W9-BRL-401

ORGANIC STRUCTURAL SPECTROSCOPY

JOSEPH B. LAMBERT
Northwestern University

HERBERT F. SHURVELL
Queen's University

DAVID A. LIGHTNER
University of Nevada at Reno

R. GRAHAM COOKS
Purdue University

Prentice Hall
Upper Saddle River, New Jersey 07458

Library of Congress Cataloging-in-Publication Data
Organic structural spectroscopy / Joseph B. Lambert . . . [et al.].
                p.   cm.
    Includes bibliographical references and index.
    ISBN 0-13-258690-8
    1. Organic compounds—Analysis.   2. Spectrum analysis.
I.   Lambert, Joseph B.
QD272.S6074   1998   97-40522
547'.122—dc21   CIP

Acquisitions Editor: *John Challice*
Editorial Assistant: *Betsy A. Williams*
Manufacturing Manager: *Trudy Pisciotti*
Total Concept Coordinator: *Kimberly Karpovich*
Art Director: *Jayne Conte*
Cover Designer: *Bruce Kenselaar*
Cover Art: *Rolando E. Corujo*
Production Supervision/Composition: *Accu-Color, Inc.*

© 1998 by Prentice-Hall, Inc.
Simon & Schuster / A Viacom Company
Upper Saddle River, New Jersey 07458

Printed in the United States of America

10 9 8 7 6 5 4 3 2

ISBN 0-13-258690-8

Prentice-Hall International (UK) Limited, *London*
Prentice-Hall of Australia Pty. Limited, *Sydney*
Prentice-Hall Canada Inc., *Toronto*
Prentice-Hall Hispanoamericana, S.A., *Mexico*
Prentice-Hall of India Private Limited, *New Delhi*
Prentice-Hall of Japan, Inc., *Tokyo*
Simon & Schuster Asia Pte. Ltd., *Singapore*
Editora Prentice-Hall do Brasil, Ltda., *Rio de Janiero*

Contents

iii

Contents

PART III ELECTRONIC ABSORPTION AND CHIROPTICAL SPECTROSCOPY

10 Introduction and Experimental Methods, 252

11 UV–Vis, CD, and ORD, 274

12 Structural Analysis, 304

Preface

The synthetic chemist wants to know the structure of a reaction product to continue a synthesis. The analytical chemist needs to identify an environmental contaminant. The forensic chemist is asked to identify a drug. In these and many other situations, the first thing the chemist does is obtain spectroscopic information. Quickly obtained proton nuclear magnetic resonance (NMR), infrared (IR), or mass spectra may provide significant progress towards identifying the reaction product, pollutant, or drug. These are only the initial steps, however, in the deductive process of elucidating chemical structure through spectroscopy. A number of other nuclei, such as carbon-13, may be examined by NMR, and more sophisticated multipulse and two-dimensional NMR experiments may be carried out. Contributions from ultraviolet spectroscopy and fragmentation information from mass spectrometry (MS) often are necessary. Certain structural problems may require the acquisition of Raman or chiroptical data as well. Thus today's chemist must have access to and knowledge of a wide range of spectroscopic techniques.

This textbook describes how an assembly of spectroscopic experiments can lead to the deduction of precise chemical structures. The methodologies apply equally well to organic, organometallic, and biological molecules. We begin discussion of each spectroscopic technique with the most elementary principles and then carry the material to the state of the art that, in our opinion, is necessary for current standard structure elucidation. This book was designed as a teaching text and offers a carefully struck balance between background theory and applied problem-solving. It also contains the following features:

- Thorough discussion of the chemical shift and coupling constant as the underlying principles of NMR
- Equal emphasis on proton and carbon-13 NMR techniques, reflecting current practice
- Description of multipulse and two-dimensional NMR techniques for signal enhancement and structure elucidation
- Extensive discussion of IR group frequencies and their relationship to structure
- Parallel inclusion of IR and Raman spectroscopies and explanation of how Raman supplements and expands IR experiments
- Fundamentals of electronic spectroscopy, including structural correlations such as the Woodward–Fieser rules
- Integration of chiroptical methods (optical rotatory dispersion and circular dichroism) for applications to chiral molecules
- Full treatment of modern ionization techniques in mass spectrometry (electrospray, MALDI, and others), which have greatly enhanced applications of MS to large molecules
- Complete coverage of structural analysis from mass spectral fragmentations
- Bibliographies that follow each chapter to provide access to additional, up-to-date sources of further information

- Extensive exercises for each spectroscopic technique (complete answers are provided in an accompanying Solutions Manual, ISBN 0-13-281403-X)
- Numerous tables of chemical shifts, coupling constants, infrared and Raman group frequencies, electronic structure correlations, and characteristics of the mass spectral ionization techniques
- Comprehensive raw spectroscopic data from all techniques for 35 unknowns with structures ranging from simple to mildly complex

The text is designed for a one-semester course, although exploring all nuances would take longer. If less time is available, the student or instructor may omit various sections without loss of understanding in others. We invite your comments. If any textual material seems unclear or incorrect, please feel free to contact us to discuss, amplify, or clarify the text.

ACKNOWLEDGMENTS

The authors were assisted by numerous individuals in assembling the text. Sharon Fateley, Nancy Olson, Irene Shurvell, and Carol Slingo provided expert word processing. Ernani Basso, Stefan Boiadjiev, Guodeng Chen, Jeffrey Denault, Bing Bing Feng, Lynn Gill, Derek Lightner, Hai Luo, Shugang Ma, Scott Miller, Pat Mulligan, Mario Nappi, Qing Ning, S.M. Quadri, Edwin Rivera, Leah Riter, Henry Rohrs, Arlene Rothwell, Portia Mahal Sabido, Catharine E. Shawl, Jianwei Shen, Srinivasan, Feng Wang, Carsten Weil, Mitch Wells, Philip Wong, Shengtian Yang, and Yong Yim recorded spectra found in the figures and exercises, drew illustrations, or provided literature searches. The authors are grateful to the following chemists for reviewing all or part of the manuscript:

Harold M. Bell, *Virginia Polytechnic Institute and State University*
Karl D. Bishop, *Michigan State University*
J. Thomas Brenna, *Cornell University*
Paul T. Buonora, *University of Scranton*
Rainer Glaser, *University of Missouri-Columbia*
John B. Grutzner, *Purdue University*
Bruce B. Jarvis, *University of Maryland at College Park*
Francis M. Klein, *Creighton University*
Iwao Ojima, *SUNY at Stony Brook*
Bruce R. Osterby, *University of Wisconsin-La Crosse*
Timothy Patten, *University of California, Davis*
Lev Ryzhkov, *Towson State University*
Daniel Singleton, *Texas A&M University*
James M. Takacs, *University of Nebraska, Lincoln*
Akos Vertes, *The George Washington University*

Joseph B. Lambert
lambert@casbah.acns.nwu.edu

Herbert F. Shurvell
shurvell@chem.queensu.ca

David A. Lightner
lightner@chem.unr.edu

R. Graham Cooks
cooks@purdue.edu

chapter 1

Introduction

1-1 THE SPECTROSCOPIC APPROACH TO STRUCTURE

"How can I determine the structure of the molecule?" This question is asked by the synthetic chemist after completing any chemical reaction. It is also foremost in the mind of the natural product chemist extracting molecules from plants or animals in hopes of developing new medicinal materials. And the forensic chemist isolating drugs or toxins from a suspect or victim. And the environmental chemist examining the effects of materials contaminating soil, bodies of water, or the atmosphere. And the archaeological chemist tracing dietary information from food residues in pottery. And the biological chemist examining enzymatic mechanisms in the body. The quest for structural information on solids, liquids, or gases, on crystalline, powdered, or glassy materials, on mixtures or pure compounds is a continuing challenge to chemists of every type.

Diffraction methods offer the ideal solution to many structural problems by providing data that often may lead to a complete structure. Equipment for neutron or electron diffraction, however, is not widely available, and these methods apply only to relatively small molecules. X-ray diffraction is restricted to materials in the solid, normally crystals, for optimal results. Crystallographic methods cannot be applied to mixtures. None of these diffraction methods can provide the quick structural information that the synthetic chemist needs to proceed to the next step or the analytical chemist needs to report to the lawyer, athlete, or geologist about the structures of specific molecules. Because crystallography is time consuming, it cannot produce in a timely fashion the sheer quantity of results often needed.

Various forms of spectroscopy can provide a wide array of structural information in the most rapid fashion possible, for all phases of matter, and on mixtures as well as on pure compounds. The equipment is routinely available, although ranging from relatively inexpensive (thousands of dollars) to expensive (millions of dollars). The process of structural elucidation by spectroscopic methods is deductive. One or more spectroscopic experiments are carried out, and structural conclusions are reached by analyzing the resulting data. This text examines the four most common and useful forms of organic structural spectroscopy. Each provides its own special kind of data that apply to molecular structure. The next section contains a brief introduction to each method.

1-2 CONTRIBUTIONS OF DIFFERENT FORMS OF SPECTROSCOPY

1-2a Nuclear Magnetic Resonance Spectroscopy

Nuclear magnetic resonance (NMR) spectroscopy provides information about the types, numbers, and connectivity of particular atoms. Thus it can show that ethanol, CH_3CH_2OH, has two types of carbons in the ratio 1/1 and three types of hydrogens in the ratio 3/2/1, and that the methyl and methylene groups are bonded together rather than separated by oxygen. Thus, for this simple molecule, the entire structure is deduced. The NMR experiment applies only to nuclei that have the quantum mechanical property of spin, through excitation of transitions between different quantized spin states. The most commonly examined nuclei are hydrogen (often referred to as the proton, ^{1}H) and carbon-13 (^{13}C). Most elements in the periodic table have NMR-active isotopes.

Examination of the spin properties of hydrogens, carbons, and other nuclei in organic, biological, organometallic, and inorganic molecules characterizes each nucleus according to a physical parameter called the chemical shift. Partial or even complete structures may be derived from analysis of chemical shifts alone. Interactions between NMR-active nuclei, as measured by the coupling constant and the relaxation time, provide information about connectivity between atoms. Because coupling and relaxation depend on the distance between nuclei within a molecule, stereochemical information also is obtainable. Furthermore, because chemical reactions move nuclei from one position to another within a molecule or even to a new molecule, NMR is used to follow the course of many kinds of reactions.

The NMR experiment may be applied to molecules in any state of matter, but routine applications are carried out on liquids. When analyzing protons, the NMR experiment may be applied to microgram quantities, and, when analyzing carbons, to milligram quantities. The samples may be mixtures. Thus NMR comprises a very general approach to structural elucidation, as described in Part I of this volume.

1-2b Vibrational Spectroscopy

Infrared (IR) and Raman spectra result from vibrations that occur naturally but predictably within molecules. The spectra provide a variety of information about a molecule's structure, particularly its symmetry and the identity of functional groups. The high sensitivity of IR spectroscopy and the ease of sample preparation contribute to its widespread use. Grating infrared spectrometers are inexpensive, compact, and so simple to use that an undergraduate student can obtain a spectrum after just a few minutes of instruction. Although Raman spectrometers are more complex and expensive, they can provide important information for structural analysis.

The infrared and Raman spectra of any compound are unique and can be used as fingerprints for purposes of identification. Most importantly, vibrations of many functional groups always give rise to features within well-defined ranges in the spectra, regardless of the overall structure of the molecule containing the group. Consequently, vibrational spectroscopy is particularly useful in the identification of functionalities such as carbonyl or nitrile. The richness of peaks in many spectra always implies that careful interpretation is necessary, as described in Part II.

1-2c Electronic Spectroscopy

Electronic absorption spectroscopy measures the energy and probability of promoting a molecule from its ground electronic state to an electronically excited state. Excitation involves moving an electron from an occupied molecular orbital to a higher, unoccupied orbital. Since an organic molecule typically has numerous occupied and unoccupied molecular orbitals, many different electronic excitations are possible. The associated transition energies normally are found in the ultraviolet (UV) and visible (vis) regions of the electromagnetic spectrum (Section 1-3). The transition energies of most importance in understanding chemical bonds are the lowest energy excitations, which involve promoting an electron from the highest occupied bonding or nonbonding orbital to the lowest unoccupied molecular orbital. Information about electronic transitions may be obtained either directly from the UV–vis spectrum or by a difference procedure known as circular dichroism (CD), which is sensitive only to chiral molecules.

UV–vis and CD spectroscopies are used qualitatively to detect certain functional groups, according to the position and intensity of the absorption band, as described in Part III. The sensitivity of the UV–vis experiment is quite high, with detectability up to

10^{-9} M. Analysis of UV–vis absorption can indicate interaction between neighboring groups, such as that occurring in conjugated ketones, and it can be used to study chemical or photochemical reactions that alter the functionality. CD can determine conformation and absolute configuration of molecules. The methods of electronic spectroscopy may be applied quantitatively to determine solute concentrations, to follow reaction kinetics, and to measure equilibrium constants.

1-2d Mass Spectrometry

Mass spectrometry (MS) examines ions in the gas phase that are produced by techniques such as the collision of molecules with electrons. The routine use and the principal contribution of MS is to determine molecular weights, both nominal (nearest integer) and exact (five or more significant figures). In addition to intact ionic versions of the molecule, fragment ions often are generated, and their structures can prove helpful in deducing the complete molecular structure. The MS experiment involves generating ions in the gas phase and measuring their mass-to-charge ratio and relative abundances. Because isotopes differ in mass, they can be recognized by MS, thus allowing compounds to be analyzed according to their isotopic makeup.

Mass spectra can be taken on samples in any state of matter, including fragile, thermally labile solids. Mixtures are examined by MS with instruments that combine mass analysis with separations based on gas chromatography (GC–MS), liquid chromatography (LC–MS), or a second stage of mass spectrometry (MS–MS). The sensitivity of MS is unmatched by other common spectroscopic techniques. Work is carried out routinely at the submicrogram level ($<10^{-6}$ g), and quantities as small as femtomoles (10^{-12} mol) are detectable.

Chemical reactions invariably occur during the MS experiment. It is possible to follow their kinetics, to measure their thermodynamic parameters, and to prepare new compounds in the mass spectrometer. Part IV describes the use of MS to obtain molecular weights, to analyze fragment ions, and to follow gas phase reactions.

1-3 THE ELECTROMAGNETIC SPECTRUM

Propagation of energy through space is characterized by both electrical and magnetic properties, and so the phenomenon is referred to as *electromagnetic radiation,* or, loosely, *light.* Such radiation has wave properties, in that its magnitude fluctuates sinusoidally over time (Figure 1-1) as it moves through space at the speed of light (velocity $c = 2.998 \times 10^8$ m s^{-1} in vacuum). The length of one full cycle is called the wavelength (λ), for example, the distance from crest to crest or trough to trough in the diagram. The number of full-cycle fluctuations that occur over some period of time (usually one second) is called the frequency (υ). Wavelength and frequency are inversely related, and the speed of light is the constant of proportionality ($\upsilon = c/\lambda$). Thus the longer the wavelength, the shorter the frequency. More energetic waves fluctuate more rapidly in space, so that the energy (ΔE) and frequency are directly related, with Planck's constant ($h = 6.624 \times 10^{-34}$ J $\cdot$ s) as the constant of proportionality ($\Delta E = h\upsilon$). The maximum excursion from the zero point is the amplitude of the wave (A in Figure 1-1) and may be thought of as its intensity.

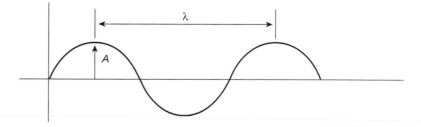

Figure 1-1

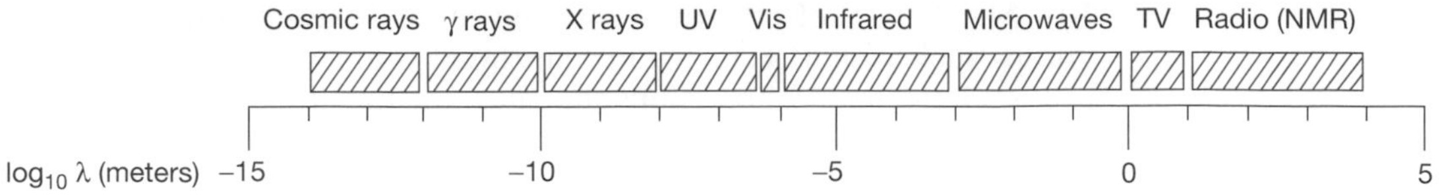

Figure 1-2 Names given to various regions of the electromagnetic spectrum; the plot is linear in the logarithm of wavelength (λ).

Light has been given various names according to its wavelength (Figure 1-2). Cosmic, γ, and X rays have the shortest wavelengths, and radiofrequency waves have the longest wavelength. Because there is a sequence of related phenomena that differ in wavelength, the whole series is called the *electromagnetic spectrum.* As drawn in Figure 1-2, wavelength increases from left to right, whereas frequency and energy increase from right to left.

NMR, vibrational, and electronic spectroscopies involve absorption of electromagnetic energy, respectively from the radiofrequency, infrared, and ultraviolet/visible regions of the electromagnetic spectrum. Each *absorption spectroscopy* promotes normal or ground state molecules into higher energy or excited states. For NMR, only the spin state of the nucleus is changed. Infrared absorption gives rise to well-defined molecular vibrations, and ultraviolet–visible absorption results in excited electronic states. The energy absorbed (ΔE) is roughly 100 kcal mol^{-1} for electronic excitation, 10 kcal mol^{-1} for vibrational excitation, and 10^{-6} kcal mol^{-1} for nuclear spin excitation. Each of these spectroscopies has its preferred units for wavelength, frequency, and energy, developed historically according to custom. Mass spectrometry involves measurement of mass numbers rather than energy absorption (hence it is a spectro*metry* rather than a spectro*scopy*) and is not associated with a region of the electromagnetic spectrum.

1-4 MOLECULAR WEIGHT AND MOLECULAR FORMULA

Determining the molecular weight and the molecular formula are important steps in structural analysis. The traditional, nonspectroscopic procedures for obtaining both include lowering the freezing or melting point, elevating the boiling point, and measuring vapor pressure changes. Although the equipment for these classic physical chemistry methods is neither expensive nor complex, it is rarely available to the synthetic chemist. Today molecular weight is obtained more reliably and easily by mass spectrometry, as described in Part IV.

High resolution mass spectrometry also can provide the molecular formula, that is, the identity and number of each atom in the molecule, for example, $C_6H_{10}O$ for cyclohexanone. More traditionally, chemists used quantitative elemental analysis to derive the *empirical formula* (the lowest integer ratio of elements present), which is related to the molecular formula by an integral factor. Thus the empirical formula of ethane is CH_3, but its molecular formula is $(CH_3)_2$ or C_2H_6; the empirical formula of cyclohexane is an uninformative CH_2 but its molecular formula is $(CH_2)_6$ or C_6H_{12}.

Almost all journals that publish synthetic chemistry research require accurate elemental analyses of new compounds as a proof of purity. The data are generally obtained by submitting samples to a commercial laboratory for analysis, although many research laboratories have their own equipment. These analyses provide the percentages of carbon, hydrogen, nitrogen, and other elements (oxygen usually is calculated by difference from 100%). These elemental percentages are converted to the empirical formula by the following procedure. Division of an atomic percentage X by the atomic weight M of the atom gives the relative number of atomic equivalents ($E = X/M$) for each element in the compound. The numbers for all atoms ($E, E', E'' \ldots$) are converted to the simplest ratio through division by the smallest value of E: E/E (which equals 1.0), E'/E, $E''/E \ldots$. These numbers are then multiplied by the lowest integer n that yields whole numbers for all the ratios: $n, nE'/E, nE''/E \ldots$. The whole numbers are the atomic ratios in the empirical formula. The procedure is illustrated as follows.

Atoms	Carbon	Hydrogen	Oxygen
Elemental analysis (X)	48.57	8.19	43.24
Relative atoms ($E = X/M$)	$\dfrac{48.57}{12.01} = 4.04$	$\dfrac{8.19}{1.008} = 8.13$	$\dfrac{43.23}{16.00} = 2.70$
Simplest ratio (E/E'')	$\dfrac{4.04}{2.70} = 1.50$	$\dfrac{8.13}{2.70} = 3.01$	$\dfrac{2.70}{2.70} = 1.00$
Lowest integer ratio (nE/E'')	$2 \times 1.50 = 3$	$2 \times 3.01 \approx 6$	$2 \times 1.00 = 2$
Empirical formula	C_3	H_6	O_2

As noted, the result of such calculations is the empirical rather than the molecular formula. Moreover, the procedure is poor at distinguishing small differences in carbon and hydrogen, it is subject to experimental error in the percentages (particularly that of oxygen), and it is very sensitive to the purity of the sample. For these reasons, high resolution mass spectrometry is preferred, although it is subject to its own set of limitations, as described in Part IV. The conservative approach is to obtain both mass spectral and elemental analytical data.

The molecular formula itself provides important structural information beyond the simple facts of elemental identity and number. Introduction of a ring or a double bond into an alkane structure reduces the number of hydrogen atoms by two. Thus the formula for the homologous series of alkanes is C_nH_{2n+2}, whereas that for alkenes or monocyclic alkanes is C_nH_{2n}. Detailed examination of the molecular formula can provide the number of sites of unsaturation. Each ring is one site, each double bond is one site, and each triple bond is two sites.

The *unsaturation number* (or index of hydrogen deficiency) (U) is given by eq. 1-1, in which C is the number of tetravalent elements (carbon, silicon, etc.), X is the

$$U = C + 1 - \tfrac{1}{2}(X - N) \qquad (1\text{-}1)$$

number of monovalent elements (hydrogen and the halogens), and N is the number of trivalent elements (nitrogen, phosphorus, etc.). The expression is an elaboration of the formula for the alkane series, whose unsaturation number is zero. Divalent elements such as oxygen and sulfur extend chains but do not alter the number of unsaturations, so they do not appear in the formula. Trivalent elements add a bond and univalent elements subtract a bond, as indicated in the last term. The unsaturation number for cyclohexanone is $6 + 1 - \tfrac{1}{2}(10) = 2$, one for the ring and one for the carbonyl group.

1-5 STRUCTURAL ISOMERS AND STEREOISOMERS

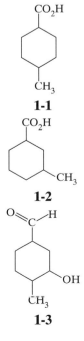

1-1

1-2

1-3

In the most detailed sense, the structure of a molecule includes all its bond lengths, valence angles, and torsional angles, which in turn define atom connectivity, stereochemistry, and conformation. This type of information normally is available only from crystallography or some forms of multidimensional nuclear magnetic resonance spectroscopy.

In the absence of quantitative knowledge of bond lengths and angles, the spectroscopic method is to deduce the structure by identifying components of the molecule, by showing that specific atoms are connected to each other, and by determining the relationship between atoms in space. The spectroscopic objective is to obtain enough such information to overdetermine the structure.

The first step usually is to determine the molecular formula, because it defines a family of molecules or *isomers* to be considered. This family is composed of molecules that differ in the connectivity of atoms (*structural isomers*) and molecules that differ only in the spatial relationships of atoms (*stereoisomers*). For example, molecules **1-1**, **1-2**, and **1-3** all have the molecular formula $C_8H_{14}O_2$, but they clearly have different structures. Molecules **1-1** and **1-2** differ only in the locational relationship of the carboxyl and methyl groups and are said to be *positional isomers*. Molecule **1-3** differs more radically because it possesses hydroxyl and aldehyde functionalities instead of the carboxyl group, so it is a *functional isomer* of the first two molecules. To distinguish among this set of molecules, spectroscopy needs (1) to show how the various carbon atoms are connected and (2) to distinguish between organic functional groups.

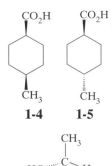

1-4   **1-5**

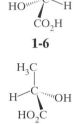

1-6

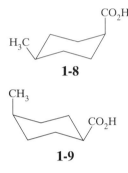

1-7

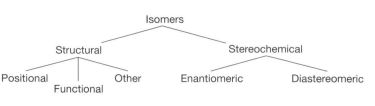

Isomers
Structural   Stereochemical
Positional   Other   Enantiomeric   Diastereomeric
Functional

Figure 1-3   The isomer tree.

A closer look at molecule **1-1** reveals that the methyl and carboxyl groups can be located on the same side or on opposite sides of the six-membered ring, as indicated by **1-4** and **1-5**. These two forms are different molecules, as bond cleavage is necessary to change one into the other. They have the same connectivity or constitution, so they are not structural isomers. Because they differ in the spatial arrangement of atoms, they are stereoisomers. Another type of stereoisomer includes pairs of molecules that are nonsuperimposable mirror images, such as **1-6** and **1-7** (two forms of lactic acid). Such molecules are called *enantiomers* and are said to be chiral. Stereoisomers that are not enantiomers, such as **1-4** and **1-5**, are called *diastereoisomers*.

When the six-membered ring of **1-5** is depicted three-dimensionally, it assumes a chair shape (**1-8**). By a series of rotations about single bonds, this molecule can switch all axial positions into equatorial positions, and vice versa, to produce the molecule **1-9**. These molecules are stereoisomers, but because they can interconvert by rotating about single bonds they are called *conformational isomers*, which comprise a special category of diastereoisomers or enantiomers. All forms of spectroscopy can contribute to assigning stereochemical properties to molecules. Figure 1-3 illustrates the relationships among various classes of isomers.

1-8

1-9

The remainder of this book describes the spectroscopic methods used to perform these structural distinctions: to obtain the molecular formula, to identify functional groups, to determine atom connectivities, and to arrange the atoms in space. The successful result provides the qualitative structure of the molecule, including the spatial arrangements of the atoms.

PROBLEMS

1-1 Specify whether the following pairs of molecules are structural isomers or stereoisomers.

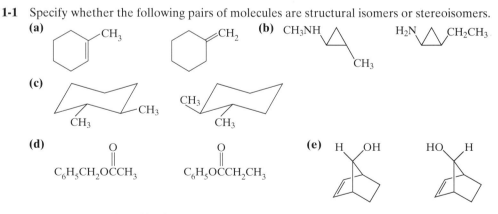

1-2 Carry out the following operations for each of the molecules in Problem **1-1.**
  (a) Write down the molecular formula and calculate elemental percentages for C, H, N, and O (when present).
  (b) Determine the unsaturation number.
  (c) Specify the functional groups present.

1-3 For the molecules in Problems **1-1a**, **1-1d**, and **1-1e**, determine the number and relative proportions of different types of carbon atoms. Do the same with the hydrogens. Atoms are said to be different if they cannot be interconverted by a symmetry operation such as reflection through a mirror plane or rotation about an axis.

1-4 Determine the empirical formula of the molecules that gave the following elemental analyses.
  (a) C, 59.89; H, 8.12; O, 31.99.
  (b) C, 66.02; H, 10.34; N, 10.96; O, 12.68.

part I

NUCLEAR MAGNETIC RESONANCE SPECTROSCOPY

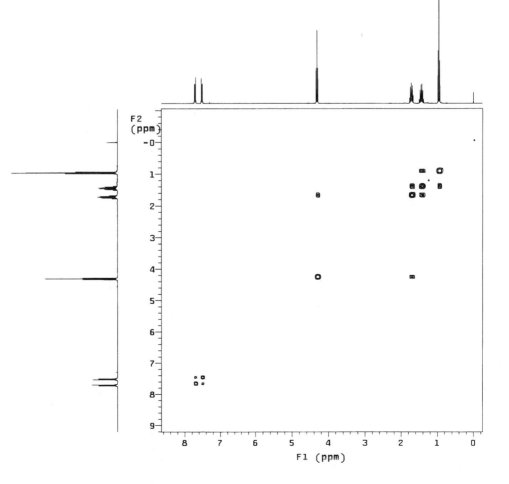

chapter 2

Introduction and Experimental Methods

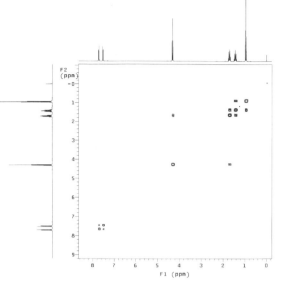

Structure determination for almost all organic and biological molecules and many inorganic molecules begins with nuclear magnetic resonance (NMR) spectroscopy. During the half-century existence of NMR spectroscopy, the field has undergone several internal revolutions, repeatedly redefining itself as an increasingly complex and effective structural tool. Aside from X-ray crystallography, which can provide the complete molecular structure of pure, crystalline materials, NMR spectroscopy is the chemist's most direct tool for identifying the structure of both pure compounds and mixtures. The process involves performing several NMR experiments to deduce the molecular structure from the magnetic properties of the atomic nuclei and the surrounding electrons.

2-1 MAGNETIC PROPERTIES OF NUCLEI

The simplest atom, hydrogen, is found in almost all organic compounds and is composed of a single proton and a single electron. Its notation is ^1H, in which the superscript signifies the sum of protons and neutrons (the atomic mass). The key aspect of the hydrogen nucleus for the purpose of NMR is that it has angular momentum properties that resemble the classic concept of a spinning particle. Because the spinning nucleus is positively charged, it generates a magnetic field and possesses a *magnetic moment* μ, just as a charge moving in a circle creates a magnetic field (Figure 2-1). The magnetic moment μ is a vector, because it has both magnitude and direction, as defined by its axis of spin in Figure 2-1. The NMR experiment exploits the magnetic properties of nuclei to provide structural information.

The spin properties of protons and neutrons in the nuclei of heavier elements combine to define the overall spin of the nucleus. When both the atomic number and the atomic mass are even, the nucleus has no magnetic properties, as signified by a zero value of a *spin quantum number*, $I = 0$ (Figure 2-2). Common nonmagnetic nuclei are carbon (^{12}C) and oxygen (^{16}O). For spinning nuclei, the spin quantum number can take on only certain values. Those with a spherical shape have a spin I of $\frac{1}{2}$, and those with a nonspherical, or quadrupolar, shape have a spin of 1 or more (in increments of $\frac{1}{2}$). Com-

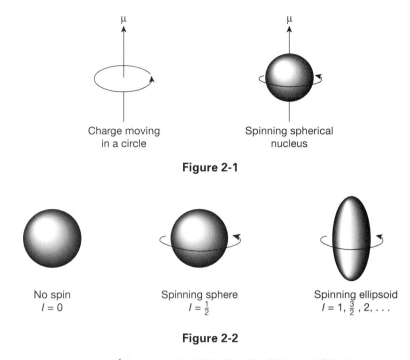

μ   μ

Charge moving   Spinning spherical
in a circle   nucleus

Figure 2-1

No spin   Spinning sphere   Spinning ellipsoid
$I = 0$ $I = \frac{1}{2}$ $I = 1, \frac{3}{2}, 2, \ldots$

Figure 2-2

mon nuclei with a spin of $\frac{1}{2}$ include ^{1}H, ^{13}C, ^{15}N, ^{19}F, ^{29}Si, and ^{31}P. Quadrupolar nuclei include ^{2}H, ^{11}B, ^{14}N, ^{17}O, ^{33}S, and ^{35}Cl. The magnitude of the magnetic moment produced by a spinning nucleus varies from atom to atom, as defined by the equation $\mu = \gamma \hbar I$. The quantity $\hbar$ is Planck's constant divided by 2π, and γ is a characteristic of the nucleus called the *gyromagnetic* or *magnetogyric ratio*. The larger the γ, the larger the magnetic moment of the nucleus.

To study nuclear magnetic properties, the experimentalist subjects nuclei to a strong laboratory magnetic field of value B_0, whose units are tesla or T (1 tesla = 10^4 gauss). In the absence of the laboratory field, all the nuclear magnets of the same isotope have the same energy. When the B_0 field is turned on along a direction designated as the z axis, the energies of the nuclei are affected. There is a slight preference for alignment in the general direction of B_0 ($+z$) over the opposite direction ($-z$). Nuclei with a spin of $\frac{1}{2}$ assume only these two arrangements. The interaction is illustrated in Figure 2-3. At the left is a magnetic moment with a $+z$ component and on the right one with a $-z$ component. The nuclear magnets are not actually lined up fully in the $+z$ or $-z$ directions. Rather, the force of B_0 causes the magnetic moment to move in a circular fashion about the $+z$ direction in the first case and about the $-z$ direction in the second, a motion called *precession*.

In terms of vector analysis, the B_0 field in the z direction operates on the x component of μ to create a force in the y direction (Figure 2-3, inset in the middle). The nuclear moment then begins to move toward the y direction. Because the force of B_0 on μ is always perpendicular to both B_0 and μ, the motion of μ describes a circular orbit around the $+z$ or the $-z$ direction, in complete analogy to the forces present in a spinning top or gyroscope.

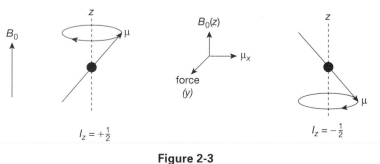

Figure 2-3

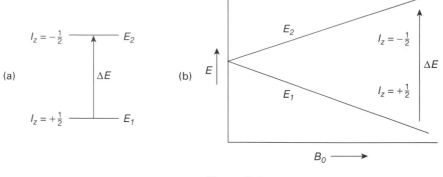

Figure 2-4

Because only two directions of precession exist for a spin $\frac{1}{2}$ nucleus, two assemblages or spin states are created, with assigned quantum numbers of $I_z = +\frac{1}{2}$ for precession with the field $(+z)$ and $-\frac{1}{2}$ for precession against the field $(-z)$. There is a slight energetic preference for the $I_z = +\frac{1}{2}$ state. In the absence of B_0, the precessional motions are absent, there is no preference for the $+I_z$ state, and all nuclei have the same energy.

The populational preference for $+z$ precession in the presence of B_0 is defined by Boltzmann's law (eq. 2-1), in which n is the population of a spin state, k is Boltzmann's

$$\frac{n(+\frac{1}{2})}{n(-\frac{1}{2})} = \exp(\Delta E/kT) \tag{2-1}$$

constant, T is the absolute temperature, and ΔE is the energy difference between the spin states. Figure 2-4a depicts the energies of the two states and the difference ΔE between them.

The precessional motion of the magnetic moment around B_0 occurs with an angular frequency ω_0, called the *Larmor frequency*, whose units are radians per second, rad s^{-1}. As B_0 increases, so does the angular frequency, that is, $\omega_0 \propto B_0$. The constant of proportionality between ω_0 and B_0 is the gyromagnetic ratio, that is, $\omega_0 = \gamma B_0$. This natural precession frequency corresponds to the frequency in Planck's relationship, $\Delta E = h\nu_0$ in linear frequency or $\hbar\omega_0$ in angular frequency ($\omega = 2\pi\nu$). In this way the energy difference between the spin states is related to the Larmor frequency (eq. 2-2). Thus as the

$$\Delta E = \hbar\omega_0 = h\nu_0 = \gamma\hbar B_0 \tag{2-2}$$

B_0 field increases, the difference in energy between the two spin states increases, as illustrated in Figure 2-4b.

These equations indicate that the natural precession frequency of a spinning nucleus depends only on the nuclear properties contained in the gyromagnetic ratio γ and on the laboratory-determined value of the magnetic field B_0. For a proton in a magnetic field B_0 of 7.05 tesla (T), the frequency of precession is 300 megahertz (MHz), and the difference in energy between the spin states is only 0.0286 cal mol^{-1} (0.120 J mol^{-1}). This value is extremely small in comparison with the energy differences between vibrational or electronic states.

In the NMR experiment the two states illustrated in Figure 2-4 are made to interconvert by applying a second magnetic field B_1. When the frequency of the B_1 field is the same as the Larmor frequency of the nucleus, energy can flow by absorption and emission between this newly applied field and the nuclei. Absorption of energy occurs as $+\frac{1}{2}$ nuclei become $-\frac{1}{2}$ nuclei, and emission occurs as $-\frac{1}{2}$ nuclei become $+\frac{1}{2}$ nuclei. Since the experiment begins with an excess of $+\frac{1}{2}$ nuclei, there is a net absorption of energy. This process is called *resonance,* and the absorption may be detected electronically and displayed as a plot of frequency vs. amount of energy absorbed. Because the resonance frequency ν_0 is highly dependent on the structural environment of the nucleus, NMR has become the structural tool of choice for chemists. Figure 2-5 illustrates the NMR spectrum for the protons in benzene. Absorption is represented by a peak upwards from the baseline.

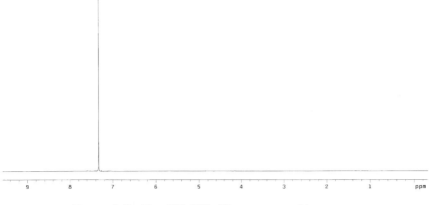

Figure 2-5   The 300 MHz ^{1}H spectrum of benzene.

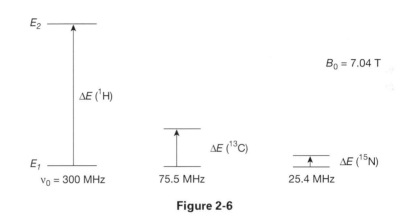

Figure 2-6

Because gyromagnetic ratios vary among elements and even among isotopes of a single element, resonance frequencies also vary. The proton has one of the largest gyromagnetic ratios, so its spin states are far apart and the value of ΔE is large. Other important nuclei such as ^{13}C and ^{15}N have much smaller gyromagnetic ratios and hence have smaller differences between the energies of the two spin states (Figure 2-6). Absorption of energy is easier to observe when the spin states are energetically farther apart. An increase in ΔE means that there are more nuclei in the lower spin state (E_1) as compared with the upper spin state (E_2), according to the Boltzmann relationship (see eq. 2-1). The increase in $+\frac{1}{2}$ nuclei at the expense of $-\frac{1}{2}$ nuclei leads to a greater net absorption of energy and a larger signal.

In summary, the NMR experiment consists of immersing magnetic nuclei in a strong field B_0 to distinguish them according to their values of I_z, followed by application of a B_1 field whose frequency corresponds to the Larmor frequency. The frequency is related to the energy difference between the I_z spin states, $\Delta E = h\nu$. This energy difference depends on the value of B_0 (Figure 2-4) and on the gyromagnetic ratio of the nucleus $\Delta E = \gamma \hbar B_0$.

When spins have values other than $\frac{1}{2}$, there are more than two available spin states. For $I = 1$ nuclei such as ^{2}H and ^{14}N, the magnetic moments may precess about three directions relative to B_0: parallel ($I_z = +1$), perpendicular (0), and opposite (-1). In general there are $2I + 1$ spin states, for example, six for $I = 5/2$ (^{17}O has this spin). The values of I_z extend from $+I$ to $-I$ in increments of 1 ($+I, +I-1, +I-2, \ldots -I$).

2-2 THE CHEMICAL SHIFT

NMR is a valuable structural tool because the observed resonance frequency ν_0 depends on the molecular environment as well as on γ and B_0. The electron cloud that surrounds the nucleus also has charge, motion, and hence a magnetic moment. The magnetic field that electrons generate alters the B_0 field in the microenvironment around the nucleus.

The actual field experienced by a given nucleus thus depends on the nature of the surrounding electrons. This electronic modulation of the B_0 field is termed *shielding*, which is represented by the Greek letter σ. The actual field experienced by the nucleus becomes B_{local} and may be expressed as $B_0(1 - σ)$, in which electronic shielding σ normally is positive. Variation of the resonance frequency with shielding has been termed the *chemical shift*.

The expression for the resonance frequency in terms of shielding is given by eq. 2-3, by substituting B_{local} into eq. 2-2. Decreased shielding results in a higher resonance

$$ν_0 = γB_0(1 - σ)/2π \qquad (2\text{-}3)$$

frequency $ν_0$ at constant B_0, since σ enters the equation with a negative sign. For example, the presence of an electron-withdrawing group in a molecule reduces the electron density around a proton, so that there is less shielding and a higher resonance frequency than in the case of a molecule that lacks the electron-withdrawing group. Thus protons in fluoromethane (CH_3F) resonate at a higher frequency than those in methane (CH_4) because the fluorine atom withdraws electrons from around the hydrogen nuclei.

Figure 2-7 shows the NMR spectra separately of the protons and the carbons of methyl acetate ($CH_3CO_2CH_3$). Although 98.9% of naturally occurring carbon is the

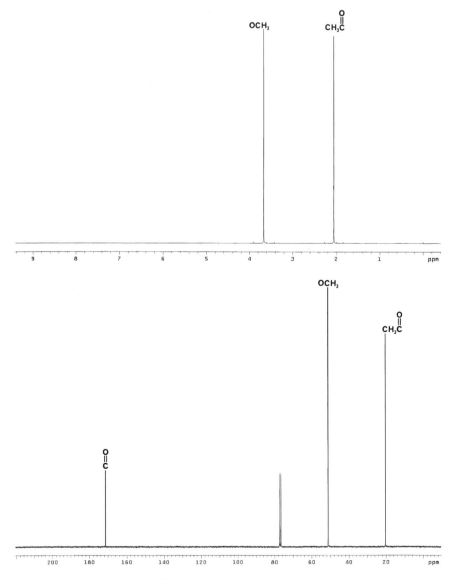

Figure 2-7 The 300 MHz ^{1}H spectrum (upper) and the 75.45 MHz ^{13}C spectrum (lower) of methyl acetate in $CDCl_3$ (the triplet at δ 77 is from solvent). The ^{13}C spectrum has been decoupled from the protons.

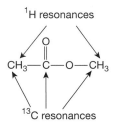

¹H resonances

CH₃—C—O—CH₃

¹³C resonances

Figure 2-8

nonmagnetic ^{12}C, the carbon NMR experiment is carried out on the 1.1% of ^{13}C, which has an I of $\frac{1}{2}$. Because of differential electronic shielding, the ^{1}H spectrum contains separate resonances for the two types of protons ($O—CH_3$ and $C—CH_3$), and the ^{13}C spectrum contains separate resonances for the three types of carbon ($O—CH_3, C—CH_3$, and carbonyl) (Figure 2-8).

The proton resonances may be assigned on the basis of the electron-withdrawing abilities (electronegativity) of the neighboring atoms. The ester oxygen is more electron withdrawing than the carbonyl group, so the $O—CH_3$ resonance occurs at higher frequency than (and to the left of) the $C—CH_3$ resonance. By convention, frequency in the spectrum increases from right to left for consistency with other forms of spectroscopy. Therefore shielding increases from left to right.

The system of units depicted in Figure 2-7 and used throughout this book has been developed to overcome the fact that chemical information is found in small differences between large numbers. An intuitive system might be absolute frequency, for example in Hz. At the common field of 7.05 T, for example, all protons resonate in the vicinity of 300 MHz. A scale involving numbers like 300.000764, however, is cumbersome and the frequencies would vary from one B_0 field to another. For each element or isotope, a reference material has been chosen and assigned a relative frequency of zero. For both protons and carbons, the substance is tetramethylsilane [$(CH_3)_4Si$, usually called TMS], which is soluble in most organic solvents, unreactive, and volatile. In addition, the low electronegativity of silicon means that the protons and carbons are surrounded by a relatively high density of electrons. Hence they are highly shielded and resonate at low frequency. Shielding by silicon in fact is so strong that the proton and carbon resonances of TMS are placed at the right extreme of the spectrum, thus providing a convenient spectral zero. In Figures 2-5 and 2-7 the position marked "0 ppm" is the hypothetical position of TMS.

To have a common unit that applies to any B_0 field, the chemical shift δ of a given nucleus i is calculated from that of the reference TMS by the equation $\delta = (\nu_i - \nu_{TMS})/\nu_{TMS}$. Because the difference between chemical shifts $\nu_i - \nu_{TMS}$ is generally on the order of Hz and the absolute frequency for TMS (ν_{TMS}) is in MHz, values of δ are in parts per million (ppm). As seen in the ^{1}H spectrum of methyl acetate (see Figure 2-7), the δ value for the $C—CH_3$ protons is 2.17 ppm and that for the $O—CH_3$ protons is 3.67 ppm. These values remain the same in spectra taken at a field of either 1.41 T (60 MHz) or 18.8 T (800 MHz), which represent the extremes of spectrometers currently in use. Chemical shifts in Hz, however, vary from field to field. Thus a resonance that is 90 Hz from TMS at 60 MHz is 450 Hz from TMS at 300 MHz but always has a δ value of 1.50 ppm ($\delta = 90/60 = 450/300 = 1.50$). Note that a resonance to the right of TMS has a negative value of δ. Since TMS is insoluble in water, other internal standards are used for this solvent, such as 3-(trimethylsilyl)-1-propanesulfonic acid [$(CH_3)_3Si(CH_2)_3SO_3Na$] or 3-(trimethylsilyl)propionic acid [$(CH_3)_3SiCH_2CH_2CO_2Na$] (sodium salts).

In the first generation of spectrometers, the range of chemical shifts, such as those in the scale at the bottom of Figures 2-5 and 2-7, was generated by varying the B_0 field while holding the B_1 field and hence the resonance frequency constant. Consideration of eq. 2-3 [$\nu_0 = \gamma B_0(1 - \sigma)2\pi$] indicates that an increase in shielding (σ) requires B_0 to be raised to keep ν_0 constant. Since more shielded nuclei resonate at the right side of the spectrum, the B_0 field in this experiment increases from left to right. Consequently, the right end came to be known as the high field or upfield end, and the left end as the low field or downfield end. This method was termed *continuous wave, field sweep*.

Modern spectrometers allow the frequency to be varied while the B_0 field is kept constant. An increase in shielding (σ) lowers the right side of eq. 2-3, so that ν_0 must decrease to maintain a constant B_0. Thus the right end of the spectrum, as noted before, corresponds to lower frequencies for more shielded nuclei. The general result is that frequency increases from right to left and field increases from left to right. Figure 2-9 summarizes the terminologies. The right end of the spectrum is still often referred to as the high field or upfield end in deference to the old field sweep experiment, although it is more appropriate to call it the low frequency or more shielded end.

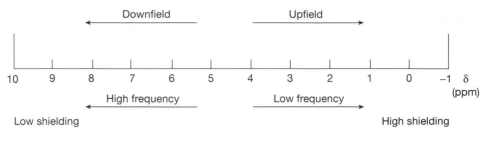

Figure 2-9

2-3   EXCITATION AND RELAXATION

To understand the NMR experiment more fully, it is useful to reconsider Figure 2-3 in terms of a collection of nuclei (Figure 2-10). At equilibrium, the $I_z = +\frac{1}{2}$ nuclei precess around the $+z$ axis and the $-\frac{1}{2}$ nuclei precess around the $-z$ axis. Only 20 spins are shown on the surface of the double cone, and the excess of $+\frac{1}{2}$ over $-\frac{1}{2}$ nuclei is exaggerated (12 to 8). The ratio of populations of the two states is given by the Boltzmann equation (see eq. 2-1). Inserting the numbers for $B_0 = 7.04$ T yields the result that for every million spins there are about 50 more with $+\frac{1}{2}$ than $-\frac{1}{2}$ spin. If the magnetic moments are added vectorially, there is a net vector in the $+z$ direction because of the excess of $+\frac{1}{2}$ over $-\frac{1}{2}$ spins. The sum of all the individual spins is called the *magnetization* (M), and the large arrow pointing along the $+z$ direction in Figure 2-10 represents the resultant $M = M_z$. Because the spins are distributed randomly around the z axis, there is no net x or y magnetization, that is, $M_x = M_y = 0$.

Figure 2-10 also shows the vector that represents the B_1 field placed along the x axis. The frequency is varied until it matches the Larmor frequency of the nuclei. When that happens, some $+\frac{1}{2}$ spins turn over and become $-\frac{1}{2}$ spins, so that M_z decreases slightly. The vector B_1 exerts a force on M whose result is perpendicular to both vectors (see Figure 2-10, inset at lower right). If B_1 is on only briefly, the magnetization vector M just tips slightly off the z axis, moving toward the y axis, which represents the mutually perpendicular direction. Figure 2-11 illustrates the result.

The 20 spins of Figure 2-10 originally had twelve spin $+\frac{1}{2}$ and eight spin $-\frac{1}{2}$ nuclei, which after application of the B_1 field are shown in Figure 2-11 as eleven spin $+\frac{1}{2}$ and nine spin $-\frac{1}{2}$ nuclei, that is, only 1 nucleus has changed its spin. The decrease in M_z is apparent, as is the tipping of the magnetization vector off the axis. The positions on the circles, or *phases,* of the 20 nuclei no longer are random, because the tipping requires

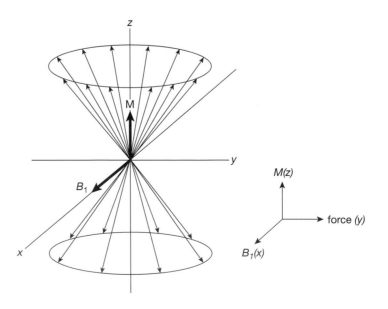

Figure 2-10

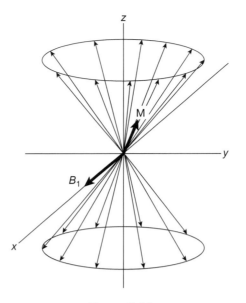

Figure 2-11

bunching of the spins toward the y axis. Thus the phases of the spins now have some coherence, and there are x and y components of the magnetization. The xy component of the magnetization is the signal detected electronically as the resonance.

The B_1 field in Figures 2-10 and 2-11 oscillates back and forth along the x axis. As Figure 2-12 illustrates from a view looking down the z axis, B_1 may be considered either (1) to oscillate linearly along the x axis at so many times per second (with frequency v) or (2) to move circularly in the x plane with angular frequency ω in radians per second ($2\pi v$). These representations are essentially equivalent. Resonance occurs when the frequency and phase of B_1 match that of the nuclei precessing at the Larmor frequency.

Figure 2-11 represents a snapshot in time, with the motion of both the B_1 vector and the precessing nuclei frozen. Another way to look at the freeze frame is to consider that the x and y axes are rotating at the frequency of the B_1 field. In terms of Figure 2-12, the axes are following the circular motion. Consequently, B_1 appears to be frozen in position and magnitude along the x axis This *rotating coordinate system* is used throughout this book to simplify magnetization diagrams. In the rotating frame, the individual nuclei no longer precess around the z axis but also are frozen, for as long as they all have the same Larmor frequency matched by B_1.

Application of B_1 at the resonance frequency results in energy absorption ($+\frac{1}{2}$ nuclei become $-\frac{1}{2}$) and emission ($-\frac{1}{2}$ nuclei become $+\frac{1}{2}$). Because initially there are more $+\frac{1}{2}$ than $-\frac{1}{2}$ nuclei, the net effect is absorption. As B_1 irradiation continues, however, this excess of $+\frac{1}{2}$ nuclei will disappear, so that the rates of absorption and emission become equal. Under these conditions the sample is said to be *saturated*. The situation is ameliorated because there are natural mechanisms whereby nuclear spins return toward equilibrium from saturation. Any process that returns the z magnetization to its equilibrium condition with the excess of $+\frac{1}{2}$ spins is called *spin–lattice* or *longitudinal relaxation* and is usually a first-order process with time constant T_1. Relaxation also is necessary to destroy magnetization created in the xy plane. Any process that returns the

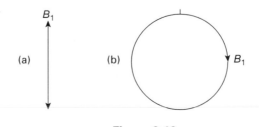

Figure 2-12

x and y magnetizations to their equilibrium condition of zero is called *spin–spin* or *transverse relaxation* and is usually a first-order process with time constant T_2.

Spin–lattice relaxation derives from the existence of local oscillating magnetic fields in the sample that correspond to the resonance frequency. The primary source of these fields is other magnetic nuclei that are in motion. As a molecule tumbles in solution in the B_0 field, each nuclear magnet generates a field caused by its motion. If this field has the Larmor frequency, excess spin energy of neighboring spins can pass to this motional energy. The resonating spins are relaxed and the absorption process can be repeated.

For effective relaxation, the tumbling magnetic nuclei must be spatially close to the resonating nucleus. For ^{13}C, attached protons provide spin–lattice relaxation. A carbonyl carbon or a carbon attached to four other carbons thus relaxes very slowly and is more easily saturated because the attached atoms are nonmagnetic (^{12}C and ^{16}O are nonmagnetic, so their motion provides no relaxation). Protons are relaxed by their nearest neighbor protons, so that CH_2 and CH_3 groups obtain relaxation by geminal protons, but CH protons must rely on vicinal or other protons.

Spin–lattice relaxation also is responsible for generating the initial excess of $+\frac{1}{2}$ nuclei when the sample is first placed in the probe. In the absence of the B_0 field, all spins have the same energy. When the sample is immersed in the B_0 field, magnetization begins to build up as spins flip through interaction with surrounding magnetic nuclei in motion, eventually creating the equilibrium ratio of $+\frac{1}{2}$ and $-\frac{1}{2}$ states.

For x and y magnetization to decay towards zero (spin–spin relaxation), the phases of the nuclear spins must become randomized (compare Figures 2-10 and 2-11). The mechanism that gives the phenomenon its name involves the interaction of two nuclei with opposite spin. The process whereby one spin goes from $+\frac{1}{2}$ to $-\frac{1}{2}$ while the other goes from $-\frac{1}{2}$ to $+\frac{1}{2}$ involves no net change in z magnetization and hence no spin–lattice relaxation. The switch in spins results in dephasing, because the new spin state has a different phase from the old one. In terms of Figure 2-11, a spin vector disappears from the upper cone surface and reappears on the lower cone surface (or vice versa) at a new phase position. As this process continues, the phases become randomized around the z axis, and xy magnetization disappears.

A similar result arises when the B_0 field is not perfectly homogeneous. Again in terms of Figure 2-11, if the spin vectors are not located in exactly identical B_0 fields, they will differ slightly in Larmor frequencies and hence precess around the z axis at different rates. As the spins move faster or more slowly relative to each other, eventually their phases become randomized. When nuclei resonate over a range of Larmor frequencies, the linewidth is broadened. The spectral linewidth at half height and the spin–spin relaxation are related by the expression $lw_{\frac{1}{2}} = 1/\pi T_2$.

2-4 PULSED EXPERIMENTS

In pulsed NMR the sample is irradiated close to the resonance frequency with an intense B_1 field for a very short time. For the duration of the pulse, the B_1 vector on the rotating x axis exerts a force on the M vector on the z axis to push the magnetization toward the y axis (Figure 2-13). This diagram simplifies Figures 2-10 and 2-11 by eliminating the indi-

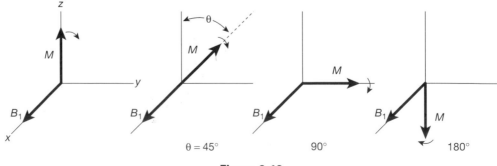

Figure 2-13

vidual spins. The force between two vectors is their cross product ($\mathbf{M} \times \mathbf{B}$ here), whose resultant is perpendicular to both vectors. As long as the strong B_1 field is on, the magnetization rotates or precesses around B_1. The angle θ of rotation increases as long as B_1 is present. A short pulse might leave the magnetization at a 45° angle relative to the z axis. A pulse twice as long (90°) aligns the magnetization along the y axis. A pulse of double this duration (180°) brings the magnetization along the $-z$ direction, meaning that there is an excess of $-\frac{1}{2}$ spins, or a population inversion. The exact angle θ caused by a pulse is determined by its duration, t_p. The angle θ (see Figure 2-13) therefore is ωt_p, in which ω is the Larmor frequency in the B_1 field, so that $\theta = \gamma B_1 t_p$.

If the B_1 irradiation is halted when the magnetization reaches the y axis (90° pulse), and the magnetization is detected over time at the resonance frequency, it would be seen to decay (Figure 2-14). Alignment of the magnetization along the y axis is a nonequilibrium situation. The x and y magnetization decays by spin–spin relaxation. At the same time, z magnetization reappears by spin–lattice relaxation. The reduction of y magnetization with time shown in Figure 2-14 is called the *free induction decay* (FID) and is a first-order process with time constant T_2.

The illustration of Figure 2-14 is artificial because it involves only a single type of nucleus with B_1 at its resonance frequency. Most samples have quite a few different types of protons or carbons, so that several resonance frequencies are involved. What happens when excitation is not carried out precisely at the resonance frequency? Imagine again the case of a single resonance, like that of benzene in Figure 2-5. After the 90° pulse (Figure 2-15), the spins are lined up along the y axis (a). If the xy coordinate system is rotating at a frequency slightly different from the Larmor frequency, then the nuclear magnet moves off the y axis within the xy plane (b). Only nuclei precessing at the same frequency as B_1 appear to be stationary in the rotating coordinate system. After additional time, the magnetization continues to rotate, reaching the $-y$ axis (c) and eventually returning to the $+y$ axis (d). The detected y magnetization during this cycle first decreases, falls to zero as it passes the $y = 0$ point, moves to a negative value in the $-y$ region (c), and then returns to positive values. The magnitude of the magnetization thus varies periodically like a cosine function. When it again is along the $+y$ axis (d), the magnetization has decreased because of spin–spin relaxation. Moreover, it has moved out of the xy plane (contrary to the diagram), as z magnetization returns through spin–lattice relaxation. The magnetization varies as a cosine function with time, as it continually passes through a sequence of events illustrated by Figure 2-15.

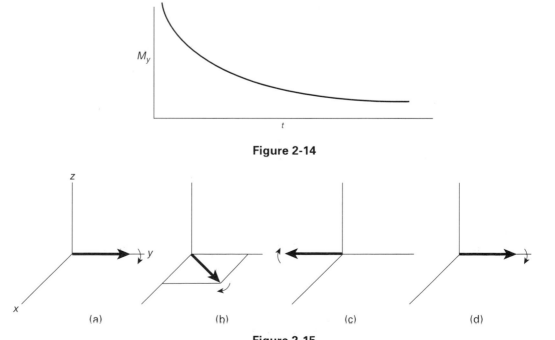

Figure 2-14

Figure 2-15

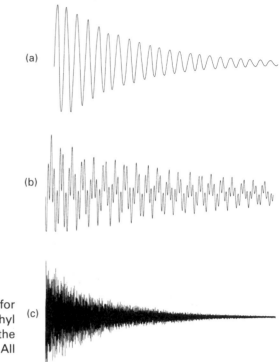

(a)

(b)

(c)

Figure 2-16 The free induction decay for the ^{1}H spectra of (a) acetone and (b) methyl acetate. (c) The free induction decay for the ^{13}C spectrum of 3-hydroxybutyric acid. All samples are without solvent.

Figure 2-16a shows what the FID looks like for the protons of acetone. The horizontal distance between each maximum is the reciprocal of the difference between the Larmor frequency and the B_1 frequency. The intensities of the maxima decrease as y magnetization is lost through spin–spin relaxation. Because the linewidth of the spectrum is determined by T_2, the FID contains all the necessary information to display a spectrum: frequency and linewidth, as well as overall intensity.

When there are two resonating nuclei, their decay patterns are superimposed, reinforcing and interfering to create a complex FID, as in Figure 2-16b for the protons of methyl acetate. By the time there are four frequencies, as in the carbons of 3-hydroxybutyric acid shown in Figure 2-16c, it is nearly impossible to unravel the frequencies visually. The mathematical process called Fourier analysis matches the FID with a series of sinusoidal curves and obtains from them the frequencies, linewidths, and intensities of each component. The FID is a plot in time (see Figures 2-14 and 2-16), so the experiment is said to occur in the *time domain*. However, the experimentalist wants a plot of frequencies, so the spectrum must be transformed to a *frequency domain* as shown in Figures 2-5 and 2-7. The Fourier transformation (FT) is carried out rapidly by computer and the experimentalist does not need to examine the FID.

2-5 THE COUPLING CONSTANT

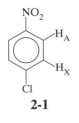

NO$_2$

H$_A$

H$_X$

Cl

2-1

The form of a resonance can be altered when there are other magnetic nuclei nearby. In 1-chloro-4-nitrobenzene **(2-1)**, for example, there are two types of protons, labeled A and X. For the time being, we will ignore any effects from identical A and X protons across the ring. Each proton has a spin of $\frac{1}{2}$ and therefore can exist in two I_z spin states, $+\frac{1}{2}$ and $-\frac{1}{2}$, which differ in population only in parts per million. Almost exactly half the A protons have $+\frac{1}{2}$ X neighbors, and half $-\frac{1}{2}$ X neighbors. The magnetic environments provided by these two types of X protons are not identical, so the A resonance is split into two peaks (Figures 2-17a and 2-18). By the same token, the X nucleus exists in two distinct magnetic environments because the A proton has two spin states. The X resonance also is split into two peaks (Figures 2-17b and 2-18). Quadrupolar nuclei such as

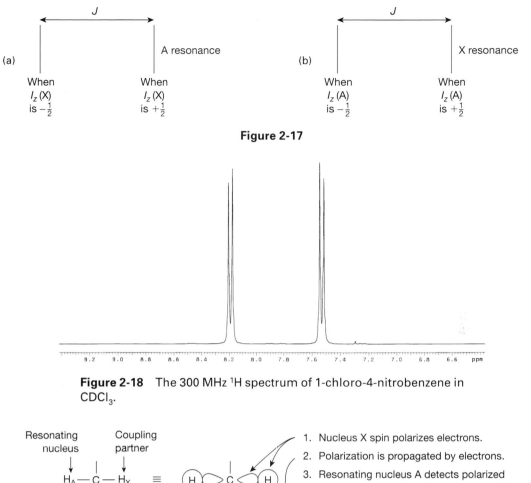

Figure 2-17

Figure 2-18   The 300 MHz ^{1}H spectrum of 1-chloro-4-nitrobenzene in CDCl$_3$.

Figure 2-19

the chlorine and nitrogen atoms in molecule **2-1** often act as if they are nonmagnetic and may be ignored in this context. Thus the proton resonance of chloroform (CHCl$_3$) is a singlet; the proton resonance is not split by chlorine.

The influence of neighboring spins on the multiplicity of peaks is called *spin–spin splitting*. The distance between the two peaks for the resonance of one nucleus split by another is a measure of how strongly the nuclear spins influence each other and is called the *coupling constant*, whose symbol is *J* and is measured in Hz. In 1-chloro-4-nitrobenzene **(2-1)** the coupling between A and X is 10.0 Hz, a relatively large value of *J*. In general, when there are only two nuclei in the coupled system, the resulting spectrum is referred to as AX. Notice that the splitting in both the A and the X portions of the spectrum is the same (Figure 2-18), since *J* is a measure of the interaction between the nuclei and must be the same for either nucleus. Moreover, *J* in Hz is independent of B_0, because the magnitude of the interaction depends only on nuclear properties and not on external quantities such as the field. Thus in 1-chloro-4-nitrobenzene **(2-1)**, *J* is 10.0 Hz when measured either at 7.05 T as in Figure 2-18 or at 14.1 T (600 MHz).

For two nuclei to couple, there must be a mechanism whereby spin information is transferred between them. The most common mechanism involves the interaction of electrons along the bonding path between the nuclei (see Figure 2-19 for an abbreviated coupling pathway over two bonds). Electrons, like protons, act like spinning particles and have a magnetic moment. The X proton very slightly influences or polarizes the spins of its surrounding electrons, making the electron spins slightly favor one I_z state.

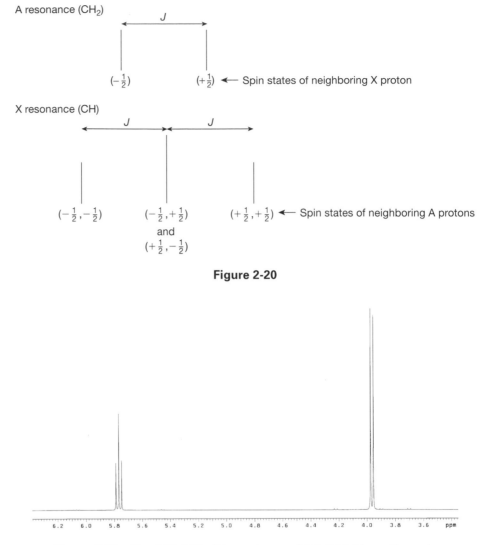

Figure 2-20

Figure 2-21   The 300 MHz ^{1}H spectrum of 1,1,2-trichloroethane in CDCl$_3$.

Thus a spin $+\frac{1}{2}$ proton polarizes the electron to be $-\frac{1}{2}$. The electron in turn polarizes the other electron of the C—H bond, and so on, finally reaching the resonating A proton. Because J normally represents a through-bond interaction, it is useful when drawing conclusions about bond properties such as bond strengths and steric arrangements.

(ClCH$_2^A$—CH^{X}Cl$_2$)

2-2

Additional splitting occurs when a resonating nucleus is close to more than one nucleus. For example, 1,1,2-trichloroethane (**2-2**) has two types of protons, which we have labeled A (CH$_2$) and X (CH). The A protons experience two different environments from the $+\frac{1}{2}$ and $-\frac{1}{2}$ spin states of the X proton and therefore are split into a 1/1 doublet. The X proton, however, experiences three different magnetic environments because the A spins now must be considered collectively: both may be $+\frac{1}{2}$ (++), both $-\frac{1}{2}$ (−−), and one may be $+\frac{1}{2}$ while the other is $-\frac{1}{2}$ (two equivalent possibilities: +− and −+). The three different A environments—(++), (+−)/(−+), (−−)—therefore result in three X peaks in the ratio 1/2/1 (Figure 2-20). Thus the spectrum of **2-2** contains a doublet and a triplet and is referred to as A$_2$X (or AX$_2$ if the labels are switched) (Figure 2-21). The value of J is found in three different spacings in the spectrum, between any two adjacent peaks (Figures 2-20 and Figure 2-21).

As the number of neighboring spins increases, so does the complexity of the spectrum. The ethyl groups in diethyl ether form an A$_2$X$_3$ spectrum (Figure 2-22). The methyl protons are split into a 1/2/1 triplet by the neighboring methylene protons. Because the methylene protons are split by three methyl protons, there are four peaks

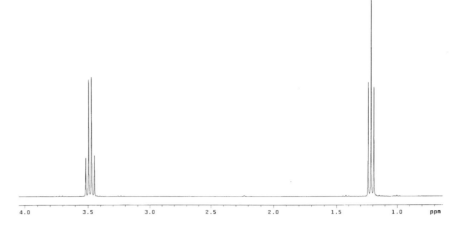

Figure 2-22   The 300 MHz ^{1}H spectrum of diethyl ether in CDCl$_3$.

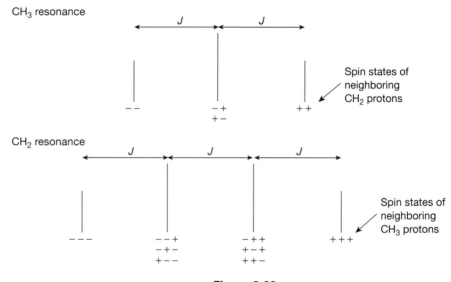

Figure 2-23

in the methylene resonance. The neighboring methyl protons can have all positive spins $(+++)$, two spins positive and one negative (three ways: $++-$, $+-+$, $-++$), one spin positive and two negative (three ways: $+--$, $-+-$, $--+$), and all negative spins $(---)$. The result is a 1/3/3/1 quartet (Figure 2-23). The triplet–quartet pattern seen in Figure 2-22 is a reliable diagnostic indicator for the presence of an ethyl group.

   The splitting patterns of larger spin systems may be deduced in a similar fashion. If a nucleus is coupled to n equivalent nuclei with $I = \frac{1}{2}$, there are $n + 1$ peaks, unless second-order effects discussed later are present. The intensity ratios to a first-order approximation correspond to the coefficients in the binomial expansion and may be obtained from Pascal's triangle (Figure 2-24), since arrangements of I_z states are statistically independent events. Pascal's triangle is constructed by summing two adjacent integers and placing the result one row lower and between the two integers. Zeros are imagined outside the triangle. The first row (1) gives the resonance multiplicity when there is no neighboring spin, the second row (1/1) when there is one neighboring spin, and so on. We have already seen that two neighboring spins give a 1/2/1 triplet and three give a 1/3/3/1 quartet. Four neighboring spins are present for the CH proton in the arrangement —CH$_2$—CHX—CH$_2$— (X is nonmagnetic) and the CH resonance is a 1/4/6/4/1 quintet. The CH resonance from an isopropyl group, —CH(CH$_3$)$_2$, is a 1/6/15/20/15/6/1 septet. Several common spin systems are given in Table 2-1.

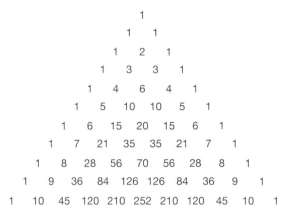

Figure 2-24   Pascal's triangle.

TABLE 2-1

Common First-Order Spin–Spin Splitting Patterns

Spin System	Molecular Substructure	A Multiplicity	X Multiplicity
AX	—CH^A—CH^X—	doublet (1/1)	doublet (1/1)
AX_2	—CH^A—CH^X_2—	triplet (1/2/1)	doublet (1/1)
AX_3	—CH^A—CH^X_3	quartet (1/3/3/1)	doublet (1/1)
AX_4	—CH^X_2—CH^A—CH^X_2—	quintet (1/4/6/4/1)	doublet (1/1)
AX_6	CH^X_3—CH^A—CH^X_3	septet (1/6/15/20/15/6/1)	doublet (1/1)
A_2X_2	—CH^A_2—CH^X_2—	triplet (1/2/1)	triplet (1/2/1)
A_2X_3	—CH^A_2—CH^X_3	quartet (1/3/3/1)	triplet (1/2/1)
A_2X_4	—CH^X_2—CH^A_2—CH^X_2—	quintet (1/4/6/4/1)	triplet (1/2/1)

Except in cases of second-order spectra (discussed in Chapter 4), coupling between protons that have the same chemical shift does not lead to splitting in the spectrum. It is for this reason that the spectrum of benzene in Figure 2-1 is a singlet, even though the protons are coupled to each other. For the same reason, protons within a methyl group normally do not split each other. Examples of unsplit spectra include those of acetone, cyclopropane, and dichloromethane.

All the coupling examples given so far have been between vicinal protons, over three bonds (H—C—C—H). Coupling over four or more bonds usually is small or unobservable. It is possible for geminal protons (—CH_2—) to split each other, provided each proton of the methylene group has a different chemical shift. Geminal splittings are observed when the carbon is part of a ring with unsymmetrical substitution on the upper and lower faces, when there is a single chiral center in the molecule, or when an alkene lacks an axis of symmetry (XYC=CH_2).

Coupling can occur between ^1H and ^{13}C as well as between two protons. Because ^{13}C is in such low natural abundance (about 1.1%), these couplings are not important in analyzing ^1H spectra. In 99 cases out of 100, protons are attached to nonmagnetic ^{12}C atoms. Small satellite peaks from the 1.1% of ^{13}C sometimes can be seen in ^1H spectra. In the ^{13}C spectrum, the carbon nuclei are coupled to nearby protons. The largest couplings occur with protons that are directly attached to the carbon. Thus the ^{13}C resonance of a methyl carbon is split into a quartet, of a methylene carbon into a triplet, and of a methinyl carbon (CH) into a doublet; a quaternary carbon is not split by one bond coupling. Figure 2-25 (upper) shows the ^{13}C spectrum of 3-hydroxybutyric acid, ($CH_3CH(OH)CH_2CO_2H$), which contains a carbon resonance with each type of multiplicity. From right to left are seen a quartet (CH_3), a triplet (CH_2), a doublet (CH), and a singlet (CO_2H). Thus the splitting pattern in the ^{13}C spectrum is an excellent diagnostic indicator for each of these types of groupings within a molecule.

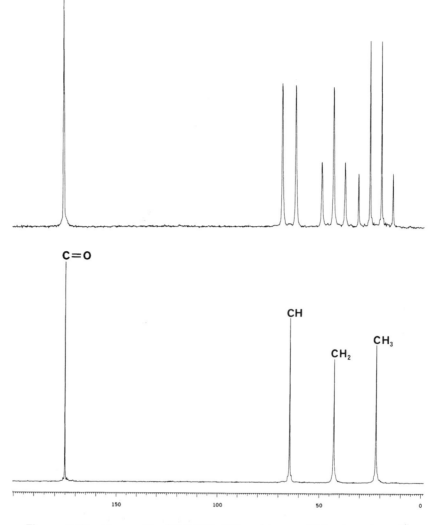

Figure 2-25 Upper: The 22.6 MHz ^{13}C spectrum of 3-hydroxybutyric acid, $CH_3CH(OH)CH_2CO_2H$, without solvent. Lower: The ^{13}C spectrum of the same compound with proton decoupling.

Instrumental procedures, called *decoupling*, are available by which spin–spin splittings may be removed. These methods are discussed in Chapter 5 and involve irradiating one nucleus with an additional field (B_2) while observing another nucleus resonating in the B_1 field. Because a second field is being applied to the sample, the experiment is called *double resonance*. This procedure was used to obtain the ^{13}C spectrum of methyl acetate in Figure 2-7 and the lower spectrum of 3-hydroxybutyric acid in Figure 2-25. It is commonly used to obtain a very simple ^{13}C spectrum, so that each carbon gives a singlet. Measurement of both decoupled and coupled ^{13}C spectra then produces in the first case a simple picture of the number and types of carbons and in the second case the number of protons to which they are attached (Figure 2-25).

2-6 QUANTITATION

The signal detected when nuclei resonate is directly proportional to the number of spins present. Thus a methyl group produces three times the proton signal of a methinyl proton. This difference in intensity can be measured through electronic integration and exploited to elucidate the molecular structure. Figure 2-26 illustrates electronic integration for the ^1H spectrum of ethyl *trans*-crotonate ($CH_3CH=CHCO_2CH_2CH_3$). The vertical displacement of the continuous line above or through each resonance provides a measure of the area under the peaks. The vertical displacements show that the dou-

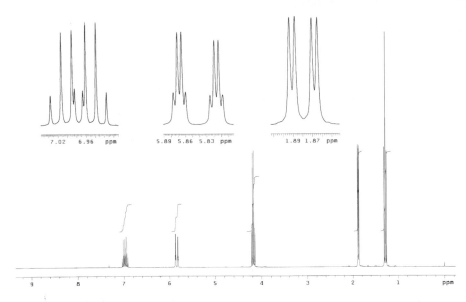

Figure 2-26 The 300 MHz ^{1}H spectrum of ethyl *trans*-crotonate in CDCl$_3$. The continuous line above or through resonances is the integral.

blet at δ 5.84, the quartet at δ 4.19, and the triplet at δ 1.28 are respectively in the ratio 1/2/3. The integration provides only relative intensity data, so that the experimentalist must choose a resonance with a known or suspected number of protons and normalize the other integrals to it. Often integrals are provided digitally by the spectrometer.

Each of the peaks in the illustrated ^{1}H spectrum of ethyl crotonate (Figure 2-26) may be assigned by examination of the integral and splitting pattern. The triplet at lowest frequency (highest field) (δ 1.28) has an integral of 3 and must come from the methyl part of the ethyl group. Its *J* value corresponds to that of the quartet in the middle of the spectrum at δ 4.19, whose integral is 2. This latter multiplet then must come from the methylene group. The mutually coupled methyl triplet and methylene quartet form the resonances for the ethyl group attached to oxygen (—OCH$_2$CH$_3$). The methylene resonance is at higher frequency than the methyl resonance because CH$_2$ is closer than CH$_3$ to the electron-withdrawing oxygen.

The remaining resonances in the spectrum come from protons coupled to more than one type of proton and deserve closer examination. The highest frequency (lowest field) resonance (δ 6.98) has an intensity of unity and comes from one of the two alkenic (—CH=) protons. This resonance is split into a doublet (*J* = 16 Hz) by the other alkenic proton, and then each member of the doublet is further split into a quartet (*J* = 7 Hz) by coupling to the methyl group on carbon with a crossover of the inner peaks. Stick diagrams are useful in analyzing complex multiplets, as in Figure 2-27, for the resonance at δ 6.98.

The resonance of unit integral at δ 5.84 is from the other alkenic proton and is split into a doublet (*J* = 16 Hz) by the proton at δ 6.98. There is a small coupling (1 Hz) over four bonds to the methyl group, giving rise to a quartet (Figure 2-28). The significance of these differences in the magnitude of couplings is discussed in Chapter 4. The resonance at δ 6.98 can be recognized as being from the proton closer to the methyl (*J* is over three rather than four bonds).

The resonance at δ 1.88 has an integral of 3 and hence comes from the remaining methyl group, on the double bond. As it is split by both alkenic protons but with unequal couplings (7 and 1 Hz), four peaks result (Figure 2-29). This grouping is called a doublet of doublets; the term quartet normally is reserved for 1/3/3/1 multiplets. The two unequal couplings in the resonance at δ 1.88 correspond precisely to the quartet splittings found, respectively, in the two alkenic resonances.

Final assignments: δ   1.9 7.0   5.8   4.2 1.3
                                 CH$_3$CH=CHCO$_2$CH$_2$CH$_3$

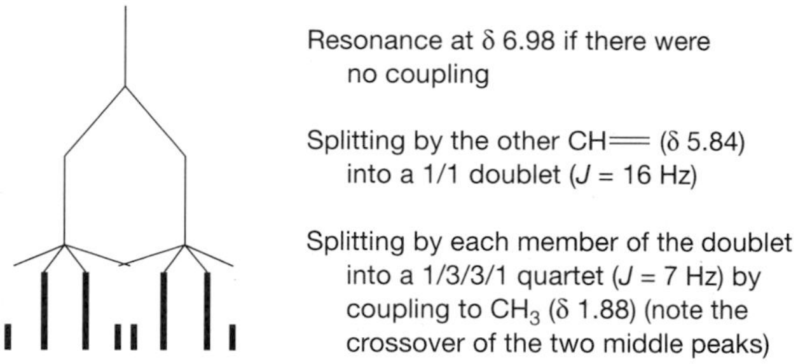

Resonance at δ 6.98 if there were
no coupling

Splitting by the other CH═══ (δ 5.84)
into a 1/1 doublet (J = 16 Hz)

Splitting by each member of the doublet
into a 1/3/3/1 quartet (J = 7 Hz) by
coupling to CH_3 (δ 1.88) (note the
crossover of the two middle peaks)

Figure 2-27

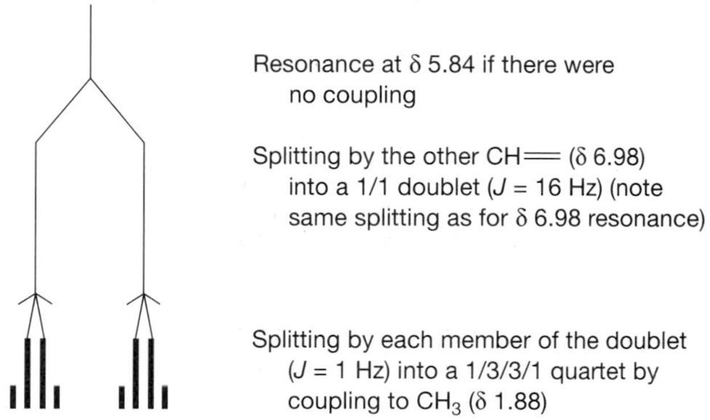

Resonance at δ 5.84 if there were
no coupling

Splitting by the other CH═══ (δ 6.98)
into a 1/1 doublet (J = 16 Hz) (note
same splitting as for δ 6.98 resonance)

Splitting by each member of the doublet
(J = 1 Hz) into a 1/3/3/1 quartet by
coupling to CH_3 (δ 1.88)

Figure 2-28

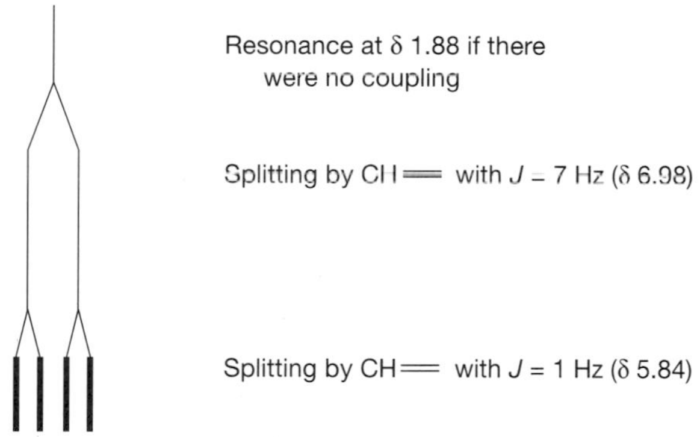

Resonance at δ 1.88 if there
were no coupling

Splitting by CH═══ with J = 7 Hz (δ 6.98)

Splitting by CH═══ with J = 1 Hz (δ 5.84)

Figure 2-29

Integration may be used as a measure of the relative amounts of components in a mixture. In this case, after normalizing for the number of protons in a grouping, the proportions of the components may be calculated from the relative integrals of protons in different molecules. An internal standard may be included with a known concentration. Comparison of other resonances to those of the standard thus can provide absolute concentration.

2-7 COMMONLY STUDIED NUCLIDES

Which nuclei (in this context, "nuclides") are useful in chemical problems? The answer depends on one's area of specialty. Certainly for the organic chemist, the most common elements are carbon, hydrogen, oxygen, and nitrogen (Table 2-2). The biochemist would add phosphorus to the list. The organometallic or inorganic chemist would focus on whichever elements are of potential use, possibly boron, silicon, tin, mercury, platinum, or even some of the low intensity nuclei such as iron and potassium. The success of the experiment depends on several factors.

TABLE 2-2
NMR Properties of Common Nuclei

Nuclide	Spin	Natural Abundance (N_a) (%)	Natural Sensitivity (N_s)(for equal numbers of nuclei) (vs. ^{1}H)	Receptivity (vs. ^{13}C) $\left(\dfrac{N_a \cdot N_s}{1.108 \cdot 0.0159}\right)$	NMR Frequency (at 7.05 T)	Reference Substance
Proton	$\frac{1}{2}$	99.985	1.00	5680	300.00	$(CH_3)_4Si$
Deuterium	1	0.015	0.00965	0.0082	46.05	$(CD_3)_4Si$
Lithium-7	$\frac{3}{2}$	92.58	0.293	1540	38.86	LiCl
Boron-11	$\frac{3}{2}$	80.42	0.165	754	96.21	$Et_2O \cdot BF_3$
Carbon-13	$\frac{1}{2}$	1.108	0.0159	1.00	75.45	$(CH_3)_4Si$
Nitrogen-14	1	99.63	0.00101	5.69	21.69	$NH_3(l)$
Nitrogen-15	$\frac{1}{2}$	0.37	0.00104	0.0219	30.42	$NH_3(l)$
Oxygen-17	$\frac{5}{2}$	0.037	0.0291	0.0611	40.68	H_2O
Fluorine-19	$\frac{1}{2}$	100	0.833	4730	282.27	CCl_3F
Silicon-29	$\frac{1}{2}$	4.70	0.00784	2.09	59.61	$(CH_3)_4Si$
Phosphorus-31	$\frac{1}{2}$	100	0.0663	377	121.44	85% H_3PO_4
Sulfur-33	$\frac{3}{2}$	0.76	0.00226	0.0973	23.04	CS_2
Chlorine-35	$\frac{3}{2}$	75.53	0.0047	20.2	29.40	NaCl (aq)
Cobalt-59	$\frac{7}{2}$	100	0.277	1570	71.19	$K_3Co(CN)_6$
Tin-119	$\frac{1}{2}$	8.58	0.517	25.2	37.29	$(CH_3)_4Sn$
Platinum-195	$\frac{1}{2}$	33.8	0.00994	19.1	64.38	Na_2PtCl_6
Mercury-199	$\frac{1}{2}$	16.84	0.00567	5.42	53.73	$(CH_3)_2Hg$

Spin. The overall spin of the nucleus is determined by the spin properties of the protons and neutrons, as discussed in Section 2-1 (see the second column in Table 2-2). By and large, spin $\frac{1}{2}$ nuclei exhibit more favorable NMR properties than quadrupolar nuclei ($I > \frac{1}{2}$). Nuclei with odd mass numbers have half-integral spins ($\frac{1}{2}$, $\frac{3}{2}$, etc.), whereas those with even mass and odd charge have integral spins (1, 2, etc.). Quadrupolar nuclei have a unique mechanism for relaxation that can result in extremely short relaxation times, as discussed in Section 5-2. The relationship between lifetime (Δt) and energy (ΔE) is given by the Heisenberg uncertainty principle: $\Delta E \Delta t \sim \hbar$. When the lifetime of the spin state as measured by the relaxation time is very short, the larger uncertainty in energies implies a larger band of frequencies, or a broadened signal, in the NMR spectrum. The relaxation time and hence the line broadening depends on the distribution of charge within the nucleus as determined by the quadrupole moment. For example, quadrupolar broadening makes ^{14}N ($I = 1$) a generally less useful nucleus than ^{15}N ($I = \frac{1}{2}$).

Natural Abundance. Nature provides us with nuclides in varying amounts (see the third column in Table 2-2). Whereas ^{19}F and ^{31}P are 100% abundant and ^{1}H nearly so, ^{13}C is present only to the extent of 1.1%. The most useful nitrogen (^{15}N) and oxygen (^{17}O) nuclei occur to the extent of much less than 1%. The NMR experiment of course is easier for nuclides with higher natural abundance. Because so little ^{13}C is present, there is a very small probability of having two ^{13}C atoms at adjacent positions in the same molecule ($0.011 \times 0.011 = 0.00012$, or about 1 in 10,000). Thus J couplings are not easily observed between two ^{13}C nuclei in ^{13}C spectra, although procedures to measure them have been developed.

Natural Sensitivity. Nuclides have differing sensitivities to the NMR experiment, as determined by the gyromagnetic ratio and the energy difference ΔE ($= \gamma \hbar B_0$) between the spin states (see Figure 2-6) (see the fourth column in Table 2-2). The larger the energy difference, the more nuclei are present in the lower spin state (see eq. 2-1) and hence are available to absorb energy. The proton is one of the most sensitive nuclei, whereas ^{13}C and

^{15}N unfortunately are rather weak. Tritium (^{3}H) is very useful to the biochemist as a radioactive label. It has $I = \frac{1}{2}$ and is highly sensitive. As a hydrogen label, deuterium also is useful, but it has very low natural sensitivity. Nuclei of interest to the inorganic chemist vary from poorly sensitive iron and potassium to highly sensitive cobalt. Thus it is important to be familiar with the natural sensitivity of a nucleus before designing a NMR experiment.

Receptivity. The signal intensity for a spin $\frac{1}{2}$ nucleus is determined by both natural abundance (in the absence of synthetic labeling) and natural sensitivity. The mathematical product of these two factors is a good measure of how amenable a specific nucleus is to the NMR experiment. Because chemists are quite familiar with the ^{13}C experiment, the product of natural abundance and natural sensitivity for a nucleus is divided by the product for ^{13}C to give the factor known as the receptivity (see the fifth column in Table 2-2). Thus the receptivity of ^{13}C by definition is 1.00. The ^{15}N experiment then is seen to be about 50 times less sensitive than that for ^{13}C, since the receptivity of ^{15}N is 0.0219.

In addition to these factors, Table 2-2 also contains the NMR resonance frequency at 7.05 T (see the sixth column). The last column contains the reference substance for each nuclide, for which $\delta = 0$.

2-8   EXPERIMENTAL METHODS

2-8a   The Spectrometer and the Sample

Although there is a wide variety of NMR instrumentation available, certain components are common to all, including a magnet to supply the B_0 field, devices to generate the B_1 pulse and to receive the NMR signal, a probe for positioning the sample in the magnet, hardware for stabilizing the B_0 field and optimizing the signal, and computers for controlling much of the operation and processing the signals.

Early NMR machines relied on electromagnets. Although a generation of chemists used them, the magnets had low sensitivity and poor stability. Less popular permanent magnets were simpler to maintain but still had low sensitivity. Most research grade instruments today use a superconducting magnet and operate at 3.5–18.8 T (150–800 MHz for protons). These magnets provide high sensitivity and stability. Possibly most important, the very high fields produced by superconducting magnets result in better separation of resonances, because chemical shifts and hence chemical shift differences increase with field strength. The superconducting magnet resembles a solid cylinder with a central axial hole. The direction of the B_0 field (z) is aligned with the axis of the cylinder.

Separate from the magnet is a console that contains, among other components, the transmitter and receiver of the NMR signal. The B_1 field of the transmitter in the pulse experiment is 1–40 mT. In spectrometers designed to record the resonances of several nuclides (multinuclear spectrometers), the B_1 field must be tunable over a range of frequencies. The console in addition contains a recorder to display the signal.

The sample is placed in the most homogeneous region of the magnetic field by means of an adjustable probe. The probe contains a holder for the sample, mechanical means for adjusting its position in the field, electronic leads for supplying the B_1 and B_2 (double resonance) fields and for receiving the signal, and devices for improving magnetic homogeneity.

Liquid samples are examined in a cylindrical tube, usually 5 mm in outer diameter for protons. The axis of the tube is placed parallel to the (z) axis of the superconducting cylinder. Usually a volume of 300–400 μl is used, but, if sample is in short supply, microtubes are available that require a much smaller volume. Wider diameter tubes are available for samples in low concentration or for low sensitivity nuclei. In practice, ^{1}H spectra may be obtained on less than 1 μg of material, although the result depends on the molecular weight.

The sample must have good solubility in a solvent, which must have no resonances in the regions of interest. For protons, $CDCl_3$, D_2O, and acetone-d_6 are traditional NMR solvents. If the spectrum is to be recorded above or below room temperature, the solvent chosen must not boil or freeze during the experiment.

2-8b Optimizing the Signal

The NMR experiment is plagued by the dual problems of sensitivity and resolution. Peak separations of <0.5 Hz may need to be resolved, so the B_0 field must be uniform to a very high degree (for a separation of 0.3 Hz at 300 MHz, field homogeneity must be better than 1 part in 10^9). Corrections to field inhomogeneity may be made for small gradients in B_0 by the use of *shim coils*. For example, the field along the z direction might be slightly higher at one point than at another. Such a gradient may be compensated for by applying a small current through a shim coil built into the probe. Shim coils are available for correcting gradients in all three cartesian coordinates as well as higher order gradients (x^2, xy, and so on).

The sample is spun along the axis of its cylinder at a rate of 20–50 Hz by an air flow to improve homogeneity within the tube. Spinning improves resolution because a nucleus at a particular location in the tube experiences a field that is averaged over a circular path. In the superconducting magnet, the axis of the tube is in the z direction (in electromagnets it is spun along the y direction). Spinning does not average gradients along the axis of the cylinder, so shimming is required primarily for z gradients for a superconducting magnet or y gradients for an electromagnet.

All magnets are subject to field drift, which can be minimized by electronically locking the field to the resonance of a substance contained in the sample. Because deuterated solvents are used quite commonly for this purpose, an *internal lock* normally is at the deuterium frequency for both ^1H and ^{13}C spectra. In some instruments, the field is locked to a sample contained in a separate tube located permanently elsewhere in the probe. This *external lock* is usually found only in spectrometers designed for a highly specific use, such as taking only ^1H spectra.

Since there is an excess of only some 50 spin $\frac{1}{2}$ nuclei per million, the NMR experiment is inherently insensitive. The sensitivity of a given experiment depends on the natural abundance and the natural sensitivity of the observed nucleus (related to the magnetic moment and gyromagnetic ratio). The experimentalist has no control over these factors. The sample size may be increased by using a larger diameter container, and the field strength may be increased (sensitivity increases with the $\frac{3}{2}$ power of B_0). Collapse of peaks through decoupling effectively enhances sensitivity because, for example, several lines in a multiplet become concentrated at a single frequency, as shown in Figure 2-25 (the baseline clearly has less noise in the lower, decoupled spectrum). Decoupling also can bring about sensitivity enhancement through the nuclear Overhauser effect (see Section 5-4). Complex manipulation of pulses can raise sensitivity, as is discussed in Section 5-6.

Finally, routine improvement of signal to noise is achieved through multiple scanning or signal averaging. In this procedure, the digitized NMR spectrum is stored in computer memory. The spectrum is recorded multiple times and is stored in the same locations. Any signal present is reinforced, but noise tends to be canceled out. If n such scans are carried out and added digitally, the theory of random processes states that the signal amplitude is proportional to n and the noise is proportional to $\sqrt{n}$. The signal to noise ratio (S/N) therefore increases by $n/\sqrt{n}$ or $\sqrt{n}$. Thus 100 scans added together theoretically enhance S/N by a factor of about 10. Multiple scanning is routine for most nuclei and necessary for many, such as ^{13}C and ^{15}N.

2-8c Spectral Parameters

In the pulsed experiment, resolution is controlled directly by the amount of time taken to acquire the signal. To distinguish two signals separated by Δv (in Hz), acquisition of data must continue for at least $1/\Delta v$ seconds. For example, a desired resolution of 0.5 Hz in a ^{13}C spectrum requires an *acquisition time* (t_a) of 1/0.5 or 2.0 s. Sampling for a longer time would improve resolution; for example, acquisition for 4.0 s could yield a resolution of 0.25 Hz. Thus longer acquisition times are necessary to produce narrow lines. The tail of the FID contains mostly noise mixed with signals from narrow lines. If sensitivity is the primary concern, either shorter acquisition times should be used or the later portions of the FID may be reduced artificially by a weighting function. In this way noise is reduced, but at the possible expense of lower resolution. Figure 2-30 illustrates this problem. Decreased weighting of the data from longer times (from the bottom to the top of the figure) results in lower resolution and broader peaks.

After acquisition is complete, a *delay time* is necessary to allow the nuclei to relax before they can be examined again for the purpose of signal averaging. For optimal results, the delay time is on the order of 3–5 times the spin–lattice relaxation time. For a typical ^{13}C relaxation time of 10 s, a total cycle thus might take up to 50 s. Such cycle times are excessive and can be reduced by using a pulse angle θ (see Figure 2-13) that is less than 90°. When fewer spins are flipped, T_1 relaxation can occur more rapidly and the delay time can be shortened. In practice, proper balance between θ and t_a (to optimize resolution) can result in essentially no delay time, and the next pulse cycle can be initiated immediately after acquisition.

The detected range of frequencies (the *spectral width*) is determined by how often the detector samples the value of the FID: the *sampling rate*. The FID is made up of a collection of sinusoidal signals (see Figure 2-16). A single, specific signal must be sampled at least twice within one sinusoidal cycle to determine its frequency. For a collection of signals up to a frequency of N, the FID thus must be sampled at a rate of $2N$ Hz. For example, for a ^{13}C spectral width of 15,000 Hz (200 ppm at 75 MHz), the signal must be sampled 30,000 times per second. Figure 2-31 illustrates this situation. The top signal (a) is sampled exactly twice per cycle (each dot is a sampling). The higher frequency signal in (b) is sampled at the same rate, but not often enough to determine its frequency. In fact, the lower frequency signal in (c) gives exactly the same collection of points as in (b). A real signal from the points in (b) is indistinguishable from a *foldover signal* with the frequency in (c). For example, if the ^{13}C spectrum is sampled only 15,000 times per second, for a spectral width of 7500 Hz (100 ppm at 75 MHz), a signal with a frequency of 150 ppm (11,250 Hz) appears as a foldover peak in the 0–100 ppm region.

If a signal is sampled 20,000 times per second, the detector spends 50 µs on each point. The reciprocal of the sampling rate is called the *dwell time*, which signifies the amount of time between sampling. Reducing the dwell time means that more data points are collected in the same period of time, so that a larger computer memory is required. If the acquisition time is 4.0 s (for a resolution of 0.25 Hz) and the sampling

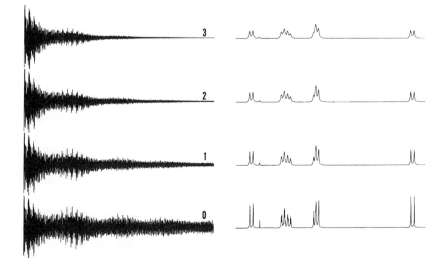

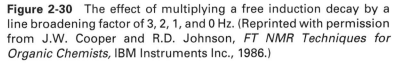

Figure 2-30 The effect of multiplying a free induction decay by a line broadening factor of 3, 2, 1, and 0 Hz. (Reprinted with permission from J.W. Cooper and R.D. Johnson, *FT NMR Techniques for Organic Chemists,* IBM Instruments Inc., 1986.)

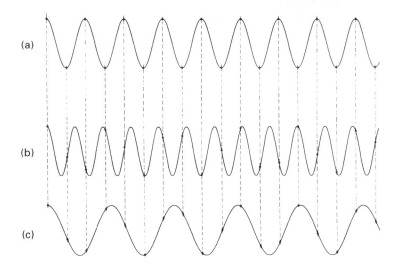

(a)

(b)

(c)

Figure 2-31   (a) Sampling of a sine wave exactly two times per cycle. (b) Sampling a sine wave less than twice per cycle. (c) The lower frequency sine wave that contains the same points as in (b) and is sampled more than twice per cycle. The frequency of (b) is not detected but appears as a foldback peak at the frequency of (c). (Reprinted with permission from J.W. Cooper and R.D. Johnson, *FT NMR Techniques for Organic Chemists,* IBM Instruments Inc., 1986.)

rate is 20,000 times per second (for a spectral width of 10,000 Hz), the computer must store 80,000 data points. Making do with fewer points because of computer limitations would require either lowering the resolution or decreasing the spectral width.

2-9 DYNAMIC EFFECTS

According to the principles outlined in the previous sections, the ^{1}H spectrum of methanol (CH_3OH) should contain a doublet of integral 3 for the CH_3 (coupled to OH) and a quartet of integral 1 for the OH (coupled to CH_3). Under conditions of high purity or low temperatures, such a spectrum is observed (Figure 2-32, lower). The presence of a small amount of acidic or basic impurity, however, can catalyze the intermolecular exchange of the hydroxyl proton. When this proton becomes detached from the molecule by any mechanism, information about its spin states is no longer available to the rest of the molecule. For coupling to be observed, the rate of exchange must be approximately slower than the magnitude of the coupling in Hz. Thus a proton could exchange a few times per second and still maintain coupling. If the rate of exchange is faster than $1/J$, no coupling is observed between the hydroxyl proton and the methyl protons. Thus at high temperatures (Figure 2-32, upper), the ^{1}H spectrum of methanol contains only two singlets. If the the temperature is lowered or the amount of acidic or basic catalyst is decreased, the exchange rate slows down. The coupling constant continues to be washed out until the exchange rate reaches a critical value at which the proton resides sufficiently long on oxygen to permit the methyl group to detect the spin states. As can be seen from Figure 2-32, the transition from *fast exchange* (upper) to *slow exchange* (lower) can be accomplished for methanol over an 80° C temperature range. Under most spectral conditions, there are minor amounts of acid or base impurities, so hydroxyl protons do not usually exhibit couplings to other nuclei. The integral is still unity for the OH group because the amount of catalyst is small. Sometimes the exchange rate is intermediate, between fast and slow exchange, and broadened peaks are observed. The situation for amino protons (NH or NH_2) is similar.

A process that averages coupling constants also can average chemical shifts. A mixture of acetic acid and benzoic acid often contains only one ^{1}H resonance for the CO_2H groups from both molecules. The carboxyl protons exchange between molecules so rapidly that the spectrum exhibits only the average of the two. Moreover, if the solvent is

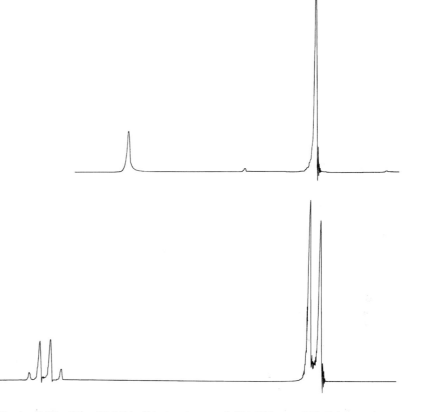

Figure 2-32 The 60 MHz ^{1}H spectrum of CH_3OH at $+50°$ C (upper) and at $-30°$ C (lower).

water, exchangeable protons such as OH in carboxylic acids or alcohols do not give separate resonances. Thus the ^{1}H spectrum of acetic acid (CH_3CO_2H) in water contains two, not three, peaks. The water and carboxyl protons appear as a single resonance whose chemical shift falls between those of the pure materials. If the rate of exchange between —CO_2H and water could be slowed sufficiently, separate resonances would be observed.

Intramolecular (unimolecular) reactions also can influence the appearance of the NMR spectrum. The molecule cyclohexane, for example, contains distinct axial and equatorial protons, yet the spectrum at room temperature exhibits only one sharp singlet. There is no splitting because all protons have the same average chemical shift. Flipping of the ring interconverts the axial and equatorial positions. When the rate of this process is greater (in s^{-1}) than the chemical shift difference between the axial and equatorial protons (in Hz, which of course is s^{-1}), the NMR experiment does not distinguish the two types of protons and only one peak is observed (fast exchange). At lower temperatures, however, the process of ring flipping is much slower. By $-100°$ C the NMR experiment can distinguish the two types of protons, so two resonances are observed (slow exchange). At intermediate temperatures, broadened peaks are observed that reflect the transition from fast to slow exchange. Figure 2-33 illustrates the spectral changes as a function of temperature for cyclohexane in which all protons but one have been replaced by deuterium to remove vicinal proton–proton couplings and simplify the spectrum (eq. 2-4).

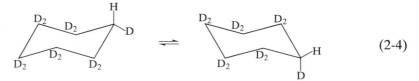

(2-4)

These processes that bring about averaging of spectral features occur reversibly, whether by acid-catalyzed intermolecular exchange or by unimolecular reorganization.

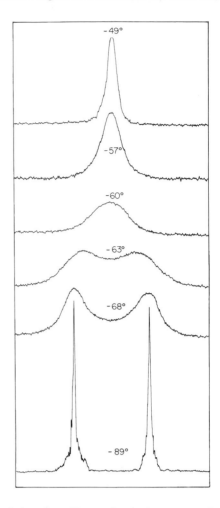

Figure 2-33 The 60 MHz ^{1}H spectrum of cyclohexane-d_{11} as a function of temperature. (Reproduced with permission from F.A. Bovey, F.P. Hood III, E.W. Anderson, and R.L. Kornegay, *J. Chem. Phys.* **41,** 2042 [1964].)

NMR provides one of the few methods for examining the effects of reaction rates when the system is at equilibrium. Most other kinetic methods require that one substance be transformed irreversibly into another. The dynamic effects of the averaging of chemical shifts or coupling constants provide a nearly unique window into processes that occur on the order of a few times per second. The subject is examined further in Section 5-2.

2-10 SPECTRA OF SOLIDS

All of the examples and spectra illustrated thus far have been for liquid samples. Certainly it would be useful to be able to take NMR spectra of solids, so why have we avoided discussion of solid samples? Under conditions normally used for liquids, the spectra of solids are broad and unresolved, providing only modest amounts of information. There are two primary reasons for the poor resolution of solid samples. In addition to the indirect spin–spin (J) interaction that occurs between nuclei through bonds, nuclear magnets also can couple through the direct interaction of their nuclear dipoles. This *dipole–dipole coupling* occurs directly through space. The resulting coupling is designated with the letter D and is much larger than J coupling. In solution, dipoles are continuously reorienting through molecular tumbling. Just as two bar magnets have no net interaction when averaged over all mutual orientations, two nuclear magnets have no net dipolar interaction because of the randomizing effect of tumbling. The indirect J coupling is not averaged to zero because tumbling cannot average out a through-bond interaction. In contrast, nuclear dipoles in the solid phase are held rigidly in position, and the dominant interaction between nuclei is the D coupling, which is on the order of several hundred to a few thousand hertz. One spin might have a dipolar coupling with several nearby spins. Such interactions override J couplings and even most chemical shifts, so that very broad signals are produced.

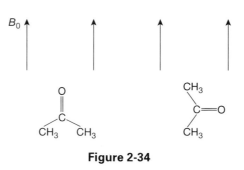

Figure 2-34

As with the J coupling, the D coupling may be eliminated by application of a strong B_2 field. Power levels for removal of D must be much higher than those for J decoupling, since D is two to three orders of magnitude larger than J. Thus high-powered decoupling is used routinely to reduce the linewidth of the spectra of solids.

The second factor that contributes to line broadening for solids is *anisotropy of chemical shielding* (the term "chemical shift anisotropy" should be avoided in this context, since, strictly speaking, chemical shift is a scalar quantity and cannot be anisotropic). In solution the observed chemical shift is the average of the shielding of a nucleus over all orientations in space, as the result of molecular tumbling. In the solid, shielding of a specific nucleus in a molecule depends on the way the molecule is oriented with respect to the B_0 field. Consider the carbonyl carbon of acetone. When the B_0 field is parallel to the C=O bond, the nucleus experiences a different shielding from when it is perpendicular to the C=O bond (Figure 2-34). The ability of electrons to circulate and give rise to shielding varies according to the arrangement of bonds in space. Differences between the abilities of electrons to circulate in the two arrangements in Figure 2-34, as well as for all other arrangements, generates a range of shieldings and hence a range of resonance frequencies.

Double irradiation does not average chemical shielding anisotropy, since the effect is entirely geometrical. The problem is largely removed by spinning the sample to mimic the process of tumbling. The effects of spinning are optimized when the axis of spinning is set at an angle of 54°44' to the direction of the B_0 field. This is the angle between the edge of a cube and the adjacent solid diagonal. Spinning along this diagonal averages each cartesian direction just as tumbling in solution does. When the sample is spun at this axis to the field, the various arrangements of Figure 2-34 average and the chemical shieldings are reduced to the isotropic chemical shift. The technique therefore has been called *magic angle spinning* (MAS). Because shielding anisotropies are generally a few hundred to several thousand hertz, the rate of spinning must exceed this range in order to average all orientations. Typical minimum spinning rates are 2–5 kHz, but rates up to 20 kHz are available.

The combination of strong irradiation to eliminate dipolar couplings and magic angle spinning to eliminate chemical shielding anisotropy results in spectra of solids that are almost as high in resolution as those of liquids. Figure 2-35 shows the ^{13}C spectrum of polycrystalline β-quinolmethanol clathrate. The broad, featureless spectrum at the top (a) is typical of solids. Strong double irradiation (b) eliminates dipolar couplings and brings out some features. Magic angle spinning in addition to decoupling (c) produces a high resolution spectrum.

Relaxation times are extremely long for solids because the motion of nuclei necessary for spin–lattice relaxation is slow or absent. Carbon-13 spectra could take an extremely long time to record because the nuclei must be allowed to relax for several minutes between pulses. This problem is solved by taking advantage of the more favorable properties of the protons that are coupled to the carbons. The same double irradiation process that eliminates J and D couplings also is used to transfer some of the proton's higher magnetization and faster relaxation to the carbon atoms. The process is called *cross polarization* (CP) and is standard for most solid spectra of ^{13}C. After the protons are moved onto the y axis by a 90° pulse, a continuous y field is applied to keep the magnetization precessing about the y axis, a process called *spin locking*. The fre-

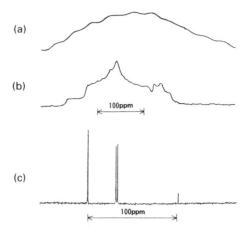

Figure 2-35 The ^{13}C spectrum of polycrystalline β-quinolmethanol clathrate (a) without dipolar decoupling, (b) with decoupling, and (c) with both decoupling and magic angle spinning. (Reproduced with permission from T. Terao, *JEOL News,* **19,** 12 [1983].)

quency of this field ($\gamma_H B_H$) is controlled by the operator. When the ^{13}C channel is turned on, its frequency ($\gamma_C B_C$) can be set equal to the ^{1}H frequency (the *Hartmann-Hahn condition,* $\gamma_H B_H = \gamma_C B_C$). Both protons and carbons then are precessing at the same frequency and hence have the same net magnetization, which for carbon is increased over the normal pulse experiment. Carbon resonances thus have enhanced intensity. When carbon achieves maximum intensity, B_C is turned off (ending the *contact time*) and carbon magnetization is acquired while B_H is retained for dipolar decoupling.

The higher resolution and sensitivity of the cross polarization and magic angle spinning experiment (CP/MAS) opened vast new areas to NMR. Inorganic and organic materials that do not dissolve may be subjected to NMR analysis. Synthetic polymers and coal were two of the first materials to be examined. Biological and geological materials such as wood, humic acids, and biomembranes became general subjects for NMR study. Problems unique to the solid state also may be examined, such as structural and conformational differences between solids and liquids.

PROBLEMS

2-1 Determine the number of different hydrogen atoms and their relative proportions in the following molecules. Do the same for carbon atoms.

(a) 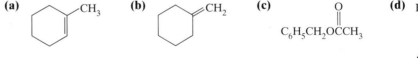 ——CH$_3$ **(b)** ——CH$_2$ **(c)** $C_6H_5CH_2OCCH_3$ (with O double-bonded)   **(d)** H OH

2-2 What is the expected multiplicity for each proton resonance in the following molecules?
(a) ClCH$_2$CH$_2$CH$_2$Cl   **(b)** BrCH(CH$_3$)$_2$ **(c)** C$_6$H$_5$OCCH$_2$CH$_3$ (with O double-bonded)   **(d)** H O Cl / H H

2-3 Predict the multiplicities for the ^{1}H and the ^{13}C resonances in the absence of decoupling for each of the following compounds. For the ^{13}C spectra, give only the multiplicities caused by coupling to attached protons. For the ^{1}H spectra, give only multiplicities caused by coupling to vicinal protons (HCCH).

(a) CH$_3$CH$_2$CH$_2$OCCH$_3$ (with O double-bonded)   **(b)** 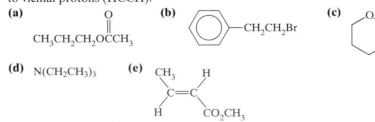 ——CH$_2$CH$_2$Br   **(c)** (ring with O)

(d) N(CH$_2$CH$_3$)$_3$ **(e)** CH$_3$ H / C=C / H CO$_2$CH$_3$

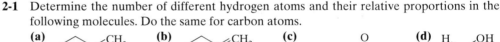

2-4 For each of the following 300 MHz ^{1}H spectra, carry out the following operations. **(i)** From the elemental formula, calculate the unsaturation number. **(ii)** Calculate the relative integrals for each group of protons. Then convert to absolute numbers by selecting one group to be of known integral. **(iii)** Assign a structure to each compound; be sure that your structure agrees with the spectrum in all aspects: number of different proton groups, integrals, splitting patterns.

(a) C_4H_9Br

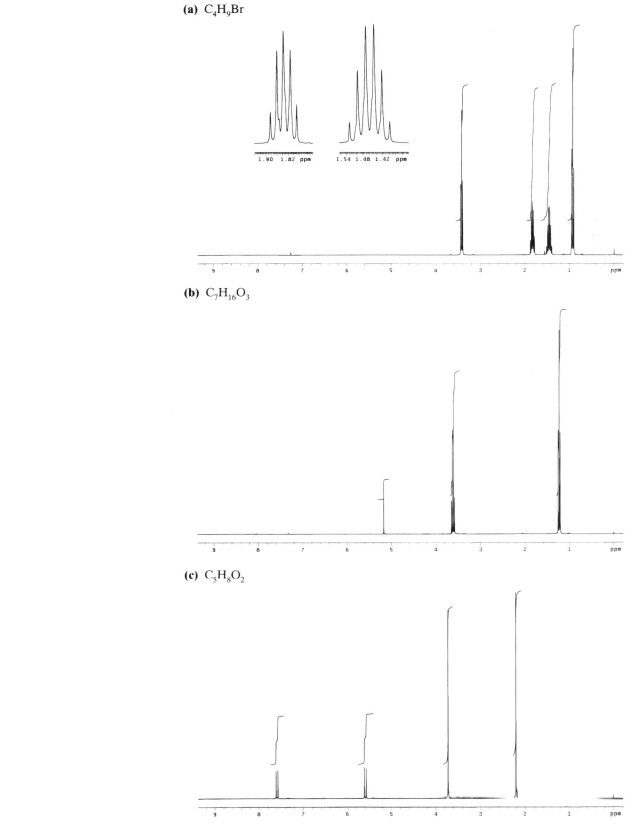

(b) $C_7H_{16}O_3$

(c) $C_5H_8O_2$

3-1 FACTORS

Progressive replacement of hydrogen with chlorine on a methane molecule moves the chemical shift to higher frequency (downfield) because of chlorine's ability to remove electron density from the remaining protons: δ 0.23 for CH_4, 3.05 for CH_3Cl, 5.30 for CH_2Cl_2, and 7.27 for $CHCl_3$. The trend for a series of methyl resonances often can be explained in the same fashion by the inductive or polar effect. The chemical shifts for the series CH_3X for which X is F, HO, H_2N, H, Me_3Si, or Li respectively are δ 4.26, 3.38, 2.47, 0.23, 0.0 (TMS, the standard), and −0.4 (this value is very solvent dependent; the minus sign indicates a lower frequency than TMS), following the electronegativity of the atom attached to CH_3.

Electron density is influenced by resonance as well as by inductive effects, as seen in unsaturated molecules such as alkenes and aromatics. Donation of electrons through resonance by a methoxy group increases the electron density at the β position of a vinyl ether (3-1) and at the para position of anisole. Thus the chemical shift of the β protons

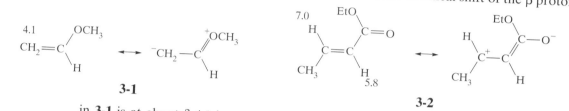

3-1

3-2

in **3-1** is at about δ 4.1 in comparison with δ 5.28 in ethene. The resonance frequency decreases, as expected with the increased shielding from electron donation. The electron-withdrawing inductive effect of CH_3O is overpowered by the resonance effect. Groups such as nitro, cyano, and acyl withdraw electrons by both resonance and induction, so they can bring about significant shifts to higher frequency (downfield). Ethyl *trans*-crotonate (**3-2**, also see Section 2-6) illustrates this effect. The electron-withdrawing group shifts the β proton strongly to high frequency (δ 7.0). Although the α proton is not subjected to this strong resonance effect, it is close enough to the electron-withdrawing carboethoxy group to be shifted slightly to higher frequency (δ 5.8) by the inductive effect.

Hybridization of the carbon to which a proton is attached also influences electron density. As the proportion of s character increases from sp^3 to sp^2 to sp orbitals, bonding electrons move closer to carbon and away from the protons, which then become deshielded. For this reason methane and ethane resonate respectively at δ 0.23 and 0.86 but ethene resonates at δ 5.28. Ethyne (acetylene) is an exception in this consideration, as we shall see. Hybridization contributes to shifts in strained molecules, such as cyclobutane (δ 1.98) and cubane (δ 4.00), for which hybridization is intermediate between sp^3 and sp^2.

Induction, resonance, and hybridization modulate electron density at the proton itself, as the result of local electron currents around the nucleus (Figure 3-1b). Purely magnetic effects of substituents in the absence of changes in electron density also can have major effects on proton shielding, but only when the groups have a nonspherical shape. The drawing in Figure 3-1c represents the combined effects of local fields and nonlocal fields. The group giving rise to the nonlocal field could be, for example, methyl, phenyl, or carbonyl, and the resonating nucleus need not be attached directly to the group. To see why a spherical or isotropic ("same in all directions") group contributes no nonlocal effect, consider a proton attached to such a group, for example, chlorine. The local effect arises from the electrons that surround the resonating proton. The electrons in the substituent, which are not around the proton, also precess in the applied field. They induce a magnetic field that opposes B_0 and can have a value at the position of the proton. The nonlocal induced field may be represented by magnetic lines of force. If the bond from the spherical substituent to the resonating proton is parallel to the direction of B_0, as in Figure 3-2a, the lines of force from the induced field oppose B_0 at the proton, shielding it. If the bond from the substituent to the proton is perpendicular to B_0, as in Figure 3-2b, the induced lines of force reinforce those of B_0, deshielding the proton. Because the group is isotropic, the two arrangements are equally probable. As the molecule tumbles in solution, the effects of the induced field cancel out. Other orientations are canceled in a similar fashion. Thus an isotropic substituent has no effect in addition to what it provides to local currents from induction or resonance.

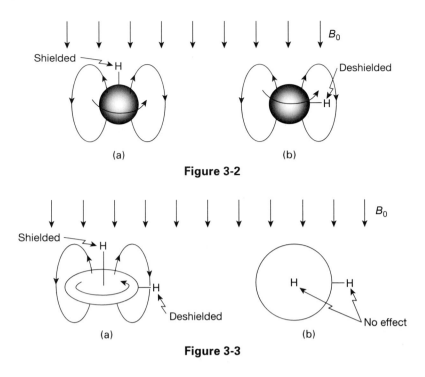

Figure 3-2

Figure 3-3

Most substituents, however, are not spherical. The flat shape of an aromatic ring, for example, resembles an oblate ellipsoid, and the elongated shape of single or triple bonds resembles a prolate ellipsoid. For a proton situated at the edge of an oblate ellipsoid such as a benzene ring, there again are two extremes (Figure 3-3). When the flat portion is perpendicular to the static field (a), a proton at the edge is deshielded, since the induced lines of force reinforce the B_0 field. For the same geometry (a), a proton situated over the middle of the ellipsoid is shielded, as the induced lines of force oppose B_0. For this geometry, the induced field is large because aromatic electrons circulate easily around the ring. When the ring is parallel to B_0 (b), induced currents would have to move from one ring face to the other. As a result, little current or field is induced from this geometry. The cancelation seen for the sphere as the molecule tumbles in solution does not occur for aromatic rings. A group that has appreciably different currents induced by B_0 from different orientations in space is said to have *diamagnetic anisotropy*. Because an oblate ellipsoid has the larger effect for the geometry shown in Figure 3-3a, a proton at the edge of an aromatic ring is deshielded and one at or over the center is shielded. It is for this reason that benzene resonates at such a high frequency (low field) (δ 7.27) compared with alkenes (compare ethene at δ 5.28).

Figure 3-4 illustrated how the diagmagnetic effect of benzene is shielding ($+$) above and below the ring but deshielding ($-$) around the edge. The effect changes sign at a null point at which the angle from the ring is 54°44', corresponding to a zero value of the expression $(3\cos^2\theta - 1)$. Although the protons of benzene reside in the deshielded portion of the cone, molecules have been constructed to explore the full range of the effect. The methylene protons of methano[10]annulene **(3-3)** are con-

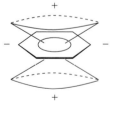

Figure 3-4

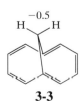

3-3

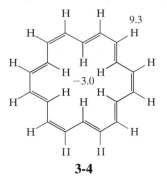

3-4

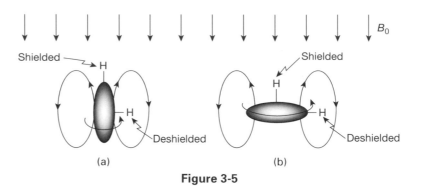

Figure 3-5

strained to positions above the aromatic 10π electron system and consequently are shielded to a position (δ -0.5) even lower frequency (higher field) than TMS. [18]Annulene **(3-4)** has one set of protons around the edge of the aromatic ring that resonates at a deshielded position of δ 9.3 and a second set located toward the center of the ring that resonates at a shielded position of δ -3.0.

The two arrangements of a prolate ellipsoid, used as a model for a chemical bond, may be considered in a similar fashion (Figure 3-5). In this case it is not always clear which arrangement has a stronger induced current. The π electrons of acetylene (ethyne) provide one clearcut case. When the axis of the molecule is parallel to the B_0 field (as in Figure 3-5a), the π electrons are particularly susceptible to circulation around the cylinder. The alternative arrangement in Figure 3-5b is ineffective for acetylene and therefore does not provide a canceling effect. The acetylenic proton is attached to the end of this array of electrons and hence is shielded. For this reason, the acetylene resonance (δ 2.88) falls between those of ethane (δ 0.86) and ethene (δ 5.28). The effects of hybridization thus are superseded by those of diamagnetic anisotropy.

Circulation of charge within a single bond is less strong than that of a triple bond, and the larger effect is seen when its axis is perpendicular to B_0 (as in Figure 3-5b). Thus a proton at the side of a single bond is more shielded than one along its end (Figure 3-6a). The axial and equatorial protons of a rigid cyclohexane ring exemplify these arrangements (Figure 3-6b). The two illustrated protons are equivalently positioned with respect to the 1,2 and 6,1 bonds, which thus produce no differential effect. The 1-axial proton, however, is in the shielding region of the further removed 2,3 and 5,6 bonds (darkened), whereas the 1-equatorial proton is in their deshielding region. In general, axial protons are more shielded and resonate at lower frequency (higher field) than equatorial protons, typically by about 0.5 ppm. The high frequency position for methine compared with methylene protons and for methylene compared with methyl protons [CH_3X to CH_3CH_2X to $(CH_3)_2CHX$] for a single X group has been attributed to the anisotropy of the additional C—C bonds (Figure 3-7), although changes in hybridization also can contribute. The highly shielded position of cyclopropane resonances may be attributed either to an aromatic-like ring current or to the anisotropy of the C—C bond that is opposite to a CH_2 group in the three-membered ring.

The anisotropy of double bonds is more difficult to assess, because they have three nonequivalent axes. Protons situated over double bonds are more shielded than those in the plane (Figure 3-8), either for alkenes (a) or carbonyl groups (b). The positions of the

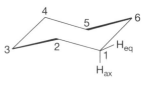

Figure 3-6

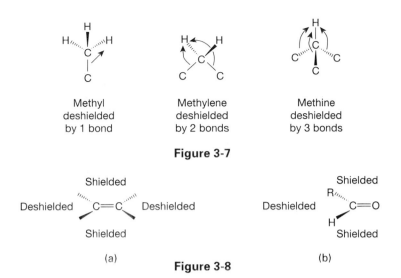

Figure 3-7

Figure 3-8

(a)   (b)

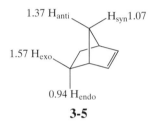

3-5

methylene protons in norbornene (3-5) may be explained in this fashion, since the syn and endo protons respectively are shielded with respect to the anti and exo protons. The highly deshielded position of aldehydes (δ 9.7) is attributed to a combination of a strong inductive effect and the diamagnetic anisotropy of the carbonyl group.

The nonspherical cloud of lone pairs of electrons also may exhibit diamagnetic anisotropy. A proton that is hydrogen bonded to a lone pair invariably is deshielded. Thus the hydroxyl proton in free ethanol (as a dilute solute in a non-hydrogen bonding solvent such as CCl_4) resonates at δ 0.7, but in pure ethanol with extensive hydrogen bonding it resonates at δ 5.3. Carboxylic protons (CO_2H) resonate at extremely high frequency (low field) because every proton is hydrogen bonded within a dimer. Lone pair anisotropy also has been invoked to explain trends in ethyl groups, CH_3CH_2X. The resonance positions of the CH_2 group attached to X are well explained by induction [for X = F (4.36), Cl (3.47), Br (3.37), I (3.16)], but the trend for the more distant methyl group is opposite (in the same order, 1.24, 1.33, 1.65, 1.86). As the value of X increases, the lone pair moves closer to the methyl group and more strongly deshields it.

Functional group effects on proton chemical shifts may be explained largely by the two general effects described above. (1) Electron withdrawal or donation by resonance or induction (including hybridization) alters the electron density around the resonating proton. Higher electron density shields the proton and moves its resonance position to a lower frequency (downfield). (2) Diamagnetic anisotropy of nonspherical substituents is largely responsible for the proton resonance positions of aromatics, acetylenes, aldehydes, cyclopropanes, cyclohexanes, alkenes, and hydrogen-bonded species.

3-2 PROTON CHEMICAL SHIFTS AND STRUCTURE

Assignment of structure based on NMR spectra requires knowledge of the relationship between chemical shifts and functional groups. Normally, both proton and carbon spectra are recorded and analyzed. This section considers the relationship between proton resonances and structure. Figure 3-9 summarizes the resonance ranges for common proton functionalities.

3-2a Saturated Aliphatics

Alkanes. Cyclopropane has the lowest frequency position (δ 0.22) of any simple hydrocarbon because of a ring current or the anisotropy of the carbon–carbon bonds. Unsubstituted methane has essentially the same chemical shift (δ 0.23). Progressive addition of carbon–carbon bonds to methane results in a shift to higher frequency (downfield), as in the series ethane (CH_3CH_3, δ 0.86), propane ($CH_3CH_2CH_3$, δ 1.33), and isobutane [$(CH_3)_3CH$, δ 1.56]. Cyclic structures other than cyclopropane and

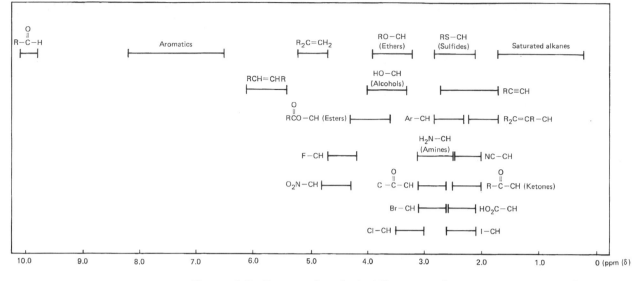

Figure 3-9 Proton chemical shift ranges for common structural units. The symbol CH represents methyl, methylene, or methine, and R represents a saturated alkyl group. The range for —CO$_2$**H** and other strongly hydrogen-bonded protons is off scale to the left. The indicated ranges are for common examples; actual ranges can be larger.

cyclobutane (δ 1.98) have resonance positions similar to those of open chain systems, for example, δ 1.43 for cyclohexane. In complex natural products such as steroids or alkaloids, a large number of structurally similar alkane protons leads to overlapping resonances in the region δ 0.8–2.0, whose analysis requires the highest possible field.

Functionalized Alkanes. The presence of a functional group alters the resonance position of neighboring protons according to the inductive effect of the group and its diamagnetic anisotropy. Ethane (δ 0.86) is a useful point of reference for methyl groups. Replacement of one methyl group in ethane with hydroxyl yields methanol, whose resonance position is δ 3.38. The electron-withdrawing effect of the oxygen atom is the primary cause of the large shift to higher frequency (lower field). Just as in unfunctionalized alkanes, methylene groups (CH$_3$**CH$_2$**OH, δ 3.56) and methine groups [(CH$_3$)$_2$**CH**OH, δ 3.85] are found at progressively higher frequency. In general, methylene and methine protons respectively resonate about 0.4 and 0.7 ppm to higher frequency than analogous methyl groups. There is considerable variation from one example to another, depending on the remainder of the structure, so that resonances for a given functionality can range over 1 ppm. Ether resonances are similar to those for alcohols (CH$_3$OCH$_3$, δ 3.24). Ester alkoxy groups, however, usually resonate at even higher frequency (**CH$_3$**O(CO)CH$_3$, δ 3.67) because the attached oxygen is more electron withdrawing as the result of ester resonance **(3-6)**.

3-6   **3-7**

Because nitrogen is not so electron withdrawing as oxygen, amines resonate at somewhat lower frequency (higher field) than ethers: δ 2.42 for methylamine (**CH$_3$**NH$_2$ in aqueous solution). Introducing a positive charge through quaternization (electron withdrawal) causes a shift to high frequency, as in (CH$_3$)$_3$N (δ 2.22) compared with (CH$_3$)$_4$N$^+$ (δ 3.33). An intermediate charge, as produced in amides through resonance, results in an intermediate shift, as for *N,N*-dimethylformamide (**3-7**, δ 2.88).

The lower electronegativity of sulfur means that sulfides are at even lower frequency (higher field), δ 2.12 for dimethyl sulfide (CH$_3$SCH$_3$). Halogens move reso-

TABLE 3-1

Substituent Parameters for Shoolery's Rule (R = H or Alkyl)

Substituent	Δ_i	Substituent	Δ_i
CH_3	0.47	C_6H_5	1.83
$CR{=}CR_2$	1.32	Br	2.33
$C{\equiv}CR$	1.44	OR	2.36
NR_2	1.57	Cl	2.53
SR	1.64	OH	2.56
CN	1.70	O(CO)R	3.01
CO—R	1.70	NO_2	3.36
I	1.82	F	4.00

nances to higher frequency according to the electronegativity of the atom: δ 2.15 for CH_3I, 2.69 for CH_3Br, 3.06 for CH_3Cl, and 4.27 for CH_3F. In all these cases, the shifts probably are affected by the anisotropy of the C—X bond, but this factor is hard to assess and is somewhat diminished by free rotation in open chain systems. Pseudohalogens also cause shifts to higher frequencies, as for example cyano in acetonitrile (CH_3CN, δ 2.00) and nitro in nitromethane (CH_3NO_2, δ 4.33). Electron-donating substituents such as silicon in TMS cause shifts to lower frequency.

Methyl groups on carbon–carbon double bonds are usually found in the region δ 1.7–2.5, as for isobutylene [$(CH_3)_2C{=}CH_2$, δ 1.70] and toluene ($C_6H_5CH_3$, δ 2.31). Methyl groups on carbon–oxygen double bonds are found in the region δ 2.0–2.7, as for acetone ($CH_3(CO)CH_3$, δ 2.07), acetic acid (CH_3CO_2H, δ 2.10), acetaldehyde (CH_3CHO, δ 2.20), and acetyl chloride (CH_3COCl, δ 2.67).

Empirical correlations between structure and proton chemical shift have been developed for common but relatively simple structural units. The earliest, called Shoolery's rule, provides the chemical shift of protons in a Y—CH_2—X group from substituent parameters Δ_i added to the chemical shift of methane (eq. 3-1 and Table 3-1).

$$\delta = 0.23 + \Delta_X + \Delta_Y \qquad (3\text{-}1)$$

The calculation is reasonably successful for CH_2XY, but additivity fails due to steric effects between groups for many CHXYZ examples. For example, the calculated shift for CH_2Cl_2 is 0.23 + (2 × 2.53) = 5.29 (obs. δ 5.30), but for $CHCl_3$ it is 0.23 + (3 × 2.53) = 7.82 (obs. δ 7.27).

3-2b   Unsaturated Aliphatics

Alkynes.   The anisotropy of the triple bond results in a relatively low frequency (upfield) position for protons on sp-hybridized carbons. For acetylene itself, the chemical shift is δ 2.88, and the range is about δ 1.8–2.9.

Alkenes.   The increased electronegativity of the sp^2 carbon and the modest anisotropy of the carbon–carbon double bond result in a high frequency (low field) position for protons on alkene carbons. The range is quite large (δ 4.5–7.0), as the exact resonance position depends on the nature of the substituents on the double bond. The value for ethene is δ 5.28. 1,1-Disubstituted hydrocarbon alkenes (vinylidenes), including exomethylene groups on rings ($={=}CH_2$), resonate at somewhat lower frequency (higher field), as in isobutylene [$(CH_3)_2C{=}CH_2$, δ 4.73]. The CH_2 part of a vinyl group, —CH=CH_2, is also usually at lower frequency than δ 5. 1,2-Disubstituted alkenes, as found for example in endocyclic ring double bonds (—CH=CH—) and trisubstituted double bonds, generally resonate at higher frequency than δ 5, as in *trans*-2-butene ($CH_3CH{=}CHCH_3$, δ 5.46). Angle strain on the double bond moves the resonance position to higher frequency, as in norbornene (**3-8**, δ 5.94). Conjugation also usually moves the resonance position to higher frequency, as in 1,3-cyclohexadiene (**3-9**, δ 5.78). The double bonds in 1,3-cyclopentadiene (**3-10**) are both strained and conjugated, so the

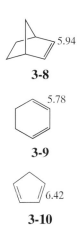

5.94

3-8

5.78

3-9

6.42

3-10

position is at even higher frequency, δ 6.42. The phenyl ring of styrene ($C_6H_5CH{=}CH_2$) withdraws electrons from the double bond by induction, so the position of the nearer CH (α) proton is moved to high frequency, δ 6.66. The more distant CH_2 (β) protons resonate at δ 5.15 and 5.63. The anisotropy of the aromatic ring is largely responsible for the difference between the β protons (the closer cis proton is shifted to higher frequency). The nonaromatic portion of styrene, $-CH{=}CH_2$, is a *vinyl* group, and the term should be restricted to that structure. The term *alkenic*, not vinylic, should be used generically for protons on double bonds.

Carbonyl groups are strongly electron withdrawing by induction and resonance. Thus the β protons on double bonds conjugated with a carbonyl group have very high frequency (downfield) resonances, for example δ 6.83 in the α,β-unsaturated ester *trans*-$CH_3CH_2O_2CCH{=}CHCO_2CH_2CH_3$. Compounds **3-11** and **3-12** illustrate the effects

3-11   **3-12**

of conjugation on alkene chemical shifts. Whereas the alkenic protons of cyclohexene resonate at a normal δ 5.59, the oxygen atom in the unsaturated ether **3-11** donates electrons to the β position by resonance and moves the β proton to lower frequency (δ 4.65). The oxygen withdraws electrons inductively from the α position, whose proton resonance moves to higher frequency (δ 6.37). In contrast, the carbonyl group in the unsaturated ketone **3-12** withdraws electrons from the β position by resonance and moves the β proton to higher frequency (δ 6.88). In this case the inductive effect of the carbonyl group causes a small shift to higher frequency for the α proton (δ 5.93).

These effects were quantified in the empirical approach of Tobey and of Pascual, Meier, and Simon, who used eq. 3-2 to provide the chemical shift of a proton on a double

$$\delta = 5.28 + Z_{\text{gem}} + Z_{\text{cis}} + Z_{\text{trans}} \qquad (3\text{-}2)$$

bond. Substituent constants Z_i (Table 3-2) for groups geminal, cis, or trans to the proton under consideration are added to the chemical shift of ethene. For example, the res-

TABLE 3-2

Substituent Parameters for the Tobey-Simon Rule

Substituent	Z_{gem}	Z_{cis}	Z_{trans}
H	0.0	0.0	0.0
Alkyl	0.44	−0.26	−0.29
CH_2O, CH_2I	0.67	−0.02	−0.07
CH_2S	0.53	−0.15	−0.15
CH_2Cl, CH_2Br	0.72	0.12	0.07
CH_2N	0.66	−0.05	−0.23
$C{=}C$	0.50	0.35	0.10
$C{\equiv}N$	0.23	0.78	0.58
$C{=}C$ (isolated)	0.98	−0.04	−0.21
$C{=}C$ (conjugated)	1.26	0.08	−0.01
$C{=}O$ (isolated)	1.10	1.13	0.81
$C{=}O$ (conjugated)	1.06	1.01	0.95
CO_2H (isolated)	1.00	1.35	0.74
CO_2R (isolated)	0.84	1.15	0.56
CHO	1.03	0.97	1.21
OR (R aliphatic)	1.18	−1.06	−1.28
OCOR	2.09	−0.40	−0.67
Aromatic	1.35	0.37	−0.10
Cl	1.00	0.19	0.03
Br	1.04	0.40	0.55
NR_2 (R aliphatic)	0.69	−1.19	−1.31
SR	1.00	−0.24	−0.04

onances of a crotonaldehyde (CH_3—CH_β=CH_α—CHO) at δ 6.87 and 6.03 may be assigned and the stereochemistry of the molecule determined. For the cis stereochemistry, the calculated shift for H_β is 6.93 (5.28 + 0.44 + 1.21) and for H_α 6.02 (5.28 + 1.03 − 0.29); for the trans stereochemistry the calculated shift for H_β is 6.69 (5.28 + 0.44 + 0.97) and for H_α 6.05 (5.28 + 1.03 − 0.26). The cis stereochemistry is supported and the resonances are assigned. Although the parameters Z_i incorporate inductive and resonance effects, steric effects can cause deviations from observed positions.

Aldehydes. The aldehydic proton is shifted to very high frequency (low field) by induction and diamagnetic anisotropy of the carbonyl group. For acetaldehyde (CH_3CHO) the value is δ 9.80, and the range is relatively small, generally δ 10 ± 0.3.

3-2c Aromatics

Diamagnetic anisotropy of the benzene ring augments the already deshielding influence of the sp² carbon atoms to yield a high frequency (low field) position for benzene, δ 7.27. Inductive and resonance effects of substituents are similar to those in alkenes. For toluene ($C_6H_5CH_3$), the electronic effect of the methyl group is small, and all five aromatic protons resonate at about δ 7.2. A narrow range thus is typical for saturated hydrocarbon substituents (arenes). Conjugating substituents, however, result in a large spread in the aromatic resonances and in spectral multiplicity from spin–spin splitting. For nitrobenzene (Figure 3-10), the inductive effect of the nitro group (3-13) moves all

3-13   **3-14**

3-15

3-16

resonances to higher frequency with respect to benzene, but the ortho and para protons are further shifted by electron withdrawal by resonance. By contrast, the methoxy group in anisole (3-14) donates electrons by resonance, so the ortho and para positions are at lower frequency than for benzene. The α protons in heterocycles generally are shifted to high frequency, as in pyridine (3-15) and pyrrole (3-16), largely because of the inductive effect of the heteroatom.

Aromatic proton resonances also may be treated empirically, provided that no two substituents are ortho to each other (steric effects). The shift of a particular proton is obtained by adding substituent parameters to the shift of benzene, as shown in eq. 3-3 and

$$\delta = 7.27 + \Sigma S_i \qquad (3\text{-}3)$$

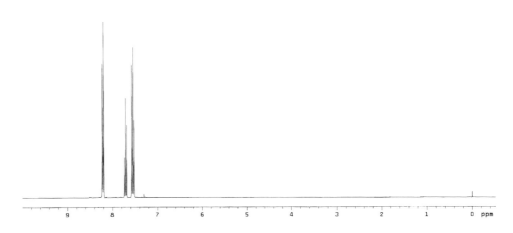

Figure 3-10 The 300 MHz ¹H spectrum of nitrobenzene in $CDCl_3$.

and excited states. Because of symmetry considerations, the $\pi \longrightarrow \pi^*$ transition is excluded. Contributions from low-lying excited states (small ΔE) make the largest contribution, since ΔE appears in the denominator. Saturated molecules, such as alkanes, typically have no low-lying excited states (large ΔE), so that σ^p is small and alkane carbon resonances are found at very low frequency (high field). (Paramagnetic shielding causes shifts to high frequency, whereas diamagnetic shielding causes shifts to low frequency.) Similarly, the nitrogen atoms in aliphatic amines and the oxygen atoms in aliphatic ethers have no low-lying excited states, so that their ^{15}N and ^{17}O resonances also are found at low frequency. Carbonyl carbons, $C{=}O$, have a low-lying excited state involving movement of electrons from the oxygen lone pair to the antibonding π orbital that generates a paramagnetic current. This $n \longrightarrow \pi^*$ transition causes the large shift to high frequency that characterizes carbonyl groups, up to 210 ppm from the zero of TMS. Even larger shifts to high frequency, to $\delta\,335$, have been observed for carbocations, R_3C^+.

The radial term in eq. 3-4 is responsible for effects related to electron density that parallel inductive effects on proton chemical shifts. Paramagnetic shielding is larger when the p electrons are closer to the nucleus. Thus substituents that donate or withdraw electrons influence the paramagnetic shift. Electron donation increases repulsion between electrons, which can be relieved by an increase in r. The paramagnetic shielding then decreases, which causes a shift to lower frequency (upfield). Similarly, electron withdrawal permits electrons to move closer to the nucleus, increasing the paramagnetic shielding and causing a shift to higher frequency. Thus placing a series of electron-withdrawing atoms on carbon results in progressively higher frequency shifts, as in the series CH_3Cl ($\delta\,25$), CH_2Cl_2 ($\delta\,54$), $CHCl_3$, ($\delta\,78$), and CCl_4 ($\delta\,97$). The situation is qualitatively similar to that for protons, but the numbers are much larger because the shift is from the paramagnetic term. Again, however, substituent effects generally follow the electronegativity of groups attached to carbon.

Electronegativity is a measure of the ability of a nucleus to attract electrons. A highly electronegative element such as oxygen attracts p electrons more than does carbon and reduces the value of r. Thus ^{17}O shifts then are correspondingly larger than ^{13}C shifts. A plot of the ^{17}O shifts of aliphatic ethers versus the ^{13}C shifts of the analogous alkanes is linear with a slope of about three. The linearity shows that oxygen and carbon chemical shifts are sensitive to the same structural factors, and the slope indicates that oxygen is more sensitive to these factors because its 2p electrons are closer to the nucleus.

The third factor in eq. 3-4, ΣQ_{ij}, is related to charge densities and bond orders and can be considered to be a measure of multiple bonding. The greater the degree of multiple bonding, the greater the shift to high frequency (low field). This term provides a rationale for the series ethane ($\delta\,6$), ethene ($\delta\,123$), and the central sp-hybridized carbon of allene ($\delta\,214$). Arene shifts are similar to those of alkenes (benzene, $\delta\,129$). The effects of diamagnetic anisotropy on carbon chemical shifts are similar in magnitude to the effects on protons but are small in relation to the range of carbon shifts. The chemical shifts of alkynes are at an intermediate position ($\delta\,72$ for acetylene), because their linear structure has zero angular momentum about the $C{\equiv}C$ axis.

Interpretation of the chemical shifts of most elements other than hydrogen is accomplished by analysis of the three factors in eq. 3-4: accessibility of certain excited states, distance of the p electrons, and multiple bonding. For carbon, the shifts of alkanes, alkenes, arenes, alkynes, and carbonyl groups and the effects of electron-donating or electron-withdrawing groups may be interpreted in this fashion. There are exceptions, the most prominent being the effect of heavy atoms. The series CH_3Br ($\delta\,10$), CH_2Br_2 ($\delta\,22$), $CHBr_3$ ($\delta\,12$), and CBr_4 ($\delta\,-29$) defies any explanation based on electronegativity, unlike the analogous series given above for chlorine. The same series with iodine is monotonic to lower frequency ($\delta\,-290$ for CI_4), that is, opposite to the chlorine series. This so-called heavy-atom effect has been attributed to a new source of angular momentum from spin–orbit coupling. These anomalous shifts to low frequency (high field) can be expected when nuclei other than hydrogen have a heavy atom substituent.

3-4   CARBON CHEMICAL SHIFTS AND STRUCTURE

Figure 3-11 illustrates general ranges for ^{13}C chemical shifts.

3-4a   Saturated Aliphatics

Acyclic Alkanes.   The absence of low-lying excited states and of π bonding minimizes paramagnetic shielding and places alkane chemical shifts at very low frequency (high field). Methane itself resonates at δ -2.1. The series ethane (CH_3CH_3), propane ($CH_3CH_2CH_3$), isobutane ($CH_3)_3CH$) follows a steady trend to higher frequency: δ 5.7, 16.1, 25.2, similar to the methyl, methylene, methine series in proton shifts. Replacement of H by CH_3 adds about 9 ppm to the chemical shift of the attached carbon (the effect is similar for replacement by saturated CH_2, CH, or C). Because the added methyl group is attached directly to the resonating carbon, the shift has been termed the α *effect* **(3-19).** The effect is not restricted to replacement of H by carbon. Any group X

that replaces hydrogen on a resonating carbon atom causes a relatively constant shift that depends primarily on the electronegativity of X.

Replacement of hydrogen by CH_3 (or by CH_2, CH, or C) at a β position **(3-20)** also causes a constant shift of about $+9$ ppm. Thus the central carbon in pentane ($CH_3CH_2CH_2CH_2CH_3$) is shifted by the α effects of the two methylene groups and the β *effects* of the two methyl groups. Replacement of a γ hydrogen **(3-21)** by CH_3 (or by

CH$_2$, CH, or C) causes a shift of about -2.5 (to low frequency or upfield). Unlike the α and β effects, this γ *effect* has an important stereochemical component. Because of these α, β, and γ effects, the alkane chemical shift range is relatively large. Methyl resonances in alkanes are typically found at δ 5–15, depending on the number of β substituents; methylene resonances are at δ 15–30; methine resonances are at δ 25–45.

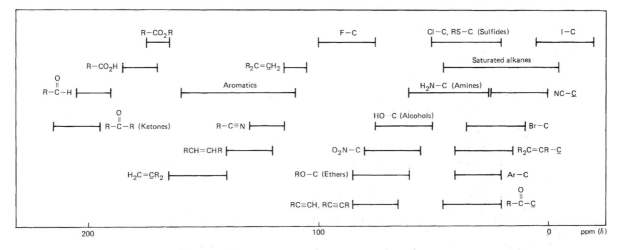

Figure 3-11 Carbon chemical shift ranges for common structural units. The symbol C represents methyl, methylene, methine, or quaternary carbon; R represents a saturated alkyl group. The indicated ranges are for common examples; actual ranges can be larger.

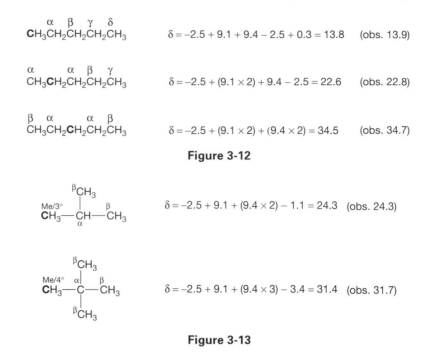

$$\overset{\alpha}{C}H_3\overset{\beta}{C}H_2\overset{\gamma}{C}H_2\overset{\delta}{C}H_2CH_3 \qquad \delta = -2.5 + 9.1 + 9.4 - 2.5 + 0.3 = 13.8 \quad (obs.\ 13.9)$$

$$\overset{\alpha}{C}H_3\overset{\alpha}{C}H_2\overset{\beta}{C}H_2\overset{\gamma}{C}H_2CH_3 \qquad \delta = -2.5 + (9.1 \times 2) + 9.4 - 2.5 = 22.6 \quad (obs.\ 22.8)$$

$$\overset{\beta}{C}H_3\overset{\alpha}{C}H_2\overset{\alpha}{C}H_2\overset{\beta}{C}H_2CH_3 \qquad \delta = -2.5 + (9.1 \times 2) + (9.4 \times 2) = 34.5 \quad (obs.\ 34.7)$$

Figure 3-12

Me/3°
$$\overset{\beta}{C}H_3$$
$$CH_3—\underset{\alpha}{CH}—\overset{\beta}{C}H_3 \qquad \delta = -2.5 + 9.1 + (9.4 \times 2) - 1.1 = 24.3 \quad (obs.\ 24.3)$$

Me/4°
$$\overset{\beta}{C}H_3$$
$$CH_3—\overset{\alpha}{C}—\overset{\beta}{C}H_3 \qquad \delta = -2.5 + 9.1 + (9.4 \times 3) - 3.4 = 31.4 \quad (obs.\ 31.7)$$
$$\overset{\beta}{C}H_3$$

Figure 3-13

Carbon-13 chemical shifts lend themselves conveniently to empirical analysis because shifts are easily measured and tend to have well-defined substituent effects. For saturated, acyclic hydrocarbons, eq. 3-5 provides an empirical measure of chemical

$$\delta = -2.5 + \Sigma A_i n_i \qquad (3\text{-}5)$$

shifts, developed by David Grant. For any resonating carbon, a substituent parameter A_i for each carbon atom in the molecule, up to a distance of five bonds, is added to the chemical shift of methane (δ −2.5). There are different substituent parameters for carbons (whether CH_3, CH_2, CH, or C) that are α (9.1), β (9.4), γ (−2.5), δ (0.3), or ϵ (0.1) to the resonating carbon. We already have alluded to the first three figures. If there is more than one α carbon, the substituent parameter is multiplied by the appropriate number n_i, and similar factors are applied for multiple substitution at other positions. Figure 3-12 illustrates the calculation for each carbon in pentane. The methyl chemical shift is calculated by adding contributions from single α, β, γ, and δ carbons to the shift (−2.5) of methane. The shift of the 2 carbon is calculated by adding contributions of two α carbons, one β carbon, and one γ carbon. The observed shifts are calculated usually to within 0.3 ppm, providing a very reliable means for spectral assignment.

There are complications. Corrections must be applied if there is branching, because eq. 3-5 applies only to straight chains. The resonance position of a methyl group is corrected for the presence of an adjacent tertiary (CH) carbon by adding −1.1 and for an adjacent quaternary carbon by −3.4. Methylene carbons have corrections of −2.5 and −7.2, respectively, for adjacent tertiary and quaternary carbons. Methine carbons have corrections of −3.7, −9.5, and −1.5 for adjacent secondary, tertiary, and quaternary carbons. Finally, quaternary carbons have corrections of −1.5 and −8.4 for adjacent primary and secondary carbons (corrections for adjacent tertiary and quaternary carbons undoubtedly are significant but are not known accurately). For example, the methyl group in isobutane (the first example in Figure 3-13) is adjacent to a tertiary carbon. The methyl chemical shift is calculated by adding contributions of one α carbon, two β carbons, and the correction of −1.1 since the methyl group is adjacent to a tertiary (3°) center. In the second example in Figure 3-13 (neopentane) the methyl group is adjacent to a quaternary (4°) center.

It is noteworthy that the γ effect of a carbon substituent is negative (−2.5). A γ carbon can be either gauche or anti (Figure 3-14) to the resonating carbon, and the pro-

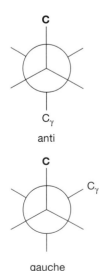

anti

gauche

Figure 3-14

portion of conformers can vary from molecule to molecule. The value -2.5 is a weighted average for open chain conformers and does not serve accurately for all situations. For a pure γ-*anti effect*, the shift is about $+1$, and for a pure γ-*gauche effect* it is about -6. The average value of -2.5 measured by Grant clearly indicates a mix of the two conformations. Hydrocarbons with unusually large deviations from the average mix may give poor results from eq. 3-5. The α and β effects are determined by fixed geometries and have no stereochemistry component.

Cyclic Alkanes.   With a resonance position of δ -2.6, cyclopropane has the lowest frequency resonance of hydrocarbons. Cyclobutane resonates at δ 23.3, and the remaining cycloalkanes generally resonate within 2 ppm of cyclohexane at δ 27.7. The fixed stereochemistry represented by cyclohexane requires an entirely new set of empirical parameters that depend on the axial or equatorial nature of the substituent, as well as on the distance from the resonating carbon. Table 3-4 lists the parameters for methyl substitution that are added to the value for cyclohexane (δ 27.7). The substituent parameter for a γ-axial methyl is large and negative (-5.4), reflecting the pure gauche stereochemistry between the perturbing and resonating carbons. A γ equatorial group represents a γ-anti effect and has little perturbation in this case. Corrections again are needed for branching. For two α methyls that are geminal (both on the resonating carbon), the correction is -3.4; for two β methyls that are geminal, it is -1.2. Thus the calculated resonance position for C2 of 1,1,3-trimethylcyclohexane **(3-22)** is

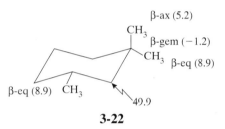

3-22

$27.7 + 5.2 + (8.9 \times 2) - 1.2 = 49.5$ (obs. 49.9). A pair of vicinal, diequatorial methyls that are respectively α and β to the resonating carbon require a correction of -2.3, and similar axial–equatorial vicinal methyls require -3.1.

Functionalized Alkanes.   Replacement of a hydrogen on carbon with a heteroatom or an unsaturated group usually results in shifts to higher frequency (downfield) because of inductive effects on the radial term. Strongly electron-withdrawing groups thus have large, positive α effects. In the halogen series CH_3X, the methyl chemical shifts are δ 75.4 for fluorine, 25.1 for chlorine, 10.2 for bromine, and -20.6 for iodine. Multiple substitution results in larger effects, δ 77.7 for $CHCl_3$. Recall that the α effect of heavy atoms such as iodine or bromine is influenced by a spin–orbit mechanism and hence does not follow the simple order of electronegativity. The general range for the α halogen effect in hydrocarbons extends from the values given for the simple CH_3X systems to about 25 ppm higher frequency (downfield) for CH_2X and CHX systems, since the α and β effects of the unspecified hydrocarbon pieces contribute to the shift to higher frequency.

Methanol (CH_3OH) resonates at δ 49.2, and the range for hydroxy-substituted carbons is δ 49–75. Dimethyl ether [$(CH_3)_2O$] resonates at δ 59.5, and the range for alkoxy-substituted carbons is δ 59–80. The ether range is translated a few ppm to higher

TABLE 3-4

Substituent Parameters for Methyl Substitution on Cyclohexane

Stereochemistry	α	β	γ	δ
Equatorial	5.6	8.9	0.0	0.3
Axial	1.1	5.2	−5.4	−0.1

frequency (downfield) from alcohols because each ether must have one additional β effect with respect to the analogous alcohol.

The lower electronegativity of nitrogen moves the amine range somewhat to lower frequency (upfield). Methylamine in aqueous solution resonates at δ 28.3, with the range for amines extending some 30 ppm to higher frequency. The amine range is larger than the alcohol range because nitrogen can carry up to three substituents, with the possibility of more α and β effects. Dimethyl sulfide resonates at δ 19.5, acetonitrile at δ 0.3, and nitromethane at δ 57.3, with the respective ranges for thioalkoxy, cyano, and nitro substitution extending some 25 ppm to higher frequency. The low frequency position for cyano substitution is related to the cylindrical shape.

Attachment of methyl to a double bond has only a small effect. The position for the methyls of *trans*-2-butene (*trans*-CH₃CH=CHCH₃) is δ 17.3, and that for the methyl of toluene (C₆H₅CH₃) is δ 21.3. The range for carbons on double bonds is about δ 10–40. Methyls on carbonyl groups are at slightly higher frequency (lower field), δ 30.2 for acetone and δ 31.2 for acetaldehyde, with a range of about δ 30–45.

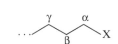

Terminal

3-23

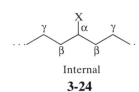

Internal

3-24

Introducing heteroatoms or unsaturation into alkanes requires completely new sets of empirical parameters that depend on the substituent, on its distance from the resonating carbon (α, β, γ), and on whether the substituent is terminal **(3-23)** or internal **(3-24)** (Table 3-5). These numbers represent the effect on a resonating carbon of replacing a hydrogen atom at the respective position with a group X. With the exception of the cyano, acetyleno, and the heavy atom iodine, the α effects are determined largely by the electronegativity of the substituent. It is interesting that the β effects are all positive and generally of similar magnitude (6 to 11) and that the γ effects are all negative and generally of similar magnitude (-2 to -5). Although the details are not entirely understood, it is clear that simple polar considerations do not dominate the β and γ effects.

To utilize the substituent parameters given in Table 3-5, add the appropriate values to the chemical shift of the carbon in the analogous hydrocarbon, rounded off to the nearest ppm. As seen in Figure 3-15, the chemical shift of the 1 carbon of 1,3-dichloropropane may be calculated from the value (16) for the methyl carbon of propane and from the figures in Table 3-5. The chemical shift of the β carbon of cyclopentanol similarly may be calculated from the value (27) for cyclopentane.

TABLE 3-5

Carbon Substituent Parameters for Functional Groups*

X	Terminal X (3-23)			Internal X (3-24)		
	α	β	γ	α	β	γ
F	68	9	−4	63	6	−4
Cl	31	11	−4	32	10	−4
Br	20	11	−3	25	10	−3
I	−6	11	−1	4	12	−1
OH	48	10	−5	41	8	−5
OR	58	8	−4	51	5	−4
OAc	51	6	−3	45	5	−3
NH₂	29	11	−5	24	10	−5
NR₂	42	6	−3			−3
CN	4	3	−3	1	3	−3
NO₂	63	4		57	4	
CH=CH₂	20	6	−0.5			−0.5
C₆H₅	23	9	−2	17	7	−2
C≡CH	4.5	5.5	−3.5			−3.5
(C=O)R	30	1	−2	24	1	−2
(C=O)OH	21	3	−2	16	2	−2
(C=O)OR	20	3	−2	17	2	−2
(C=O)NH₂	22		−0.5	2.5		−0.5

*From F.W. Wehrli, A.P. Marchand, and S. Wehrli, *Interpretation of Carbon-13 NMR Spectra,* 2nd ed., John Wiley & Sons Ltd., Chichester, UK, 1988.

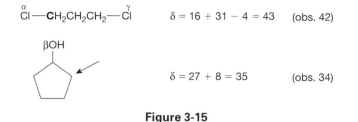

$$\delta = 16 + 31 - 4 = 43 \quad \text{(obs. 42)}$$

$$\delta = 27 + 8 = 35 \quad \text{(obs. 34)}$$

Figure 3-15

3-4b   Unsaturated Compounds

The effects of diamagnetic anisotropy are of similar magnitude for carbon as for proton, but the much larger paramagnetic shielding renders the phenomenon relatively small for carbon. Thus benzene (δ 128.7) and cyclohexene (δ 127.2) have almost identical carbon resonance positions, in contrast to the case for their protons. The full range of alkene and aromatic resonances is about δ 100–170.

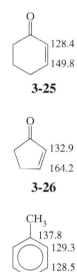

3-25

3-26

Alkenes.   Alkenic carbons that bear no substituents (=CH$_2$) resonate at low frequency (high field), such as isobutylene [(CH$_3$)$_2$C=CH$_2$] at δ 107.7, and have a range for hydrocarbons of about δ 104–115. Alkenic carbons that have one substituent (=CHR), like those in *trans*-2-butene (δ 123.3), resonate in the range δ 120–140. Finally, disubstituted alkenic carbons (=CRR'), like that in isobutylene (δ 146.4), resonate at highest frequency (δ 140–165). Polar substituents on double bonds, particularly those in conjugation with the bond, can alter the resonance position appreciably. α,β-Unsaturated ketones, such as **3-25** and **3-26,** have lower frequency α resonances and higher frequency β resonances (the effect is reduced in acyclic molecules). Electron donation, as in enol ethers, reverses the effect, as in CH$_2$=CHOCH$_3$, ($\delta(\alpha)$ 153.2, $\delta(\beta)$ 84.2).

Alkene chemical shifts may be estimated from substituent parameters added to the shift for ethene (δ 123.3). For α, β, and γ carbons on the same end of the double bond as the resonating carbon, respective increments of 10.6, 7.2, and -1.5 are added. For α', β', and γ' carbons on the opposite end of the double bond from the resonating carbon, respective increments of -7.9, -1.8, and -1.5 are added. An increment of -1.1 is added if any pair of substituents are cis to each other. Thus the shift of the unsubstituted carbon in 1-butene (CH$_3$CH$_2$CH=CH$_2$) is calculated to be 123.3 $-$ 7.9 $-$ 1.8 = 113.6 (obs. 113.3), and the shift of the substituted carbon (CH$_3$CH$_2$CH=CH$_2$) is 123.3 + 10.6 + 7.2 = 141.1 (obs. 140.2).

3-27

Alkynes and Nitriles.   An alkyne carbon that carries a hydrogen substituent (≡CH) generally resonates in the narrow range δ 67–70. An alkyne carbon that carries a carbon substituent (≡CR) resonates at slightly higher frequency (downfield) (δ 74–85), because of α and β effects from the R group. Effects of conjugating, polar substituents expand the total range to δ 20–90. Nitriles resonate in the range δ 117–130 (acetonitrile is at δ 117.2). The $n \longrightarrow \pi^*$ transition pushes the range to high frequency.

3-28

Aromatics.   Alkyl substitution, as in toluene **(3-27)**, has its major (α) effect on the ipso carbon. Because this carbon has no attached proton, its relaxation time is much longer than those of the other carbons and its intensity is usually lower. Conjugating substituents like nitro **(3-28)** have strong perturbations on the aromatic resonance positions, as the result of a combination of traditional α, β, and γ effects and changes in electron density through resonance. A similar interplay of effects is seen in the resonance positions of pyridine **(3-29)** and pyrrole **(3-30).**

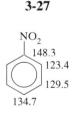

3-29

Aromatic resonances may be calculated empirically by adding increments to the benzene chemical shift (δ 128.7) for each substituent that is ipso, ortho, meta, or para to the resonating carbon (Table 3-6).

3-30

The facility and accuracy of these calculations have been exploited through computer programs for general predictions of ^{13}C chemical shifts. The results are only as good as the data set used in their creation. The programs assume that the effects of mul-

TABLE 3-6

Carbon Substituent Parameters for Aromatic Systems*

X	Ipso	Ortho	Meta	Para
CH_3	8.9	0.7	−0.1	−2.9
CH_2OH	13.3	−0.8	−0.6	−0.4
$CH{=}CH_2$	9.5	−2.0	0.2	−0.5
CN	−19.0	1.4	−1.5	1.4
CO_2CH_3	1.3	−0.5	−0.5	3.5
CHO	9.0	1.2	1.2	6.0
$CO{-}CH_3$	7.9	−0.3	−0.3	2.9
F	35.1	−14.1	1.6	−4.4
Cl	6.4	0.2	1.0	−2.0
Br	−5.4	3.3	2.2	−1.0
I	−32.0	10.2	2.9	1.0
NH_2	19.2	−12.4	1.3	−9.5
OH	26.9	−12.6	1.8	−7.9
OCH_3	30.2	−15.5	0.0	−8.9
SCH_3	10.2	−1.8	0.4	−3.6
NO_2	19.6	−5.3	0.8	6.0

*From J.B. Stothers, *Carbon-13 NMR Spectroscopy,* Academic Press, New York, 1973.

tiple substitution are additive unless specific corrections have been incorporated. Unconsidered or nonadditive phenomena, such as conformational and other steric effects, can cause unexpected deviations between observed and calculated chemical shifts. No such programs are available for protons. The small range of chemical shifts and the large number of shielding factors have made proton correlations very limited.

3-4c Carbonyl Groups

Carbonyl groups have no direct representation in proton NMR spectra, so carbon NMR provides unique information for their analysis. The entire carbonyl chemical shift range, δ 160–220, is well removed from that of almost all other functional groups. Like aromatic ipso carbons and nitriles, carbonyl carbons other than in aldehydes carry no attached protons and hence relax more slowly and tend to have low intensities.

Aldehydes resonate toward the middle of the carbonyl range, about δ 190–205, with acetaldehyde (CH_3CHO) at δ 199.6. Unsaturated aldehydes, in which the carbonyl group is conjugated with a double bond or phenyl ring, are shifted to lower frequency (upfield): benzaldehyde (C_6H_5CHO) at δ 192.4 and $CH_2{=}CHCHO$ at δ 192.2. The α, β, and γ effects of substituents on ketones add to the carbonyl chemical shift, and hence are found at the high frequency end of the carbonyl range. Their overall range is δ 195–220: acetone at δ 205.1, cyclohexanone at δ 208.8. Again unsaturation shifts the resonances to lower frequency.

Carboxylic derivatives fall in the range δ 155–185. The resonances for the series carboxylate (CO_2^-), carboxyl (CO_2H), ester (CO_2R) often are well defined, for example for sodium acetate (δ 181.5), acetic acid (δ 177.3), and methyl acetate (δ 170.7). The range for esters is about δ 165–175 and that for acids is δ 170–185. Acid chlorides are at slightly lower frequency (higher field): δ 160–170, with δ 168.6 for acetyl chloride ($CH_3(CO)Cl$). Anhydrides have a similar range: δ 165–175, with δ 167.7 for acetic anhydride ($CH_3(CO)O(CO)CH_3$). Lactones overlap the ester range, with the six-membered lactone at δ 176.5. Amides also have a similar range: δ 160–175, with δ 172.7 for acetamide ($CH_3(CO)NH_2$). Oximes have a larger range, extending from δ 145–165. The central carbon of allenes ($R_2C{=}C{=}CR_2$) falls in the ketonic range, δ 200–215, although the outer carbons have a much lower frequency range, δ 75–95.

3-5 SUMMARY AND TABLES OF CHEMICAL SHIFTS

Structural analysis of an unknown organic material normally begins with examination of the ^{1}H and ^{13}C spectra. Resonance positions are analyzed if possible with the benefit of knowledge of the molecular formula and structural information based on synthetic precursors. Representative chemical shifts are given in Tables 3-7 to 3-11, drawn from references at the end of this chapter.

Observed chemical shifts depend not only on the structure of the compound but also on the solvent. Saturated hydrocarbons such as cyclohexane, pentane, and hexane interact with solutes only through van der Waals forces and have a minimal effect on the chemical shift. Polar molecules such as acetone, chloroform, diethyl ether, dimethyl sulfoxide, or acetonitrile, or even nonpolar molecules with polar substituents, such as carbon tetrachloride, can have electrostatic interactions with a polar substrate. Because such substrates interact more strongly with the solvent than with the reference, TMS, there is a net effect on chemical shifts, sometimes several tenths of a ppm for protons.

Solvent effects on proton resonances can be particularly large for solvents with large diamagnetic anisotropy. Flat molecules such as benzene, toluene, nitrobenzene, nitromethane, pyridine, or acetone, interact with many solutes differently from the way they do with TMS and hence give rise to significant solvent shifts. Rod-shaped molecules such as acetonitrile, carbon disulfide, or sulfur dioxide also are anisotropic, but their solvent shifts are opposite to those of flat solvents. Thus the position of the CH_2 protons in propargyl chloride ($HC{\equiv}CCH_2Cl$) moves from δ 3.87 in noninteracting cyclohexane to 4.17 in the rod-like acetonitrile and to 3.42 in the disc-like benzene. Differential solvent effects within complex molecules, resulting from local structure, can be exploited to separate overlapping proton resonances at any field strength, so that it may be useful to employ two or more solvents, say pyridine, acetone, and acetonitrile for polar solutes, in order to separate all resonances.

TABLE 3-7
Methyl and Methylene Groups

	$\delta(^1H)$		$\delta(^{13}C)$			$\delta(^1H)$		$\delta(^{13}C)$	
	CH_2	CH_3	CH_2	CH_3		CH_2	CH_3	CH_2	CH_3
CH_3Li		−0.4		−13.2	$(CH_3)_2NCHO$		2.88		36.0
CH_3CH_3		0.86		5.7			2.97		30.9
$(CH_3)_3CH$		0.89		25.2	CH_3Cl		3.06		25.1
$(CH_3)_4C$		0.94		31.7	$(CH_3)_2O$		3.24		59.5
$(CH_3)_3COH$		1.22		29.4	$(CH_3)_4N^+$		3.33		55.6
$CH_3CH{=}CH_2$		1.72		18.7	CH_3OH		3.38		49.2
$CH_3C{\equiv}CH$		1.80		−1.9	$CH_3CO_2\mathbf{CH_3}$		3.67		51.0
$(CH_3)_3P{=}O$		1.93		18.6	$CH_3OC_6H_5$		3.73		54.8
CH_3CN		2.00		0.3	CH_3F		4.27		75.4
$\mathbf{CH_3}CO_2CH_3$		2.01		18.7	CH_3NO_2		4.33		57.3
$CH_3(CO)CH_3$		2.07		30.2	$(CH_3CH_2)_2S$	2.49	1.25	26.5	15.8
CH_3CO_2H		2.10		18.6	$CH_3CH_2NH_2$	2.74	1.10	36.9	19.0
$(CH_3)_2S$		2.12		19.5	$CH_3CH_2C_6H_5$	2.92	1.18	29.3	16.8
CH_3I		2.15		−20.6	CH_3CH_2I	3.16	1.86	0.2	23.1
CH_3CHO		2.20		31.2	CH_3CH_2Br	3.37	1.65	28.3	20.3
$(CH_3)_3N$		2.22		47.3	CH_3CH_2Cl	3.47	1.33	39.9	18.7
$CH_3C_6H_5$		2.31		21.3	$(CH_3CH_2)_2O$	3.48	1.20	67.4	17.1
CH_3NH_2		2.42		30.4	CH_3CH_2OH	3.56	1.24	57.3	15.9
$CH_3(SO)CH_3$		2.50		40.1	CH_3CH_2F	4.36	1.24	79.3	14.6
$CH_3(CO)Cl$		2.67		32.7	$CH_3CH_2NO_2$	4.37	1.58	70.4	10.6
CH_3Br		2.69		10.2	$BrCH_2CH_2Br$	3.63		32.4	
$(CH_3)_4P^+$		2.74		11.3	$HOCH_2CH_2OH$	3.72		63.4	
$CH_3(SO_2)CH_3$		2.84		42.6	$ClCH_2CH_2Cl$	3.73		51.7	

TABLE 3-8

Saturated Ring Systems

		^1H	^{13}C			^1H	^{13}C
Cyclopropane		0.22	−2.6	Pyrrolidine	(α)	2.75	47.4
Cyclobutane		1.98	23.3		(β)	1.59	25.8
Cyclopentane		1.51	26.5	Piperidine	(α)	2.74	47.5
Cyclohexane		1.43	27.7		(β)	1.50	27.2
Cycloheptane		1.53	29.4		(γ)	1.50	25.5
Cyclopentanone	(α)	2.06	37.0	Thiirane		2.27	18.9
	(β)	2.02	22.3	Tetrahydrothiophene	(α)	2.82	31.7
Cyclohexanone	(α)	2.22	40.7		(β)	1.93	31.2
	(β)	1.8	26.8	Sulfolane	(α)	3.00	51.1
	(γ)	1.8	24.1		(β)	2.23	22.7
Oxirane		2.54	40.5	1,4-Dioxane		3.70	66.5
Tetrahydrofuran	(α)	3.75	69.1				
	(β)	1.85	26.2				
Oxane (tetrahydropyran)	(α)	3.52	68.0				
	(β)	1.51	26.6				
	(γ)	—	23.6				

TABLE 3-9

Alkenes

		^1H	^{13}C	
CH_2=CHCN	(α)	5.5–6.4	107.7	
	(β)		137.8	
CH_2=CHC$_6$H$_5$	(α)	6.66	112.3	
	(β)	5.15, 5.63	135.8	
CH_2=CHBr	(α)	6.4	115.6	
	(β)	5.7–6.1	122.1	
CH_2=CHCO$_2$H	(α)	6.5	128.0	
	(β)	5.9–6.5	131.9	
CH_2=CH(CO)CH$_3$	(α)	5.8–6.4	138.5	
	(β)		129.3	
CH_2=CHO(CO)CH$_3$	(α)	7.28	141.7	
	(β)	4.56, 4.88	96.4	
CH_2=CHOCH$_2$CH$_3$	(α)	6.45	152.9	
	(β)	3.6–4.3	84.6	
$\overset{4}{C}H_3CH$=$\overset{3}{C}CH_3$—$\overset{2}{C}H$=$\overset{1}{C}H_2$	(1)	5.02		
	(2)	6.40		
	(4)	5.70		
$(CH_3)_2C$=CHCO$_2$CH$_3$	(α)	—	114.8	
	(β)	5.62	155.9	
Cyclopentene		5.60	130.6	
Cyclohexene		5.59	127.2	
1,3-Cyclopentadiene		6.42	132.2, 132.8	
1,3-Cyclohexadiene		5.78	124.6, 126.1	
2-Cyclopentenone	(α)	6.10	132.9	
	(β)	7.71	164.2	
2-Cyclohexenone	(α)	5.93	128.4	
	(β)	6.88	149.8	
exo-Methylenecyclohexane	(=CH$_2$)	4.55	106.5	
	(C=)	—	149.7	
Allene	(=CH$_2$)	4.67	74.0	
	(=C=)	—	213.0	

TABLE 3-10

Aromatics

	¹H			¹³C			
	o	m	p	i	o	m	p
$C_6H_5CH_3$	7.16	7.16	7.16	137.8	129.3	128.5	125.6
$C_6H_5CH{=}CH_2$	7.24	7.24	7.24	138.2	126.7	128.9	128.2
$C_6H_5SCH_3$	7.23	7.23	7.23	138.7	126.7	128.9	124.9
C_6H_5F	6.97	7.25	7.05	163.8	114.6	130.3	124.3
C_6H_5Cl	7.29	7.21	7.23	135.1	128.9	129.7	126.7
C_6H_5Br	7.49	7.14	7.24	123.3	132.0	130.9	127.7
C_6H_5OH	6.77	7.13	6.87	155.6	116.1	130.5	120.8
$C_6H_5OCH_3$	6.84	7.18	6.90	158.9	113.2	128.7	119.8
$C_6H_5O(CO)CH_3$	7.06	7.25	7.25	151.7	122.3	130.0	126.4
$C_6H_5(CO)CH_3$	7.91	7.45	7.45	136.6	128.4	128.4	131.6
$C_6H_5CO_2H$	8.07	7.41	7.47	130.6	130.0	128.5	133.6
$C_6H_5(CO)Cl$	8.10	7.43	7.57	134.5	131.3	129.9	136.1
C_6H_5CN	7.54	7.38	7.57	109.7	130.1	127.2	130.1
$C_6H_5NH_2$	6.52	7.03	6.63	147.9	116.3	130.0	119.2
$C_6H_5NO_2$	8.22	7.48	7.61	148.3	123.4	129.5	134.7

	¹H			¹³C		
	α	β	Other	α	β	Other
Naphthalene	7.81	7.46	—	128.3	126.1	—
Anthracene	7.91	7.39	8.31	130.3	125.7	132.8
Furan	7.40	6.30	—	142.8	109.8	—
Thiophene	7.19	7.04	—	125.6	127.4	—
Pyrrole	6.68	6.05	—	118.4	108.0	
Pyridine	8.50	7.06	7.46	150.2	123.9	135.9

TABLE 3-11

Carbonyl Compounds

	¹H(CH₃)	¹H(other)	¹³C(C=O)
$H(CO)OCH_3$	3.79	8.05 (HCO)	160.9
$CH_3(CO)Cl$	2.67	—	168.6
$CH_3(CO)OCH_2CH_3$	2.02 (CH₃CO)	4.11 (CH₂), 1.24(CH₃C)	169.5
$CH_3(CO)N(CH_3)_2$	2.10 (CH₃CO)	6.94, 7.04 (CH₃N)	169.6
CH_3CO_2H	2.10	1.37 (HO)	177.3
$CH_3CO_2^-Na^+$	—	—	181.5
$CH_3(CO)C_6H_5$	2.62	—	196.0
$CH_3(CO)CH{=}CH_2$	2.32	5.8–6.4 (CH=CH₂)	197.2
$H(CO)CH_3$	2.20	9.80 (HCO)	199.6
$CH_3(CO)CH_3$	2.07	—	205.1
2-Cyclohexenone	—	5.93, 6.88 (CH$_α$=CH$_β$)	197.1
2-Cyclopentanone	—	6.10, 7.71 (CH$_α$=CH$_β$)	208.1
Cyclohexanone	—	1.7–2.5	208.8
Cyclopentanone	—	1.9–2.3	218.1

PROBLEMS

3-1 From Shoolery's rule, calculate the expected resonance position for the CH_2 resonance in **(a)** CH_3CH_2I **(b)** $NC{-}CH_2CH{=}CH_2$ **(c)** $CH_3OCH_2C_6H_5$ **(d)** $CH_3C{\equiv}CCH_2Br$

3-2 The proton resonance positions of the *cis*- and *trans*-1,2-dibromoethenes are δ 6.65 and 7.03. Which comes from the cis isomer and which from the trans isomer?

3-3 A trisubstituted benzene possessing one bromine and two methoxy substituents exhibits three aromatic resonances, at δ 6.40, 6.46, and 7.41. What is the substitution pattern?

3-4 Calculate the expected ¹³C resonance positions for all the carbon atoms in the following molecules.

(a) $CH_3CH_2CH(CH_3)CH(CH_3)_2$ **(c)** $CH_3CH_2CH(NO_2)CH_3$
(b) $ICH_2CH_2CH_2Br$ **(d)** $(CH_3)_3CCN$

3-5 Should the α or the β proton of naphthalene resonate at higher frequency? Why? Compare both resonance positions with that of benzene.

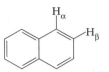

3-6 The —OH proton resonance is found at δ 5.80 for phenol in dilute $CDCl_3$ and at δ 10.67 for 2-nitrophenol in dilute $CDCl_3$. Explain.

3-7 Derive the structures of the compounds that have the following ^1H (300 MHz) and ^{13}C (75 MHz) spectra. The 1/1/1 triplet at δ 78 in the ^{13}C spectra is from the solvent $CDCl_3$ (used in all cases except **f**).

(a) $C_4H_6O_2$

(b) $C_4H_8Cl_2$

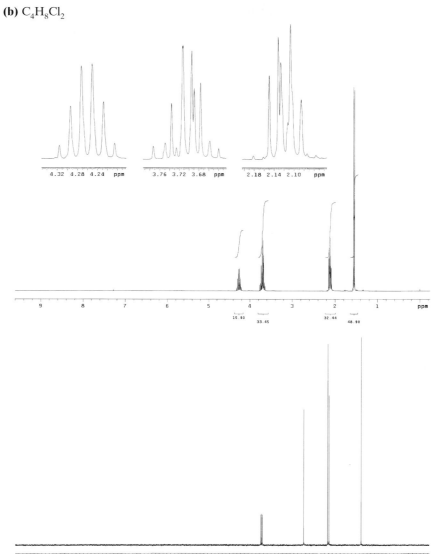
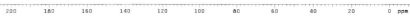

(c) C_5H_9OCl

(d) $C_4H_8O_2$

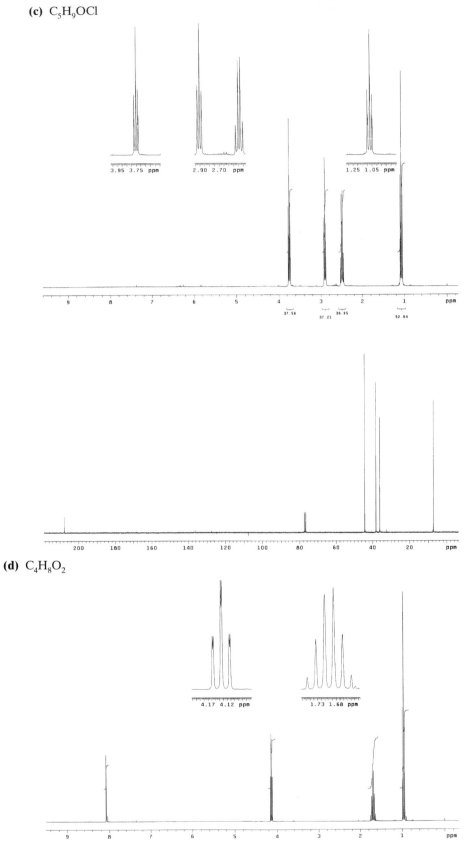

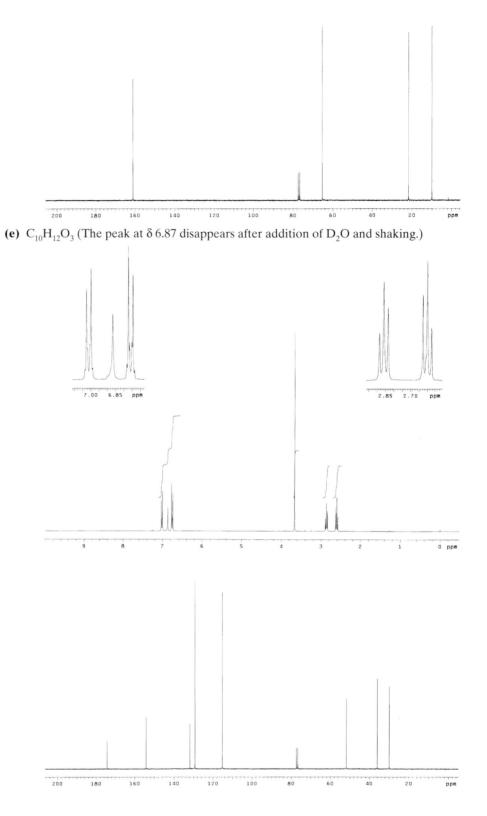

(e) $C_{10}H_{12}O_3$ (The peak at δ 6.87 disappears after addition of D_2O and shaking.)

chapter 4

The Coupling Constant

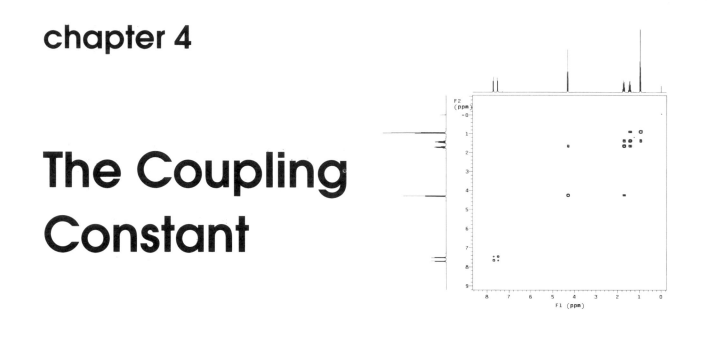

4-1 FIRST- AND SECOND-ORDER SPECTRA

Most spectra illustrated up to this point are said to be *first order*. For a spectrum to be first order, the frequency difference (Δv) between the chemical shifts of any given pair of nuclei must be much larger than the value of the coupling constant J between them, approximately $\Delta v/J > 10$. In addition, an important symmetry condition discussed in the next section must hold. First-order spectra exhibit a number of useful and simple characteristics:

- Multiplicities that result from coupling reflect the $n + 1$ rule for $I = \frac{1}{2}$ nuclei exactly ($2nI + 1$, in general). Thus two neighboring protons split the resonating nucleus into three peaks.
- The intensities of spin–spin multiplets correspond to the coefficients of the binomial expansion given by Pascal's triangle for spin $\frac{1}{2}$ nuclei (see Figure 2-24).
- Nuclei with the same chemical shift do not split each other, even when the coupling constant between them is nonzero.
- Spacings between adjacent components of a spin–spin multiplet are equal to J.
- Spin–spin multiplets are centered on the resonance frequency.

When the chemical shift difference is less than about 10 times J ($\Delta v/J \lesssim 10$), *second-order* effects appear in the spectrum, including deviations in intensities from the binomial pattern and other exceptions from the above characteristics. By the Pople notation, nuclei that have a first-order relationship are represented by letters that are far apart in the alphabet (AX), and those that have a second-order relationship are represented by adjacent letters (AB). Figure 4-1 illustrates the progression for two spins from AB to almost AX. When $\Delta v/J$ is 0.4, the spectrum is practically a singlet. Intensity distortions increase peak heights toward the center of the multiplet. A second-order multiplet typically leans toward the resonances of its coupling partner. The peak intensities within a multiplet are not equal even when $\Delta v/J = 15$. With the wide availability of proton frequencies of 300 MHz and higher, first-order spectra have become common but by no means exclusive.

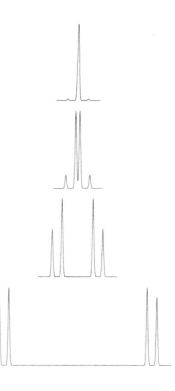

Figure 4-1 The two-spin spectrum with $\Delta\nu/J$ values of 0.4 (top), 1.0, 4.0, and 15.0.

4-2 CHEMICAL AND MAGNETIC EQUIVALENCE

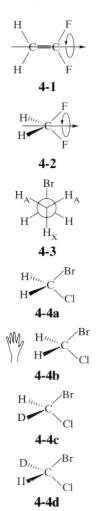

4-1

4-2

4-3

4-4a

4-4b

4-4c

4-4d

In addition to the requirements concerning chemical shift differences compared with coupling constants $\Delta\nu/J$, first-order spectra must pass a symmetry test. Any pair of chemically equivalent nuclei must have the same coupling constant to any other nucleus. Nuclear pairs that fail this test are said to be *magnetically nonequivalent*. To apply this test, it is useful to understand the role of symmetry in the NMR spectrum.

Nuclei are *chemically equivalent* if they can be interchanged by a symmetry operation of the molecule. Thus the two protons in 1,1-difluoroethene **(4-1)** or in difluoromethane **(4-2)** may be interchanged by a 180° rotation. Nuclei that are interchangeable by rotational symmetry are said to be *homotopic*. Rotation about carbon–carbon single bonds is so rapid that the chemist rarely considers the fact that the three methyl protons in CH_3CH_2Br are not in fact equivalent (compare nuclei A and X in **4-3**). Rapid C—C rotation, however, results in an average environment in which they are equivalent. Dynamic effects are considered more thoroughly in Section 5-2.

Nuclei related by other symmetry operations such as a plane are called *enantiotopic*, provided that there is no rotational axis of symmetry. For example, the protons in bromochloromethane **(4-4a)** are chemically equivalent and enantiotopic because they are related by the plane of symmetry containing C, Br, and Cl. If the molecule is placed in a chiral environment, this statement no longer holds true. Such an environment may be created by using a solvent composed of an optically active material or by placing the molecule in the active site of an enzyme. Such an environment may be represented as shown in **4-4b,** in which bromochloromethane has a small hand placed to one side. The protons are no longer equivalent because the hand is a chiral object. As the plane of symmetry is lost in the presence of the hand (or any chiral environment), the nuclei are not enantiotopic but have become chemically nonequivalent.

The term enantiotopic was invented because replacement of one proton of the pair by another atom or group, such as deuterium, produces the enantiomer (nonsuperimposable mirror image, **4-4c**) of the molecule that results when the other proton is replaced by the same group **(4-4d).** A pair of homotopic nuclei treated in this fashion produce identical molecules (superimposable mirror images). Enantiotopic or homotopic protons need not be on the same carbon atom. Thus the alkenic protons in cyclopropene **(4-5)** are homotopic, but those in 3-methylcyclopropene **(4-6)** are enan-

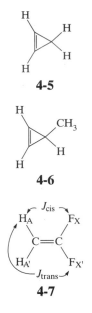

4-5

4-6

4-7

4-8

4-9

tiotopic. Chemically equivalent nuclei (either homotopic or enantiotopic) are represented by the same letter in the spectral shorthand of Pople. Cyclopropene **(4-5)** is A_2X_2, as is difluoromethane **(4-2),** since the two fluorine atoms have spins of $\frac{1}{2}$. The ring protons of 3-methylcyclopropene **(4-6)** constitute an AX_2 group.

To be *magnetically equivalent,* nuclei that already are chemically equivalent must have the same coupling constant to any nucleus. This is a more stringent test than chemical equivalence, because it is necessary to go beyond considering just the symmetry of the overall molecule. The first two molecules discussed in this chapter provide contrasting results. In difluoromethane **(4-2)** each of the two hydrogens has the same coupling to a specific fluorine atom because they have the same spatial relationship to the fluorine. Consequently, the protons are magnetically equivalent. By the same token, the two fluorine atoms also are magnetically equivalent by reference to coupling to either proton, and the spin system is labeled A_2X_2.

In 1,1-difluoroethene **(4-7),** however, the two protons do not have the same spatial relationship to a given fluorine. Therefore, they have different couplings, one a J_{cis} and the other a J_{trans} **(4-7),** and are said to be magnetically nonequivalent. These spins are represented by the notation AA'XX', so that the two couplings may be denoted by J_{AX} and $J_{AX'}$ (in contrast, an A_2X_2 system such as difluoromethane **(4-2)** or cyclopropene **(4-5)** has only one coupling, J_{AX}). In an AA'XX' system, J_{AX} and $J_{A'X'}$ are the same, as are $J_{AX'}$ and $J_{A'X}$. Any spin system that contains nuclei that are chemically equivalent but magnetically nonequivalent is by definition second order. Moreover, raising the magnetic field cannot alter basic structural relationships between nuclei, so that the spectrum remains second order at the highest accessible fields.

The AA'XX' notation may be interpreted as follows. The chemical shifts of the A and X nuclei are very far from each other (opposite ends of the alphabet). The A and A' nuclei are chemically equivalent but magnetically nonequivalent, as are the X and X' nuclei. Figure 4-2 illustrates the proton AA' part of the spectrum of 1,1-difluoroethene, in which 10 peaks are visible. This appearance is quite different from the simple 1/2/1 triplet expected in the first-order case. The multiplicity of peaks in Figure 4-2 in fact permits measurement of $J_{AA'}$, the coupling between the equivalent protons.

Magnetic nonequivalence is not uncommon. The spin systems for both para- and ortho-disubstituted benzene rings are AA'XX' (or AA'BB' if the chemical shifts are close). Figure 4-3 illustrates the proton spectrum of 1,2-dichlorobenzene **(4-8),** which is AA'XX' and relatively complex. Constraints of a ring frequently convey magnetic nonequivalence, as for example in butyrolactone **(4-9).** Even open chain systems such as 2-chloroethanol ($ClCH_2CH_2OH$, Figure 4-4) contain magnetically nonequivalent spin systems, although they are understandable only by examination of the contribut-

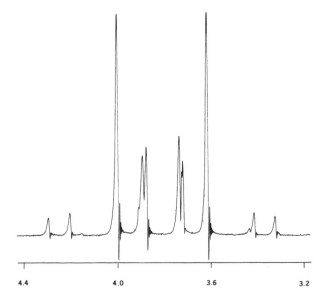

Figure 4-2   The 90 MHz ^1H spectrum of 1,1-difluoroethene in $CDCl_3$.

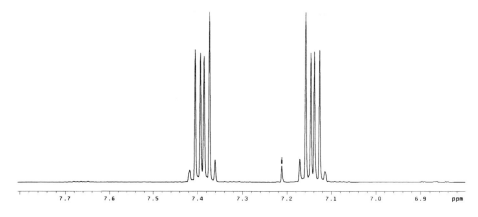

Figure 4-3 The 300 MHz ^{1}H spectrum of 1,2-dichlorobenzene in CDCl$_3$. An impurity is signified by the letter *i*.

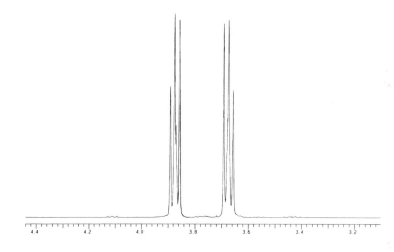

Figure 4-4 The 300 MHz ^{1}H spectrum of 2-chloroethanol (methylene resonances only) in CDCl$_3$.

ing rotamers (see Problems at the end of the chapter). Butyrolactone, chloroethanol, and both *o*- and *p*-dichlorobenzene thus all give AA'XX' (or AA'BB') spectra (if the hydroxyl proton is ignored in the alcohol).

Magnetically equivalent nuclei must be chemically equivalent and hence have the same chemical shift, that is, they must be *isochronous*. When protons are on different carbons, it usually is straightforward to determine whether they are chemically equivalent through examination of symmetry operations. Geminal protons (those on the same carbon, CH$_2$), sometimes are more subtle. Consider the protons of ethylbenzene (C$_6$H$_5$CH$_2$CH$_3$) and of its β-bromo-β-chloro derivative (C$_6$H$_5$CH$_2$CHClBr). Rotation about the saturated C—C bond creates three rotamers for each molecule, which may be represented by the Newman projections shown in **4-10** and **4-11**. In the rotamer **4-10a,** H$_A$ and H$_{A'}$ are chemically equivalent and enantiotopic by reason of the plane

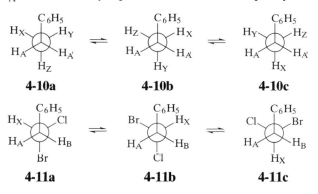

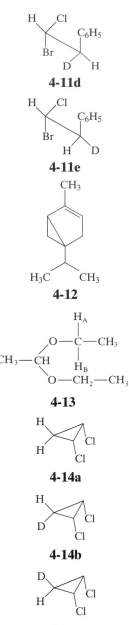

4-11d

4-11e

4-12

4-13

4-14a

4-14b

4-14c

of symmetry. When methyl rotation is slow, H_A and $H_{A'}$ are magnetically nonequivalent, because each would couple unequally with either H_X or H_Y. The plane of symmetry actually requires that H_Y should be labeled $H_{X'}$, but we retain the different lettering to illustrate the effect of methyl rotation. Thus the frozen structure **4-10a** contains an AA'XX'Z spectrum. Rapid methyl rotation averages the X, Y, and Z environments, so that the three methyl protons become chemically equivalent on average, and the A and A' proton have equal couplings to the methyl protons on average and hence become magnetically equivalent. On average there is only one coupling constant, and the spectrum is A_2X_3, if the aromatic protons are ignored.

Molecule **4-11** contains a chiral or stereogenic center in place of the methyl group, so that the three rotamers are now distinct **(4-11a-c).** Moreover, no symmetry operation in any of them relates H_A to H_B. Consequently, even with rapid C—C rotation, H_A and H_B have different chemical shifts and exhibit a mutual coupling constant. The spin system is ABX. The AB protons in **4-11** exemplify a particular type of chemically nonequivalent nuclei that are termed *diastereotopic*. Diastereoisomers are stereoisomers other than enantiomers. Replacement of H_A by deuterium gives **4-11d**, a diastereoisomer of **4-11e**, which is formed when H_B is replaced by deuterium. The deuterated derivative has two stereogenic centers. In general the protons of a saturated methylene group are diastereotopic when there is a stereogenic center elsewhere in the molecule because there is no symmetry operation that relates the two protons. The protons in **4-4b** are diastereotopic because the hand provides the stereogenic center. Accidental degeneracy can occur when the chemical shift difference is small or unobservable, so that diastereotopic protons can appear to be equivalent in the spectrum.

Methyl groups in an isopropyl group can be diastereotopic when stereogenic centers are present in the molecule, as in α-thujene **(4-12).** The proton resonance then appears as a pair of doublets (coupled to the methine proton) and the carbon resonance appears as two singlets (proton decoupled). A stereogenic center is not necessary for methylene protons to be diastereotopic. The diethyl acetal of acetaldehyde **(4-13)** contains diastereotopic protons because the symmetry axis of the molecule is not a symmetry axis for the CH_2 protons. This situation may be understood by examining the rotamers or by replacing H_A with deuterium. This latter operation creates two stereogenic centers, —OCHD(CH_3) and —OCH(CH_3)O—, and the resulting molecule is a diastereoisomer of the molecule in which H_B is replaced with deuterium. The methylene protons in *cis*-1,2-dichlorocyclopropane **(4-14a)** are diastereotopic because **4-14b** and **4-14c** are diastereoisomers. The axial and equatorial protons on a single carbon in ring-frozen cyclohexane are diastereotopic, because cyclohexane-axial-*d* and cyclohexane-equatorial-*d* are diastereoisomers. When ring flipping is fast on the NMR time scale, the geminal protons become equivalent on average. Thus the diastereotopic nature of protons can depend on the rate of molecular interconversions.

4-3 SIGNS AND MECHANISMS: ONE BOND COUPLINGS

Spin–spin coupling arises because information about nuclear spin is transferred from nucleus to nucleus via the electrons. Exactly how does this process occur? Several mechanisms have been considered, but the most important is the *Fermi contact mechanism*. According to this model, an electron in a bond X—Y, in which both X and Y are magnetic, spends a finite amount of time at the same point in space as, say, nucleus X. The electron and nucleus are then said to be in contact. If nucleus X has a spin of $I_z = +\frac{1}{2}$, then by the Pauli exclusion principle the spin of the electron must be opposite $(-\frac{1}{2})$ so that the two spins can occupy the same space at the same time. In this way the nuclear spin polarizes the electron spin (makes it to prefer one spin state). The electron in turn shares an orbital in the X—Y bond with another electron, which must have a spin of $+\frac{1}{2}$ when the spin of the first electron is $-\frac{1}{2}$. This second $(+\frac{1}{2})$ electron occupies the same point in space as nucleus Y only when Y has a spin of $-\frac{1}{2}$. Thus whenever X has a spin of $+\frac{1}{2}$, a spin of $-\frac{1}{2}$ is slightly favored for Y, as shown in Figure 4-5 for a ^{13}C—^1H coupling (an upward-pointing arrow represents a $+\frac{1}{2}$ spin and a downward-pointing arrow

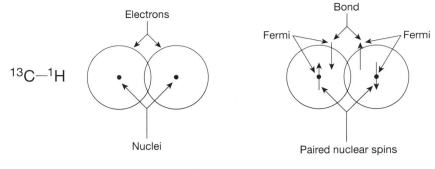

^{13}C—^{1}H

Figure 4-5

a $-\frac{1}{2}$ spin). Since the bonding electrons are used to pass the spin information, the contact term is not averaged to zero by molecular tumbling.

When one spin slightly polarizes another spin oppositely, as in the above model for coupling across X—Y, the coupling constant J between the spins by convention is said to have a positive sign. A negative coupling occurs when spins polarize each other in the same (parallel) direction. Qualitative models analogous to that shown in Figure 4-5 indicate that coupling over two bonds, as in H—C—H, is negative and coupling over three bonds, as in H—C—C—H, is positive. There are numerous exceptions to this qualitative approach, but it is useful in understanding that J has sign as well as magnitude.

High-resolution NMR spectra normally are not dependent on the absolute sign of coupling constants. Reversal of the sign of every coupling constant in a spin system results in an identical spectrum. Many spectra, however, depend on the *relative signs* of component couplings. For example, the general ABX spectrum is determined in part by three couplings, J_{AB}, J_{AX}, and J_{BX}. Different spectra can be obtained when J_{AX} and J_{BX} have the same sign (both positive or both negative) from when they have opposite signs (one positive and the other negative), even when the magnitudes are the same.

The usual convention for referring to a coupling constant is to denote the number of bonds between the coupled nuclei by a superscript on the left of the letter J and any other descriptive material by a subscript on the right or parenthetically. A two bond *(geminal)* coupling between protons then is $^2J_{HCH}$ or $^2J(HCH)$, and a three bond *(vicinal)* coupling between a proton and a carbon is $^3J_{HCCC}$ or $^3J(HCCC)$. Beyond three bonds, couplings between protons are said to be *long range.*

The one bond coupling between carbon-13 and protons is readily measured from the ^{13}C spectrum when the decoupler is turned off. Although usually unobserved because of decoupling, this coupling can provide useful information and illustrates several important principles. Because a p orbital has a node at the nucleus, only electrons in s orbitals can contribute to the Fermi coupling mechanism. For protons, all electrons reside in the 1s orbital, but for other nuclei only that proportion of the orbital that has s character can contribute to coupling. When a proton is attached to an sp^3 carbon atom (25% s character), $^1J(^{13}C–^1H)$ is about half as large as that for a proton attached to an sp carbon atom (50%). The alkenic CH (sp^2) coupling is intermediate. The values of ^1J for methane (sp^3), ethene (sp^2), benzene (sp^2), and ethyne (sp) are 125, 157, 159, and 249 Hz, respectively. These numbers define a linear relationship between the percentage of s character of the carbon orbital and the one bond coupling (eq. 4-1). The zero intercept

4-15

4-16

4-17

$$\% \ s(C—H) = 0.2J(^{13}C—^1H) \tag{4-1}$$

of this equation indicates that there is no coupling when the s character is zero, in agreement with the Fermi contact model.

The one bond CH coupling ranges from about 100 to 320 Hz, and much of this variation may be interpreted in terms of the J-s relationship. The coupling constant in cyclopropane (162 Hz) demonstrates that the carbon orbital to hydrogen is approximately sp^2 hybridized. Intermediate values in hydrocarbons may be interpreted in terms of fractional hybridization. The J = 144 Hz for the indicated CH bond in tricyclopentane **(4-15)** corresponds to 29% s character (sp$^{2.4}$), 160 Hz in cubane **(4-16)** to 32% s character (sp^2), and 179 Hz in quadricyclane **(4-17)** to 36% s character (sp$^{1.8}$).

of the other two substituents. In cycloalkenes the value varies from 1.3 Hz in cyclopropene to 8.8 Hz in cyclohexene. In acyclic alkenes, J_{trans} has a range of 10–24 Hz and J_{cis} of 2–19 Hz. Because the ranges overlap, the distinction between cis and trans isomers is fully reliable only when both isomers are in hand. When bonds are intermediate between single and double bonds, ^3J is proportional to the overall bond order, as in $^3J_{12} = 8.6$ Hz and $^3J_{23} = 6.0$ Hz in naphthalene.

The ortho coupling in benzene derivatives varies over the relatively small range of 6.7–8.5 Hz, depending on the resonance and inductive effects of the substituents. The presence of heteroatoms in the ring expands the range at the lower end down to 2 Hz, because of the effects of electronegativity (pyridines) and of smaller rings (furans, pyrroles). When one carbon is sp^3 and one is sp^2 (H—C—C(=X)—H) the range is 5–8 Hz for freely rotating acyclic hydrocarbons (X = CR$_2$) and 1–5 Hz for aldehydes (X = O). The value varies in hydrocarbon rings from −0.8 Hz in cyclobutene to +3.1 Hz in cyclohexene and +5.7 Hz in cycloheptene. For the central bond in dienes (**H**—C(=X)—C(=Y)—**H**), the range is 10–12 Hz for transoid systems (X,Y = CR$_2$). When constrained to rings, the pathway is cisoid and the coupling is 1.9 Hz in cyclopentadiene and 5.1 Hz in 1,3-cyclohexadiene. In α, β-unsaturated aldehydes (X = O, Y = CR$_2$), the coupling is about 8 Hz if transoid and 3 Hz if cisoid.

The H–C–C–C, H–C–C–F, and C–C–C–C couplings also follow Karplus-like relationships. The ^3J(C–C–C–C) couplings have a range of values (3–15 Hz) that is larger than the two bond case (the range for ^2J(C–C–C) is 1-10 Hz). The F–C–C–F and H–C–C–P couplings appear not to follow the Karplus pattern.

4-6 LONG-RANGE COUPLINGS

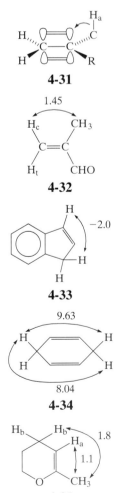

4-31

4-32

4-33

4-34

4-35

Coupling between protons that occurs over more than three bonds is said to be long range. Sometimes coupling between ^{13}C and protons over more than one bond also is called long range, but the term is inappropriate for ^2J(CCH) and ^3J(CCCH). Long-range coupling constants between protons normally are less than 1 Hz and frequently are unobservably small. In two structural circumstances, however, such couplings commonly become significant.

Interactions of C—H bonds with π electrons of double and triple bonds and aromatic rings along the coupling pathway often increase the magnitude of the coupling constant. One such case is the four bond allylic coupling, **HC—C=CH**, with a range of about +1 to −3 Hz and typical values close to −1 Hz. Larger values are observed when the saturated C—H$_a$ bond **(4-31)** is parallel to the π orbitals. This σ–π overlap enables coupling to be transmitted more effectively. When the C—H$_a$ bond is orthogonal to the π orbitals, there is no σ–π contribution and couplings are small (< 1 Hz). In acyclic systems, the dihedral angle is averaged over both favorable and unfavorable arrangements, so an average ^4J is found, as in 2-methylacrylein (**4-32**, ^4J = 1.45 Hz). Ring constraints can freeze bonds into the favorable arrangement, as in indene (**4-33**, ^4J = -2.0 Hz).

The five bond doubly allylic coupling (also called homoallylic), **HC—C=C—CH**, depends on the orientation of two C—H bonds with respect to the π orbitals. For acyclic systems such as the 2-butenes, ^5J typically is 2 Hz, with a range of 0–3 Hz. When both protons are well aligned, the coupling can be quite large, as in the planar 1,4-cyclohexadiene **(4-34)**, for which the cis coupling is 9.63 Hz and the trans coupling is 8.04 Hz. It is not unusual for the doubly allylic coupling to be larger than the allylic, as in **4-35** (^{4}J(CH$_3$—H$_a$) = 1.1 Hz, ^5J = 1.8 Hz).

Coupling constants are particularly large in alkynic and allenic systems, in which σ–π overlap can be very effective. In allene itself (CH$_2$=C=CH$_2$), ^4J is −7 Hz. In 1,1-dimethylallene, ^5J decreases to 3 Hz. In both propyne (methylacetylene, ^4J = 2.9 Hz) and 2-butyne (dimethylacetylene, ^5J = 2.7 Hz), the long-range coupling is enhanced because the triple bond imposes no steric limitations on σ–π overlap. Appreciable long-range couplings have been observed over up to seven bonds in polyalkynes.

Conjugated double bonds provide a more complicated situation. In butadiene, there are two four-bond (−0.86, −0.83 Hz) and three five-bond (+0.60, +1.30, +0.69) couplings. In aromatic rings, the meta coupling is a ^4J (range 1–3 Hz) and the para coupling is a ^5J (range 0–1 Hz). In benzene itself, $^3J_{\text{ortho}}$ is 7.54 Hz, $^4J_{\text{meta}}$ is 1.37 Hz, and $^5J_{\text{para}}$ is 0.69 Hz. None of these couplings in butadiene and benzene involves σ-π overlap. Protons on saturated carbon atoms attached to an aromatic ring ($\mathbf{CH_3}$—$\mathbf{C_6H_5}$) couple with all three types of protons on the ring. These benzylic couplings depend on the σ–π interaction between the substituent C—H bonds and the aromatic π electrons, much like the allylic coupling ($^4J_{\text{ortho}}$ = 0.6–0.9 Hz, $^5J_{\text{meta}}$ = 0.3–0.4 Hz, $^6J_{\text{para}}$ = 0.5–0.6 Hz). A doubly benzylic coupling can take place between protons on different saturated carbons directly attached to the benzene ring ($\mathbf{CH_3}$—$\mathbf{C_6H_4}$—$\mathbf{CH_3}$), as in xylenes ($^5J_{\text{ortho}}$ = 0.3–0.5 Hz).

In the second major category of long-range coupling, enhanced values often are observed between protons that are related by a planar W or zigzag pathway. This geometry is seen, for example, in the 1,3-diequatorial arrangement between protons in six-membered rings (**4-36**, ^4J = 1.7 Hz). The norbornane framework (**4-37**) contains several

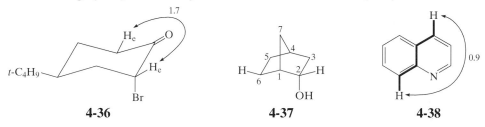

| **4-36** | **4-37** | **4-38** |

W arrangements, including that illustrated between the 2 and 6 exo protons, but also between the bridgehead protons (1 and 4) and between 3-endo and 7-anti protons.

In the planar, zigzag arrangement, there is favorable overlap between parallel C—H and C—C bonds, analogous to the optimal vicinal coupling at ϕ = 180°. The zigzag pathway is entirely within the σ framework but is important for many π systems, including aromatic meta couplings (hence the enhanced ^4J = 1.37 Hz in benzene). Five-bond zigzag pathways similarly can give rise to enhanced long-range couplings, such as the ^5J = +1.3 Hz in 1,3-butadiene ($\mathbf{H}$—C═C—C═C—$\mathbf{H}$) and the ^5J = 0.9 Hz coupling between the indicated protons in quinoline (**4-38**).

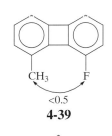

4-39

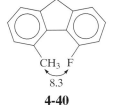

4-40

Although coupling information is always passed via electron-mediated pathways, in some cases part of the through bond pathway may be skipped, as in allylic and benzylic couplings with σ–π overlap. Two nuclei that are within van der Waals contact in space over any number of bonds can interchange spin information if at least one of the nuclei possesses lone pair electrons. These so-called through space couplings are found most commonly, but not exclusively, in H–F and F–F pairs. The six bond CH_3–F coupling is negligible in **4-39** (H–F distance 2.84 Å) but is 8.3 Hz in **4-40** (1.44 Å) (the sum of the H and F van der Waals radii is 2.55 Å). In the latter case coupling information is probably passed from the proton through the lone pair electrons to the fluorine nucleus. Such a mechanism very likely is important in the geminal F–C–F coupling, which is unusually large. Values of $^2J(\text{FCF})$ are larger for sp^3 CF_2 than for sp^2 CF_2, as the smaller tetrahedral angle brings the fluorine atoms closer together.

4-7 SPECTRAL ANALYSIS

We have not said much about how coupling constants are extracted from spectra. Measurement is straightforward when the spectrum is first order, as chemical shifts correspond to the midpoint of a resonance multiplet. The midpoint falls between the components of a doublet from coupling to one other spin, it is coincident with the middle peak of a triplet from coupling to two other spins, and so on. The coupling constant corresponds to the distance between adjacent peaks in the resonance multiplet. These simple characteristics may fail in second-order spectra. Because most nuclei other than the proton have very large chemical shift ranges and because these nuclei often are in low natural abundance and hence do not show coupling to each other, second-order analysis is primarily a consideration for proton spectra alone. For protons, spectra measured above 500 MHz are usually first order from the Δv/J criterion. Magnetic nonequiva-

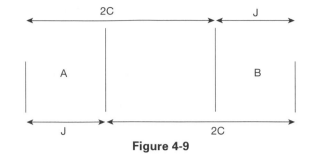

Figure 4-9

lence (Section 4-2), however, is independent of field and produces second-order spectra such as AA'XX' even with the most expensive superconducting magnet.

The AX spectrum consists of two doublets, all components with equal intensities (the spectra in Figure 2-18 and at the bottom of Figure 4-1 are very close to first order). The doublet spacing is J_{AX} and the midpoints of the doublets are ν_A and ν_X. The second-order, two spin (AB) system also contains four lines, but the inner peaks are always more intense than the outer peaks (see Figures 4-1 and 4-9). The coupling constant (J_{AB}) still is obtained directly from the doublet spacings, but no specific peak position or simple average corresponds to the chemical shifts. The A chemical shift (ν_A) occurs at the weighted average position of the two A peaks (and similarly for ν_B). The chemical shift difference $\Delta\nu_{AB}$ is most easily calculated from eq. 4-4, in which $2C$ is the spacing

$$\Delta\nu_{AB} = (4C^2 - J^2)^{\frac{1}{2}} \tag{4-4}$$

between alternate peaks (see Figure 4-9). The values of ν_A and ν_B are then readily determined by adding and subtracting $\frac{1}{2}\Delta\nu_{AB}$ to the midpoint of the quartet. The ratio of intensities of the larger inner peaks to the smaller outer peaks is given by the expression $(1 + J/2C)/(1 - J/2C)$.

Analysis of three spin systems can be carried out readily by inspection only in the first-order cases AX_2 and AMX. The second-order AB_2 spectrum can contain up to nine peaks, four from spin flips of the A proton alone, four from spin flips of the B protons alone, and one from simultaneous spin flips of both A and B protons. The ninth peak is called a *combination line* and is ordinarily forbidden and of low intensity. Although these patterns may be analyzed by inspection, recourse normally is made to computer programs. The other second-order three spin systems (ABB', AXX', ABX, and ABC) and almost all second-order systems of four spins (AA'BB', AA'XX', ABXY, etc.) or larger are seldom able to be analyzed by inspection, and recourse is made to computer methods.

Most spectrometers today contain the software for spectral calculation for up to seven spins. The first step is a trial and error procedure of approximating the chemical shifts and coupling constants in order to match the observed spectrum through computer simulation. Chemical shifts are varied until the widths and locations of the observed and calculated multiplets agree approximately. Then the coupling constants or their sums and differences are varied systematically until a reasonable match is obtained. This method is relatively successful for three and four spins but is difficult for larger systems.

Refinements of direct calculations or of this trial and error procedure utilize iterative computer programs. The program of Castellano and Bothner-By (LAOCN3) iterates on peak positions but requires assignments of peaks to specific spin flips. The program of Stephenson and Binsch (DAVINS) operates directly on unassigned peak positions.

Second-order spectra are characterized by peak spacings that do not correspond to coupling constants, by nonbinomial intensities, by chemical shifts that are not at resonance midpoints, and by resonance multiplicities that do not follow the $n + 1$ rule (see Figure 4-1). Even when the spectrum has the appearance of being first order, it may not be. Lines can coincide in such a way that the spectrum assumes a simpler appearance than seems consistent with the actual spectral parameters *(deceptive simplicity)*. In the ABX spectrum, the X nucleus is coupled to two closely coupled nuclei (A and B). When $\Delta\nu_{AB}/J$ is extremely small, the A and B spin states are fully mixed, and X responds as if the nuclei were equivalent. Thus the ABX spectrum resembles an A_2X spectrum, as if

$J_{AX} = J_{BX}$. Figure 4-10 illustrates this situation. When Δv_{AB} = 3.0 Hz (Figure 4-10a), the calculated example looks like a first-order A_2X spectrum with one coupling constant. When Δv_{AB} = 8.0 Hz (Figure 4-10b), a typical ABX spectrum is obtained. Deceptive simplicity sometimes, but not always, can be removed by use of a higher field. When the spectrum is deceptively simple, only sums or averages of coupling constants may be measured.

The AA'XX' spectrum often is observed as a deceptively simple pair of triplets, resembling A_2X_2. In this case it is the A and A' nuclei that are closely coupled ($\Delta v_{AA'}$ = 0 Hz and $J_{AA'}$ is large). This type of deceptive simplicity is not eliminated by raising the field because A and A' are chemically equivalent. The chemist should beware of the pair of triplets that falsely suggests magnetic equivalence (A_2X_2) and equal couplings ($J_{AX} = J_{AX'}$) when the molecular structure suggests AA'XX'. Sometimes the couplings between A and X may be observed by lowering the field to turn the AA'XX' spectrum into AA'BB' with a larger number of peaks that may permit a complete analysis.

A particularly subtle example of second-order complexity occurs in the ABX spectrum when A and B are very closely coupled, J_{AX} is large, and J_{BX} is zero. With no coupling to B, the X spectrum should be a simple doublet from coupling to A. Since A and B are closely coupled, however, the spin states of A and B are mixed, and the X spectrum is perturbed by the B spins (the phenomenon has been termed *virtual coupling*, which is something of a misnomer since B is not coupled to X). As an example in a slightly larger but analogous spin system, the CH and CH$_2$ protons of β-methylglutaric acid **(4-41)** are closely coupled. Although the CH$_3$ group is coupled only to the CH

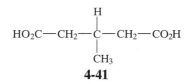

4-41

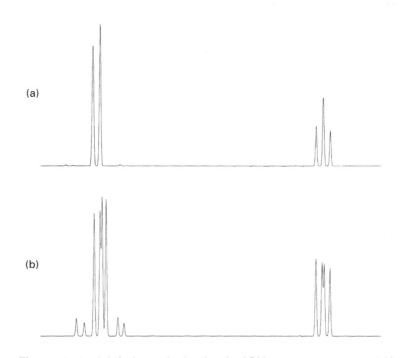

Figure 4-10 (a) A deceptively simple ABX spectrum: v_A = 0.0 Hz, v_B = 3.0 Hz, v_X = 130.0 Hz; J_{AB} = 15.0 Hz, J_{AX} = 5.0 Hz J_{BX} = 3.0 Hz. (b) The same parameters, except v_B = 8.0 Hz. The larger value of Δv_{AB} removes the deceptive simplicity and produces a typical ABX spectrum.

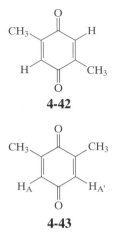

4-42

4-43

proton, its resonance is much more complicated than a simple doublet (Figure 4-11a). The CH and CH$_2$ protons are closely coupled, so their spin states are mixed and the CH$_3$ group interacts with a mixture of CH and CH$_2$ spin states even though $J = 0$ between CH$_3$ and CH$_2$. This problem is eliminated at a higher field, at which the CH and CH$_2$ resonances are well separated. Methyl, unmixed with the CH$_2$ spin states, then couples cleanly with CH.

The dimethylbenzoquinones provide a further example. The proton spectrum of the 2,5-dimethyl isomer (**4-42,** Figure 4-11b) contains a first-order methyl doublet and an alkene quartet. The spectrum of the 2,6 isomer (**4-43,** Figure 4-11c) is much more complicated. The alkenic protons in both molecules are equivalent (AA'). In **4-42** they are coupled only to the methyl protons ($J_{AA'} = 0$ Hz), but in **4-43** they are closely coupled to each other because of the zigzag pathway. The multiplicity of the methyl resonance is perturbed not only by the adjacent alkenic proton but also by the one on the opposite side of the ring. In other words, **4-42** is (AX$_3$)$_2$ but **4-43** is AA'X$_3$X'$_3$. This effect is not altered at a higher field because A and A' are chemically equivalent.

Some institutions still have available only iron core 60 MHz spectrometers, whose proton spectra are largely second order. Some clarification of these spectra may be obtained by the use of *paramagnetic shift reagents.* These molecules contain unpaired spins and form Lewis acid–base complexes with dissolved substrates. The unpaired spin exerts a strong paramagnetic shielding effect (hence to higher frequency or downfield) on nuclei close to it. The effect drops off rapidly with distance, so that those nuclei in the substrate that are closest to the site of acid–base binding are affected more. Consequently, the shift to higher frequency varies through the substrate and hence leads to greater separation of peaks. Two common shift reagents contain lanthanides: tris(dipivalomethanato)europium(III)·2(pyridine) (called Eu(dpm)$_3$ without pyridine) and 1,1,1,2,2,3,3-heptafluoro-7,7-dimethyloctanedionatoeuropium(III) (or Eu(fod)$_3$). Shift reagents are available with numerous rare earths as well as other elements. Almost all organic functional groups that are Lewis bases have been found to respond to these

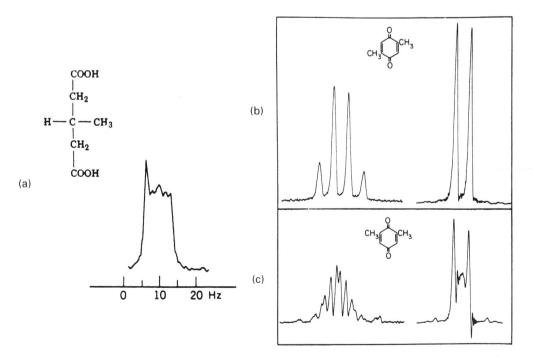

Figure 4-11 (a) The 60 MHz methyl ^{1}H resonance of β-methylglutaric acid. (b) and (c) The 60 MHz ^{1}H spectra of the 2,5 - and 2,6-dimethylbenzoquinones. ([a] Reproduced with permission from F.A.L. Anet, *Can. J. Chem.,* **39,** 2267 [1961]; [b] and [c] Reproduced with permission from E.D. Becker, *High Resolution NMR,* 2nd ed., Academic Press, Orlando, FL, 1980, p. 166.)

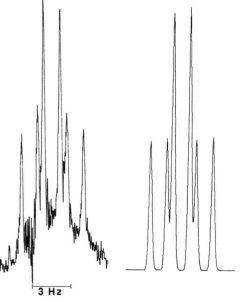

Figure 4-12 The 90 MHz ^{1}H spectrum of cyclopropene, showing the observed (left) and calculated (right) high frequency ^{13}C satellite of the alkenic protons. (Reproduced with permission from J.B. Lambert, A.P. Jovanovich, and W.L. Oliver, Jr., *J. Phys. Chem.* **74**, 2221 (1970). Copyright 1970 American Chemical Society.)

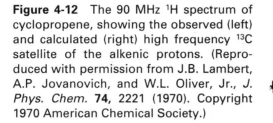

reagents. When the shift reagent is chiral, it can complex with the enantiomers and generate separate resonances from which enantiomeric ratios may be obtained.

Spectral analysis can sometimes be facilitated by taking advantage of dilute spins present in the molecule. Earlier in this chapter cyclopropene (**4-5**) was mentioned as an example of an A_2X_2 spectrum, and in Section 4-5 the vicinal coupling between the protons on the double bond (J_{AA}) was quoted as being 1.3 Hz. How was such a coupling constant between two chemically equivalent protons measured? Its small value prohibits the use of deuterium, as J_{HD} would be only 0.2 Hz. For 1.1% of the molecules the double bond spin system is H–^{12}C–^{13}C–H. The proton on ^{12}C resonates at almost the same position as the molecules with no ^{13}C. The large one bond ^{13}C–^{1}H coupling produces multiplets, called *satellites*, on either side of the centerband and separated from it by about $\frac{1}{2}J(CH)$. The separation of each satellite from the centerband serves as an effective chemical shift difference, so that the H–H coupling between H—^{12}C and H—^{13}C is present in the satellite. Figure 4-12 shows the satellite spectrum of the alkenic protons of cyclopropene. The satellite is a doublet of triplets, since the alkenic proton on ^{13}C is coupled to the other alkenic proton and to the two methylene protons. Other dilute spins produce satellite spectra that are commonly observed in proton spectra, including ^{15}N, ^{29}Si, ^{77}Se, ^{111}Cd, ^{113}Cd, ^{117}Sn, ^{119}Sn, ^{125}Te, ^{195}Pt, and ^{199}Hg.

The most general and effective method for analyzing complex proton spectra involves the use of two dimensions, as described in Chapter 6. Even this method, however, has limitations imposed on it by the presence of second-order relationships.

4.8 TABLES OF COUPLING CONSTANTS

The following tables (Tables 4-1 to 4-5) summarize values of coupling constants by class of structure, extracted from the references found at the end of the chapter. Further examples may be obtained by examination of these references.

TABLE 4-1
One Bond Couplings

		Hz			Hz
$^{13}C - {}^1H$	CH_3CH_3	125	$^{13}C - {}^{19}F$	CH_2F_2	235
	$(CH_3)_4Si$	118		CF_3I	345
	CH_3Li	98		C_6F_6	362
	$(CH_3)_3N$	132	$^{13}C - {}^{31}P$	CH_3PH_2	9.3
	CH_3CN	136		$(CH_3)_3P$	−13.6
	$(CH_3)_2S$	138		$(CH_3)_4P^+$ I⁻	56
	CH_3OH	142	$^{13}C - {}^{15}N$	CH_3NH_2	−4.5
	CH_3F	149		$C_6H_5NH_2$	−11.4
	CH_3Cl	150		$CH_3(CO)NH_2$	−14.8
	CH_2Cl_2	177		$CH_3C\equiv N$	−17.5
	$CHCl_3$	208		Pyridine	+0.62
	Cyclohexane	125		$CH_3HC\equiv N-OH$ (E,Z)	−4.0, −2.3
	Cyclobutane	136	$^{15}N - {}^1H$	CH_3NH_2	−64.5
	Cyclopropane	162		CH_3CONH_2	−89
	Tetrahydrofuran (α,β)	145, 133		Pyridinium	−90.5
	Norbornane (C1)	142		$HC\equiv N^+H$	−134
	Bicyclo[1.1.1]pentane (C1)	164		$(C_6H_5)_2C\equiv NH$	−51.2
	Cyclohexene (Cl)	157	$^{15}N - {}^{15}N$	Azoxybenzene	12.5
	Cyclopropene (C1)	226		Phenylhydrazine	6.7
	Benzene	159	$^{15}N - {}^{31}P$	$C_6H_5NHP(CH_3)_2$	53.0
	1,3-Cyclopentadiene (C2)	170		$C_6H_5NH(PO)(CH_3)_2$	−0.5
	$CH_2\equiv CHBr$ (gem)	197		$[(CH_3)_2N]_3P\equiv O$	−26.9
	Acetaldehyde (CHO)	172	$^{13}C - {}^{13}C$	CH_3CH_3	35
	Pyridine (α,β,γ)	177, 157, 160		$CH_3(CO)CH_3$	40
				CH_3CO_2H	57
	Allene	168		$CH_2\equiv CH_2$	68
	Propyne (≡CH)	248		$CH\equiv CH$	171
	$(CH_3)_2C^+H$ (⁺CH)	164	$^{31}P - {}^1H$	$C_6H_5(C_6H_5CH_2)(PO)H$	474
	$HC\equiv N$	269	$^{31}P - {}^{31}P$	$(CH_3)_2P-P(CH_3)_2$	−179.7
	Formaldehyde	222		$(CH_3)_2(PS)(PS)(CH_3)_2$	18.7
	Formamide	191			

TABLE 4-2
Geminal Proton–Proton (H—C—H) Couplings

	Hz		Hz
CH_4	−12.4	Oxirane	+5.5
$(CH_3)_4Si$	−14.1	$CH_2\equiv CH_2$	+2.3
$C_6H_5CH_3$	−14.4	$CH_2\equiv O$	+40.22
$CH_3(CO)CH_3$	−14.9	$CH_2\equiv NOH$	9.95
CH_3CN	−16.9	$CH_2\equiv CHF$	−3.2
$CH_2(CN)_2$	−20.4	$CH_2\equiv CHNO_2$	−2.0
CH_3OH	−10.8	$CH_2\equiv CHOCH_3$	−2.0
CH_3Cl	−10.8	$CH_2\equiv CHBr$	−1.8
CH_3Br	−10.2	$CH_2\equiv CHCl$	−1.4
CH_3F	−9.6	$CH_2\equiv CHCH_3$	2.08
CH_3I	−9.2	$CH_2\equiv CHCO_2H$	1.7
CH_2Cl_2	−7.5	$CH_2\equiv CHC_6H_5$	1.08
Cyclohexane	−12.6	$CH_2\equiv CHCN$	0.91
Cyclopropane	−4.3	$CH_2\equiv CHLi$	7.1
Aziridine	+1.5	$CH_2\equiv C\equiv C(CH_3)_2$	−9.0

TABLE 4-3

Vicinal Proton–Proton (H—C—C—H) Couplings

	Hz		Hz
CH_3CH_3	8.0	$CH_2=CH_2$ (cis,trans)	11.5, 19.0
$CH_3CH_2C_6H_5$	7.62	$CH_2=CHLi$ (cis,trans)	19.3, 23.9
CH_3CH_2CN	7.60	$CH_2=CHCN$ (cis,trans)	11.75, 17.92
CH_3CH_2Cl	7.23	$CH_2=CHC_6H_5$ (cis,trans)	11.48, 18.59
$(CH_3CH_2)_3N$	7.13	$CH_2=CHCO_2H$ (cis,trans)	10.2, 17.2
CH_3CH_2OAc	7.12	$CH_2=CHCH_3$ (cis,trans)	10.02, 16.81
$(CH_3CH_2)_2O$	6.97	$CH_2=CHCl$ (cis,trans)	7.4, 14.8
CH_3CH_2Li	8.90	$CH_2=CHOCH_3$ (cis,trans)	7.0, 14.1
$(CH_3)_2CHCl$	6.4	$ClHC=CHCl$ (cis,trans)	5.2, 12.2
$ClCH_2CH_2Cl$ (neat)	5.9	Cyclopropene (1–2)	1.3
$Cl_2CHCHCl_2$ (neat)	3.06	Cyclobutene (1–2)	2.85
Cyclopropane (cis,trans)	8.97, 5.58	Cyclopentene (1–2)	5.3
Oxirane (cis,trans)	4.45, 3.10	Cyclohexene (1–2)	8.8
Aziridine (cis,trans)	6.0, 3.1	Benzene	7.54
Cyclobutane (cis,trans)	10.4, 4.9	C_6H_5Li (2–3)	6.73
Cyclopentane (cis,trans)	7.9, 6.3	$C_6H_5CH_3$ (2–3)	7.64
Tetrahydrofuran (α–β: cis,trans)	7.94, 6.14	$C_6H_5CO_2CH_3$ (2–3)	7.86
Cyclopentene (3–4: cis,trans)	9.36, 5.72	C_6H_5Cl (2–3)	8.05
Cyclohexane (av.: cis,trans)	3.73, 8.07	$C_6H_5OCH_3$ (2–3)	8.30
Cyclohexane (ax–ax)	12.5	$C_6H_5NO_2$ (2–3)	8.36
Cyclohexane (eq–eq and ax–eq)	3.7	$C_6H_5N(CH_3)_2$ (2–3)	8.40
Piperidine (av. α–β: cis,trans)	3.77, 7.88	Naphthalene (1–2, 2–3)	8.28, 6.85
Oxane (av. α–β: cis,trans)	3.87, 7.41	Furan (2–3, 3–4)	1.75, 3.3
Cyclohexanone (av. α–β: cis,trans)	5.01, 8.61	Pyrrole (2–3, 3–4)	2.6, 3.4
Cyclohexene (3–4: cis,trans)	2.95, 8.94	Pyridine (2–3, 3–4)	4.88, 7.67

TABLE 4-4

Carbon Couplings Other Than $^1J(^{13}C-^1H)$

	Hz		Hz
$\mathbf{C}H_3\mathbf{C}H_3$	−4.8	$\mathbf{C}H_3\mathbf{C}H_3$	34.6
$\mathbf{C}H_3CH_2Cl$	2.6	$\mathbf{C}H_3\mathbf{C}H_2OH$	37.7
$Cl_2\mathbf{C}H-\mathbf{C}HCl_2$	+1.2	$\mathbf{C}H_3\mathbf{C}HO$	39.4
Cyclopropane (^2J)	−2.6	$\mathbf{C}H_3\mathbf{C}\equiv N$	56.5
$(\mathbf{C}H_3)_2CH\mathbf{C}H_2CH(CH_3)_2$	5.	$\mathbf{C}H_3\mathbf{C}O_2C_2H_5$	58.8
$(\mathbf{C}H_3)_2\mathbf{C}=O$	5.9	$\mathbf{C}H_2=\mathbf{C}H_2$	67.2
$\mathbf{C}H_3(\mathbf{C}O)H$	26.7	$\mathbf{C}H_2=\mathbf{C}HCN$	74.1
$\mathbf{C}H_3CH=\mathbf{C}(CH_3)_2$	4.8	$\mathbf{C}_6H_5\mathbf{C}N$ (ipso)	80.3
$\mathbf{C}H_2=\mathbf{C}H_2$	−2.4	$\mathbf{C}_6H_5NO_2$ (1,2)	55.4
$\mathbf{C}HCl=\mathbf{C}HCl$ (cis,trans)	16.0, 0.8	$H\mathbf{C}\equiv \mathbf{C}H$	170.6
$\mathbf{C}H_2=\mathbf{C}HBr$ (cis,trans)	−8.5, +7.5	$(\mathbf{C}H_3CH_2)_3\mathbf{P}$	+14.1
Benzene [$^2J(CH), ^3J(CH)$]	+1.0, +7.4	$(\mathbf{C}H_3CH_2)_4\mathbf{P}^+$ Br$^-$	−4.3
$\mathbf{C}H_3\mathbf{C}\equiv CH$ ($CH_3, \equiv CH$)	−10.6, +50.8	$(\mathbf{C}H_3O)_3\mathbf{P}$	+10.05
$\mathbf{C}F_3\mathbf{C}F_3$	46.0	$(\mathbf{C}H_3O)_3\mathbf{P}=O$	−5.8
$\mathbf{C}H_3(\mathbf{C}O)\mathbf{F}$	59.7	$(\mathbf{C}H_3)_3\mathbf{P}=S$	+56.1
$Cl_2\mathbf{C}=\mathbf{C}F_2$	44.2	$\mathbf{C}H_3(\mathbf{C}H_3O)_2\mathbf{P}=O$	+142.2

TABLE 4-5

Nitrogen-15 Couplings Beyond One Bond

	Hz		Hz
CH$_3$NH$_2$	−1.0	CH$_3$CH$_2$CH$_2$NH$_2$	1.2
Pyrrole (**HNCH**)	−4.52	CH$_3$CONH$_2$	9.5
Pyridine (**NCH**)	−10.76	CH$_3$C≡N	3.0
Pyridinium (**HNCH**)	−3.01	Pyridine (**NCC**)	+2.53
(CH$_3$)$_2$NCHO (CH$_3$,CHO)	+1.1, −15.6	Pyridinium (**HNCC**)	+2.01
H—C≡N	8.7	Aniline (**NCC**)	−2.68
H$_2$N(CO)CH$_3$	1.3	Pyrrole (**HNCC**)	−3.92
Pyrrole (**HNCCH**)	−5.39	CH$_3$CH$_2$CH$_2$NH$_2$	1.4
Pyridine (**NCCH**)	−1.53	Pyridine (**NCCC**)	−3.85
Pyridinium (**HNCCH**)	−3.98	Pyridinium (**HNCCC**)	−5.30
CH$_3$—C≡N	−1.7	Aniline (**NCCC**)	−1.29

PROBLEMS

4-1 Characterize the indicated protons as being homotopic, enantiotopic, or diastereotopic; magnetically equivalent or nonequivalent.

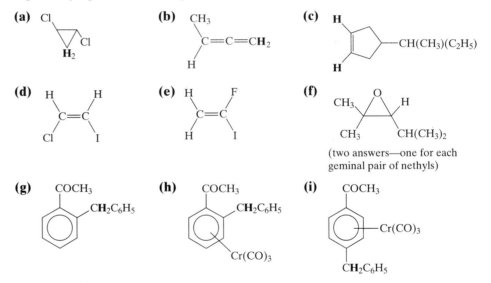

(a) Cl, Cl, **H$_2$**

(b) CH$_3$, C=C=CH$_2$, H

(c) H, —CH(CH$_3$)(C$_2$H$_5$), H

(d) H, H, C=C, Cl, I

(e) H, F, C=C, H, I

(f) CH$_3$, O, H, CH$_3$, CH(CH$_3$)$_2$
(two answers—one for each geminal pair of nethyls)

(g) COCH$_3$, CH$_2$C$_6$H$_5$

(h) COCH$_3$, CH$_2$C$_6$H$_5$, Cr(CO)$_3$

(i) COCH$_3$, Cr(CO)$_3$, CH$_2$C$_6$H$_5$

4-2 What is the spin notation for each of the following molecules (AX, AMX, AA'XX', etc.)? Consider only major isotopes.

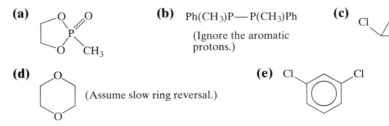

(a) O, O, P, O, CH$_3$

(b) Ph(CH$_3$)P—P(CH$_3$)Ph
(Ignore the aromatic protons.)

(c) Cl, Cl

(d) O, O
(Assume slow ring reversal.)

(e) Cl, Cl

4-3 Write out the rotamers of 2-chloroethanol (ClCH$_2$CH$_2$OH). What is the spin notation at slow rotation for each rotamer and at fast rotation for the average?

4-4 Below is the ^{1}H-decoupled ^{31}P spectrum* of the platinum complex with the illustrated structure (the resonances of the anion are omitted). Explain all the peaks and give the spin notation. What should the ^{195}Pt spectrum look like?

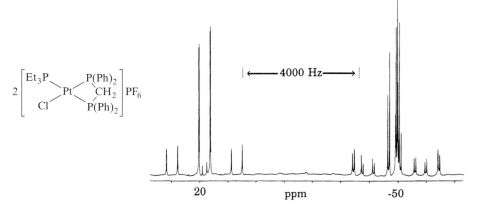

4-5 Eliminating four moles of HBr from the molecule below should give the indicated cyclopropane. The $^1J(^{13}C—^1H)$ for the bridge CH$_2$ group in the isolated product was measured to be 142 Hz. Explain in terms of product structures.

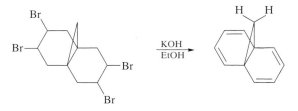

4-6 There are two isomers of thiane 1-oxide, **(a)** and **(b).** The observed geminal coupling constant between the α protons is −13.7 Hz in one isomer and −11.7 Hz in the other. Which coupling belongs to which isomer and why?

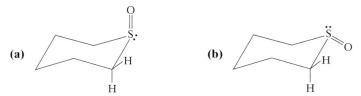

4-7 The ^{1}H spectrum of 1,3-dioxane (below) at slow ring reversal contains three multiplets with the following geminal couplings: −6.1, −11.2, and −12.9 Hz. Without reference to any chemical shift data, assign the resonances.

4-8 Does the angular methyl group in *trans*-decalins **(a)** or in *cis*-decalins **(b)** have the larger linewidth? Explain?

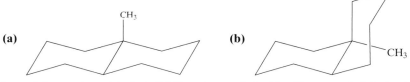

4-9 In cycloheptatriene **(a)**, J_{23} is 5.3 Hz, whereas in its bistrifluoromethyl derivative **(b)**, J_{23} is 6.9 Hz. Explain.

*From D.E. Berry, *J. Chem. Educ.*, **71**, 899-902 (1994). Copyright 1994 American Chemical Society. Reprinted by permission of the American Chemical Society.

4-10 The following four 300 MHz ^{1}H spectra are of lutidines (dimethylpyridines). From the chemical shifts and coupling patterns, deduce the placement of methyl groups on each molecule. Assume the spacings are first order.

(a)

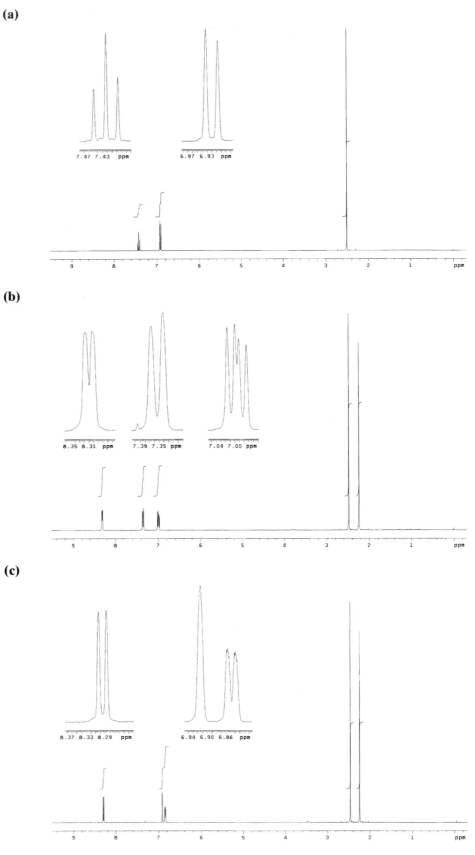

(d)

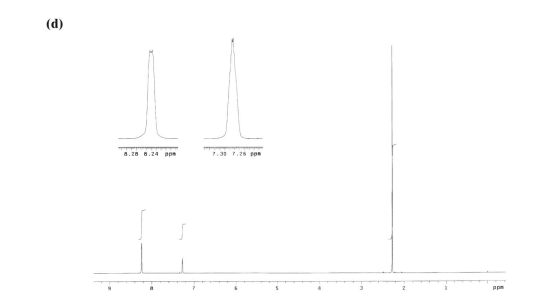

4-11 Proceed as in Problem **4-10** with the following four 300 MHz ^{1}H spectra of dichlorophenols.

(a)

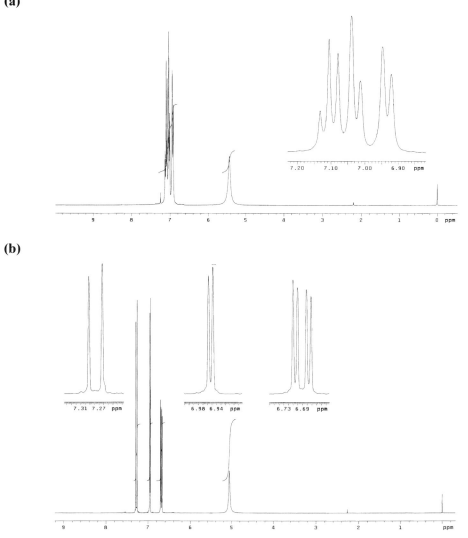

(b)

(c)

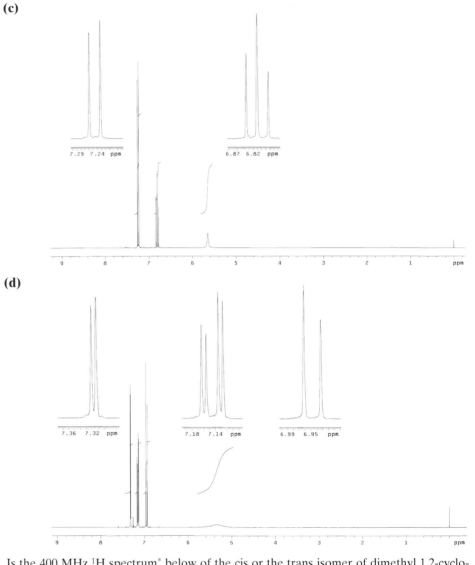

(d)

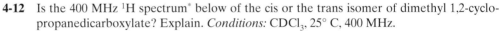

4-12 Is the 400 MHz ^{1}H spectrum* below of the cis or the trans isomer of dimethyl 1,2-cyclo-propanedicarboxylate? Explain. *Conditions:* CDCl$_3$, 25° C, 400 MHz.

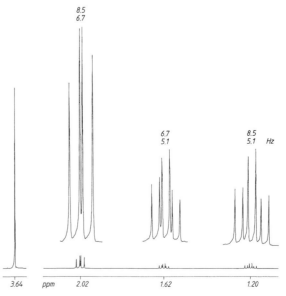

*From E. Breitmaier, *Structure Elucidation by NMR in Organic Chemistry,* John Wiley & Sons Ltd., Chichester, UK, 1993, p. 71. Copyright 1993 John Wiley & Sons Ltd. Reprinted by permission of John Wiley & Sons Ltd.

4-13 The ^{1}H spectrum of 2-hydroxy-5-isopropyl-2-methylcyclohexanone below has a ${}^3J_{56} = 3$ Hz in benzene-d_6 but 11 Hz in CD$_3$OD. Explain.

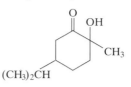

4-14 Analyze the following ^{1}H spectrum* of a thujic ester. The CH$_3$ resonances are not shown. Assign the resonances to specific protons and give very approximate coupling constants. Explain your chemical shift assignments.

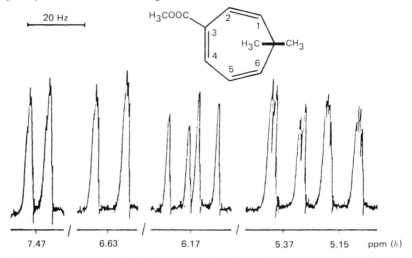

4-15 Deduce the structure (with relative stereochemistry) of the compound C$_6$H$_{12}$O$_6$ having the following 300 MHz ^{1}H NMR spectrum in D$_2$O (the peaks at δ 2.9 are from the reference, 3-(trimethylsilyl)propionic acid). Hydroxyl resonances are not shown. The triplet (a) at δ 4.04 has integral 1 and $J = 2.8$ Hz. The second-order triplet (b) at δ 3.61 has integral 2 and $J = 9.6$ Hz. The second-order doublet of doublets (c) at δ 3.52 has integral 2 and $J = 2.8$, 9.6 Hz. The triplet (d) at δ 3.26 has integral 1 and $J = 9.6$ Hz. The ^{13}C NMR spectrum shows four resonances, all between δ 71 and 75.

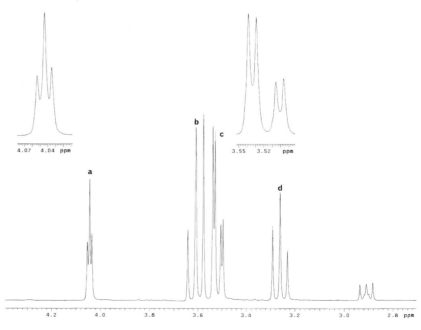

*From H. Günther et al., *Org. Magn. Reson.*, **6**, 388 (1974). Copyright 1974 John Wiley & Sons Ltd. Reprinted by permission of John Wiley & Sons Ltd.

BIBLIOGRAPHY

4.1. Coupling (general): I. Ando and G.A. Webb, *Theory of NMR Parameters,* Academic Press, London, 1983; *Nuclear Magnetic Resonance,* A Specialist Periodical Report, The Chemical Society, London, reviewed in each issue.

4.2. Magnetic equivalence: K. Mislow and M. Raban, *Top. Stereochem.,* **1,** 1 (1966); W.H. Pirkle and D.J. Hoover, *Top. Stereochem.,* **13,** 263 (1982).

4.3. One bond couplings: W. McFarlane, *Quart. Rev.,* **23,** 187 (1969); C.J. Jameson and H.S. Gutowsky, *J. Chem. Phys.,* **51,** 2790 (1969); J.H. Goldstein, V.S. Watts, and L.S. Rattet, *Progr. NMR Spectrosc.,* **8,** 103 (1971).

4.4. Geminal, vicinal, and long-range ^{1}H—^{1}H couplings: S. Sternhell, *Rev. Pure Appl. Chem.,* **14,** 15 (1964); A.A. Bothner-By, *Advan. Magn. Reson.,* **1,** 195 (1965); M. Barfield and B. Charkrabarti, *Chem. Rev.,* **69,** 757 (1969); S. Sternhell, *Quart. Rev.,* **23,** 236 (1969); V.F. Bystrov, *Russ. Chem. Rev.,* **41,** 281 (1972); J. Hilton and L.H. Sutcliffe, *Progr. NMR Spectrosc.,* **10,** 27 (1975); M. Barfield, R.J. Spear, and S. Sternhell, *Chem. Rev.,* **76,** 593 (1976).

4.5. Carbon-13 couplings: J.B. Stothers, *Carbon-13 NMR Spectroscopy,* Academic Press, New York, 1973; J.L. Marshall, D.E. Müller, S.A. Conn, R. Seiwell, and A.M. Ihrig, *Acc. Chem. Res.,* **7,** 333 (1974); D.F. Ewing, *Ann. Rep. NMR Spectrosc.,* **6A,** 389 (1975); R.E. Wasylishen, *Ann. Rep. NMR Spectrosc.,* **7,** 118 (1977); P.E. Hansen, *Org. Magn. Reson.,* **11,** 215 (1978); V. Wray, *Progr. NMR Spectrosc.,* **13,** 177 (1979); G.C. Levy, R.L. Lichter, and G.L. Nelson, *Carbon-13 Nuclear Magnetic Resonance Spectroscopy,* 2nd ed., Wiley–Interscience, New York, 1980; V. Wray and P.E. Hansen, *Ann. Rep. NMR Spectrosc.,* **11A,** 99 (1981); P.E. Hansen, *Ann. Rep. NMR Spectrosc.,* **11A,** 65 (1981); P.E. Hansen, *Progr. NMR Spectrosc.,* **14,** 175 (1981); W.H. Pirkle and D.J. Hoover, *Top. Stereochem.,* **13,** 263 (1982); J.L. Marshall, *Carbon–Carbon and Carbon–Proton NMR Couplings,* Verlag Chemie, Deerfield Beach, FL, 1983; L.B. Krivdin and G.A. Kalabia, *Progr. NMR Spectrosc.,* **21,** 293 (1989); L.B. Krivdin and E.W. Della, *Progr. NMR Spectrosc.,* **23,** 301 (1991).

4.6. Fluorine-19 couplings: J.M. Emsley, L. Phillips, and V. Wray, *Progr. NMR Spectrosc.,* **10,** 82 (1977).

4.7. Phosphorus-31 couplings: E.G. Finer and R.K. Harris, *Progr. NMR Spectrosc.,* **6,** 61 (1970).

4.8. Spectral analysis: J.D. Roberts, *An Introduction to the Analysis of Spin–Spin Splitting in High-Resolution Nuclear Magnetic Resonance Spectra,* W.A. Benjamin, New York, 1961; K.B. Wiberg and B.J. Nist, *The Interpretation of NMR Spectra,* W.A. Benjamin, New York, 1962; R.J. Abraham, *The Analysis of High Resolution NMR Spectra,* Elsevier Science Inc., Amsterdam, 1971; R.A. Hoffman, S. Forsén, and B. Gestblom, *NMR Basic Princ. Progr.,* **5,** 1 (1971); C.W. Haigh, *Ann. Rep. NMR Spectrosc.,* **4,** 311 (1971); P. Diehl, H. Kellerhals, and E. Lustig, *NMR Basic Princ. Progr.,* **6,** 1 (1972).

chapter 5

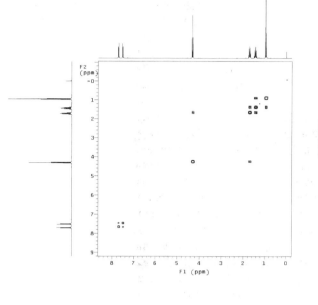

Further Topics in One-Dimensional NMR

Although the chemical shift and the coupling constant are the two fundamental measurables in NMR spectroscopy, several other phenomena may be studied in a single NMR time dimension. In this chapter we first examine the processes of spin–lattice and spin–spin relaxation, whereby the system moves toward spin equilibrium (see Section 2-3). Relaxation times or rates provide a third important measurable related to both structural and dynamic factors. Second, we explore in greater detail the structural changes that occur on the NMR time scale (see Section 2-9). The temporal dependence of chemical shifts and coupling constants influences both lineshapes and intensities and can be used to generate rate constants for reactions, a fourth NMR measurable. Third, we describe the family of experiments that utilize a second irradiation frequency, B_2. Double irradiation can give rise to spectral simplification, intensity perturbations, structural inferences, and information about rate processes. Finally, we expand on the technique of using several pulses, rather than only a single 90° pulse, sometimes separated by defined time periods, to improve sensitivity, simplify spectral patterns, measure relaxation times and coupling constants, draw structural conclusions, and improve the accuracy of pulse definitions.

5-1 SPIN–LATTICE AND SPIN–SPIN RELAXATION

Application of the B_1 field at the resonance frequency results in energy absorption and the conversion of some $+\frac{1}{2}$ spins into $-\frac{1}{2}$ spins. Thus magnetization in the z direction (M_z) decreases. Spin–lattice, or longitudinal, relaxation returns the system to equilibrium with time constant T_1. Such relaxation occurs because of the presence of natural magnetic fields in the sample that fluctuate at the Larmor frequency. Because of the frequency match, excess spin energy can flow into the molecular surroundings, sometimes called the lattice, and $-\frac{1}{2}$ spins can return to the $+\frac{1}{2}$ state.

The major source of these magnetic fields is magnetic nuclei in motion. Like the model of a charge moving in a circle, a magnetic dipole in motion creates a magnetic field, whose frequency depends on the rate of motion and on the magnetic moment of the dipole. For appropriate values of these parameters, the resulting magnetic field can fluctuate at the same frequency as the resonance (Larmor) frequency of the nucleus in question, permitting energy to flow from excited spins to the lattice. Such a process is called *dipole–dipole relaxation* ($T_1(DD)$) because it involves the interaction of the nuclear magnetic dipole with the dipole of the fluctuating field of the lattice. The resulting relaxation time or rate (R_1) depends on nuclear properties of both resonating and moving nuclei, on the distance between them, and on the rate of motion of the moving nucleus. The dependence takes the form of eq. 5-1 for the case of ^{13}C relaxed by protons

$$R_1(DD) = \frac{1}{T_1(DD)} = n\,\gamma_C^2\gamma_H^2\hbar^2 r_{CH}^{-6}\tau_c \tag{5-1}$$

in motion. The nuclear properties as usual are represented by the gyromagnetic ratios (this time to the second power). The symbol n stands for the number of protons that are nearest neighbors to the resonating carbon and hence are most effective at relaxing it. The rapid falloff with distance is indicated by the inverse of the nearest neighbor C—H distance r_{CH} to the sixth power. The motional properties of the protons are described by the effective correlation time τ_c, which is the time required for the molecule to rotate one radian and is typically in the nanosecond to picosecond range for organic molecules in solution.

   Thus carbon relaxation is faster (and the relaxation time is shorter) when there are more attached protons, when the internuclear distance is short, and when rotation in solution is slow. A quaternary carbon has a long relaxation time because it lacks an attached proton and r to other protons is large. The ratio of the carbon relaxation time of methyl to methylene to methinyl is 6/3/2 ($1/\frac{1}{2}/\frac{1}{3}$) because of differences in the number of attached protons, other things being equal. Because the rate of molecular tumbling in solution slows as molecular size increases, larger molecules relax more rapidly (for example, cholesteryl chloride > phenanthrene > benzene). Eq. 5-1 is an approximation to a more complete equation and represents what is called the *extreme narrowing limit*. Because the frequency of motion of the moving nuclear magnet must match the resonance frequency of the excited nuclear magnet, dipolar relaxation becomes ineffective for both rapidly moving small molecules and slowly moving large molecules. Many molecules of interest to biochemists fall into the latter category, for which eq. 5-1 does not apply. Rapid internal rotation of methyl groups in small molecules also can reduce the effectiveness of dipole–dipole relaxation. The optimal correlation times (τ_c) for dipolar relaxation lie in the range of about 10^{-7} to 10^{-11} s (the inverse of the resonance frequency). Because the resonance frequency depends on the value of B_0, this range also depends on B_0.

   When dipolar relaxation is slow, other mechanisms of relaxation become important. Fluctuating magnetic fields also can arise (1) from the interruption of the motion of rapidly rotating small molecules or groups within a molecule: spin rotation relaxation, $T_1(SR)$; (2) from tumbling of molecules with anisotropic chemical shielding at high fields: $T_1(CSA)$; (3) from scalar coupling constants that fluctuate through chemical exchange or through quadrupolar interactions: $T_1(SC)$; (4) from tumbling of paramagnetic molecules: $T_1(P)$ (unpaired electrons have very large magnetic dipoles); and (5) from the tumbling of quadrupolar nuclei: $T_1(Q)$. In the absence of quadrupolar nuclei or paramagnetic species, these alternative mechanisms often are unimportant. A major exception is relaxation of methyl carbons by spin rotation.

   The actual value of T_1 must be known at least approximately in order to decide how long to wait between pulses for return of the system to equilibrium (the delay time). In addition, $T_1(DD)$ offers structural information because of its dependence on r_{CH} and dynamic information because of its dependence on τ_c. For these reasons, convenient methods have been developed to measure T_1, the commonest of which is called *inversion recovery*. The strategy is to create a nonequilibrium distribution of spins and then to follow their return to equilibrium as a first-order rate process. Inverting the spins through application of a 180° pulse creates a maximum deviation from equilibrium (see Figure 5-1b). If a very short amount of time τ is allowed to pass (c) and a 90° pulse is applied to move the spins into the xy plane for observation, the nuclear magnets are aligned along the $-y$ axis (d) and an inverted peak is obtained. During the time

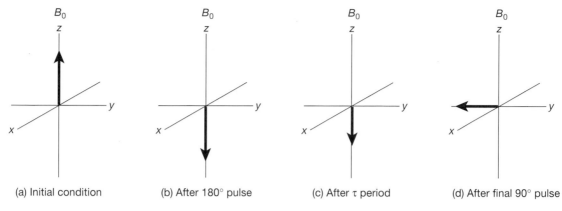

(a) Initial condition   (b) After 180° pulse   (c) After τ period   (d) After final 90° pulse

Figure 5-1

τ, some T_1 relaxation occurs (compare Figure 5-1b and c). The z magnetization at the end of the time τ (c) is shorter than at the beginning of τ (b). The peak produced after the 90° pulse then is smaller than if the 180 and 90° pulses had been combined as a 270° pulse (that is, if $\tau = 0$). The inversion recovery pulse sequence is summarized as 180°–τ–90°(acquire) and is an example of a simple multipulse sequence.

A series of such experiments with increasingly longer values of τ results in further relaxation between the 180° and 90° pulses. After the 90° pulse (Figure 5-1d), the resulting peak changes from negative through zero to positive as τ increases, until complete relaxation occurs, when τ is very long. Figure 5-2 shows a stack of such experiments for the carbons of chlorobenzene. Because the carbon ipso to chlorine has no directly attached proton, much longer values of τ are needed for the inverted peak to turn over. Relaxation is not complete even by $\tau = 80$ s. A plot of τ vs. the logarithm of the differences between intensities of each peak I_t and of the peak at equilibrium I_∞ [$\log(I_\infty - I_t)$] gives T_1 from the slope $(-1/T_1)$.

Relaxation in the xy plane, or spin–spin (transverse) relaxation (T_2), might be expected to be identical to T_1, because movement of the magnetization from the xy plane back onto the z axis restores z magnetization at the same rate as it depletes xy magnetization. There are, however, other mechanisms of xy relaxation that do not affect z magnetization. We have already seen in Section 2-3 that inhomogeneity of the B_0 mag-

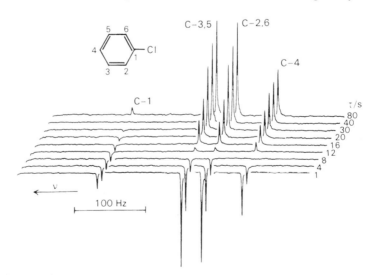

Figure 5-2 A stack plot for the inversion recovery experiment of the carbon-13 resonances of chlorobenzene at 25 MHz. The time τ in the pulse sequence $180° - \tau - 90°$ is given in seconds at the right. (From R.K. Harris, *Nuclear Magnetic Resonance Spectroscopy,* Pitman Publishing Ltd., London, 1983, p. 82. Reproduced with permission of Addison Wesley Longman Ltd.)

netic field randomizes phases in the *xy* plane and hastens *xy* relaxation. As a result, T_2 is always less than or equal to T_1. Additionally, *xy* relaxation can occur when two nuclei mutually exchange their spins, one going from $+\frac{1}{2}$ to $-\frac{1}{2}$ and the other from $-\frac{1}{2}$ to $+\frac{1}{2}$. This spin–spin, double flip, or flipflop mechanism is most significant in large molecules. The process can result in *spin diffusion*. The excitation of a specific proton changes the magnetization of surrounding protons as flipflop interactions spread around the molecule. The interpretation of the spectra of large molecules such as proteins must take such a process into consideration.

Proton relaxation times depend on the distance between the resonating nucleus and the nearest neighbor protons. The closer the neighbors, the faster the relaxation and the shorter the T_1. The two isomers **5-1a** and **5-1b** [Bz = Ph(C=O)], may be

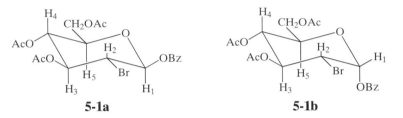

5-1a   **5-1b**

distinguished by their proton relaxation times. In the isomer in **5-1a,** H_1 is axial and close to the 3 and 5 axial protons, resulting in a T_1 of 2.0 s. In the isomer in **5-1b,** H_1 is equatorial and has more distant nearest neighbors, resulting in a T_1 of 4.1 s. In this way the structure of the anomers may be distinguished. The remaining values of T_1 may be interpreted in a similar fashion. For example, H_2 in isomer **5-1a** has only the H_4 axial proton as a nearest neighbor, so its T_1 is a relatively long 3.6 s. In isomer **5-1b,** H_2 has not only the axial H_4 but also the vicinal H_1 as nearest neighbors, so T_1 is faster, 2.1 s.

When a molecule is rigid and rotates equally well in any direction (isotropically), all the carbon relaxation times (after correction for the number of attached protons) should be nearly the same. The nonspherical shape of a molecule, however, frequently leads to preferential rotation in solution around one or more axes (anisotropic rotation). For example, toluene prefers to rotate around the long axis that includes the methyl, ipso, and para carbons. As a result, on average these carbons (and their attached protons) move less in solution than the ortho and meta carbons, because atoms on the axis of rotation remain stationary during rotation. The more rapidly moving ortho and meta carbons thus have a shorter effective correlation time τ_c and hence by eq. 5-1 a longer T_1. The actual values are shown in structure **5-2.** The longer value for the ipso carbon arises because it lacks a directly bonded proton and r_{CH} in eq. 5-1 is very large.

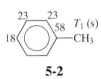

5-2

When molecules are not rigid, the more rapidly moving pieces relax more slowly because their τ_c is shorter. Thus in decane **(5-3)** the methyl carbon relaxes most slowly,

$$CH_3CH_2CH_2CH_2CH_2CH_2CH_2CH_2CH_2CH_3$$
$$nT_1 \quad 26.1 \quad 13.2 \quad 11.4 \quad 10.0 \quad 8.8$$

5-3

followed by the ethyl carbon and so on to the fifth carbon in the middle of the chain. Structure **5-3** gives the values of nT_1 (*n* is the number of attached protons), so that the figures may be compared for all carbons without consideration of substitution patterns. These values indicate the relative rates of motion of each carbon.

The inversion recovery experiment used to measure T_1 also may be exploited to simplify spectra. In Figure 5-2, the spectrum for $\tau = 40$ s entirely lacks a resonance for the ipso carbon (C-1). Similarly, for a τ of about 10 s, all the other ring carbons are nulled and only the negative peak for C-1 is obtained. Such *partially relaxed spectra* can be used not only to obtain partial spectra in this fashion but also to eliminate specific peaks. When deuterated water (D_2O) is used as the solvent, the residual HOD peak is undesirable. An inversion recovery experiment can reveal the value of τ for which the water peak is nulled. The rest of the protons will have positive or negative intensities at that τ, depending on whether they relax more rapidly or more slowly than water. The experiment may be refined by applying the 180° pulse selectively only at the resonance

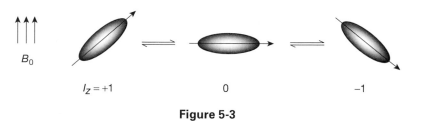

$$I_z = +1 \qquad\qquad 0 \qquad\qquad -1$$

Figure 5-3

position of water. Selection of τ for nulling of this peak then produces a spectrum that lacks the water peak but otherwise is quite normal for the remaining resonances. Such a procedure is an example of *peak suppression.*

The dominant mode of spin–lattice relaxation of nuclei with spins greater than $\frac{1}{2}$ results from the quadrupolar nature of such nuclei. These nuclei are considered to have an ellipsoidal rather than a spherical shape. When $I = 1$, as for ^{14}N or ^2H, there are three stable orientations in the magnetic field: parallel, orthogonal, and antiparallel, as shown in Figure 5-3. When these ellipsoidal nuclei tumble in solution within an unsymmetrical electron cloud of the molecule, they produce a fluctuating electric field that can bring about relaxation.

The mechanism is distinguished from dipole–dipole relaxation in two ways. First, it does not require a second nucleus in motion; the quadrupolar nucleus creates its own fluctuating field by moving in the unsymmetrical electron cloud. Second, because the mechanism is extremely effective when the quadrupole moment of the nucleus is large, T_1 can become very short (milliseconds or less). For such cases, the uncertainty principle applies, whereby the product of ΔE (the spread of energies of the spin states as measured by the linewidth $\Delta\upsilon$) and Δt (the lifetime of the spin state as measured by the relaxation time) must remain constant ($\Delta E \Delta t \sim$ Planck's constant). Thus when the relaxation time is very short, the linewidth becomes very large. Nuclei with large quadrupole moments often exhibit very large linewidths, for example about 20,000 Hz for the ^{35}Cl resonance of CCl_4. The common nuclides ^{17}O and ^{14}N have smaller quadrupolar moments and exhibit sharper resonances, typically tens of Hz. The small quadrupole moment of deuterium results in quite sharp peaks, typically a Hz or a few Hz. The linewidth also depends on the symmetry of the molecule, which controls how unsymmetrical the electron cloud is. π Systems are more unsymmetrical and give broader lines (as in amides and pyridines for ^{14}N). Spherical or tetrahedral systems have no quadrupolar relaxation, since the electron cloud is symmetrical, and exhibit very sharp linewidths like those of spin $-\frac{1}{2}$ nuclei (^{14}N in $^+NH_4$, ^{10}B in $^-BH_4$, ^{35}Cl in Cl^-, ^{33}S in $SO_4{}^{2-}$).

Very important to the organic chemistry is the effect of quadrupolar nuclei on the resonances of nearby protons. When quadrupolar relaxation is extremely rapid, a neighboring nucleus experiences only the average spin environment of the quadrupolar nucleus, so that no spin coupling is observed. Thus protons in chloromethane produce a sharp singlet even though ^{35}Cl and ^{37}Cl have spins of 3/2 and actually exist in four spin states. Chemists have come to think of the halogens (other than fluorine) as being nonmagnetic, although they only appear so because of their rapid quadrupolar relaxation. At the other extreme, deuterium has such a weak quadrupole moment that neighboring protons exhibit normal couplings to ^2H. Thus nitromethane with one deuterium (CH_2DNO_2) shows a 1/1/1 triplet since the protons are influenced by the three spin states ($+1, 0, -1$) of deuterium (analogous to Figure 5-3). Nitromethane with two deuteriums (CHD_2NO_2) shows a 1/2/3/2/1 quintet from coupling to the various combinations of the three spin states ($++; +0,0+; +-,00,-+; -0,0-; --$). This quintet is often observed in deuterated solvents such as acetone-d_6, acetonitrile-d_3, or nitromethane-d_3, because incomplete deuteration results in an impurity containing one proton in place of a deuterium.

The ^{14}N nucleus falls between these extremes. In highly unsymmetrical cases such as the interior nitrogen in biuret, $NH_2(CO)NH(CO)NH_2$, quadrupolar relaxation is rapid enough to produce the average singlet for the attached proton. The protons of the ammonium ion on the other hand give a sharp 1/1/1 triplet with full coupling between ^1H and ^{14}N because quadrupolar relaxation is absent. When relaxation is at an inter-

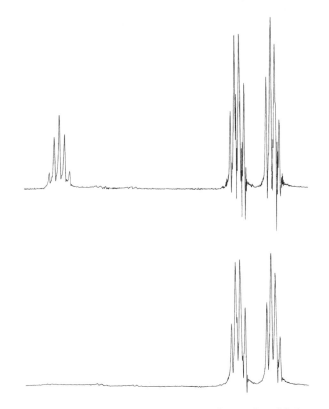

Figure 5-4   The 90 MHz proton spectrum of pyrrole with (upper) and without (lower) ^{14}N decoupling.

mediate rate, it is possible to observe three broadened peaks, one broadened average peak, or broadening to the point of invisibility. Irradiation at the ^{14}N frequency removes the ^{14}N—^{1}H interaction (Section 5-3), so that ^{14}N appears to be nonmagnetic. Figure 5-4 shows the normal spectrum of pyrrole at the bottom, containing only the AA'BB' set from the CH protons and no visible NH resonance because the line is extremely broad. Irradiation at the ^{14}N frequency decouples the NH proton from ^{14}N and results in a quintet NH resonance from coupling to the four α CH protons.

5-2   REACTIONS ON THE NMR TIME SCALE

NMR is an excellent tool for following the kinetics of an irreversible reaction through the disappearance or appearance of peaks over time periods of minutes to hours. Thus the spectrum may be obtained several times at specific time intervals, with rate constants calculated in the usual fashion from changes in peak intensities. These molecular changes take place on a time scale much longer than pulse or acquisition times of the NMR experiment. More importantly, NMR has a unique capability for the study of the kinetics of reactions that occur at equilibrium and affect the lineshapes, usually with activation energies in the range 4.5 to 25 kcal mol^{-1} (Section 2-9). This range corresponds to rates in the range 10^0 to 10^4 s^{-1}, which is the same as the spacings between the exchanging resonances in Hz.

A series of spectra for the interchange of axial and equatorial protons in cyclohexane-d_{11} as a function of temperature is illustrated in Figure 2-33. When the interchange of two such chemical environments occurs faster than the NMR time scale, the result is a single peak, reflecting the average environment *(fast exchange)*. Keep in mind that these exchanges occur reversibly, and the system remains at equilibrium. When the interchange is slower than this time scale, the NMR result is two distinct peaks *(slow exchange)*. When the interchange occurs within the NMR time scale, both extremes sometimes may be reached by altering the temperature of the experiment. At temper-

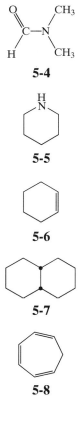

5-4

5-5

5-6

5-7

5-8

atures between the extremes, broadened lines are obtained. Bimolecular reactions such as the acid-catalyzed interchange of protons also may be studied, as in the case of the hydroxy proton of methanol (see Figure 2-32).

Normally, rotation around single bonds has a barrier below 5 kcal mol^{-1} and occurs faster than the NMR time scale. Rotation around the double bond of alkenes, on the other hand, has a barrier normally above 50 kcal mol^{-1} and is slow on the NMR time scale. There are numerous examples of intermediate bond orders, whose rotation occurs within the NMR time scale. Hindered rotation about the C—N bond in amides such as N,N-dimethylformamide (5-4) provides a classic example of site exchange.

At room temperature, exchange is slow and two methyl resonances are observed, whereas above 100° C exchange is fast and a single resonance is observed. Hindered rotation occurs on the NMR time scale for numerous other systems with partial double bonds, including carbamates, thioamides, enamines, nitrosamines, alkyl nitrites, diazoketones, aminoboranes, and aromatic aldehydes. Steric congestion about single bonds can raise the barrier to coincide with the NMR range, as in biphenyls, trypticenes, and *tert*-butyl compounds. Polyhalogenated alkanes such as 2,2,3,3-tetrachlorobutane also have barriers in the NMR range, as do molecules containing two adjacent heteroatoms possessing lone pair electrons, such as hydrazines, disulfides, sulfenamides, and aminophosphines. Hindered rotation has been observed in a few simple acyclic amines at the lowest extremes of accessible temperatures, corresponding to barriers close to or even slightly below 5 kcal mol^{-1}.

Axial–equatorial interconversion through ring reversal has been studied in a wide variety of systems in addition to cyclohexane, including heterocycles such as piperidine (5-5), unsaturated rings such as cyclohexene (5-6), fused rings such as *cis*-decalin (5-7), and rings of other than six members such as cycloheptatriene (5-8). Trisubstituted atoms with a lone pair, such as amines, may undergo the process of pyramidal atomic inversion on the NMR time scale. The resonances of the two methyls in the aziridine 5-9 become equivalent at elevated temperatures through rapid nitrogen inversion. This

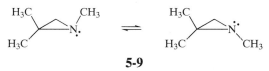

5-9

barrier is particularly high (18 kcal mol^{-1}) because of angle strain in the three-membered ring, which is higher in the transition state than in the ground state. When the nitrogen atom is attached to an electron-withdrawing atom, as in oxazolidines or N-chloroamines, the barrier is raised. Inversion about other atoms, such as phosphorus, oxygen, and sulfur, also has been studied by NMR. Inversion about nitrogen in trigonal systems such as imines provides a method of studying syn-anti interconversions.

Many other types of unimolecular reactions may be studied by NMR, including valence tautomerizations (such as the Cope rearrangement of 3,4-homotropilidene, **5-10**) and carbocation rearrangements (such as hydride shifts in the norbornyl cation, **5-11**).

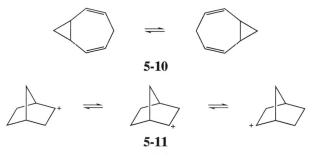

5-10

5-11

The observation of lineshape changes associated with these rate processes can be exploited for the determination of rate constants. For the simple case of two equally populated sites that do not exhibit coupling (such as cyclohexane-d_{11} in Figure 2-33 and the amide **5-4**), the rate (k_c) at the point of maximum peak broadening (the *coalescence temperature* T_c, approximately −60° C in Figure 2-33) is $\pi\Delta\nu/\sqrt{2}$, in which $\Delta\upsilon$ is the distance in Hz between the two peaks at slow exchange. The free energy of activation then may be calculated as $\Delta G_c^{\ddagger} = 2.3RT_c[10.32 + \log(T_c/k_c)]$. This result is extremely accurate and certainly easy to obtain, but the equation is limited in application.

To include spin–spin coupling, unequal populations, and more than two exchange sites, it is necessary to use computer programs such as DNMR3 that can simulate the entire lineshape at several temperatures. Such a procedure generates Arrhenius plots from which enthalpic and entropic activation parameters may be obtained, but is more susceptible to systematic errors involving inherent linewidths and spacings than is the coalescence temperature method. Consequently, it is always a good idea to use both lineshape fitting and coalescence temperature methods as an internal check.

The proportionality between k_c and $\Delta\upsilon$ ($k_c = \pi\Delta\nu/\sqrt{2}$) means that the rate is dependent on the field strength (B_0). Thus a change in field from 300 to 600 MHz alters the rate constant at T_c. At a given field strength, two nuclides such as ^{1}H and ^{13}C have different values of $\Delta\nu$. Thus the rates of exchange at T_c between protons is different from those of carbons, as in the methyl protons and methyl carbons of N,N-dimethyl-formamide (5-4).

Alternative procedures not requiring peak coalescence have been developed to expand the kinetic dynamic range of NMR. In many cases coalescence and fast exchange are never attained. The system may exchange too slowly on the NMR time scale at the highest available temperatures (as determined by the temperature range of the spectrometer or solvent or by the stability of the sample). An alternative technique called *saturation transfer* or *magnetization transfer* can provide rate constants without peak coalescence, that is, at the slow exchange limit. Continuous, selective irradiation of one slow exchange peak may partially saturate the other peak. Some of the nuclei from the first site turn into nuclei of the second type by the exchange process. The intensity of the second peak then is reduced because the newly transformed nuclei had been saturated in their previous form. This reduction in intensity is related to the rate constant of interchange and the relaxation time. Saturation transfer is observed for rates in the range 10^{-3} to 10^1 s^{-1}, which extends the NMR range on the slow exchange end by about three orders of magnitude.

Rates that are fast on the NMR lineshape time scale (no decoalescence) sometimes may be measured by observation at a different resonance frequency. Normally nuclear spins precess around the B_0 field at their Larmor frequency. Application of the usual 90° pulse in the x direction places the spins in the xy plane, along the y axis (see Figure 2-14). Continuous B_1 irradiation along the y axis (not a pulse) forces magnetization to precess around that axis (spin locking, as in the cross polarization experiment of Section 2-10). The spins are said to be locked onto the y axis. Because the spins are precessing at a lower frequency (γB_1), they are sensitive to a different range of rate processes, one corresponding to about 10^2 to 10^6 s^{-1}, which extends the NMR range on the fast exchange end by about two orders of magnitude. Rates are obtained by comparing the relaxation time while spin-locked ($T_{1\rho}$) with the usual spin–lattice relaxation time (T_1) and analyzing any differences.

Through lineshape, saturation, and spin-lock methods, the entire range of rates accessible to NMR is about 10^{-3} to 10^6 s^{-1}. Thus NMR has become an important method for studying the kinetics of reactions at equilibrium over a very large dynamic range.

5-3   MULTIPLE RESONANCE

Special effects may be routinely and elegantly created by using sources of radiofrequency energy in addition to the observing frequency ($\nu_1 = \gamma B_1$) ($\gamma = \gamma/2\pi$). The technique is called multiple irradiation or multiple resonance and requires the presence of a second transmitter coil in the sample probe to provide the new irradiating frequency ($\nu_2 = \gamma B_2$). When the second frequency is applied, the experiment is termed *double resonance* and is widely available on modern spectrometers. Less often, a third frequency ($\nu_3 = \gamma B_3$) also is provided to create a *triple resonance* experiment. We have already seen several examples of multiple irradiation experiments, including removal of proton couplings from ^{13}C spectra, sharpening of NH resonances by irradiation of ^{14}N, studying rate processes by saturation transfer, and elimination of solvent peaks by peak suppression.

One of the oldest and most generally applicable double resonance experiments is the irradiation of one proton resonance (X) and observation of the effects on the AX coupling present in another proton resonance (A). The traditional and intuitive explanation for the resulting spectral simplification, known as *spin decoupling,* is that the irradiation shuttles the X protons between the $+\frac{1}{2}$ and $-\frac{1}{2}$ spin states so rapidly that the A protons no longer can distinguish their independent existence. As a result, the A resonance collapses to a singlet. This explanation is only a rationalization, as it fails to account for phenomena at weak decoupling fields (spin tickling) and even some phenomena at very strong decoupling fields.

The actual experiment involves getting the coupled nuclei to precess about orthogonal axes. The magnitude of the coupling interaction between two spins is expressed by the scalar or dot product of their magnetic moments and is proportional to the expression $J\mu_1 \cdot \mu_2 = J\mu_1\mu_2 \cos\phi$. The quantity ϕ is the angle between the vectors. So long as both sets of nuclei precess around the same *(z)* axis, ϕ is zero ($\cos 0° = 1$) and full coupling is observed. The geometrical relationship between the spins may be altered by subjecting one of the spins to a B_2. Imagine observing ^{13}C nuclei as they precess around the z axis at the frequency of B_1. When the attached protons are subjected to a strong B_2 along the x axis, they then will precess around that axis. The angle ϕ between the ^{13}C and ^{1}H nuclear vectors then is 90°, as they respectively precess around the z and x axes. As a result, their spin–spin interaction goes to zero because the dot product is zero ($\cos 90° = 0$). The nuclei are then said to be decoupled.

Spin decoupling has been useful in identifying coupled pairs of nuclei. Figure 5-5 provides such an example for the molecule ethyl *trans*-crotonate (ethyl *trans*-but-2-enoate). The alkenic protons split each other and both are split by the allylic methyl group to form an ABX_3 spin system. Irradiation at the methyl resonance frequency (X) produces the upper spectrum in the inset for the alkenic protons, which have become a

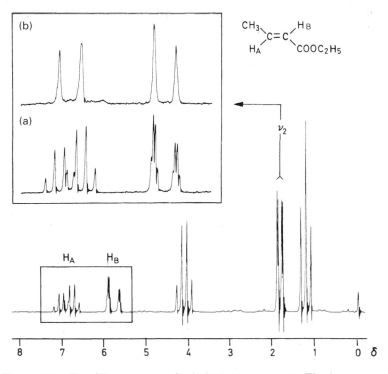

Figure 5-5 The ^{1}H spectrum of ethyl *trans*-crotonate. The inset contains an expansion of the alkenic range without (a) and with (b) decoupling of the methyl resonance at δ 1.8. (Reproduced with permission from H. Günther, *NMR Spectroscopy,* John Wiley & Sons Ltd., Chichester, UK, 1980, p. 289.)

simple AB quartet. A more complex example is illustrated in Figure 5-6. The bicyclic sugar mannosan triacetate, whose structure is given on the left of the figure, has a nearly first-order spectrum with numerous coupling partners. Irradiation of H_5 (δ 4.62) produces simplification of the resonances of its vicinal partners H_4, $H_{6/1}$, and $H_{6/2}$, as well as its long-range zigzag partner H_3.

With complex molecules, it is useful to record the difference between coupled and decoupled spectra. Features that are not affected by decoupling are subtracted out and do not appear. Figure 5-7 shows the ^1H spectrum of 1-dehydrotestosterone. The complex region between δ 0.9 and 1.1 contains the resonances of four protons. Comparison of the coupled (Figure 5-7a in the inset) and decoupled (b) spectra from irradiation of the 6α resonance shows little change as the result of double irradiation. The *difference decoupling spectrum* (c) is the result of subtracting (a) from (b). The unaffected overlapping peaks are gone. The original resonances of the affected protons are observed as

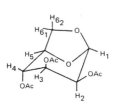

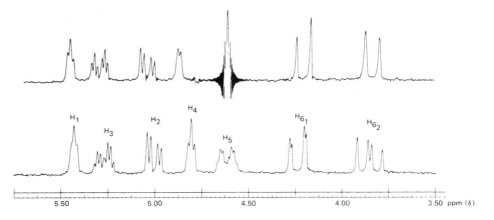

Figure 5-6 The 100 MHz ^1H spectrum of mannosan triacetate in $CDCl_3$ without decoupling (lower) and with double irradiation at δ 4.62 (upper). (Reproduced with the permission of Varian Associates.)

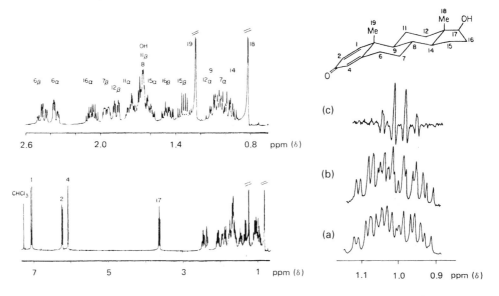

Figure 5-7 The 400 MHz ^1H spectrum of 1-dehydrotestosterone. The complete spectrum and an expansion of the low frequency region are given on the left. On the right are given (a) the coupled spectrum for the δ 0.9–1.1 region, (b) the same region decoupled from the 6α proton, and (c) the difference spectrum from subtraction of (b) from (a). (Reproduced with permission from L.D. Hall and J.K.M. Sanders, *J. Am. Chem. Soc.,* **102,** 5703 [1980]. Copyright 1980 American Chemical Society.)

negative peaks with coupling, and the simpler decoupled resonances of the same protons are present as positive peaks. The resonances must be due to the 7α protons. This procedure provides coupling relationships when spectral overlap is a serious problem. This and other simple spin decoupling experiments have been largely superseded by two-dimensional experiments (see Chapter 6).

Experiments in which both irradiated and observed nuclei are protons are called *homonuclear double resonance* and are represented by the notation $^1H\{^1H\}$. The irradiated nucleus is denoted by braces. When the observed and irradiated nuclei are different nuclides, as in proton-decoupled ^{13}C spectra, the experiment is *heteronuclear double resonance* and is denoted, for example, $^{13}C[^1H]$.

Double resonance experiments also may be classified according to the intensity or bandwidth of the irradiating frequency. If irradiation is intended to cover only a portion of the resonance frequencies, the technique is known as *selective irradiation/decoupling*. The decoupling experiments shown in Figures 5-5 and 5-6, peak suppression described in Section 4-2, and magnetization transfer discussed in Section 5-2 are examples of selective double irradiation. In the two decoupling experiments, only couplings to the selectively irradiated proton are removed. When all frequencies of a specific nuclide are irradiated, the experiment is termed *nonselective irradiation* or *broadband decoupling*. Figure 2-25 illustrates the ^{13}C spectrum of 3-hydroxybutyric acid both with and without broadband proton double irradiation. The invention of this technique was instrumental in the development of ^{13}C NMR spectroscopy as a routine tool. The double irradiation field B_2 was traditionally centered at about δ 5 of the ^1H range. To cover all the ^1H frequencies, B_2 was modulated with white noise, and the technique often was called *noise decoupling*.

The broadband decoupling experiment removes coupling patterns that could indicate the number of protons attached to a given carbon atom. The *off-resonance decoupling* method was developed to retain this information and still provide some of the advantages of the decoupling experiment. Irradiation above or below the usual 10 ppm range of ^1H frequencies leaves residual coupling given by the approximate formula $J_{res} = 2\pi J\Delta v/\gamma B_2$, in which J is the normal coupling, γ is the gyromagnetic ratio of the irradiated nucleus, and Δv is the difference between the decoupler frequency and the resonance frequency of a proton coupled to a specific carbon. Because carbon multiplicities remain intact, this technique is useful for determining, with minimal peak overlap, whether carbons are methyl (quartet), methylene (triplet), methine (doublet), or quaternary (singlet). If methylene protons are diastereotopic, methylene carbons can appear as two doublets. The outer peaks of the off-resonance decoupled triplets and quartets usually are weaker than expected from the binomial coefficients. As a result, doublets and quartets sometimes are difficult to distinguish. Figure 5-8 shows the spectrum of vinyl acetate with full decoupling and with off-resonance decoupling. In complex molecules, peak overlap and ambiguities with regard to quartets often make assignments by this technique difficult. It has been superseded by the editing experiments described in Section 5-5.

In early spin decoupling experiments, the irradiation frequency was left on continuously while observing the resonating nuclei. There are two significant problems with this method. First, application of rf energy at the decoupling frequency generates heat. As B_0 fields were raised from 100 to 800 MHz, higher decoupling intensities were required. The resulting heating was unacceptable for biological samples and for many delicate organic or inorganic samples. Secondly, with higher field strengths it became increasingly more difficult for B_2 to cover the entire range of ^1H frequencies, which had been about 1000 Hz at 100 MHz but became 5000 Hz at 500 MHz.

To overcome these problems, instead of irradiating continuously, modern decoupling methods arrange for the frequencies of an observed A nucleus associated respectively with spin-up and spin-down X nuclei to coincide at the moment of sampling by judicious timing of an additional pulse. In a $^{13}C\{^1H\}$ experiment (Figure 5-9 for two spins, ^{13}C—^1H), when a 90° B_1 pulse is applied to the observed ^{13}C nuclei along the x direction, magnetization from carbon coupled to both spin-up and spin-down protons moves into the xy plane along the y axis [(a) —→ (b)]. The carrier frequency is consid-

quencies of W_2 are in the MHz range (represented by the large spacing in Figure 5-10), whereas those of W_0 are much smaller, kHz or Hz (represented by the small but exaggerated spacing). Small molecules tumbling in solution produce fields in the MHz range and hence can provide W_2 relaxation. On the other hand, large molecules tumbling in the Hz–kHz range can provide W_0 relaxation.

Double irradiation of A in small molecules thus enhances the X intensity, provided the two nuclei are close enough for W_2 relaxation (less than about 5Å). This circumstance corresponds to what we previously referred to as the extreme narrowing limit. For larger molecules, certainly those with molecular weights over 10,000, W_0 dominates and reduction in peak intensity or inverse peaks occur. At some intermediate size the effect disappears as the crossover between regimes occurs. In general the nuclear Overhauser effect (NOE) is defined by eq. 5-2, in which the Greek letter η (eta) stands

$$\eta = (I - I_0)/I_0 \qquad (5\text{-}2)$$

for the effect, I_0 is the intensity without double irradiation, and I is the intensity with irradiation. For small molecules the maximum increment in intensity η_{max} is $\gamma_{irr}/2\gamma_{obs}$, so that a unit intensity increases up to $(1 + \eta_{max})$ (in our example A was irradiated ["irr"] and X observed ["obs"]). The increase can be less than the maximum if nondipolar relaxation mechanisms are present or if the observed nucleus is relaxed by nuclei other than the irradiated nucleus.

Whenever the two nuclei are the same nuclide (both are protons, for example), the gyromagnetic ratios cancel, η_{max} becomes 0.5, and the maximum intensity enhancement $(1 + \eta_{max})$ is a factor of only 1.5, or 50%. For the common case of broadband ^{1}H irradiation with observation of ^{13}C, η_{max} is 1.988, so the enhancement is a factor of up to 2.988, or about 200%. Certain nuclei have negative gyromagnetic ratios, so that η_{max} becomes negative and a negative peak can result. For irradiation of ^{1}H and observation of ^{15}N, $\eta_{max} = -4.92$. The maximum negative intensity thus is 3.92 times that of the original peak, or an increase of 292% but as an inverse peak. If dipolar relaxation is only partial, the ^{15}N{^{1}H} NOE can result in a completely nulled resonance. Silicon-29 also has a negative gyromagnetic ratio, so similar complications ensue. The NOE is entirely independent of spectral changes that arise from the collapse of spin multiplets through spin decoupling. The NOE does not require that nuclei A and X be spin coupled, only that they be mutually relaxed through a dipolar mechanism.

Measurement of the NOE may be carried out either directly or by difference. In the direct experiment, the spectrum is recorded twice, with and without the NOE. Figure 5-11 illustrates the relative timing of the ^{13}C pulse, the ^{1}H decoupling field, and acquisition of the ^{13}C signal. In the normal experiment with continuous broadband decoupling (a), the decoupling B_2 field is turned on and left on. It must be on not only during acquisition to ensure decoupling but also during the recovery time, when relaxation occurs and the NOE builds up. By gating the decoupler off during the recovery period, as in (b), but keeping it on during acquisition, the operator obtains decoupling but no NOE. Without irradiation during the recovery period, there is insufficient time for the NOE to build up and normal intensities are obtained. In practice, the double resonance frequency is not actually turned off but is moved far off resonance. Figure 5-11c illustrates an alternative gated experiment, in which the B_2 field is gated off during acquisition but on during the recovery period. Such an experiment provides no decoupling but generates the NOE, so it is useful for measuring ^{1}H—^{13}C couplings with enhanced intensity.

For the homonuclear proton NOE experiment just described (^{1}H{^{1}H}), it has traditionally been considered that the NOE (in percentage: 100 η) must exceed about 10% to be accepted as experimentally significant. The *difference NOE experiment,* however, can reliably measure enhancements to below 1%. By this procedure the spectra of types shown in Figure 5-11a and b are alternatively recorded and subtracted. Unaffected resonances disappear and NOEs are signified by residual peaks. Figure 5-12 illustrates the difference NOE spectrum for a portion of the ^{1}H spectrum of progesterone **(5-12),** in which the 19 methyl group has been irradiated *(arrow).* The unirradiated spectrum

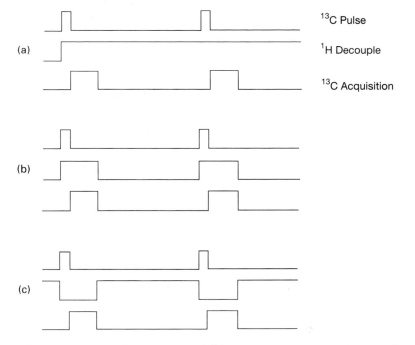

Figure 5-11 (a) Observation of ^{13}C with continuous double irradiation of ^1H (decoupling and NOE). (b) Double irradiation applied during acquisition but gated off during the wait period (decoupling, no NOE). (c) Double irradiation applied only during the wait period (NOE, no decoupling). The pulse widths are not to scale. The scheme is shown for two cycles.

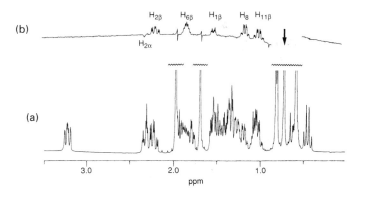

Figure 5-12 The 400 MHz ^1H spectrum (in part) of progesterone (a) without double irradiation and (b) with irradiation of the CH_3 –19 resonance displayed as a difference spectrum. (Reproduced with permission from J.K.M. Sanders and B.K. Hunter, *Modern NMR Spectroscopy*, Oxford University Press, Oxford, UK, 1993, p. 191.)

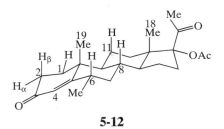

5-12

is given at the bottom and the difference spectrum at the top. Enhancements are seen by difference for five nearby protons. In general, for molecules in the extreme narrowing limit, the NOE difference experiment is preferred to the direct experiment. Proton

H$_{2\alpha}$ (the equatorial 2 proton) is not close to the 19 methyl group, but its resonances show a small negative NOE. This observation is the result of a three-spin effect (A is relaxed by B and B by C). Irradiation at A increases the Boltzmann population for B and enhances the intensity of B. This enhanced intensity of B has the opposite effect on C, decreasing the Boltzmann population and the intensity. As a result, C appears as a negative peak in the difference NOE spectrum. In the present example, A is Me-19, B is H$_{2\beta}$, and C is H$_{2\alpha}$.

The NOE experiment has three distinct uses. For heteronuclear examples, the foremost use is the increase in sensitivity. This increase combines with the collapse of multiplets through decoupling to provide the standard ^{13}C spectrum comprised of a singlet for each carbon. Because most carbons are relaxed almost entirely by their attached protons, the NOE commonly attains the maximum value of about 200%. Quaternary carbons, with more distant nearest neighbors, do not enjoy this large enhancement.

Second, interpretation of ^{13}C spin–lattice relaxation routinely requires a quantitative assessment of the dipolar component, T_1(DD). Because the NOE results from dipolar relaxation, its size is related to the dipolar percentage of overall relaxation. If the maximum or full NOE for ^{13}C{^{1}H} of 200% is observed, then T_1(obs) = T_1(DD). When other relaxation mechanisms contribute to ^{13}C relaxation, the enhancement is less than 200%. The dipolar relaxation for ^{13}C{^{1}H} may be calculated from the expression T_1(DD) = ηT_1(obs)/1.988, in which η is the observed NOE and 1.988 is the maximum NOE (η_{max}). It is then possible to discuss T_1(DD) in terms of structure, according to eq. 5-1.

In the third application, the dependence of the NOE on internuclear distances can be exploited to determine structure, stereochemistry, and conformation. Enhancements are expected when nuclei are close together. The adenosine derivative shown in **5-13** (2',3'-isopropylidene adenosine) can exist in the conformation shown with the purine ring lying over the sugar ring (syn) or in an extended form with the proton on C8 lying over the sugar ring (anti). Saturation of the H1' resonance brings about a 23% enhancement of the H8 resonance, and saturation of H2' produces an enhancement of H8 of 5% or less. Thus H8 must be positioned most closely to H1', as in the syn form shown. Structural and stereochemical distinctions frequently are possible by determination of relative orientations of protons. The synthetic penicillin derivative **5-14** could have the spiro

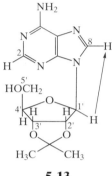

5-13

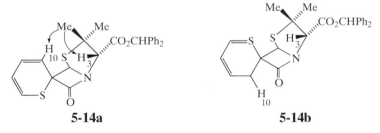

5-14a   **5-14b**

sulfur heterocycle oriented either as shown in (a) or with the sulfur atom and (CH)$_{10}$ switched as in (b). Irradiation of the methyl protons brings about enhancement of H$_{10}$ as well as H3 and clearly demonstrates that the stereochemistry is as shown in **5-14a**.

Despite the considerable advantages of the NOE experiment, its limitations must be appreciated. First, three-spin effects or spin diffusion may cause intensity perturbations when the third spin is not close to the irradiated nucleus (H$_{2\alpha}$ in Figure 5-12). Second, the size of the molecule can lead to NOE effects that are positive, negative, or null. Third, nuclei with negative gyromagnetic ratios can give diminished positive peaks, no peak, or negative peaks with diminished or enhanced intensity. Fourth, chemical exchange can cause intensity perturbation analogous to the three-spin effect. Irradiation of a nucleus can lead to intensity changes at another nucleus, which can alter its chemical identity through a dynamic exchange such as a bond rotation or valence tautomerization. The NOE can then be observed in the product nucleus, if chemical exchange is faster than relaxation of the NOE effects. Fifth, unintentional paramagnetic impurities can alter the NOE through intermolecular dipole–dipole relaxation. These considerations must be taken into account when interpreting NOE experiments.

5-5   SPECTRAL EDITING

For deducing the structure of organic molecules, one of the most useful pieces of information is a compilation of the substitution pattern of all the carbons, that is, a census of which carbons are methyl, methylene, methine, or quaternary. We have already seen that the off-resonance decoupling procedure provides such information, although with less than ideal results. Through the choice of appropriate pulses and timing, the chemist may accomplish the same task by eliminating some of the resonances from the spectrum or by reversing their polarization. Such an experiment is called *spectral editing* and includes solvent suppression for example.

Most spectral editing procedures are based on the *spin echo* experiment devised by Hahn, Carr, Purcell, Meiboom, and Gil in the 1950s largely to measure spin–spin relaxation times (T_2). An example of this experiment has already been seen in Figure 5-9, in which the 180° pulse brought the vectors from spin–spin interactions back together on the y axis as an echo. Such a procedure also refocuses dispersion in the chemical shift caused by magnetic inhomogeneity. With reference to Figure 5-13, in the absence of J, a resonance (b) fans out over a range of frequencies (c), because not every nucleus of a given type has exactly the same resonance frequency in an inhomogeneous field. The 180° pulse refocuses all the magnetization back onto the y axis after 2τ, as in (e). Chemical shift differences also may be eliminated in this fashion. Repetition of the 180° pulse every 2τ produces a train of peaks whose intensities die off with time constant T_2 (the notation $T_2{}^*$ sometimes is used to denote transverse relaxation that includes the effects of inhomogeneity). This relaxation time provides a measure of spin–spin interactions alone.

Although developed to measure T_2, this pulse sequence after a single cycle is able to improve resolution or eliminate coupling constants or chemical shifts. Moreover, it may be modified to achieve other effects. To obtain information about how many protons are attached to a carbon, the coupling information must be manipulated in a different fashion from that shown in Figure 5-5. If the protons are subjected to a 180° pulse at the same time as the carbons (Figure 5-14), the vectors from spin–spin coupling con-

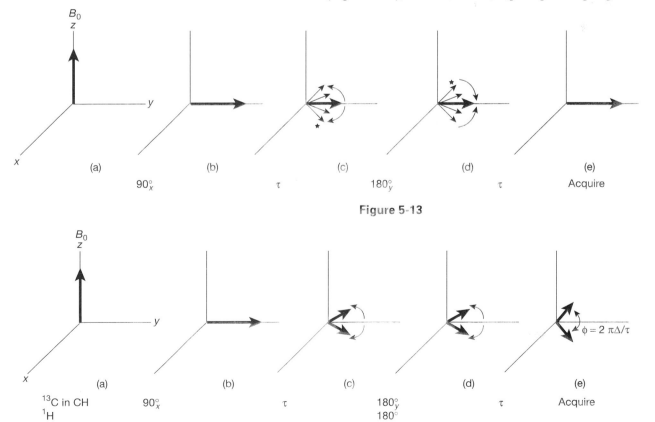

Figure 5-13

Figure 5-14

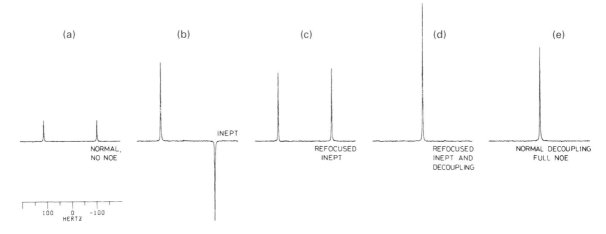

Figure 5-21 The ^{13}C spectrum of chloroform (a) without double irradiation, (b) with ^{1}H irradiation to achieve the INEPT enhancement, (c) with ^{1}H irradiation to achieve refocused INEPT enhancement, (d) with ^{1}H irradiation to achieve refocused INEPT enhancement and decoupling, and (e) with normal decoupling to achieve only the NOE. (Reproduced with permission for A.E. Derome, *Modern NMR Techniques for Chemistry Research,* Pergamon Press, Oxford, UK, 1987, p. 137.)

null signal. Methylene triplets give $-1/0/1$ INEPT intensities and methyl quartets give $-1/-1/1/1$ intensities and hence also would give null signals on decoupling. The *refocused INEPT* pulse sequence was designed to get around this problem by repeating the INEPT pulses a second time, during which polarization transfer is retained and only the negative intensity is reversed. Application of decoupling during the acquisition period then results in an enhanced peak free of coupling. Figure 5-21 compares the various experiments for chloroform.

Although INEPT is rarely used as an attached proton test, the related sequence DEPT (**D**istortionless **E**nhancement by **P**olarization **T**ransfer), has already been mentioned as being the method of choice. Similar to refocused INEPT, DEPT has a pair of $\tau\ (= 1/2J)$ periods followed by a single variable pulse θ:

$$
\begin{aligned}
S \quad &90^\circ_x - 1/2J - 180^\circ_y - 1/2J - \quad \theta_y - 1/2J - \text{Decouple} \\
I \quad &\qquad\qquad 90^\circ_x - 1/2J - 180^\circ_x - 1/2J - \text{Acquire}
\end{aligned}
$$

However, DEPT has two major differences from INEPT. First, the initial set of pulses (up to 2τ) does not result in negative peaks (hence "distortionless"), but rather 1/2/1 triplets and 1/3/3/1 quartets without decoupling. Second, the last pulse length θ is variable but independent of J. Modulation of this pulse length ($\theta = \pi/4, \pi/2, 3\pi/4$) results in generation of the various edited spectra as seen in Figure 5-18. Thus the DEPT sequence provides signal enhancement as well as a means of determining proton count on carbons, with a less stringent dependence on the exact value of J.

5-7 CARBON CONNECTIVITY

The one bond ^{13}C—^{13}C coupling potentially contains a wealth of structural information, as it indicates carbon–carbon linkage. Unfortunately, only one in about ten thousand pairs of carbon atoms contains two ^{13}C atoms and hence displays a ^{13}C—^{13}C coupling in the ^{13}C spectrum. These resonances can be detected as very low intensity satellites on either side of the centerband that is derived from molecules containing only one ^{13}C. For bonded pairs of ^{13}C, ^1J is about 30–50 Hz and the satellites are separated from the centerband by half this amount. Coupling also may be present over two or three bonds (^2J, ^3J) in the range of about 0–15 Hz. Not only are these satellites low in intensity and possibly obscured by the centerband, but spinning sidebands, impurities, and other resonances also may get in the way.

The pulse sequence INADEQUATE (**I**ncredible **N**atural **A**bundance **D**oubl**E** **QUA**ntum **T**ransfer **E**xperiment) was developed by Freeman to suppress the usual (single quantum) resonances and exhibit only the satellite (double quantum) resonances. The pulse sequence is $90_x^\circ - \tau - 180_y^\circ - \tau - 90_x^\circ - \Delta - 90_\phi^\circ$. The 180° pulse refocuses field inhomogeneities but allows the vectors from different $^{13}C-^{13}C$ coupling arrangements to continue to diverge. If the carrier frequency coincides with the centerband of a carbon resonance, the centerband spins remain on the y axis after the first 90° pulse. The delay time τ is set to $1/4J$, so that the vectors for the two satellites from the coupled $^{13}C-^{13}C$ system diverge by 180° after 2τ $[\phi = 2\pi(\Delta\nu)t = 2\pi J(2/4J) = \pi]$ and lie on the $+x$ and $-x$ axes. The second 90_x° pulse then rotates the centerband spins to the $-z$ axis but leaves the satellite spins aligned along the x axis. Thus the centerband is not available for detection in the xy plane but the satellites are. The phase of the final 90° pulse (90_ϕ°) is cycled through a series of directions represented by ϕ $(+x, +y, -x, -y)$.

This pulse sequence is an example of a class of experiments detecting *multiple quantum coherence*. The vector diagrams used throughout this book illustrate only the coherence of the spins of a single nucleus. Spins in the xy plane are said to be coherent when they have an ordered relationship between their phases, so that they all precess around one axis and can be depicted by a vector along that direction. Spins rotating with random phase in the xy plane are said to be incoherent. Simultaneous coherence of two spins, as created in the INADEQUATE experiment, is not well depicted by the vector diagrams, so that the reason for the final 90° pulse is not well represented. The special phase properties of double quantum coherence are employed in the ϕ pulse to suppress the centerband (single quantum coherence).

Figure 5-22 contains the INADEQUATE spectrum for piperidine. The double quantum (satellite) peaks are antiphase, so each $^{13}C-^{13}C$ coupling constant is represented by a pair of peaks: one up, one down $(+1/-1)$. The spectrum for C-4 of piperidine thus contains two such doublets, a large one for $^1J_{34}$ and a small one for $^2J_{24}$. For C-3, there are two large doublets because the one bond couplings $^1J_{23}$ and $^1J_{34}$ to the adjacent carbons are slightly different. There also is a small $^3J_{23'}$ between C-2 and the nonadjacent C-3. The spectrum for C-2 shows $^1J_{23}$, $^2J_{24}$, and $^3J_{23'}$.

Although more distant couplings are observable, the most important are the one bond couplings, which vary slightly for every carbon–carbon bond. Thus a match of $^1J(^{13}C-^{13}C)$ for any two carbons strongly suggests that they are bonded to each other.

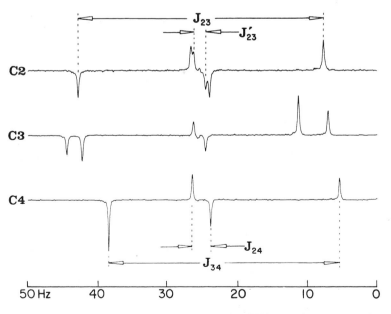

Figure 5-22 The one-dimensional INADEQUATE spectrum for the carbons of piperidine. (Reproduced with permission form A. Bax, R. Freeman, and S.P. Kempsell, *J. Am. Chem. Soc.*, **102**, 4849 (1980). Copyright 1980 American Chemical Society.)

Even in complex molecules, there is sufficient variability of couplings that INADE-QUATE can be used to map the complete connectivity of the carbon framework, provided that it is not broken by a heteroatom. The major drawback to the INADEQUATE experiment is its extremely low sensitivity, as it uses only 0.01% of the carbons in the molecule.

5-8 PHASE CYCLING AND COMPOSITE PULSES

We have used 90° and 180° pulses extensively to carry out a variety of experiments. In each case, it is important that the length of the pulse provide the desired angle of rotation with accuracy. Various artifacts can arise because of imperfections in the pulses. Figure 5-23 illustrates the effect on the inversion recovery experiment ($180_x^\circ - \tau - 90_x^\circ$, Figure 5-1) used to determine T_1, but with the initial inverting pulse not quite 180°. The magnetization after the pulse is slightly off the z axis (b), so that there is a small amount of transverse (xy) magnetization present at the start of the τ period (the y component is shown in Figure 5-23b). After the τ period the z magnetization has decreased, but the component of magnetization in the xy plane caused by the pulse imperfection persists (c). Following the final 90° pulse, the z magnetization is moved into the xy plane for detection (d). The pulse imperfection in the drawing causes a reduction in intensity, but the spectral phase also can be altered. Almost certainly, there would be errors in the 90° pulse as well, but these are not under consideration here.

Such errors may be largely eliminated by alternating the relative phase of the 180° pulse. If the second 180° rotation is carried out counterclockwise instead of clockwise about the x axis ($-x$ or $-180°$), the result of inversion is illustrated in (f). The unwanted transverse magnetization now appears along the $-y$ axis. After time τ (g) and the final 90° pulse (h), the imperfection is still present but now has the opposite effect on the z magnetization from that in (d). When the two results are added, as in (i), the effect of the imperfection cancels out. The pulse therefore is alternated between x and $-x$. Such a procedure is called *phase cycling,* a technique that permeates modern NMR spectroscopy.

Phase cycling has improved procedures for broadband heteronuclear decoupling. As described in Section 5-3, modern methods use repeated 180° pulses rather than continuous irradiation. Imperfections in the 180° pulse, however, would accumulate and

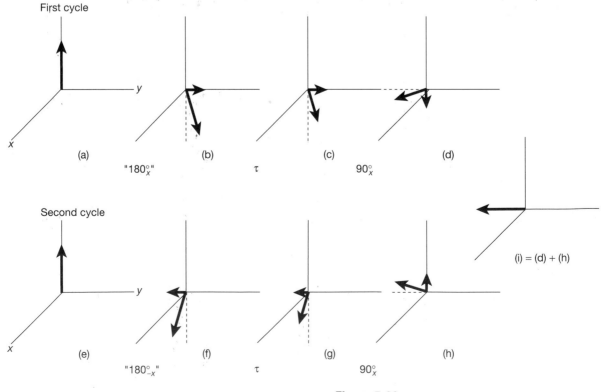

Figure 5-23

render the method unworkable. Consequently, phase cycling procedures have been developed to cancel out the imperfections. The most successful to date is the WALTZ method of Freeman, which uses the sequence $90^\circ_x - 180^\circ_{-x} - 270^\circ_x$ in place of the 180° pulse $(90 - 180 + 270 = 180)$, with significant cancelation of imperfections. The expanded WALTZ-16 sequence cycles through various orders of the simple pulses and achieves a very effective decoupling result.

A third example of phase cycling is used to place the reference frequency in the middle of the spectrum instead of off to one side. The NMR experiment as described heretofore is sensitive only to the difference $\Delta\omega$ between a signal and the reference. This situation necessitates placing the reference frequency to one side of all the resonances, so that there is no confusion of two signals that are respectively to higher and lower frequency than the reference by the exact same amount ($+\Delta\omega$ and $-\Delta\omega$). Such sideband detection, however, always contains signals from noise on the signal-free side of the reference. Placement of the reference in the middle of the spectrum avoids this unnecessary noise but requires a method to distinguish between signals with $+\Delta\omega$ and $-\Delta\omega$. *Quadrature detection* accomplishes this task by splitting the signal in two and detecting it twice, using reference signals with the same frequency but 90° out of phase. Signals with the same absolute value of $\Delta\omega$ but opposite signs are distinguished in this experiment (as obtaining θ by knowing both $\sin\theta$ and $\cos\theta$). In this case, systematic errors can arise if the two reference frequencies are not exactly 90° out of phase. The resulting signal artifacts, called *quad images,* can appear as low intensity peaks. The CYCLOPS cycle involves four steps that move the 90° pulse and the axis of detection from $+x$ to $+y$ to $-x$ to $-y$, and changes the way the two receiver channels are added, with the result that imperfections in the phase difference cancel out.

Phase cycling not only can remove artifacts from pulse or phase imperfections but also can assist in selection of coherence pathways. The inversion recovery experiment can be described with a slightly different vocabulary to illustrate this process. When spins are aligned entirely along the z axis, the order of coherence is said to be zero (phases around the xy plane are random). An exact 90° pulse creates maximum single quantum coherence by lining the spins up along, for example, the y direction. Phase cycling in the inversion recovery experiment (Figure 5-23) removes undesired single quantum coherence (transverse or xy magnetization) and leaves coherence of order zero until the end of the π period, at which time the final 90° pulse creates single quantum coherence. In this way phase cycling selects the desired degree of coherence. Double quantum coherence, involving the relationship between two spins, is not well illustrated in these vector diagrams. The INADEQUATE experiment involves selection of double quantum over single quantum coherence (elimination of the centerband and retention of the satellites) in part through phase cycling in the final 90° pulses, whose subscript ϕ refers to a sequence of pulses with different phases.

Imperfections in pulses also may be corrected by using *composite pulses* instead of single pulses. The 180° pulse that inverts longitudinal magnetization for measurement of T_1 or other purposes may be replaced by the series $90^\circ_x - 180^\circ_y - 90^\circ_x$, which results in the same net 180° pulse angle but reduces the error from as much as 20% to as little as 1%. As Figure 5-24 shows, the 180° pulse precisely compensates for what-

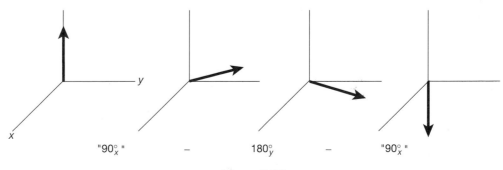

Figure 5-24

ever imperfection existed in the 90° pulse (180° normally is taken as double the optimized 90° pulse, so errors in one are present in the other). The three components of the WALTZ-16 method described above ($90^\circ_x - 180^\circ_{-x} - 270^\circ_x$) comprise a composite pulse for 180°_x.

PROBLEMS

5-1 Give the spectral notation (AB, ABX, etc.) for the following substituted ethanes, first at slow C—C rotation, then at fast rotation. Draw all stable conformations. The spectral notation for each frozen form gives the slow rotation answer. Then imagine free rotation about the C—C bond. The identity of certain protons may average for the fast rotation answer.
   (a) CH_3CCl_3 **(b)** CH_3CHCl_2 **(c)** CH_3CH_2Cl **(d)** $CHCl_2CH_2Cl$

5-2 Ring reversal in 7-methoxy-7,12-dihydropleiadene can be frozen out at −20°C. Two conformations are observed in the ratio 2/1. When the low field part of the 12-CH_2 AB quartet in the minor isomer is doubly irradiated, the intensity of the 7-methine proton is enhanced by 27%. Double irradiation of the same proton in the major isomer has no effect on the spectrum. What are the two conformational isomers and which is more abundant?

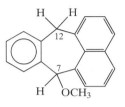

5-3 Permethyltitanocene reacts with an excess of nitrogen below −10° C to form a 1/1 complex.
$$[C_5(CH_3)_5]_2Ti + N_2 \rightleftharpoons [C_5(CH_3)_5]_2TiN_2$$
The methyl resonance of the complex is a sharp singlet above −50° C. Below −72° C the resonance splits reversibly into two peaks of not quite the same intensity. If the nitrogen is doubly labeled with ^{15}N, the ^1H-decoupled ^{15}N spectrum contains a singlet and an AX quartet ($J(^{15}N-^{15}N) = 7$ Hz), of not quite the same overall intensity at low temperatures. Explain these observations in terms of structures.

5-4 **(a)** The resonance of the methylene protons of $C_6H_5CH_2SCHClC_6H_5$ in $CDCl_3$ is an AB quartet at room temperature. Why?

   (b) The AB spectrum coalesces at high temperatures to an A_2 singlet with a $\Delta G^\ddagger$ of 15.5 kcal mol^{-1}. The rate is independent of concentration in the range 0.0190–0.267 M. Explain in terms of a mechanism.

5-5 No coupling is observed between CH_3 and ^{14}N in acetonitrile ($CH_3-C\equiv N$), but there is a coupling in the corresponding isonitrile ($CH_3-N\equiv C$). Explain. This is not a distance effect. The phenomenon is general for nitriles and isonitriles.

5-6 Comment on the following ^{14}N linewidths.

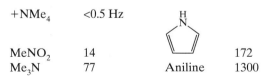

$+NMe_4$	<0.5 Hz		
$MeNO_2$	14		172
Me_3N	77	Aniline	1300

5-7   The inversion-recovery ($180° - \tau - 90°$) spectral stack* for the aromatic carbons of *m*-xylene is given below. Assign the resonances and explain the order of T_1 (look at the nulls).

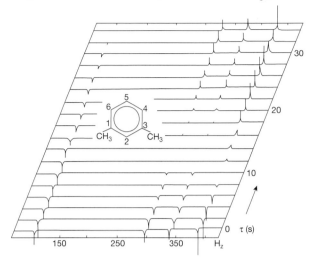

5-8   1-Decanol has the following carbon T_1 values (in s). Explain the order.

$$CH_3\!-\!CH_2\!-\!CH_2\!-\!CH_2\!-\!(CH_2)_5\!-\!CH_2\!-\!OH$$

3.1	2.2	1.6	1.1	0.8-0.83	0.65

5-9   In ribo-*C*-nucleosides the base is attached to C1' by carbon. The α and β forms (C1' epimers) may be distinguished by T_1 studies.

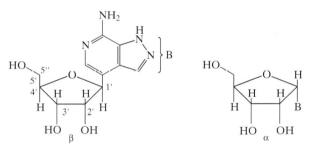

(a)   Consider the following proton T_1(s) data.

	H1'	H3'	H5'	H5''
Isomer 1	1.60	1.31	0.45	0.45
Isomer 2	3.33	1.37	0.40	0.40

Which isomer (1 or 2) is α and which is β? Why are the H1' T_1 values different for isomers 1 and 2 but the H3', H5', and H5'' values are about the same? Why is T_1 for H5' and H5'' smaller than the other values? Use the equation for dipolar relaxation in your reasoning.

(b)   Suggest another (not T_1) NMR method for distinguishing these α and β forms.

BIBLIOGRAPHY

Relaxation Phenomena

5.1.   General: D.A. Wright, D.E. Axelson, and G.C. Levy, *Magn. Reson. Rev.*, **3,** 103 (1979); G.H. Weiss and J.A. Ferretti, *Progr. NMR Spectrosc.*, **20,** 317 (1988); D.J. Wink, *J. Chem. Educ.*, **66,** 810 (1989).

5.2.   Carbon-13 relaxation: J.R. Lyerla, Jr., and G.C. Levy, *Top. Carbon-13 Spectrosc.*, **1,** 79 (1974); F.W. Wehrli, *Top. Carbon-13 Spectrosc.*, **2,** 343 (1976); D.J. Craik and G.C. Levy, *Top. Carbon-13 Spectrosc.*, **4,** 241 (1983).

*Reprinted with permission of R. Freeman from W. Bremser, H.P.W. Hill, and R. Freeman, *Messtechnick,* **79,** 14 (1971).

5.3. Nuclear Overhauser effect: J.H. Noggle and R.E. Schirmer, *The Nuclear Overhauser Effect,* Academic Press, New York, 1971; R.A. Bell and J.K. Saunders, *Top. Stereochem.,* **7,** 1 (1973); J.K. Saunders and J.W. Easton, *Determ. Org. Struct. Phys. Meth.,* **6,** 271 (1976); K.E. Kövér and G. Batta, *Progr. NMR Spectrosc.,* **19,** 223 (1987); D. Neuhaus and M. Williamson, *The NOE in Stereochemical and Conformational Analysis,* VCH, New York, 1989.

Reactions on the NMR Time Scale

5.4. General: G. Binsch, *Top. Stereochem.,* **3,** 97 (1968); *Dynamic Nuclear Magnetic Resonance Spectroscopy,* L.M. Jackman and F.A. Cotton, eds., Academic Press, New York, 1975; A. Steigel, *NMR Basic Princ. Progr.,* **15,** 1 (1978); J.I. Kaplan and G. Fraenkel, *NMR of Chemically Exchanging Systems,* Academic Press, New York, 1980; G. Binsch and H. Kessler, *Angew. Chem., Int. Ed. Engl.,* **19,** 411 (1980); J. Sändstrom, *Dynamic NMR Spectroscopy,* Academic Press, London, 1982; *Applications of Dynamic NMR Spectroscopy to Organic Chemistry,* M. Ōki, ed., VCH, Deerfield Beach, FL, 1985.

5.5. Carbon-13 applications: *Progr. NMR Spectrosc.,* **11,** 95 (1977).

5.6. Hindered rotation: H. Kessler, *Angew. Chem., Int. Ed. Engl.,* **9,** 219 (1970); W.E. Stewart and T.H. Siddall, *Chem. Rev.,* **70,** 517 (1970); M. Ōki, *Top. Stereochem.,* **14,** 1 (1983); M.L. Martin, X.Y. Sun, and G.J. Martin, *Ann. Rep. NMR Spectrosc.,* **16,** 187 (1985); C.H. Bushweller, in *Acyclic Organonitrogen Stereodynamics,* J.B. Lambert and Y. Takeuchi, eds., VCH, New York, 1992, pp. 1-55; M. Raban and D. Kost, in *Acyclic Organonitrogen Stereodynamics,* J.B. Lambert and Y. Takeuchi, eds., VCH, New York, 1992, pp. 57-88; S.F. Nelsen, in *Acyclic Organonitrogen Stereodynamics,* J.B. Lambert and Y. Takeuchi, eds., VCH, New York, 1992, pp. 89-121; B.M. Pinto, in *Acyclic Organonitrogen Stereodynamics,* J.B. Lambert and Y. Takeuchi, eds., VCH, New York, 1992, pp. 149-175.

5.7. Ring reversal and cyclic systems: H. Booth, *Progr. NMR Spectrosc.,* **5,** 149 (1969); J.B. Lambert and S.I. Featherman, *Chem. Rev.,* **75,** 611 (1975); H. Günther and G. Jikeli, *Angew. Chem., Int. Ed. Engl.,* **16,** 599 (1977); E.L. Eliel and K.M. Pietrusiewicz, *Top. Carbon-13 Spectrosc.,* **3,** 171 (1979); F.G. Riddell, *The Conformational Analysis of Heterocyclic Compounds,* Academic Press, London, 1980; A.P. Marchand, *Stereochemical Applications of NMR Studies in Rigid Bicyclic Systems,* VCH, Deerfield Beach, FL, 1982.

5.8. Atomic inversion: J.B. Lambert, *Top. Stereochem.,* **6,** 19 (1971); A. Rauk, L.C. Allen, and K. Mislow, *Angew. Chem., Int. Ed. Engl.,* **9,** 400 (1970). W.B. Jennings and D.R. Boyd, in *Cyclic Organonitrogen Stereodynamics,* J.B. Lambert and Y. Takeuchi, eds., VCH, New York, 1992, pp. 105-158; J.J. Delpuech, in *Cyclic Organonitrogen Stereodynamics,* J.B. Lambert and Y. Takeuchi, eds., VCH, New York, 1992, pp. 169-252.

5.9. Organometallics: K. Vrieze and P.W.N.M. Vanleeuwen, *Progr. Inorg. Chem.,* **14,** 1 (1971); B.E. Mann, *Ann. Rep. NMR Spectrosc.,* **12,** 263 (1982); K.G. Orrell and V. Šik, *Ann. Rep. NMR Spectrosc.,* **19,** 79 (1987).

5.10. Rates from relaxation times: J.B. Lambert, R.J. Nienhuis, and J.W. Keepers, *Angew. Chem., Int. Ed. Engl.,* **20,** 487 (1981).

Multiple Irradiation and One-Dimensional Multipulse Methods

5.11. General multiple resonance: R.A. Hoffman and S. Forsén, *Progr. NMR Spectrosc.,* **1,** 15 (1966); V.J. Kowalewski, *Progr. NMR Spectrosc.,* **5,** 1 (1969); W. McFarlane, *Determ. Org. Struct. Phys. Meth.,* **4,** 150 (1971); W. von Philipsborn, *Angew. Chem., Int. Ed. Engl.,* **10,** 472 (1971); W. MacFarlane, *Ann. Rep. NMR Spectrosc.,* **5A,** 353 (1972); R.L. Micher, *Magn. Reson. Rev.,* **1,** 225 (1972); L.R. Dalton, *Magn. Reson. Rev.,* **1,** 301 (1972); W. McFarlane and D.S Rycroft, *Ann. Rep. NMR Spectrosc.,* **9,** 320 (1979); **16,** 293 (1985).

5.12. Difference spectroscopy: J.K.M. Sanders and J.D. Merck, *Progr. NMR Spectrosc.,* **15,** 353 (1982).

5.13. Broadband decoupling: M.H. Levitt, R. Freeman, and T. Frenkiel, *Advan. Magn. Reson.,* **11,** 47 (1983); A.J. Shaka and J. Keeler, *Progr. NMR Spectrosc.,* **19,** 47 (1987).

5.14. General multipulse methods: R. Benn and H. Günther, *Angew. Chem., Int. Ed. Engl.,* **22,** 350 (1983); C.J. Turner, *Progr. NMR Spectrosc.,* **16,** 311 (1984); G.A. Morris, *Magn. Reson. Chem.,* **24,** 371 (1986); D.L. Turner, *Ann. Rep. NMR Spectrosc.,* **21,** 161 (1989); K. Nakanishi, *One-Dimensional and Two-Dimensional NMR Spectra by Modern Pulse Techniques,* University Science Books, Mill Valley, CA, 1990; R.R. Ernst and G. Bodenhausen, *Principles of Nuclear Magnetic Resonance in One and Two Dimensions,* Oxford, UK, 1990.

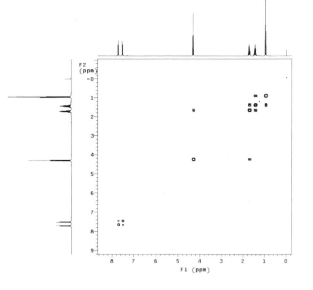

chapter 6

Two-Dimensional NMR

NMR always has been multidimensional. In addition to frequency and intensity, which serve as the axes of the standard one-dimensional spectrum, reaction rates and relaxation times have provided additional dimensions, often presented as stacked plots (see Figures 2-33 and 5-2). The second dimension of modern NMR, however, refers to an additional frequency axis. This concept was first suggested in a lecture by Jeener in 1971 and reached wide application in the 1980s when instrumentation caught up with theory. In terms of structure, we can think of the first frequency dimension as the traditional characterization of nuclei in terms of chemical shifts and couplings. In the second frequency dimension, we further consider magnetic interactions between nuclei through structural connectivity, spatial proximity, or kinetic interchange.

6-1 PROTON–PROTON CORRELATION THROUGH COUPLING

In the single-pulse experiments described up to this point, a 90° pulse is followed by a period of time during which the free induction decay is acquired (Figure 6-1a). Fourier transformation of the time-dependent magnetic information into a frequency dimension provides the familiar spectrum of δ values, now to be called a one-dimensional (1D) spectrum. If the 90° pulse is preceded by another 90° pulse (Figure 6-1b), useful relationships between spins can evolve prior to acquisition. Figure 6-2 illustrates what happens in terms of magnetization vectors.

Consider a sample that contains only one type of nucleus without any coupling partners, for example the ^{1}H spectrum of tetramethylsilane. Although the final result of the particular experiment we are about to describe may seem trivial or even pointless at first, it will take on fuller meaning when we introduce relationships with other nuclei. The isolated nucleus of Figure 6-2 begins with magnetization aligned along the z axis (a), and then along the y axis after application of the 90° pulse (b). If the coordinate system rotates at the carrier frequency and the nucleus resonates at a slightly higher frequency, the spin vector picture begins to evolve. After a short amount of time, the vector M moves to a new position in the xy plane, for example in (c). We ignore longitudinal

117

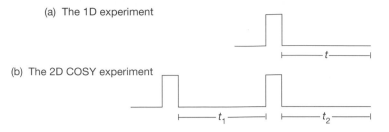

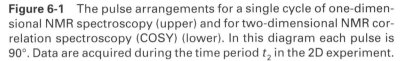

Figure 6-1 The pulse arrangements for a single cycle of one-dimensional NMR spectroscopy (upper) and for two-dimensional NMR correlation spectroscopy (COSY) (lower). In this diagram each pulse is 90°. Data are acquired during the time period t_2 in the 2D experiment.

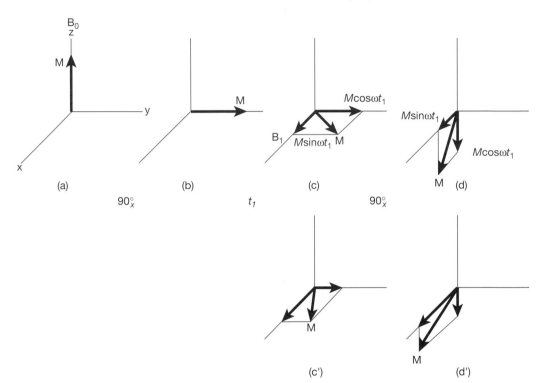

Figure 6-2 The pulse sequence for the COSY experiment. For (c') the magnetization M is allowed to evolve a longer time from (b) than for (c) before the final $90°_x$ pulse is applied to give (d').

relaxation (T_1) to simplify the drawings. The evolving magnetization vector may be decomposed into a y component ($M_y = M\cos\omega t_1$) and an x component ($M_x = M\sin\omega t_1$), in which ω is the difference between the frequency of the carrier and of the resonating nucleus and t_1 is the time elapsed since the 90° pulse.

If at this point the second 90° pulse of Figure 6-1b is applied, again along the x axis, the result is different for the two illustrated magnetization components (Figure 6-2d). The x component is unaffected, but the y component is transferred to the negative z axis. If observation is in the xy plane, the only detectable part of the magnetization is M_x. This quantity appears as a free induction decay during the time period t_2 after the second pulse. Fourier transformation of the FID as a function of t_2 yields a signal at the resonance frequency (ν_A). The intensity of this signal is determined by the quantity $M\sin\omega t_1$. The use of the subscripts is necessary to distinguish the evolution period t_1 from the acquisition period t_2 (Figure 6-1b). If t_1 is relatively short, M_x ($= M\sin\omega t_1$) is small, little x magnetization has developed (d), and the peak is small. A slightly longer value of t_1 yields a larger x component (c' and d') as $M\sin\omega t_1$ grows.

Figure 6-3 shows the result of a whole series of such experiments, with buildup of M_x ($= M\sin\omega t_1$) as t_1 increases, reaching a maximum when the spin vector M is lined up

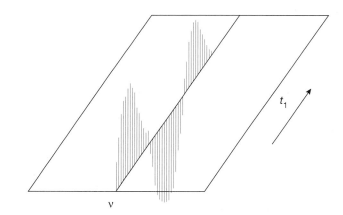

Figure 6-3 The middle diagonal serves as the baseline for a series of ^{1}H spectra of chloroform according to the COSY pulse sequence for a series of values of t_1. Each peak is one period of $90°-t_1-90°$ followed by Fourier transformation during t_2 of Figure 6-1. Fourier transformation in the t_1 dimension has not been carried out.

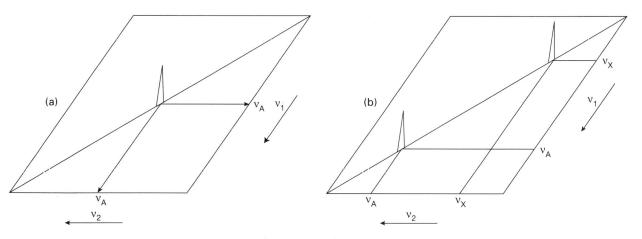

Figure 6-4 (a) The result of the COSY experiment after double Fourier transformation for a single, isolated nucleus such as that in Figure 6-3. (b) The result of the COSY experiment for two uncoupled nuclei.

along the x axis. The peak height then decreases as the vector moves to the left of the x axis, reaching zero intensity when it is lined up along the negative y axis. As it passes behind the y axis, the intensity becomes negative, reaching a negative maximum when the vector is aligned along the $-x$ axis. This negative maximum would be slightly smaller than the initial positive maximum because of T_2 relaxation. It is clear from Figure 6-3 that this family of experiments generates a sine curve when M_x is plotted as a function of t_1. Frequency (from Fourier transformation of t_2 to give v_2) is along the horizontal dimension and the time t_1 is along the vertical dimension. The type of data generated from stepping t_1 in this fashion in fact comprises a free induction decay that also may be Fourier transformed. Because the frequency ω represented by the sine curve in t_1 is the same as the frequency from the initial Fourier transformation in t_2, the result of the second Fourier transformation is a single peak at the coordinates (v_A,v_A) when plotted in two frequency dimensions (Figure 6-4a). This is the trivial result previously alluded to.

The utility of this experiment becomes evident when two coupled nuclei are treated in this fashion. Two uncoupled nuclei yield the trivial result of two peaks respectively at (v_A,v_A) and v_X,v_X), as in Figure 6-4b. These peaks are on the diagonal of the two-dimensional representation. Satisfying complications arise when the two nuclei are coupled. Figure 6-5 illustrates the possible spin states for nuclei A and X, as for example the

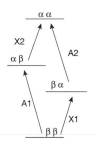

Figure 6-5

alkenic protons in β-chloroacrylic acid, ClCH=CHCO$_2$H. The AX spectrum contains four peaks due to scalar *(J)* coupling. The diagram is intended to indicate the four different frequencies, from the highest (A1) to the lowest (X2). It is useful first to consider population perturbations during an old-style, one-dimensional selective decoupling CW experiment. Irradiation, for example, of only transition A1 tends to bring the αβ and ββ states closer together in population. Consequently, there is a direct effect on the intensities of the connected transitions X1 and X2, which propagates as a secondary effect on the intensity of the A2 transition. With respect to A1, X2 is called a *progressive transition* (going on to a higher spin state), X1 is called a *regressive transition,* and A2 is called a *parallel transition.*

In the pulse experiment, energy absorption at frequency A1 has similar effects, which bring about magnetization or population transfer. The first 90° pulse serves to label all the magnetization with the 1D frequencies during period t_1: A1, A2, X1, X2. The second 90° pulse results in a population inversion for any given transition (Figure 6-2d). This perturbation causes population changes in all connected transitions of the resonances to which the nucleus is coupled, and secondarily to parallel transitions. Part of the magnetization at frequency A1, for example, is transferred by the second pulse during period t_2 to each of the other three transitions. This transferred magnetization has a new frequency. Such transfers occur from all four of the transitions to each of the other three. Thus the magnetization observed at frequency A1 during t_2 contains components modulated at frequencies A2, X1, and X2. That portion of the magnetization, for example, that has frequency A1 during t_1, but frequency X2 during t_2, is observed as a peak off of the diagonal, at frequency (v_{A1}, v_{X2}). In the absence of *J* couplings, the A and X transitions of Figure 6-5 are not connected and magnetization is not transferred.

Figures 6-6 and 6-7 are two representations of the two-dimensional experiment for β-chloroacrylic acid. In Figure 6-6 the stack representation contains several hundred complete 1D experiments, whose closely packed horizontal lines are barely distinguishable. On the diagonal from the lower left to the upper right (as drawn into the plots in Figure 6-4) are the four peaks that arise directly from resonance without magnetization transfer, that is, the components of magnetization that possess the same frequency in t_1 and t_2. These four peaks comprise the normal, four-peak 1D spectrum. All the peaks off of the diagonal represent magnetization transfer by scalar *(J)* coupling.

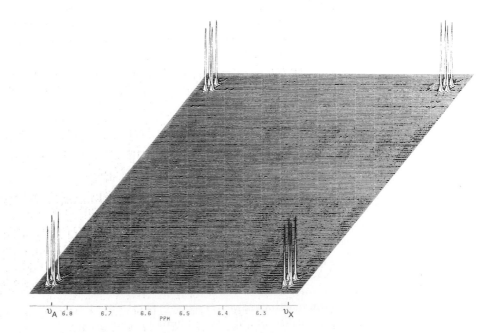

Figure 6-6   The stacked representation of the COSY experiment for two coupled nuclei (β-chloroacetic acid). (Reproduced from A.E. Derome, *Modern NMR Techniques for Chemistry Research,* Pergamon Press, Oxford, UK, 1987, p. 189.)

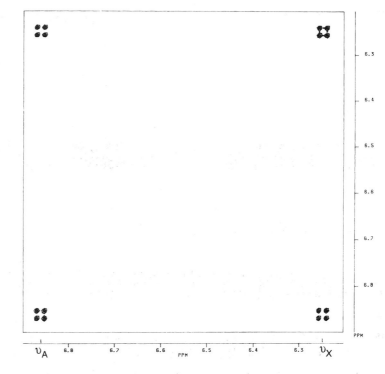

Figure 6-7 The contour representation of the COSY experiment for two coupled nuclei (β-chloroacetic acid). (Reproduced from A.E. Derome, *Modern NMR Techniques for Chemistry Research,* Pergamon Press, Oxford, UK, 1987, p. 191.)

For example, transfer between the parallel transitions A1 and A2 is found as symmetrical peaks just above and below the diagonal at the lower left. One peak represents transfer from A1 to A2 and the other from A2 to A1. Because of this reciprocal relationship, all off-diagonal peaks appear in pairs reflected across the diagonal. Normally, the off-diagonal peaks between parallel transitions are more nuisance than use and can be reduced or deleted by special techniques. The important information results from magnetization transfer between the A and the X nuclei, whose peaks are the clusters in the upper left and lower right of Figure 6-6, representing all the possible transfers between the A and X transitions: A1 to X1, X2 to A1, and so on, eight in total, including the mirror image pairs (A to X and X to A) on either side of the diagonal.

Figure 6-7 is the alternative *contour representation* of the same data, in which the distracting baselines are removed and only the peak bases remain, as if the spectator is viewing the spectrum from directly above it. By convention, the original diagonal usually is from lower left to upper right. The Jeener experiment commonly is given the semiacronym COSY, for **CO**rrelation **S**pectroscop**Y**. Since most 2D experiments involve spectral correlations, the name is not apt. Alternative terms such as 90° COSY, COSY90, H,H-COSY, or homonuclear HCOSY have been subsumed by public acceptance of the term COSY. The experiment itself has become an essential part of the analysis of complex proton spectra.

Figure 6-8 is the COSY experiment for the indicated annulene, from the work of H. Günther. The 1D spectrum is shown along both horizontal and vertical axes, and the resonances are labeled α, β, and A to F. The aromatic protons that are ortho and meta to the ring fusion provide an isolated spin system, and their coupling is represented by the off-diagonal (or cross) peak labeled α,β. The presence of a cross peak normally indicates that the protons giving the connected resonances on the diagonal are geminally or vicinally coupled. Long-range couplings do not usually provide significant cross peaks. Exceptions, however, can be expected, since long-range couplings can be large.

The COSY analysis of the remainder of the spectrum in Figure 6-8 provides the peak assignments and confirms the structure. Protons A and F are the only ones split by

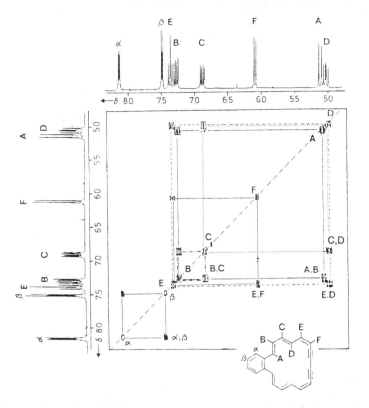

Figure 6-8 The COSY experiment for the illustrated annulene. (Reproduced from R. Benn and H. Günther, *Angew. Chem., Int. Ed. Engl.,* **22,** 350 [1983]).

a single neighbor. The coupling of A with B is trans and should be larger than the cis coupling between E and F. The two doublets then may be assigned as F (smaller coupling) at δ 6.1 and A (larger coupling) at δ 5.2. It usually is essential in a COSY analysis to be able to make an initial assignment through traditional considerations of chemical shifts and coupling constants (see Chapters 3 and 4). The COSY analysis then consists of moving from the known diagonal peak, to a cross peak, and back to the diagonal for the assignment of a new peak. Only A and F have single cross peaks (one coupling partner). All remaining resonances in the large ring have two cross peaks, which provide the means for assignment. We can start with either A or F. Dropping down from A leads to the cross peak A,B, and horizontal movement to the left leads to a diagonal peak and the assignment of proton B. The horizontal path passes through another cross peak, which must be between B and its other coupling partner, C. Moving up from B,C then leads to a diagonal peak and assignment of proton C. Horizontal movement to the right leads to the cross peak C,D, and return upward to the diagonal assigns proton D. Dropping back down from D and passing through C,D leads to the other cross peak from D, labeled E,D. Returning to the diagonal to the left assigns proton E and passes through the other cross peak from E, labeled E,F. Return upward from E,F to the diagonal completes the assignment with proton F.

A group at IBM has provided a useful example of a more complex COSY analysis with the tripeptide Pro-Leu-Gly (**6-1**). The three carbonyl groups disrupt vicinal

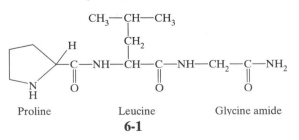

Proline   Leucine   Glycine amide

6-1

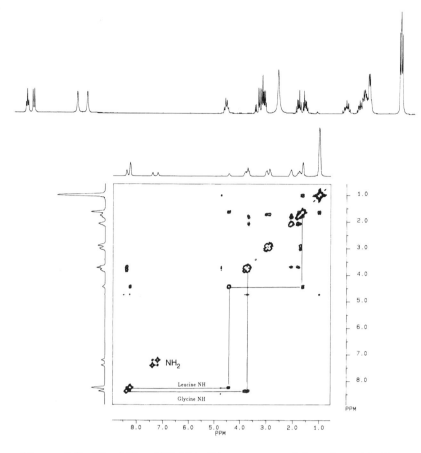

Figure 6-9 (Top) The 300 MHz ^{1}H spectrum of the tripeptide Pro-Leu-Gly in DMSO. (Bottom) COSY spectrum of Pro-Leu-Gly with connectivities of the NH protons. (Courtesy of IBM Instruments, Inc.)

connectivities, so the molecule consists of four independent spin systems: proline, leucine, glycine, and the terminal amide. The 1D ^{1}H spectrum is given at the top of Figure 6-9 without assignment. The high frequency (low field) peaks (δ 7.0–8.3) are from the protons on nitrogen, and the broad peak at δ 3.3 is from HOD. Figure 6-9 at the bottom contains the COSY spectrum with connectivities drawn in for the amide resonances. The nonequivalent terminal NH_2 resonances are immediately assigned as δ 7.0 and 7.2 because they have no external connectivities and hence no cross peaks other than between themselves. The Gly NH proton is assigned at δ 8.2 because it is a triplet (next to a CH_2) and has only the single connectivity with the CH_2 group at δ 3.6 (completing the Gly portion of the molecule). The remaining NH resonance at δ 8.1 is from Leu. It is a doublet (next to a CH) and has a cross peak with the resonance at δ 4.3, which has other connectivities. There is no third NH resonance, so the Pro NH must be quadrupolar-broadened or exchanging with HOD.

The upper portion of Figure 6-10 completes the COSY analysis of the Leu portion and confirms that the NH at δ 8.1 is part of Leu rather than Pro. The expected Leu connectivity is $NH \longrightarrow CH \longrightarrow CH_2 \longrightarrow CH \longrightarrow CH_3$. Cross peaks with the following connectivities are observed (starting with NH): δ 8.1 $\longrightarrow$ 4.3(q or dd) $\longrightarrow$ 1.5 (m) $\longrightarrow$ 0.9(dd). Apparently two of the proton resonances coincide, mostly likely those from CH_2 and the isopropyl CH. The CH_3 resonance as expected is at the lowest frequency and cannot be from any Pro group. Its higher multiplicity (dd, Figure 6-9, top) arises because the two methyl groups are diastereotopic due to the chiral center to which the butyl group is attached.

The lower portion of Figure 6-10 provides the Pro connectivity. The highest frequency resonance (δ 3.7) should be from the CH group adjacent (α) to the carbonyl group. The resonance at δ 3.7 actually is an overlap of this Pro CH (higher frequency portion) and the Gly CH$_2$ (lower frequency portion). The Pro CH has two cross peaks with the diastereotopic β protons at δ 1.7 and 1.9, which are mutually coupled and have their own cross peak. Unfortunately, the γ protons are nearby (δ 1.6), but their cross peak with the δ protons at δ 2.8 completes the assignment of the spectrum. The fully assigned 1D spectrum and structure are given in Figure 6-11.

False peaks and lack of symmetry around the diagonal are common in the COSY experiment and can arise for several reasons. (1) Differences in digital resolution in the two time periods, t_1 and t_2, prevent perfect symmetry. (2) Incorrect pulse lengths or (3) incomplete transverse relaxation during the delay time can create false cross peaks. (4) We ignored the possibility of longitudinal relaxation. Magnetization in the z direction does not precess and therefore is rotated by the second 90° pulse into the position rec-

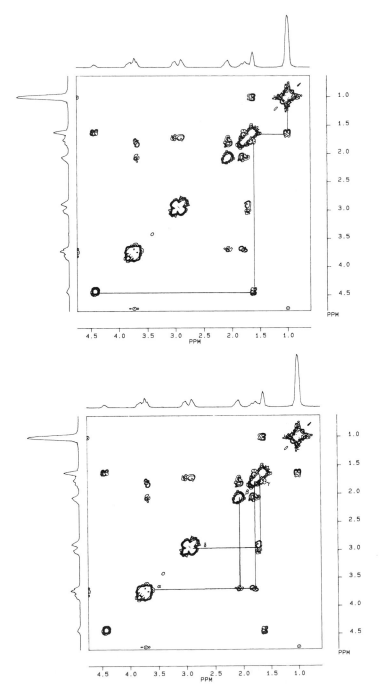

Figure 6-10 (Top) Connectivity within the leucine portion of Pro-Leu-Gly by COSY. (Bottom) Connectivity within the proline portion of Pro-Leu-Gly by COSY. (Courtesy of IBM Instruments, Inc.)

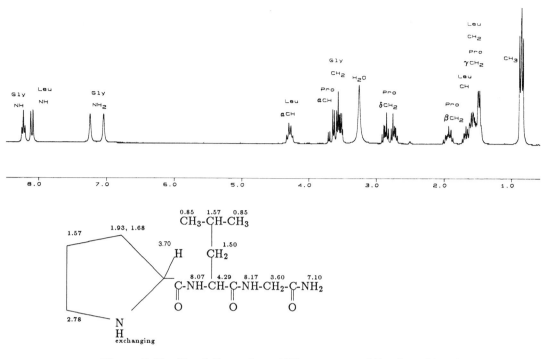

Figure 6-11 The fully assigned ^{1}H spectrum of Pro-Leu-Gly. (Courtesy of IBM Instruments, Inc.)

ognized as $\nu = 0$ (the position of the reference frequency). A stream of lines, called axial peaks, can occur at this frequency in the 2D plot. (5) Folding can occur in two dimensions and can give rise to off-position diagonal peaks and even cross peaks. All these artifacts may be minimized by optimizing pulse lengths, allowing sufficient time for transverse relaxation, use of phase cycling, and symmetrization. Axial peaks may be suppressed largely by alternating $+90°$ and $-90°$ for the second pulse, thus canceling z magnetization. The more complex CYCLOPS procedure provides suppression of axial peaks as well as elimination of other artifacts such as quad images (see Section 5-8). *Symmetrization* is a procedure for imposing bilateral symmetry around the diagonal. Most artifacts are conveniently eliminated by this procedure, but not all. For example, if two resonances have streams of axial lines, a point on one stream can occur at the precise mirror position (with respect to the diagonal) of a point on the other stream. Although the streams are largely eliminated, the two peaks at the symmetrical positions are retained and appear as handsome cross peaks. Usually common sense can reject them.

There are many variants of the standard COSY experiment that either improve on its basic aims or provide new information. We shall consider several of them without appreciable attention to the details of the pulse sequences.

COSY45. The large size of diagonal peaks sometimes can be a deterrent to understanding the significance of nearby cross peaks. The problem is aggravated by the presence of cross peaks from parallel transitions (see above). The COSY45 experiment reduces the intensities of the diagonal peaks and the cross peaks from parallel transitions. Figure 6-12 compares the COSY90 and COSY45 experiments for 2,3-dibromopropionic acid ($CH_3CHBrCHBrCO_2H$). The COSY45 experiment clarifies cluttered regions close to the diagonal but also provides information on the signs of coupling constants. The name derives from alteration of the second pulse length: $90° - t_1 - 45° - t_2$.

Long-Range COSY (LRCOSY). The normal assumption in the COSY experiment is that two- or three-bond (geminal or vicinal) couplings provide the dominant magnetization transfer to create cross peaks. Information from longer range couplings, however, also can be useful. Introducing a fixed delay Δ during the evolution and detection periods ($90° - t_1 - \Delta - 90° - \Delta - t_2$) enhances magnetization transfer from small cou-

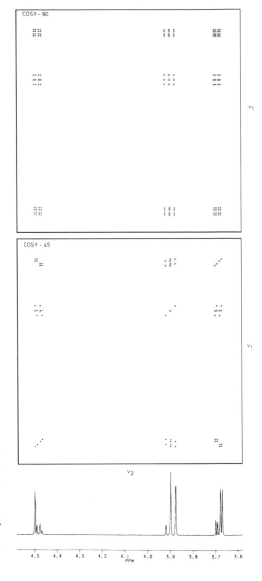

Figure 6-12 The COSY90 (top) and COSY45 (bottom) spectra of 2,3-dibromopropionic acid. The 1D spectrum at the bottom is a cross section through the COSY45 spectrum. (Reproduced from A.E. Derome, *Modern NMR Techniques for Chemistry Research,* Pergamon Press, Oxford, UK, 1987, p. 228.)

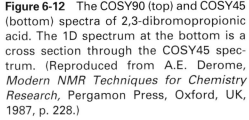

plings at the expense of large couplings. Figure 6-13 compares the COSY and LRCOSY experiments for a polynuclear aromatic compound. In the COSY spectrum, cross peaks occur only between ortho neighbors (a-b and e-f). In the LRCOSY spectrum, additional cross peaks arise between peri neighbors (c-d and d-e). Information on the connectivity between fused aromatic rings thus becomes available in the LRCOSY case.

Phase-Sensitive COSY.   Fourier transformation involves building up the signal from the sum of sine and cosine curves. Every point in the spectrum has both sine and cosine contributions, which are 90° out of phase. These contributions sometimes are called the imaginary and real terms and lead mathematically to the *dispersion mode* and *absorption mode* spectra. An in-phase or absorption signal has the familiar form of a positive peak. The dispersion signal, commonly used for electron spin resonance spectra, has a sideways S shape with a portion below and a portion above the baseline. It produces both negative and positive maxima for a given peak and a value of zero at the resonance frequency as the sign changes. NMR experiments normally are tuned to the absorption mode by the process of phasing, but the two signals also may be combined mathematically to produce what is called a *magnitude spectrum.*

Many of the COSY experiments displayed so far used magnitude representations because of phase differences between various peaks in the pure modes. Magnetization that is not transferred (and thus appears on the diagonal) and magnetization that is transferred to parallel transitions undergo no phase shift. Cross peaks, however, expe-

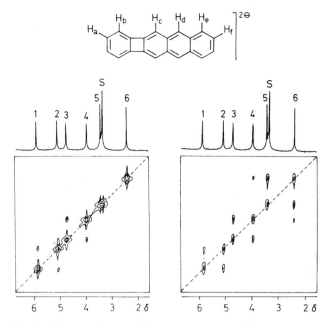

Figure 6-13   The 400 MHz COSY (left) and LRCOSY (right) spectra of naphthobiphenylene dianion (the signal S is from solvent). (Reproduced from H. Günther, *NMR Spectroscopy,* 2nd ed., John Wiley & Sons, Ltd., Chichester, UK, 1995, p. 300.)

rience phase shifts. Transfer between progressive transitions (A1 to X2 in Figure 6-5) shifts the phase −90°, and transfer between regressive transitions (A1 to X1) shifts it +90°. Because absorption and dispersion modes differ by 90°, phasing the diagonal peak to absorption results in dispersive cross peaks, or vice versa. Moreover, cross peaks from progressive and regressive transitions are always out of phase by 180° (if one is positive absorption, the other is negative absorption; or if one begins a dispersive signal negatively, the other begins positively). The magnitude or absolute value mode is used to eliminate all phase differences and produce absorption-like peaks. The resulting peaks tend to be broad and often are distorted. In small molecules with little peak overlap, there may be no problem with the use of magnitude spectra, but larger molecules such as proteins, polysaccharides, or polynucleotides may produce unacceptable overlap. Use of the phase-sensitive COSY experiment then can tune the cross peaks to a purely absorption (real) mode. This experiment not only provides enhanced resolution but also enables coupling constants to be read more easily from the cross peaks.

Because the 2D method involves two time domains, transformation in both t_1 and t_2 generates real and imaginary components. As a result, the phase-sensitive 2D signal has four modes rather than two. These phase modes or quadrants correspond to both frequency signals being real, both being imaginary, or one being real while the other is imaginary. Figure 6-14 illustrates the four modes. The real/real (RR) mode produces the familiar peak with a contour shaped like a four-pointed star at the base. Figure 6-15 illustrates what the COSY spectrum for two spins looks like when the diagonal and parallel components are tuned dispersively (both imaginary) and the progressive and regressive cross peaks are tuned absorptively (both RR but 180° out of phase). This common phase sensitive representation provides straightforward identification of the cross peaks derived from coupling and hence the determination of J.

Multiple Quantum Filtration.   The one-dimensional INADEQUATE pulse sequence suppresses the centerband singlet in order to measure ^{13}C–^{13}C couplings from the satellites (see Section 5-7). The procedure involves creating double quantum coherences. A similar procedure may be used in two dimensions to suppress singlets, which are single quantum coherences. Such singlets may arise from solvent or from uncoupled methyl resonances, both of which can constitute major impediments in locating highly

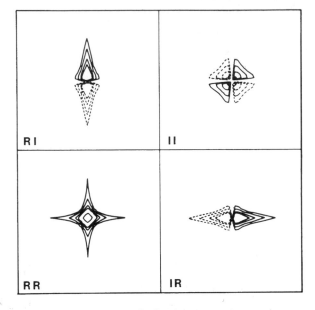

Figure 6-14 The four types of 2D phase quadrants, corresponding to frequency modes that are real/real (RR), imaginary/real (IR), real/imaginary (RI), and imaginary/imaginary (II). (Reproduced from A.E. Derome, *Modern NMR Techniques for Chemistry Research,* Pergamon Press, Oxford, UK, 1987, p. 207.)

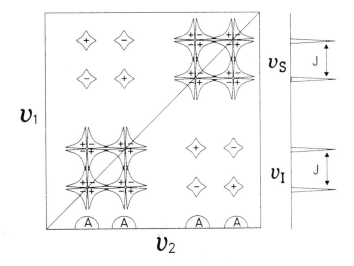

Figure 6-15 Phase-sensitive COSY diagram for two spins, with the diagonal peaks in dispersion mode and the cross peaks in antiphase absorption mode. The 1D spectrum is on the right. (Reproduced from F.J.M. van de Ven, *Multidimensional NMR in Liquids,* VCH, New York, 1995, p. 171.)

split resonances in a complex spectrum. In a double quantum filtered COSY (DQF–COSY) experiment, an extra 90° pulse is added after the second 90° COSY pulse, and phase cycling converts multiple quantum coherences into observable magnetizations. The resulting 2D spectrum lacks all singlets along the diagonal (for example, the left spectrum in Figure 6-16). An important feature of the phase-sensitive DQF–COSY experiment is that double quantum filtration allows both diagonal and cross peaks to be tuned into pure absorption at the same time. This feature reduces the size of all the diagonal signals and permits analysis of cross peaks close to the diagonal. The only disadvantage of DQF–COSY is reduction in sensitivity by a factor of two. The triple quantum filtered COSY experiment (TQF–COSY) removes both singlets and AB or AX quartets, providing greater spectral simplification. It is rarely used because of increased loss of sensitivity.

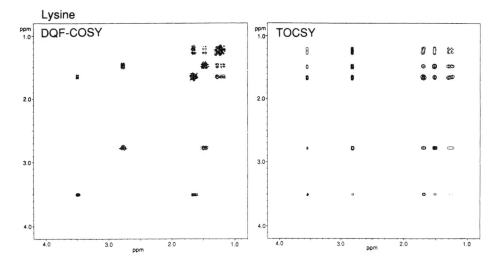

Lysine

Figure 6-16 The DQF–COSY and TOCSY spectra of lysine. (Reproduced from J.N.S. Evans, *Biomolecular NMR Spectroscopy,* Oxford University Press, Oxford, UK, 1995, p. 428.)

Total Correlation Spectroscopy (TOCSY). The connectivity within an entire spin system such as in a butyl group ($CH_3CH_2CH_2CH_2$—), must be mapped out in the standard COSY experiment from proton to proton via a series of cross peaks. By spin locking the protons during the second COSY pulse, the chemical shifts of all the protons may be brought essentially into equivalence. Recall that resonance frequencies of protons and carbons are made equal through cross polarization for solids by achieving the Hartmann-Hahn condition (see Section 2-10). In the 2D variant of this experiment, the initial 90° pulse and the t_1 period occur as usual, but the second pulse locks the magnetization along the y axis so that all protons have the spin lock frequency. All coupled spins within a spin system then become closely coupled and magnetization is transferred from one spin to all other members. Figure 6-16 on the right shows the result for lysine [$NH_2CH(CH_2 CH_2CH_2CH_2NH_2)CO_2H$]. The methylene group at the lowest frequency (upper right corner) exhibits four TOCSY cross peaks, one with each of the other three methylene protons and one with the methine proton. The TOCSY experiment, a variation of which is called the **HO**monuclear **HA**rtmann-**HA**hn or HOHAHA experiment, has particular advantages for large molecules. It has enhanced sensitivity, and both diagonal and cross peaks may be phased to the absorption mode. The process of identifying resonances within specific amino acid or nucleotide residues is considerably simplified by this procedure. Each residue can be expected to exhibit cross peaks among all its protons and none with protons of other residues.

Relayed COSY. An alternative but less general method for displaying extended levels of connectivity is provided by relayed coherence transfer (RCT). The normal COSY experiment for an AMX system with $J_{AX} = 0$ produces cross peaks between A and M and between M and X. It is not unusual for the key diagonal peak for M to be coincident with a resonance from another spin system, making it difficult to follow the connectivity path (for example, recall the Leu portion of the COSY spectrum of Pro-Leu-Gly in Figure 6-10). The RCT experiment generates a cross peak between the A/M and M/X cross peaks, eliminating the ambiguity. The RCT pulse sequence is $90°-t_1-90°-\tau-180°-\tau-90°-t_2$, in which the sequence after the second 90° pulse permits relay of coherence to the next spin. The result is shown diagrammatically in Figure 6-17 for AMX and A'M'X' systems whose M and M' resonances coincide. The COSY experiment contains the expected four cross peaks. The RCT experiment additionally contains two cross peaks connecting the cross peaks of the individual spin systems, labeled (A,M,X) and (A',M',X'). The connectivity of A → M → X and of A' → M' → X' is then rendered unambiguous.

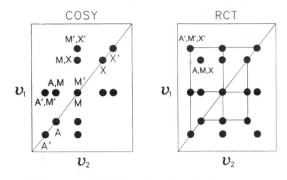

Figure 6-17 Diagram of COSY and relayed coherence transfer (RCT) experiments for two three-spin systems (AMX and A'M'X') whose M and M' portions overlap. (Adapted from F.J.M. van de Ven, *Multidimensional NMR in Liquids,* VCH, New York, 1995, p. 233.)

J-Resolved Spectroscopy. In Chapter 5 we saw how the spin echo experiment can isolate or remove characteristics of chemical shifts or coupling constants. Spin echoes may be used in two dimensions to generate one frequency dimension representing chemical shifts and another coupling constants. The sequence $90° - \frac{1}{2}t_1 - 180° - \frac{1}{2}t_1 - t_2$ for example uses the 180° pulse to refocus chemical shifts during t_1. The result for a glucose derivative is shown in Figure 6-18. The ^{1}H frequencies are found on the horizontal axis (v_2), with the normal 1D spectrum displayed at the top (a). The vertical axis ("f_1" = v_1) contains only proton–proton coupled multiplets, each centered about a zero frequency point, that is, all multiplets occur at the same chemical shift in v_1. Thus the multiplet at highest frequency (lowest field) from H-3 is a quartet with further splitting when viewed from the vertical axis. By taking a projection at an angle (45°) that causes each of the members of the individual multiplets to overlap when viewed from the horizontal axis, as at the bottom (b), a display is obtained that in essence is a proton-proton decoupled proton spectrum. Resonances are present at each frequency devoid of any couplings. This projection is a novel way to examine ^{1}H spectra, although it has not seen widespread use because it reveals no connectivities.

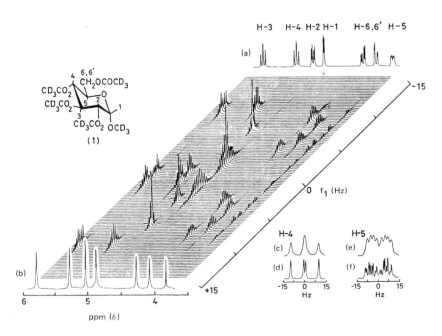

Figure 6-18 The 270 MHz 2D *J*-resolved ^{1}H spectrum of 2,3,4,6-tetrakis-*O*-trideuteroacetyl-α-D-glucopyranoside. (Reproduced from L.D. Hall, S. Sukumar, and G.R. Sullivan, *J. Chem. Soc., Chem. Commun.,* 292 [1979]).

The pulse sequence, as a variant of the spin echo experiment, also refocuses the spread of frequencies caused by field inhomogeneity, so that some improvement in resolution is obtained. The inset at the lower right of Figure 6-18 shows the normal 1D spectra of H-4 and H-5 at the top (c and e) and the unrotated projection of the 2D J-resolved spectra at the bottom (d and f, extracted from the projected spectrum (a) at the top of the 2D display). The much higher resolution of the 2D resonances is clearly evident. Thus the procedure is an effective way to measure J accurately, particularly when J is poorly resolved in the 1D spectrum. The experiment fails for closely coupled nuclei (second order spectra).

In addition to resolving small couplings that may be absent in the 1D spectrum, the J-resolved procedure can be used to distinguish homonuclear from heteronuclear couplings. The vertical axis in Figure 6-18 displays couplings only between spins that experienced the 180° pulse. Thus only $^1H-^1H$ couplings appear on this axis. Couplings to heteronuclei are not phase modulated and consequently appear as spacings along the horizontal axis. In this way, $^1H-^{19}F$ and $^1H-^{31}P$ couplings may be distinguished from $^1H-^1H$ couplings, which are removed in the rotated spectrum such as that shown in (b) at the bottom of Figure 6-18.

6-2   PROTON–HETERONUCLEUS CORRELATION

Cross peaks in the COSY experiment are generated through magnetization transfer that arises from scalar *(J)* coupling between protons or between any identical nuclides. Coupling from a proton to a different nuclide such as carbon-13 should be able to generate a similar response. Analogous cross peaks then would provide very useful information about which carbons are bonded to which protons. Thus assignment of a proton resonance would automatically lead to the assignment of the resonance of the carbon to which it is bonded, and vice versa. The simplest extension of the COSY sequence to include magnetization transfer to carbon takes the form given in Figure 6-19. The pulse sequence is very reminiscent of the one-dimensional INEPT sequence, and manipulation of magnetization is much the same (see Figure 5-19). The initial 90° ^1H pulse generates y magnetization. For the simplest case of one carbon bonded to one proton (as in $CHCl_3$), ^1H magnetization evolves during the period t_1 according to its Larmor frequency. Two ^1H vectors diverge due to coupling with ^{13}C. The second 90° ^1H pulse generates nonequilibrium z magnetization that is transferred to ^{13}C as shown in Figure 5-19. The single 90° ^{13}C pulse then provides the ^{13}C free induction decay that is acquired during t_2. The 2D spectrum then has one axis in ^1H frequencies (ν_1) and one in ^{13}C frequencies (ν_2).

For the simple AX example the 2D spectrum contains two peaks when projected onto either the ^1H or the ^{13}C axis (the A and X portions of the AX spectrum). Thus the ^1H resonances that are detected actually are the ^{13}C satellites of the usual ^1H spectrum. The 2D display contains four peaks: two along a diagonal and two symmetrically off the diagonal. Moreover, the peaks are in antiphase for each nuclide, since the INEPT spectrum without decoupling generates one peak up and one peak down. Application of decoupling would result in algebraic summing of the peaks to zero.

To bring about decoupling, an additional pulse and two fixed time periods are added (Figure 6-20). The first ^1H pulse allows chemical shifts and coupling constants to evolve during t_1. The 180° ^{13}C pulse refocuses H–C coupling constants in the ^1H dimension (^1H is then decoupled from ^{13}C). The fixed time Δ_1 allows the ^1H vectors to obtain

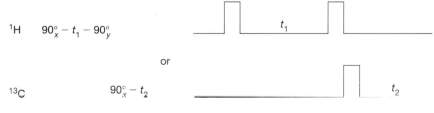

Figure 6-19

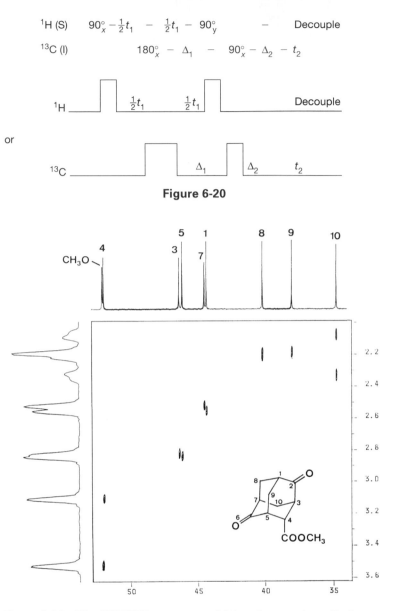

$$^1\text{H (S)} \quad 90^\circ_x - \tfrac{1}{2}t_1 \quad - \quad \tfrac{1}{2}t_1 - 90^\circ_y \qquad\qquad - \qquad \text{Decouple}$$

$$^{13}\text{C (I)} \qquad\qquad\quad 180^\circ_x - \Delta_1 \quad - \quad 90^\circ_x - \Delta_2 - t_2$$

Figure 6-20

Figure 6-21 The HETCOR spectrum of 4-(methoxycarbonyl)adamantane-2,6-dione. (Reproduced from H. Duddeck and W. Dietrich, *Structure Elucidation by Modern NMR,* Steinkopff Verlag, Darmstadt, Germany, 1989, p. 22.)

the antiphase (180° out of phase) relationship illustrated in Figure 5-19. The second 90° ^{1}H pulse moves the vectors in antiphase relationship onto the z axis and provides polarization transfer to ^{13}C, also in antiphase relationship. The 90° ^{13}C pulse is for observation. The second fixed time Δ_2 restores phase alignment and permits ^{13}C to be decoupled from ^{1}H during ^{13}C acquisition.

Figure 6-21 illustrates this procedure for an adamantane derivative. The ^{1}H frequencies are on the vertical axis and the ^{13}C frequencies are on the horizontal axis. The respective spectra are illustrated on the left and at the top. The 2D spectrum is composed only of cross peaks, each one relating a carbon to its directly bonded proton(s). Quaternary carbons are invisible to the technique, as the fixed times Δ_1 and Δ_2 normally are set to values for one bond couplings. This experiment often is a necessary component in the complete assignment of ^{1}H and ^{13}C resonances. Its name, **HET**eronuclear chemical shift **COR**relation, usually is abbreviated as HETCOR, but other acronyms, HSC (**H**eteronuclear **S**hift **C**orrelation) and H,C COSY, also are used. The method may be applied to protons with many other nuclei, such as ^{15}N, ^{29}Si, and ^{31}P, as well as ^{13}C.

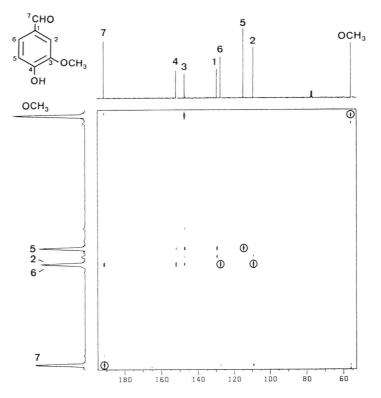

Figure 6-22 The COLOC spectrum of vanillin. (Reproduced from H. Duddeck and W. Dietrich, *Structure Elucidation by Modern NMR*, Steinkopff Verlag, Darmstadt, Germany, 1989, p. 24.)

$$^1\text{H} \quad 90° - \Delta_1 - \tfrac{1}{2}t_1 - 180° - \tfrac{1}{2}t_1 - \Delta_2 - t_2$$

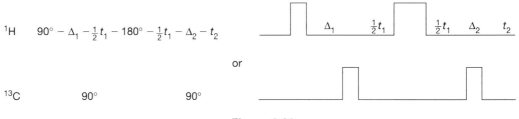

$$^{13}\text{C} \qquad 90° \qquad\qquad 90°$$

Figure 6-23

To focus on longer range H—C couplings, the fixed times Δ_1 and Δ_2 are lengthened accordingly. Loss of magnetization due to transverse relaxation then reduces sensitivity significantly. The COLOC pulse sequence (**CO**rrelation spectroscopy via **LO**ng-range **C**oupling) avoids this problem by incorporating the ^{1}H evolution period t_1 inside the Δ_1 delay period. Figure 6-22 shows a COLOC spectrum for vanillin. The circled cross peaks are residues from one bond couplings. The only long-range coupling of the methoxy group is with C-3, which indicates that methoxy is connected at that point. Other long range couplings are seen, for example, between C-1 and H-5, C-3 and H-5, and C-2 and H-7.

A major drawback to the HETCOR experiment is the low sensitivity that results from detection of the X nucleus (usually ^{13}C). The HMQC experiment (**H**eteronuclear correlation through **M**ultiple **Q**uantum **C**oherence) uses inverse detection, whereby ^{13}C responses are observed in the ^{1}H spectrum. The pulse sequence is given in Figure 6-23 and represents coherence rather than polarization transfer. The sequence is initiated by the 90° ^{1}H pulse, which creates ^{13}C polarization through the ^{1}H–^{13}C coupling constant during the fixed Δ_1 period. The remainder of the sequence is designed to select double quantum coherence (the ^{13}C satellites in the ^{1}H spectrum) over single quantum coherence (the ^{1}H centerbands), in a process similar to the 1D INADEQUATE experiment in Section 5-7. The 2D representation, as in Figure 6-24 for camphor, normally still includes the ^{1}H–^{13}C coupling information in the ^{1}H dimension, although ^{13}C irradiation can be carried out during the ^{1}H t_2 acquisition period.

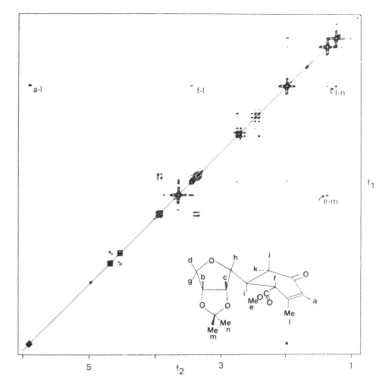

Figure 6-26 The ¹H NOESY spectrum for the indicated compound. (Courtesy of Bruker Instruments, Inc.)

At least three factors complicate the analysis of NOESY spectra. First, COSY signals may be present from scalar couplings and may confuse an analysis intended to be based entirely on interproton distances (vicinal couplings, for example, are largest when they are furthest apart, found in the antiperiplanar geometry). COSY signals may be eliminated through phase cycling or by a statistical variation of τ_m by about 20% (the NOESY signals grow monotonically but the COSY signals are sinusoidal).

Second, in small molecules the NOE tends to build up slowly and attain a theoretical maximum of only 50%, as noted earlier in the 1D context (see Section 5-4). Because a single proton may be relaxed by several neighboring protons, the actual maximum normally is much less than 50% (the same problem of course exists in the 1D NOE experiment). Moreover, as molecular size increases and behavior departs from the extreme narrowing limit, the maximum NOE decreases to zero and becomes negative. Thus for many small to medium-sized molecules the NOESY experiment may fail. For larger molecules, whose relaxation is dominated by the W_0 term, not only is the maximum NOE −100% rather than +50%, but also the NOE buildup occurs more rapidly. The NOESY experiment thus has been of particular utility in the analysis of the structure and conformation of large molecules such as proteins and polynucleotides.

Third, in addition to direct transfer of magnetization from one proton to an adjacent proton, magnetization may be transferred by spin diffusion. In this mechanism, already described in the 1D experiment (see Section 5-4), magnetization is transferred through the NOE from one spin to a nearby second spin and then to a third spin. The third spin is close to the second spin but not necessarily to the first spin. These multistep transfers thus can produce NOESY cross peaks between protons that are not close together. Spin diffusion can even occur through two or more intermediate spins, but the process becomes increasingly less efficient. Direct magnetization transfer and transfer by spin diffusion sometimes may be distinguished by examination of the NOE buildup rate, as illustrated in Figure 6-27. In this hypothetical plot of the NOE intensity as a function of the fixed time τ_m for irradiation at the A frequency of a system D–A-B-C, AA is the intensity of the diagonal peak, and the other lines represent intensities of cross peaks. The NOE between two close protons (A and B, separated by 2 Å in the

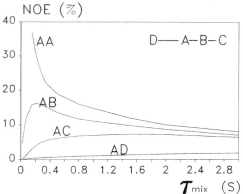

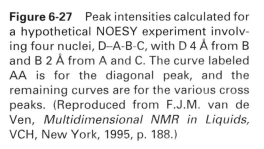

Figure 6-27 Peak intensities calculated for a hypothetical NOESY experiment involving four nuclei, D–A-B-C, with D 4 Å from B and B 2 Å from A and C. The curve labeled AA is for the diagonal peak, and the remaining curves are for the various cross peaks. (Reproduced from F.J.M. van de Ven, *Multidimensional NMR in Liquids*, VCH, New York, 1995, p. 188.)

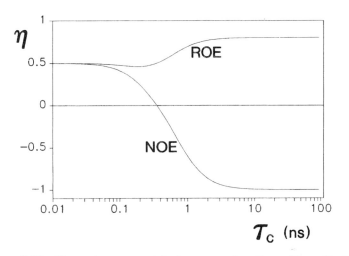

Figure 6-28 The enhancement factor η as a function of the effective correlation time τ_c for the standard nuclear Overhauser experiment (NOE) and for the spin lock or rotating coordinate variant (ROE). The curves were calculated for an interproton distance of 2.0 Å and a spectrometer frequency of 500 MHz. (Reproduced from F.J.M. van de Ven, *Multidimensional NMR in Liquids,* VCH, New York, 1995, p. 251.)

model) rises most rapidly. Protons A and D are 4 Å apart and show a very small NOE. Protons A and C also are 4 Å apart (2 Å between A and B and another 2 Å between B and C), but the intensity of the AC cross peak rises steadily through spin diffusion with B as the intermediary. The model shows that spin diffusion provides the major contribution to the AC cross peak for a distance of 4 Å, but the buildup is slower than for the direct transfer for the AB cross peak.

The rotating frame NOESY (ROESY) experiment provides some advantages for small and medium-sized, as well as large molecules. Whereas the NOE decreases to zero and becomes negative as the mean correlation time τ_c for molecular rotation increases (larger molecules move more slowly), in the rotating frame the maximum NOE remains positive and even increases from 50% to 67.5% (Figure 6-28). The pulse sequence for ROESY (previously called CAMELSPIN) is similar to TOCSY or HOHAHA, although the period of spin locking is chosen to optimize magnetization transfer through the NOE (dipolar interactions) rather than through scalar couplings. In addition to greater signal enhancement, the ROESY experiment also decreases spin diffusion, so that it offers advantages for large molecules. Just as COSY artifacts may be present in the NOESY spectrum, so can TOCSY artifacts be present in the ROESY spectrum, and steps must be taken to remove them. Use of a weak static spin-lock pulse can reduce the TOCSY peaks.

When magnetization is transferred via chemical exchange in the EXSY experiment, it may be necessary to perform several preliminary experiments to optimize the

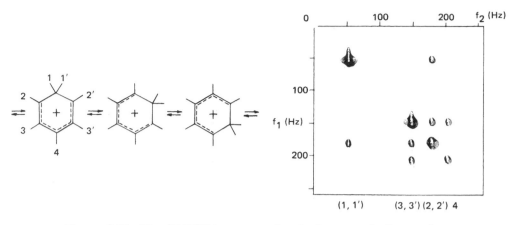

Figure 6-29 The ¹H EXSY spectrum for the heptamethylbenzenium ion. (Reproduced with permission from R.H. Meier and R.R. Ernst, *J. Am. Chem. Soc.*, **101**, 6441 [1979]). Copyright 1979 American Chemical Society.)

value of τ_m, which should be approximately $1/k$. Figure 6-29 contains the EXSY experiment from an early example by Ernst, in which the diagonal peaks run nontraditionally from upper left to lower right. At fast exchange the 1D ¹H spectrum of the heptamethylbenzenium ion contains only one methyl resonance, as the methyl group moves around the ring. At slow exchange there are distinct resonances for the four types of methyls labeled on the left in the figure. The EXSY experiment shows which methyls interchange with which. One can imagine 1,2, 1,3 or 1,4 shifts, but the EXSY experiment agrees only with the 1,2 mechanism. Each off-diagonal peak indicates magnetization transfer between two diagonal peaks. Thus the 1 methyls have a cross peak only with (and hence exchange only with) the 2 methyls; the 2 methyls exchange with the 1 and 3 methyls; the 3 methyls exchange with the 2 and 4 methyls; and the 4 methyls exchange only with the 3 methyls. This is the pattern expected for 1,2 shifts.

The intensity of the cross peaks depends on the rate constant for exchange. For the case of exchange between equally populated sites lacking spin–spin coupling (for example the two methyls of N,N-dimethylformamide, $H(CO)N(CH_3)_2$), the rate constant k is related to the mixing time τ_m, the intensity of the cross peak I_c, and the intensity of the diagonal peaks I_d by eqs. 6-1 and 6-2.

$$I_d/I_c ~\sim~ (1 - k\tau_m)/k\tau_m \tag{6-1}$$

$$k ~\sim~ 1/[\tau_m(I_d/I_c + 1) \tag{6-2}$$

6-4 CARBON–CARBON CORRELATION

The 1D INADEQUATE experiment provides a method for measuring ¹³C–¹³C coupling constants and for determining carbon–carbon connectivity by establishing coupling magnitudes that are common to two carbon atoms (see Section 5-7). In practice, application to solving connectivity problems is complicated not only by the inherently low sensitivity of detecting two dilute nuclei but also by the similarity of many ¹³C–¹³C couplings. Duddeck and Dietrich have pointed out that all the one-bond carbon–carbon couplings in cyclooctanol fall in the narrow region 34.2–34.5 Hz, except for C-1–C-2, which is 37.5 Hz. This latter problem may be largely alleviated by translating the experiment into two dimensions. The original INADEQUATE experiment (see Section 5-7) can be adapted directly to two dimensions by incrementing the fixed time Δ as the t_1 domain: $90^\circ_x - 1/4J_{CC} - 180^\circ_y - 1/4J_{CC} - 90^\circ_x - t_1 - 90^\circ_\phi - t_2$.

The time period t_1 is used to encode the double quantum frequency domain. The resulting 2D display contains a horizontal axis in ν_2 (the normal ¹³C frequencies) and a vertical axis that is a double quantum domain represented by the sum of the frequencies of coupled ¹³C nuclei ($\nu_2 = \nu_A + \nu_X$). The latter frequencies are referenced to a transmitter frequency at zero.

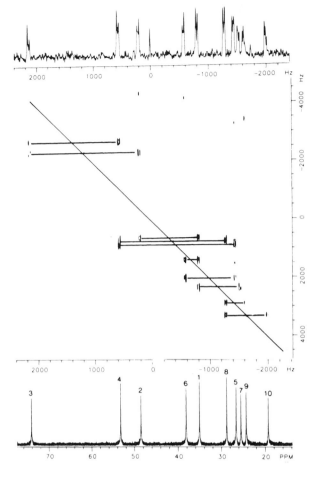

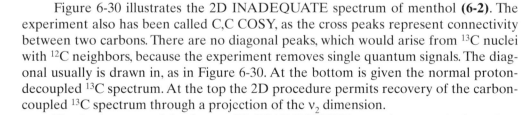

Figure 6-30 The 2D INADEQUATE spectrum of menthol, with the ¹H-decoupled ¹³C spectrum at the bottom and the ¹³C-coupled ¹³C spectrum at the top. (Reproduced from G.E. Martin and A.S. Zektzer, *Two-Dimensional NMR Methods for Establishing Molecular Connectivity,* VCH, New York, 1988, p. 362.)

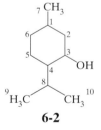

6-2

Figure 6-30 illustrates the 2D INADEQUATE spectrum of menthol **(6-2)**. The experiment also has been called C,C COSY, as the cross peaks represent connectivity between two carbons. There are no diagonal peaks, which would arise from ¹³C nuclei with ¹²C neighbors, because the experiment removes single quantum signals. The diagonal usually is drawn in, as in Figure 6-30. At the bottom is given the normal proton-decoupled ¹³C spectrum. At the top the 2D procedure permits recovery of the carbon-coupled ¹³C spectrum through a projection of the v_2 dimension.

To obtain connectivity from a 2D INADEQUATE experiment, a single assignment is made and the remainder of the structure is mapped. Only a gap caused by the presence of a heteroatom, C—X—C, prevents mapping the entire skeleton. For menthol, the oxygen-substituted C-3 resonates at highest frequency (the far left). Horizontal lines are drawn between coupled carbons in the 2D spectrum (Figure 6-30), passing through the diagonal at their midpoints. There are two cross peaks at the C-3 frequency, corresponding to connectivities to C-2 and to C-4. Of these, the secondary C-2 should be at lower frequency (higher field). The connectivity then may be followed: C-2 ⟶ C-1 ⟶ C-6 (and from C-1 to the C-7 methyl) ⟶ C-5 ⟶ C-4 ⟶ C-8 (and from C-4 to the original C-3) ⟶ C-9 and C-10.

The major disadvantage to this experiment is its extremely low sensitivity. Although numerous technical refinements have been applied to 2D INADEQUATE to ameliorate this and other disadvantages, it is unlikely that the experiment will be widely used. Alternative connectivity experiments that utilize proton sensitivity through relay methods or inverse detection may be preferred.

6-5 *HIGHER DIMENSIONS*

The enormous complexity of spectra of large biomolecules such as proteins, polynucleotides, or polysaccharides has led to the development of three- and four-dimensional experiments. Two independently incremented evolution periods (t_1 and t_2) in conjunction with three separate Fourier transformations of these and the acquisition period t_3 result in a cube of data with three frequency coordinates.

From a study by van de Ven, Figure 6-31 illustrates the complexity of the 2D NOESY spectrum of a DNA-binding protein of phage Pf3, consisting of 78 amino acids. The vertical line at δ 9.35 highlights the problems at a single resonance position in the NH region. The NH proton in a given peptide unit —CHR′—CO—NH—CHR—CO— could have one cross peak with its own CHR proton and another with the neighboring CHR′ protein, but the NOESY spectrum contains more than a dozen cross peaks at the one frequency at δ 9.35. Thus more than one NH must be generating cross peaks at this frequency.

The nitrogen HMQC experiment provides connectivity information about nitrogens and their attached protons. For proteins, use of HMQC normally requires isotopic enrichment of ^{15}N, which is obtained by growing an organism in a medium containing a single nitrogen source such as ^{15}NH$_4$Cl (similarly, ^{13}C enrichment may be obtained by use of a medium containing ^{13}C-labeled glucose). The normal 2D HMQC spectrum (^{15}N vs. ^{1}H) for this same protein is given in Figure 6-32, and two connectivities are seen

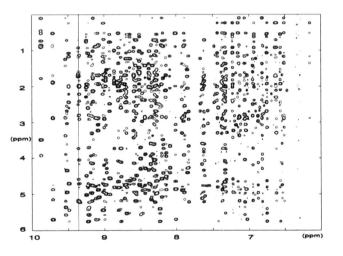

Figure 6-31 A portion of the NOESY spectrum of a DNA-binding protein of phage Pf3 containing cross peaks between NH and aliphatic protons. (Reproduced from F.J.M. van de Ven, *Multidimensional NMR in Liquids,* VCH, New York, 1995, p. 296.)

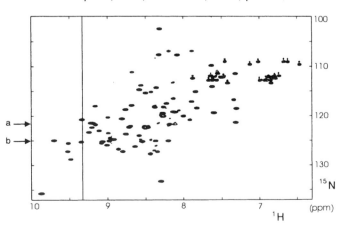

Figure 6-32 The ^{1}H,^{15}N HMQC spectrum of ^{15}N-labeled Pf3. The arrows are explained in Figure 6-35. (Reproduced from F.J.M. van de Ven, *Multidimensional NMR in Liquids,* VCH, New York, 1995, p. 296.)

at the ^{1}H frequency of δ 9.35 (vertical line). Thus there are two NH resonances (or more if there are coincidences) at δ 9.35.

The 3D experiment takes the 2D experiments in Figures 6-31 and 6-32 into an additional dimension. The 3D procedure illustrated in Figure 6-33 labels each NOESY peak with the ^{15}N frequency through the HMQC method, thus combining NOESY and HMQC data. The pulses and time delays comprise the standard NOESY sequence through the third 90° ^{1}H pulse. The pulses and delays thereafter are the standard HMQC sequence, which ends with inverse detection of ^{15}N at the ^{1}H frequencies in t_3. The totality of data requires a cube for representation, as diagrammatically in Figure 6-34, in which the flat dimensions are the NOESY data in ^{1}H frequencies and the vertical axis is the ^{15}N frequencies from HMQC. In practice, horizontal planes (single ^{15}N frequencies) are selected for analysis, as in Figure 6-35 for δ 120.7 and 124.9 (see the arrows labeled a and b in Figure 6-32). The vertical lines at δ 9.35 each show two dominant cross peaks for the NH NOEs to the inter- and intraresidue CHR. Note that both ^{15}N frequencies show a cross peak for δ 9.35 at a CHR frequency of δ 5.2, so that indeed there is overlap in Figure 6-32.

This type of heteronuclear 3D experiment is called NOESY–HMQC (in Figure 6-34 there are two ^{1}H dimensions and one ^{15}N dimension). Most 3D experiments use high-sensitivity methods and displays that are particularly effective for large molecules. Thus COSY is not often used, but TOCSY–HMQC is a useful method

^{1}H $90° - t_1 - 90° - \tau_m - 90°$ $-$ $180°$ $-$ t_3

^{15}N $180°$ $-$ $\Delta - 90° - t_2 - 90° - \Delta -$ Decouple

Figure 6-33

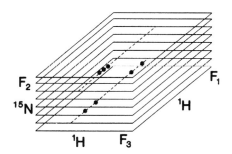

Figure 6-34 Diagram of the 3D ^{1}H,^{15}N NOESY–HMQC spectrum of Pf3 in three frequency dimensions: F_1, F_2, F_3 (ν_1, ν_2, ν_3). (Reproduced from F.J.M. van de Ven, *Multidimensional NMR in Liquids*, VCH, New York, 1995, p. 299.)

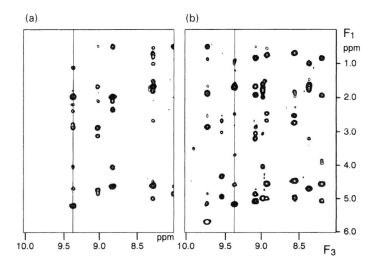

Figure 6-35 Two $\nu_1–\nu_3$ ($F_1–F_3$) planes taken from the 3D ^{1}H,^{15}N NOESY–HMQC spectrum of Pf3, corresponding to the frequencies indicated by arrows in Figure 6-32. (Reproduced from F.J.M. van de Ven, *Multidimensional NMR in Liquids*, VCH, New York, 1995, p. 300.)

to separate ^1H–^1H coupling connectivities into a ^{13}C or ^{15}N dimension. The homonuclear 3D experiment NOESY–TOCSY (all three dimensions are ^1H) separates through-space connectivities from coupling connectivities by the simple pulse sequence $90° - t_1 - 90° - \tau_m - t_2 - (\text{spin lock}) - t_3$. The three dimensions may each represent a different nuclide, as in $^1\text{H}/^{13}\text{C}/^{15}\text{N}$, and are considered as variants of the HETCOR experiment. The nuclides usually are selected to explore specific connectivities in biomolecules. The H–N–CO experiment looks at the connection ^1H—^{15}N—(C)—^{13}C=O in the peptide unit —NH—CHR—CO— and requires double labeling of ^{15}N and ^{13}C to provide sufficient sensitivity in proteins. The 3D cross peaks connect the **HN** proton in the first dimension, the HN nitrogen in the second, and the intraresidue carbonyl carbon in the third. An analogue in nucleotide analysis is the H–C–P experiment, in which the third dimension is ^{31}P. Numerous variations of these triple resonance experiments exist. In particularly complex cases a fourth time domain t_4 may be introduced to produce 4D experiments.

These higher dimensions of NMR require a computer with powerful storage and graphics capabilities. When through-bond connectivity experiments are combined with the spatial information from buildup rates of NOESY cross peaks, proton–proton distances can be obtained by referencing to known bond lengths. The result can be the complete three-dimensional structure of biomolecules. Such solution-phase structures complement solid-phase information from X-ray crystallography. In this way, NMR spectroscopy has become a structural tool for obtaining detailed molecular geometries of complex molecules.

6-6 PULSED FIELD GRADIENTS

Field inhomogeneity has been mentioned as the primary contributor to transverse relaxation (T_2) (Sections 2-3 and 5-1). Transverse *(xy)* magnetization arises because the phases of individual magnetic vectors became coherent rather than random (respectively, Figures 2-10 and 2-11). In a perfectly homogeneous field, this coherence would relax only through spin–spin interactions. In an inhomogeneous field, however, the existence of slightly unequal Larmor frequencies permits vectors to move faster or more slowly than the average, thereby randomizing their phases and destroying transverse magnetization, as described in Section 2-3.

There are several instances in which transverse magnetization is unwanted and may be eliminated by application of a pulsed field gradient (PFG) (also called a gradient pulse). An example of a PFG is illustrated in Figure 6-36. This gradient is along the direction *(z)* of the B_0 field. On application, nuclei with different positions in the sample (different z coordinates) resonate at different frequencies. Such spatial encoding of frequency information is the fundamental principle of magnetic resonance imaging (MRI). For the present context, it may be viewed as a method for inducing transverse relaxation very quickly by rapid dephasing. In a typical 2D pulse sequence, a delay time is necessary between repetitions of the pulse sequence in order for relaxation to occur. If repetition occurs before transverse magnetization has relaxed to zero, artifacts may occur in the 2D spectrum. Consequently, application of a PFG at the beginning of the sequence avoids this problem. For the NOESY experiment, the following pulse sequence may be used: $G1 - 90° - t_1 - 90° - \tau_m, G2 - 90° - t_2$.

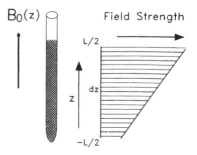

Figure 6-36 Diagram of a B_0 field gradient along the z direction. (Reproduced from F.J.M. van de Ven, *Multidimensional NMR in Liquids*, VCH, New York, 1995, p. 212.)

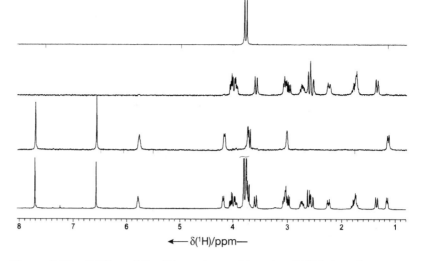

Figure 6-37 Editing of the ¹H spectrum of brucine **(6-3)** into subspectra for CH₃, CH₂, and CH (top to bottom). The complete ¹H spectrum is given at the bottom. (Reproduced with permission from T. Parella, F. Sánchez-Ferrando, and A. Virgili, *J. Magn. Reson. A,* **117,** 80 [1995]).

The first PFG (G1) destroys residual transverse magnetization from prior pulses by dephasing the magnetic vectors. A second PFG (G2) is applied during the mixing period τ_m. Only the effects on longitudinal (z) magnetization are of interest during this period. Application of the PFG eliminates false cross peaks that can arise from transverse magnetization.

In addition to dephasing transverse magnetization, PFGs are used to select coherence order. Use of phase cycling to select coherence order inevitably involves multiple scans, by which pulse sequences move through 4 or 16 or 64 variations with switching of x, $-x$, y, and $-y$, for example. Full exploitation of phase cycling thus is time consuming. The development of zero, single, or double quantum coherence depends on the rate of various dephasing processes. Proper use of PFGs permits selection of coherence order without the repetitive scans of phase cycling. For example, in the inverse detection HETCOR experiment, the single quantum coherence signal for protons attached to ¹²C (or ¹⁴N) must be suppressed while selecting the double quantum coherence signal for protons attached to ¹³C (or ¹⁵N). By phase cycling, this selection is achieved by measuring the difference between two strong signals. The PFG method selects and measures the small signal directly in a single scan.

PFG procedures have been developed to implement most of the 1D and 2D experiments discussed in the last two chapters. They may be used for solvent suppression, INADEQUATE, all common 2D experiments, and spectral editing. A PFG combination of DEPT and HMQC results in editing of proton spectra according to carbon substitution patterns. A PFG-based multiple quantum filtration leads to evolution of double, triple, or quadruple quantum coherence, respectively leading to proton spectra containing only CH, CH₂, or CH₃ resonances. Broadband decoupling removes coupling to ¹³C, while proton-proton couplings remain. Figure 6-37 illustrates the result for brucine **(6-3)**.

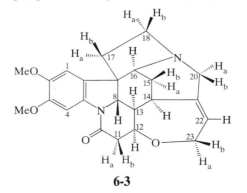

6-3

6-7 SUMMARY OF TWO-DIMENSIONAL METHODS

There is a bewildering array of two-dimensional methods available to the NMR spectroscopist today. This chapter has described a number of the most widely used experiments. A routine structural assignment begins with recording the one-dimensional ^1H and ^{13}C spectra. Many resonances may be assigned according to the principles outlined in Chapters 3 and 4 on chemical shifts and coupling constants. Normally recourse is made to 2D methods only if this traditional approach is insufficient. Some type of spectral editing for determining the number of protons attached to each carbon, such as DEPT, is helpful in completing the ^{13}C assignments (Section 5-5). The HETCOR experiment then provides correlations between the ^{13}C and ^1H resonances, possibly with completion of the ^1H assignments.

Further 2D methods are necessary if the structure is not solved in the process of making spectral assignments for hypothetical or expected structures. The COSY experiment lays the groundwork for structure determination through ^1H–^1H connectivities based on J couplings. For small molecules, there may not be enough vicinal or geminal couplings for the method to be useful. For molecules of medium complexity, COSY may be sufficient to provide the entire structure by confirming expected ^1H–^1H connectivities based on vicinal and geminal couplings. The analogous experiment based on long range couplings (LRCOSY) may be necessary to assign connections between molecular pieces that do not involve vicinal protons, for example between two rings, for substituents on a ring, or over a heteroatom or carbonyl group.

Additional 2D experiments may be necessary for larger molecules. As peaks accumulate along the diagonal, the COSY45 or DQF–COSY experiment may be used to simplify that region and uncover cross peaks that are close to the diagonal. For even larger molecules, the TOCSY or relayed COSY experiments may be necessary. Further connectivities between protons and carbons may be explored through long range couplings (COLOC). The HMQC experiment may replace HETCOR if higher sensitivity is required or if ^1H–^{13}C couplings are needed. If ^1H–^1H couplings need to be measured, either the J-resolved method or phase sensitive COSY may be carried out.

The NOESY experiment provides information about the proximity of protons and hence is used primarily for distinguishing structures that have clear stereochemical differences. For larger molecules the ROESY experiment may offer some advantages because of its lower tendency to exhibit spin diffusion. The related EXSY experiment is used only when chemical exchange is being investigated.

The 2D INADEQUATE experiment requires additional spectrometer time. It may be an experiment of last resort, although specific structures may be particularly amenable to this technique, as for example when there are several quaternary carbons that prevent COSY analysis.

PROBLEMS

For a large selection of relatively straightforward 2D spectra, refer to the early examples of the Integrated Problems (see Chapter 16). The following problems involve molecules of medium to high complexity, although none is so complex as to require 3D methods.

6-1 Below is the 300 MHz COSY spectrum* of a molecule with the formula $C_{14}H_{20}O_2$. The 1D spectrum is given on either edge. In addition to the illustrated resonances, the ^1H spectrum contains a broad singlet at δ 7.3 with integral 5. What is the structure? Show your reasoning.

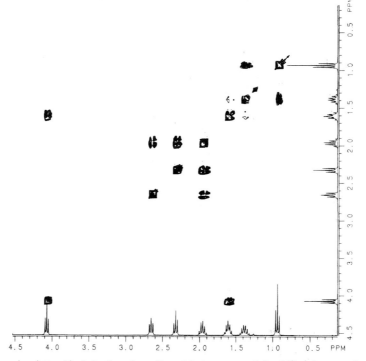

6-2 The trimerization of indole-5-carboxylic acid gives one of the following two isomers.

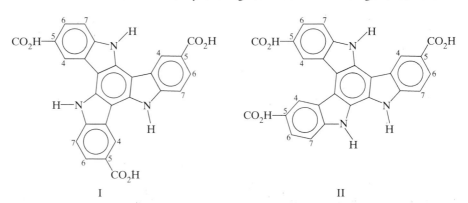

I   II

On the following page are the 360.1 MHz ^1H spectrum† of the product and the COSY spectrum (with a blow-up of the δ 7.8–8.2 region) in DMSO-d_6. The signal marked * is an impurity. The ^1H signals are labeled A to N. Signals A to D were removed with the addition of D_2O. Signal A is a broad peak at the base of B. The following 1D NOE results were obtained. Irradiation of B affected F/G and N. Irradiation of C and D affected M and L, respectively. Irradiation of F/G affected B.

---

*Reproduced with permission from S.E. Branz, R.G. Miele, R.K. Okuda, and D.A. Straus, *J. Chem. Educ.*, **72**, 659-661 (1995). Copyright 1995 American Chemical Society.
†J.G. Mackintosh, A.R. Mount, and D. Reed, *Magn. Reson. Chem.*, **32**, 559-560 (1994). Reproduced with permission from John Wiley & Sons Ltd.

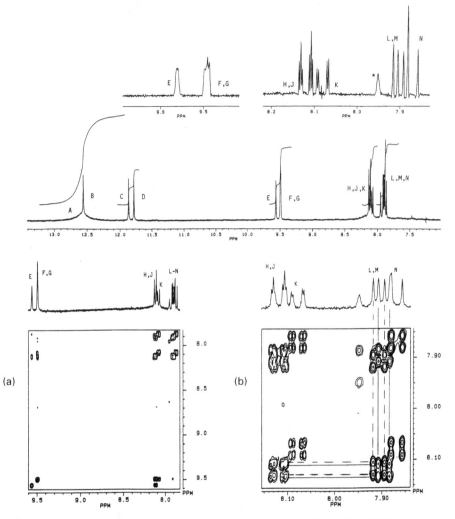

(a) From the overall appearance of the spectrum, is the trimer I or II? Explain.

(b) Using peak multiplicities, the NOE experiments, and the COSY spectrum, assign all the resonances. Discuss your reasoning in a step-by-step fashion. You should end up with an assignment of peaks A to N to specific protons.

6-3   The 500 MHz ^{1}H spectrum of the illustrated sugar derivative (extracted from a bean) is given below, aside from hydroxy resonances.

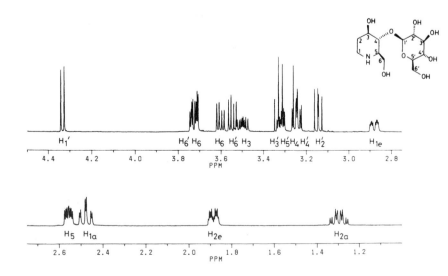

(a) The complete COSY spectrum* is given below with the assignment of one resonance. Complete the assignment for protons in the sugar ring.

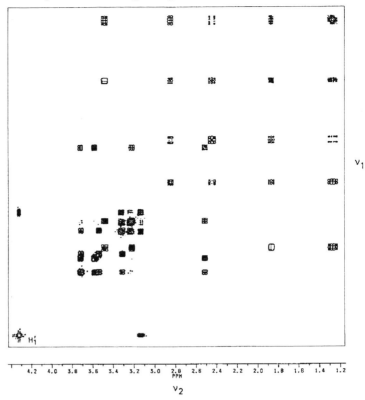

(b) A slightly expanded version of the low-frequency portion of the COSY spectrum is given below, again with the assignment of one resonance. Complete the assignment for the protons of the piperidine ring. First, it is advisable to draw out the chair conformation.

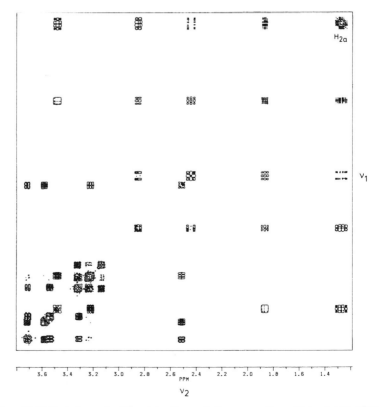

*A.E. Derome, *Modern NMR Techniques for Chemistry Research,* Pergamon Press, Oxford, UK, 1987, pp. 191-194.

6-4   Given below is the 600 MHz COSY45 spectrum* for a mixture of 3-*endo*- and 3-*exo*-fluo-rocamphor. The two high-frequency peaks are from the 3 protons respectively on the exo-F isomer ("3n," an endo proton) and the endo-F isomer ("3x," an exo proton). The remaining protons have been assigned to position but not to isomer. Using the COSY45 spectrum, indicate which peaks come from the exo and which from the endo isomer. Show your reasoning by drawing lines on the spectrum.

6-5   **(a)** From the following 300 MHz ¹H spectrum† of phenanthro[3,4-b]thiophene, identify the resonances of H-1 and H-11, as a pair. How are these distinguished from the remaining resonances?

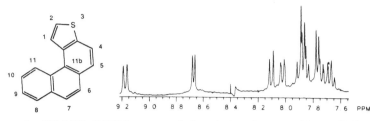

(b) From the 300 MHz COSY spectrum† given below, how would you distinguish H-1 and H-11? Assign resonances to H-2, H-8, H-9, and H-10.

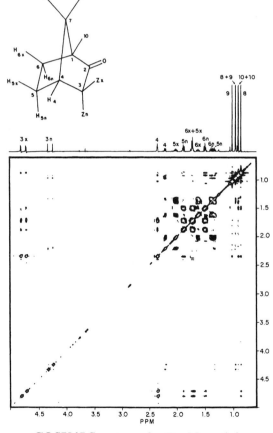

COSY45 Spectrum for Problem 6-4

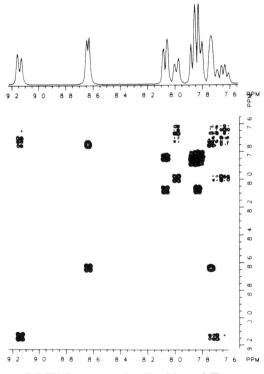

COSY Spectrum for Problem 6-5b

*C.R. Kaiser, R. Rittner, and E.A. Basso, *Magn. Reson. Chem.*, **32**, 503-508 (1994). Reproduced with permission of John Wiley & Sons Ltd.

†Reproduced from G.E. Martin and A.S. Zektzer, *Two-Dimensional NMR Methods for Establishing Connectivity,* VCH, New York, 1988, pp. 75-78.

6-6 With only the 2D INADEQUATE carbon spectrum* given below, derive the structure of the unknown. You will have to suggest a functionality attached to the highest frequency carbon, marked with an **a** in the spectrum.

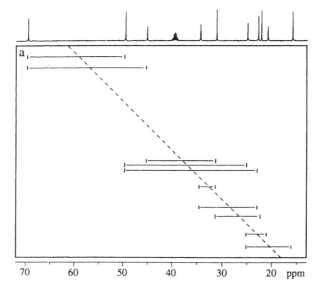

6-7 Upjohn scientists isolated a potent inhibitor of the cholesteryl ester transfer protein, U-106305. The high-resolution mass spectrum indicated the formula to be $C_{28}H_{41}NO$, and the ^{13}C spectrum had 27 distinct resonances (including one pair of equivalent carbons at δ 20.02). DEPT spectra indicated the multiplicities given in the table below.

(a) The infrared spectrum showed intense bands at 1630 and 1558 cm^{-1}. What functional group is suggested? What NMR peak confirms this assignment?

(b) What substructures are suggested by the ^{13}C peaks at δ 120–150?

(c) The HETCOR spectrum gave full ^1H assignments (see table below). From the ^1H resonance correlated with the δ 120–150 ^{13}C peaks, what else can you tell about the substructures from **(b)**? Use the magnitudes of J (values in parentheses), the values of the four ^1H and ^{13}C chemical shifts, and the proton multiplicities. In particular, note that δ 120 is d rather than dd.

(d) The UV–vis spectrum showed a strong band at 215 nm. Using the functional groups you have already deduced, what is the chromophore?

(e) Note the six low-frequency ^{13}C triplets. Each is correlated with a very low frequency (high field) pair of protons (δ < 0.7). What grouping is suggested here that is present six times?

*Reproduced with permission from N.C. Nielsen, H. Thøgersen, and O.W. Sørensen, *J. Am. Chem. Soc.,* **117,** 11365-11366 (1995). Copyright 1995 American Chemical Society.

^{13}C Chemical Shift and Multiplicities	^1H Chemical Shift and Coupling Constants	^{13}C Chemical Shift and Multiplicities	^1H Chemical Shift and Coupling Constants
7.6 (T)	0.07, 0.09 (dt: 8.43, 4.85)	20.0 (D)	1.00 (m)
7.6 (T)	0.12, 0.16 (dt: 8.39, 4.90)	20.02 (Q)	0.90 (d: 6.8)
8.0 (T)	0.08 (not first order)	20.7 (D)	1.29 (m)
11.4 (T)	0.32, 0.34 (dt: 8.20, 4.77)	21.8 (D)	0.68 (m)
13.4 (T)	0.65 (dt: 8.59, 4.87)	22.4 (D)	0.94 (m)
14.8 (T)	0.34, 0.43 (dt: 8.33, 4.60)	24.0 (D)	1.01 (m)
14.8 (D)	0.63 (m)	28.5 (D)	1.77 (h: 6.8)
17.9 (D)	0.57 (dq: 13.27, 4.93)	46.7 (T)	3.20 (d: 6.8)
18.0 (D)	0.58 (m)	120.0 (D)	5.91 (d: 15.2)
18.2 (D)	0.49 (m)	130.4 (D)	4.98 (dd: 15.5, 7.2)
18.2 (D)	0.51 (m)	131.0 (D)	4.98 (dd: 15.5, 7.7)
18.4 (D)	0.53 (m)	148.8 (D)	6.24 (dd: 15.2, 9.8)
18,41 (Q)	1.02 (d: 6.0)	166.0 (S)	
18.8 (D)	0.60 (m)		

(f) Now count up your unsaturations. You should have accounted for them all. Enumerate them.

(g) The DQF–COSY spectrum was given by the authors for one substructure as follows: the integral-6 ^{1}H resonance at δ 0.90 (d: 6.8) was linked to the integral-1 resonance at 1.77 (heptet: 6.8), which was linked to the integral-2 resonance at δ 3.20 (d: 6.8). What substructure is suggested by this 2D evidence?

(h) How is this substructure linked to a previously determined functionality? Look at the chemical shifts.

(i) You actually have almost all the structure now. The remaining unassigned ^{13}C resonances are twelve doublets and one quartet. Locate these carbons (without specific assignment) on your previous substructures and comment on the chemical shifts of the attached protons.

(j) These protons are all found in the DQF–COSY spectrum, except for the δ 1.29 resonance. Even at 600 MHz, there is severe overlap, so here are the connectivities derived from this experiment.

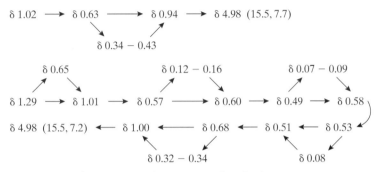

Write out the entire structure. Some stereochemistries are not set.

(k) Look at the available multiplicities and J values for the protons associated with the six high frequency triplets (all are dt with J either ~8.4 or ~4.8 Hz). This information sets the final stereochemistries. Give the full structure.

BIBLIOGRAPHY

Also see general texts referenced in previous chapters.

6.1. R.R. Ernst, G. Bodenhausen, and A. Wokaun, *Principles of Nuclear Magnetic Resonance in One and Two Dimensions,* Oxford University Press, Oxford, UK, 1987.

6.2. G.E. Martin and A.S. Zektzer, *Two-Dimensional NMR Methods for Establishing Molecular Connectivity,* VCH, New York, 1988.

6.3. K. Nakanishi, *One-Dimensional and Two-Dimensional NMR Spectra by Modern Pulse Techniques,* University Science Books, Mill Valley, CA, 1990.

6.4. *Two-Dimensional NMR Spectroscopy,* R.R. Croasmun and R.M.K. Carlson, eds., 1st and 2nd ed., VCH, New York, 1987, 1994.

6.5. H. Friebolin, *Basic One- and Two-Dimensional NMR Spectroscopy,* VCH, Weinheim, Germany, 1993.

6.6. F.J.M. van de Ven, *Multidimensional NMR in Liquids,* VCH, New York, 1995.

6.7. J.N.S. Evans, *Biomolecular NMR Spectroscopy,* Oxford University Press, Oxford, UK, 1995.

6.8. Atta-ur-Rahman and M.I. Choudhary, *Solving Problems with NMR Spectroscopy,* Academic Press, San Diego, CA, 1996.

6.9. COSY: A. Kumar, *Bull. Magn. Reson.,* **10,** 96-118 (1988).

6.10. Long-range HETCOR: G.E. Martin and A.S. Zektzer, *Magn. Reson. Chem.,* **26,** 631-652 (1990).

6.11. EXSY: R. Willen, *Progr. NMR Spectrosc.,* **20,** 1-94 (1987); C.L. Perrin and T.J. Dwyer, *Chem. Rev.,* **90,** 935-967 (1990).

6.12. Multiple quantum methods: T.J. Norwood, *Progr. NMR Spectrosc.,* **24,** 295-375 (1992).

6.13. Composite pulses: M.H. Levitt, *Progr. NMR Spectrosc.,* **18,** 61-122 (1986).

6.14. Pulsed field gradients: W.S. Price, *Ann. Rev. NMR Spectrosc.,* **32,** 51-142 (1996); T. Parella, *Magn. Reson. Chem.,* **34,** 329-347 (1996).

part II

VIBRATIONAL SPECTROSCOPY

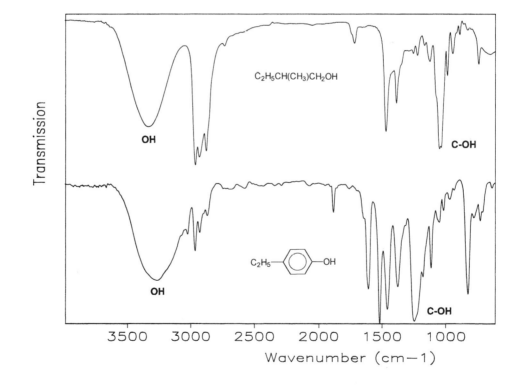

chapter 7

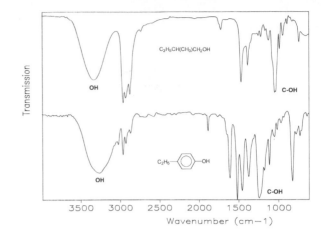

Introduction and Experimental Methods

7-1 INTRODUCTION

A rapid and simple method for obtaining preliminary information on the identity or structure of an organic molecule is to record an infrared (IR) absorption spectrum of the compound. The spectrum is a plot of the percentage of infrared radiation that passes through the sample (% transmission) vs. the wavelength (or wavenumber) of the radiation (Figure 7-1). The positions and relative sizes of the absorption peaks (also called bands) give clues to the structure of the molecule, as indicated on the figure.

The instrument that produces the spectrum is known as an IR spectrophotometer (usually shortened to spectrometer). Modern IR spectrometers are based on the Michelson interferometer (see Section 7-5b) and incorporate a fast, powerful computer. The absorption spectrum is obtained by means of a Fourier transformation of an interferogram. Hence the instruments are known as Fourier transform infrared (FT–IR) spectrometers.

Earlier instruments known as dispersive infrared spectrometers are based on monochromators that disperse the radiation from an infrared source into its component wavelengths (see Section 7-5a). The spectrum is obtained by measuring the amount of radiation absorbed by a sample as the wavelength is varied. Dispersive IR spectrometers are still in use and the early literature and spectral libraries are based on spectra recorded on these instruments.

It is relatively simple to prepare a sample and record a spectrum such as that shown in Figure 7-1. Sample handling skills can be learned quickly, and some tips on this procedure are given in Section 7-7. Recording an FT–IR spectrum consists of placing the sample in the sample compartment of the spectrometer, choosing a file name and a number of scans, and initiating the data collection. The spectrum is displayed on the computer screen and, if satisfactory, it can be saved and plotted. To record a spectrum using a dispersive instrument, the scan speed and resolution parameters are chosen and a single scan is started. The spectrum is recorded on a chart as the scan proceeds.

What an infrared spectrum actually is and how a molecule gives a spectrum are explained in Section 7-3. How IR spectrometers work is explained in Section 7-5.

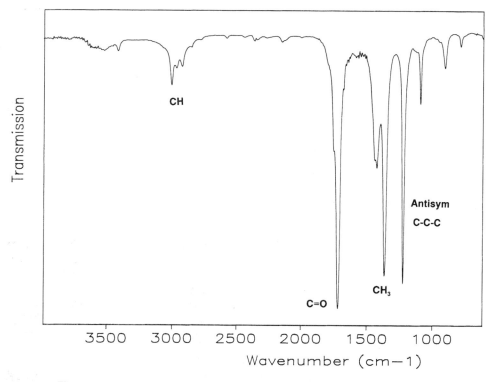

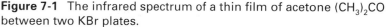

Figure 7-1 The infrared spectrum of a thin film of acetone $(CH_3)_2CO$ between two KBr plates.

Raman spectroscopy provides information complementary to that obtained from infrared spectroscopy. This technique is discussed in Sections 7-3 and 7-6. Again, there are two types of instruments: Fourier transform and dispersive.

Information on the structure of a molecule is obtained from a detailed study of those bands in the spectrum that are characteristic of certain functional groups. Some such bands are indicated on Figure 7-1. Characteristic group frequencies are discussed in detail in Chapter 8, and the interpretation of infrared and Raman spectra is the subject of Chapter 9.

7-2 VIBRATIONS OF MOLECULES

Infrared and Raman spectroscopy give information on molecular structure through the frequencies of the normal modes of vibration of the molecule. A normal mode of vibration is one in which each atom executes a simple harmonic oscillation about its equilibrium position. All atoms move in phase with the same frequency while the center of gravity of the molecule does not move. A model of the molecule can be made using balls to represent the atoms and springs to represent the bonds. Vibrations of the model involve stretching and bending the springs together with motions of the balls.

According to classical mechanics, the frequency of vibration ν (s^{-1}) of two balls of mass m (kg) connected by a spring with force constant k (N m^{-1}) can be calculated from eq. 7-1. The force constant is a measure of the resistance to stretching of the

$$\nu = \frac{1}{\pi}\sqrt{\frac{k}{2m}} \qquad (7\text{-}1)$$

spring. The force needed to displace a mass m by a distance x is $F = kx$. The vibrational motions and frequencies of a structure containing several balls (atoms) of various masses connected by springs (bonds) with different force constants can also be studied using the methods of classical mechanics. The results of these calculations are very important because they form the basis for the interpretation of vibrational spectra.

There are $3N - 6$ normal modes of vibration of a molecule, where N is the number of atoms. Each atom has three degrees of motional freedom, which can be thought of as motions in the x, y, and z directions. Thus N atoms have $3N$ independent motions. However, when the atoms are connected together in a molecule, the motions are no longer independent. Three motions become translations of the molecule, where all atoms move simultaneously in the x, y, or z directions. Another three are rotations, where all atoms rotate in phase about the x, y, and z axes. This leaves $3N - 6$ motions, in which internuclear distances and bond angles change, but the center of gravity of the molecule does not move. This number is increased by one (to $3N - 5$) if the molecule is linear. These are the normal modes of vibration of the molecule.

The methods of classical mechanics used to calculate the frequencies and forms of the normal modes of vibration of a ball-and-spring model also apply to molecules. However, appropriate atomic masses, molecular dimensions, and force constants must be used, and certain rules of quantum mechanics must be applied. This subject will be discussed further in Section 8-1.

7-3 VIBRATIONAL SPECTRA: INFRARED AND RAMAN

An infrared spectrum is obtained when a sample *absorbs* radiation in the region of the electromagnetic spectrum known as the infrared (see Figure 1-2 in Section 1-3). The expression *absorption band* is used to denote a feature observed in the spectrum. If the absorption band is quite narrow and sharp, the word *peak* is used. Thus we have the more or less interchangeable expressions: absorption band and absorption peak, or simply band, peak, or absorption.

In infrared absorption, energy is transferred from the incident radiation to the molecule, and a quantum mechanical transition occurs between two vibrational energy levels, E_1 and E_2. The difference in energy (joules) between the two vibrational energy levels is directly related to the frequency ν (s^{-1}) of the electromagnetic radiation, as shown in eq. 7-2, in which h is Planck's constant (6.624×10^{-34} J $\cdot$ s). The quantity of

$$\Delta E = E_2 - E_1 = h\nu \qquad (7-2)$$

energy, $h\nu$, is known as a photon. The frequency of vibration of the molecule corresponds directly to the frequency of infrared radiation absorbed.

The most important transitions are from the ground state (all vibrational quantum numbers $v_i = 0$) to the first excited levels (each $v_i = 1$). These allowed* transitions are known as fundamentals and they usually give rise to strong absorption bands in the infrared. A transition from the ground state to a level with one $v_i = 2$ is known as an overtone. A transition to a level for which $v_i = 1$ and $v_j = 1$ (where i and j are two different vibrations) is known as a combination. Overtones and combinations are forbidden by the simple harmonic oscillator theory of molecular vibrations, but they become weakly allowed when anharmonicity is taken into account. Many weak absorptions in an infrared spectrum can be attributed to overtones and combinations. The material for study will usually be in the form of a solid, liquid, or solution. However, gas or vapor phase spectra can also be obtained. Molecules in the gas phase can undergo changes in rotational energy at the same time as the vibrational transition, so that some rotational structure may be observed on a vibrational band. In liquids and solids, rotation is quenched by collisions and the rotational structure becomes part of the overall broadness of the band.

A Raman spectrum is produced by a *scattering* process. Monochromatic incident radiation from a laser is scattered by the sample. In dispersive instruments, the scattered (visible) light is usually observed instrumentally in a direction at 90° to the incident radiation. Since early Raman spectra were recorded on photographic plates, the fea-

---

*A transition between two energy levels is allowed when the quantum mechanical probability of the transition is not zero.

tures appeared as *lines* and *bands* on the plates; both terms are still used when discussing Raman spectra. In Fourier transform–Raman instruments the sample replaces the source lamp of a *near-IR* spectrometer. Radiation from a neodymium-doped yttrium aluminum garnet (Nd:YAG) laser is used to excite the Raman scattering, which is usually collected at 180° to the incident laser beam.

Raman spectra result from inelastic collisions of photons with molecules. In an inelastic collision, some energy is transferred either from the photon to the molecule or from the molecule to the photon, as illustrated in Figure 7-2. In the former case, the molecule will be left in a higher energy level (giving rise to the so-called *Stokes lines*). In the latter case, the molecule must already be in an excited state so that it can return to a lower state after giving up energy to the photon (giving rise to the so-called *anti-Stokes lines*). Since most molecules are in their ground vibrational state at normal temperatures, only the Stokes lines are important, and it is these that comprise the Raman spectrum of interest.

A typical Raman spectrum is shown in Figure 7-3. The zero on the abscissa corresponds to the frequency of the laser line (v_L) used to excite the spectrum. The positions of the peaks correspond to differences or shifts between v_L and the observed scattered frequencies (v_{obs}). Because the frequencies are very large, it is customary to

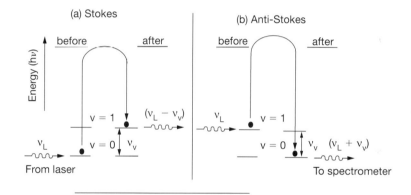

Figure 7-2 The mechanism of Raman scattering.

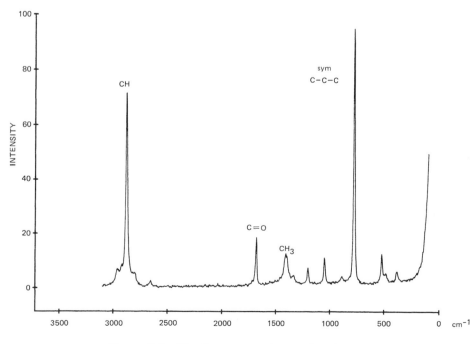

Figure 7-3 The Raman spectrum of acetone.

divide ($v_L - v_{obs}$) by c, the velocity of light (3×10^{10} cm s^{-1}) to obtain the wavenumber unit (cm^{-1}) (see Section 7-4). A Raman spectrum is then directly comparable with the infrared spectrum of the same compound. For example, compare Figure 7-3 with Figure 7-1.

The goal of this section of the book is to show the student how vibrational spectra can provide information on molecular structure. This process requires some knowledge of molecular vibrations, symmetry, and group frequencies, as well as factors that affect group frequencies. The infrared or Raman spectrum can also be used as a *fingerprint* for molecular identification. A computer can be used to search a data bank to find the spectra that most closely match that of the unknown. The final identification is then made by consultation of an atlas of spectra.

Vibrations of certain functional groups, such as OH, NH, CH$_3$, C=O, C$_6$H$_5$, etc., always give rise to bands in the infrared and Raman spectra within well-defined frequency ranges regardless of the molecule containing the functional group. The exact position of the group frequency within the range gives further information on the environment of the functional group. A parallel can be drawn here with the NMR experiment, where the resonance frequency of a nucleus depends on its electronic environment. For example, we might take the carbonyl stretching band ($v_{C=O}$) of simple aliphatic aldehydes or ketones as the standard (1730 cm^{-1}). Carboxylic acid (monomers), acid chlorides, and esters have their $v_{C=O}$ bands at higher frequencies, whereas amides and aromatic ketones have lower carbonyl stretching frequencies (Figure 7-4).

The observation of a band in the spectrum within an appropriate frequency range can indicate the presence of one *or more* different functional groups in the molecule because there is considerable overlap of the ranges of many functional groups. It is therefore necessary to examine other regions of the spectrum for confirmation of a particular group. Examples of this procedure are given in Chapter 9.

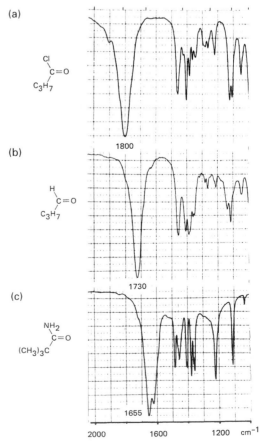

Figure 7-4 The carbonyl stretching bands of (a) butyryl chloride, (b) butyraldehyde, and (c) trimethylacetamide. (Note that the NH$_2$ bending band in (c) overlaps the C=O stretching band to give a doublet.) (Reproduced with permission of Aldrich Chemical Co., Inc., from *The Aldrich Library of FT–IR Spectra.*)

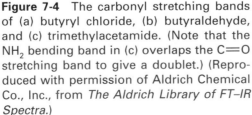

7-4 *UNITS AND NOTATION*

As Figure 7-1 shows, a spectrum is recorded graphically with the wavelength, frequency, or wavenumber (cm^{-1}) as the abscissa and the amount of absorption as the ordinate. Several units are used for both ordinate and abscissa scales. The unit of frequency ν is s^{-1} (vibrations per second). For molecular vibrations this number is very large (of the order of $10^{13}\ s^{-1}$) and inconvenient. A more convenient unit, $\bar{\nu}$ (the wavenumber), is obtained (eq. 7-3) by dividing the frequency by c, the velocity of light. A vibration of frequency

$$\bar{\nu} = \frac{\nu}{c} \tag{7-3}$$

$3 \times 10^{13}\ s^{-1}$ has a corresponding wavenumber of $1000\ cm^{-1}$ as shown in eq. 7-4.

$$\frac{3 \times 10^{13}\ s^{-1}}{3 \times 10^{10}\ cm\ s^{-1}} = 10^3\ cm^{-1} \tag{7-4}$$

Although the wavenumber is the frequency divided by the velocity, it is common practice to refer to $1000\ cm^{-1}$ as a "frequency" of $1000\ cm^{-1}$ with the division by c understood. The three accepted ways of saying cm^{-1} are "centimeters to the minus one," "reciprocal centimeters," or "wavenumbers." Infrared spectroscopists formerly used wavelength in micrometers (10^{-6} m) as the abscissa scale for their spectra. Wavelength (λ) is related to frequency (ν) or wavenumber ($\bar{\nu}$) as shown in eq. 7-5. Other older units and their equivalent SI units are given in Table 7-1.

$$\frac{1}{\lambda} = \frac{\nu}{c} = \bar{\nu} \tag{7-5}$$

In many older texts and papers on infrared spectroscopy, the positions of the infrared absorption bands are given in micrometers (formerly called microns), and most of the earlier instruments produced a spectrum whose abscissa was linear in wavelength. Thus many of the early collections of reference spectra also were published in this format. However, all modern infrared instruments record spectra with a linear wavenumber format, and all recent collections of spectra present the spectra in this format.

The positions of Raman lines cannot be expressed in units of wavelength because the lines are measured as *shifts* from the incident or exciting laser line. Hence the wavelength of a Raman line depends on the laser used. Since infrared and Raman spectra are used together to give information on molecular structure, it is convenient to use a common unit, the wavenumber (cm^{-1}). There also is strong support on theoretical grounds for using cm^{-1}, since this unit is related directly to energy ($E = h\nu = hc\bar{\nu}$).

Several units are used to measure the intensity of an infrared absorption peak. Transmittance (T) or percent transmittance (%T) is the most common, but absorbance (A) also will be encountered. Transmittance is the ratio of the radiant power or intensity (I) transmitted by a sample to the incident intensity (I_0) and can be expressed as shown in eq. 7-6, and percent transmittance as shown in eq. 7-7. Absorbance is defined in the several ways given in eq. 7-8.

$$T = \frac{I}{I_0} \tag{7-6}$$

$$\% T = 100\ \frac{I}{I_0} \tag{7-7}$$

$$A = \log_{10} \frac{I_0}{I} = \log_{10} \frac{1}{T} = \log_{10} \frac{100}{\%T} \tag{7-8}$$

In solution spectra, the intensity of absorption can be related to the concentration and the pathlength by the Beer-Lambert law, eq. 7-9, in which C is the concentration in mol L^{-1} and l is the pathlength in cm. The constant is the molar absorption coefficient,

$$A = \varepsilon Cl \tag{7-9}$$

TABLE 7-1
Wavelength Units

Older Unit	SI Unit
1 micron (μ) = 10^{-4} cm	1 micrometer (μm) = 10^{-6} m
1 millimicron (mμ) = 10^{-7} cm	1 nanometer (nm) = 10^{-9} m
1 angstrom (Å) = 10^{-8} cm	1 angstrom (Å) = 10^{-10} m

with units L mol^{-1} cm^{-1}. Thus ε is the absorbance produced by a solution of concentration 1.0 M in a cell with a pathlength of 1.0 cm. Other names for ε include extinction coefficient and molar absorptivity.

Intensities in Raman spectra are much less quantitative because the height of a peak depends on factors such as the laser power, the wavelength of the exciting radiation,* the detector, and the amplification system used. Thus quantitative results can only be obtained if an internal standard is used to determine the amount of sample actually in the laser beam and giving rise to scattering.

The intensity of a Raman line is a *linear* function of concentration, whereas the intensity of an infrared absorption band is a logarithmic function of concentration. Thus doubling the concentration of a solution should double the intensities of all Raman lines for identical instrumental settings, whereas the apparent effect on the infrared peak heights depends on the peak. For example, a weak infrared band appears to be affected much more than a strong absorption, since doubling the concentration almost doubles the intensity of a weak band but changes that of a strong band only about 10%. It is clear that caution must be exercised in discussing both infrared and Raman band intensities.

7-5   INFRARED SPECTROSCOPY: DISPERSIVE AND FOURIER TRANSFORM

Infrared spectroscopy is a well-established technique, and commercial instruments have been available since the late 1940s. Over the years a very large number of infrared spectra have accumulated in the literature and collections of reference spectra are commercially available. This makes infrared spectroscopy a very useful tool for determination of molecular structure. Direct information about the presence of functional groups is immediately available. Comparison of the infrared spectrum of an unknown material with a reference spectrum, or with the spectrum of a known compound, can provide absolute proof of the identity of the unknown substance.

For the present purposes the normal infrared range is taken to be 4000–400 cm^{-1}. However, some infrared spectrometers cover a somewhat wider range, overlapping the far infrared to 200 cm^{-1}. The region below 200 cm^{-1} is not readily accessible by infrared spectroscopy, but vibrational spectra can be obtained in this region by Raman spectroscopy (see Section 7-6).

Infrared spectra can be obtained by either dispersive or interferometric methods. Here an analogy can be drawn to NMR spectroscopy. Earlier it was seen that an NMR spectrum can be obtained by either continuous wave or Fourier transform methods. In the case of infrared spectra, dispersive instruments record the spectrum in the frequency domain, whereas interferometers record the spectrum in the time domain. The latter result is an interferogram, which must be transformed to the frequency domain by means of a Fourier transformation to obtain the infrared spectrum.

7-5a   Dispersive Infrared Spectrometers

Although instrument manufacturers no longer manufacture dispersive IR spectrometers, these instruments are still in use and are important in the historical development of infrared spectroscopy. A dispersive IR spectrometer consists of three basic parts:

---

* The scattered intensity depends on the fourth power of the frequency.

(1) a source of continuous infrared radiation, (2) a monochromator to disperse the radiation into its spectrum, and (3) a sensitive detector of infrared radiation. The sample is usually placed between the source and the monochromator.

The source of infrared radiation is usually a coil of wire with high resistance, such as nichrome, or a rod of partially conductive material, such as silicon carbide, heated by passing an electrical current through it. Temperatures of about 1200° C give the optimum yield of energy in the infrared. The monochromator contains a grating that disperses the continuous radiation into its spectrum of monochromatic components. A mechanical scanning device passes the component frequencies sequentially and continuously to the detector. In this way, the detector can sense which frequencies have been absorbed or partially absorbed by the sample and which frequencies have been unaffected. The radiation enters the monochromator through a slit and, after dispersion, leaves through another slit. The width of the entrance slit determines how much energy enters the monochromator, and the width of the exit slit determines the width of the narrow band of frequencies simultaneously reaching the detector.

The ability of the instrument to distinguish between absorptions at closely similar frequencies is known as the resolution. For most applications, a resolution of 4 cm^{-1} is adequate.

The detector is a very sensitive thermocouple. The radiation from the exit slit of the monochromator is focused by means of a mirror onto the thermocouple junction. The small voltage from the thermocouple is amplified and is used to produce the recorded spectrum. The reader is referred to page 143 of reference 7.1 for further details on dispersive IR spectrometers.

7-5b   Fourier Transform Infrared Spectrometers

FT–IR spectrometers are based on the Michelson interferometer. This instrument (Figure 7-5) consists of two plane mirrors, M1 and M2, mounted at 90° to each other and a semireflecting beam splitter (BS). One of the mirrors (M1) is fixed; the other (M2) can be moved very precisely and reproducibly through a distance (δ) of a few millimeters. The beam splitter transmits 50% of the incident radiation to one mirror and reflects 50% to the other. After reflection at M1, 50% of the radiation travels back through the BS and recombines with 50% of the radiation returned from mirror M2 and reflected by the BS. The optical path difference between the beams is known as the retardation x ($x = 2\delta$); unless $x = 0$, the recombined beams will interfere. With this arrangement, 50% of the radiation returns to the source and the other 50% passes through the sample to a detector.

Interferometers for the mid-IR region make use of the same sources as those used in dispersive instruments. However, thermocouples have a long response time and cannot be used as detectors for interferometers because the scan time is very short. One type of IR detector with a response time fast enough for an interferometer is the pyro-

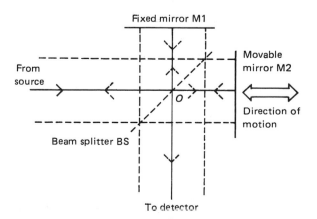

Figure 7-5 A schematic diagram of a Michelson interferometer.

electric bolometer. The best pyroelectric material available at present is deuterated triglycine sulfate (DTGS). The beam splitter is usually a very thin layer (typically 0.4 μm) of germanium deposited on an optically flat KBr plate. For the far-IR, the beam splitter is usually a thin film (typically 12.5 μm) of polyethylene terephthalate (Mylar). The thickness of the Mylar film controls the spectral range covered.

To understand how an interferometer can be used to measure an infrared spectrum, consider first a monochromatic beam of radiation of wavelength λ passing through the instrument. When $x = 0$ or $n\lambda$ (n is an integer), the recombined beams are exactly in phase, so the signal at the detector is a maximum. As mirror M2 is moved, the beams interfere and the signal falls to zero when $x = \lambda/2$. As mirror M2 continues to move at a constant velocity, the signal intensity $I(x)$ varies according to a cosine function (eq. 7-10). In eq. 7-10 $I(\nu)$ is the intensity of the source at frequency (ν cm^{-1}) and $\nu = \lambda^{-1}$. The factor of 0.5 in eq. 7-10 occurs because only one half of the incident radiation reaches the detector. The other half is reflected back towards the source. A graph of $I(x)$ vs. x is known as an interferogram.

$$I(x) = 0.5\, I(\nu) \cos 2\pi\nu x \qquad\qquad (7\text{-}10)$$

When a continuous source of infrared radiation is used, an infinite number of wavelengths simultaneously pass through the interferometer and only when $x = 0$ are all wavelengths in phase. At any other position of mirror M2 a very complex interference pattern results, giving rise to an interferogram like the one shown in Figure 7-6a. To obtain this interferogram, the mirror M2 was moved from a negative x through the zero path difference position to positive x. At $x = 0$, all wavelengths interfere construc-

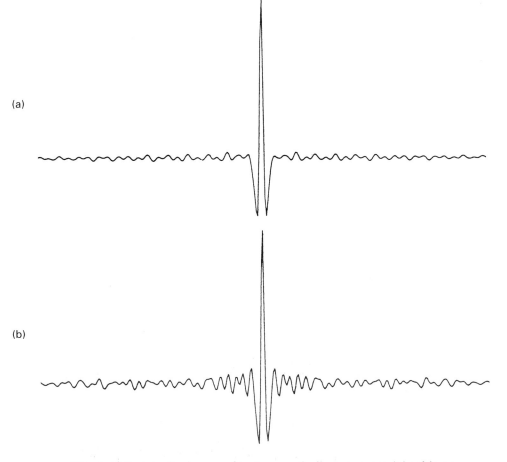

(a)

(b)

Figure 7-6 Interferograms from a continuous source: (a) with no sample; (b) with a 0.05 mm film of polystyrene placed between the interferometer and the detector.

tively and produce the very strong central signal. The intensity $I(x)$ of the radiation reaching the detector at any other retardation is the sum of the intensities of all the interfering wavelengths at this mirror position.

A sample placed between the interferometer and the detector reduces the intensity of radiation at any frequency at which the sample absorbs. Thus the infrared absorption spectrum is contained in the resulting interferogram. Figure 7-6b shows the interferogram of Figure 7-6a modified by the absorption of a 0.05 mm thick polystyrene film placed between the interferometer and the detector. The interferograms are stored digitally by the computer as data files of $I(x)$ vs. x.

There is clearly a difference in the interferograms of Figures 7-6a and 7-6b, but to obtain quantitative information about the absorption from the interferograms we need the data in the form $I(v)$ vs. v. To obtain this information, a mathematical procedure known as a Fourier transformation must be performed on both interferograms, as with FT–NMR. An FT–IR spectrum of the polystyrene sample used to obtain the interferogram of Figure 7-6b is shown in Figure 7-7.

The two main advantages of interferometers over dispersive spectrometers for infrared spectroscopy are speed and sensitivity. These advantages result from the increased energy throughput (Jacquinot's advantage) and higher signal-to-noise ratio (Fellgett's advantage) available from an interferometer. Limiting values for these advantages can be calculated (see Chapter 1 of reference 7.2). For mid-IR spectra ($4000–400$ cm^{-1}) at a resolution of 2 cm^{-1} the theoretical values are approximately 20 for Jacquinot's advantage and 40 for Fellgett's. However, sample size, detector sensitivity, and other instrument factors can reduce the magnitude of these advantages.

The resolution of an FT–IR spectrometer depends on the reciprocal of the retardation x. Hence, for a resolution of 4 cm^{-1}, the moving mirror must move 0.25 cm. The Fourier transform is computed from the digitized values of the interferogram sampled at equal intervals (Δx) of retardation. The spectral range covered is inversely proportional to the sampling interval. For a spectral range of $4000–400$ cm^{-1}, x is of the order of 10^{-4} cm. These sampling points must be separated by precisely equal intervals; otherwise noise will be introduced into the computed spectrum. This precision is possible when sampling points are triggered by the zero crossings of an interferogram generated by a helium-neon laser.

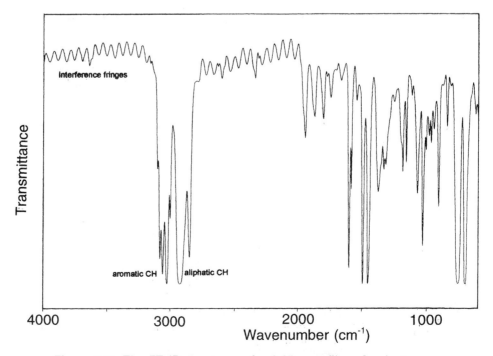

Figure 7-7 The FT–IR spectrum of a 0.05 mm film of polystyrene obtained from the interferogram shown in Fig. 7-6b.

Co-addition of several scans results in an improved signal-to-noise ratio (S/N) because the signal increases directly with the number of scans (n), while the noise, because of its statistical nature, only increases with $\sqrt{n}$. Thus, in theory, 100 scans should improve the S/N by a factor of 10 over that of a single scan. In practice, instrumental restrictions and imperfections limit the improvement in S/N ultimately obtainable by co-addition of interferograms. It should also be noted that the spectrum cannot be plotted until all of the interferograms have been recorded and averaged and the Fourier transformation carried out.

This section is not intended to be a comprehensive coverage of FT–IR spectroscopy. Consult references 7.2, 7.3, and 7.4 for further information on instrumentation and applications of this important technique.

7-6 SAMPLING METHODS FOR INFRARED TRANSMISSION SPECTRA

The very simplest sampling techniques often give quite satisfactory results. Sampling of pure liquids is quite simple. Solutions and solid samples, on the other hand, require more care. In all transmission studies, except the pressed-pellet method, discs or plates transparent to infrared radiation are needed to support the sample. Gas cells and cryostats for low-temperature work also need infrared transmitting windows.

Several infrared transmitting materials are in common use. A summary of the physical properties of the more important of these is given in Table 7-2; most can be purchased as polished windows or as sawn crystal blanks ready for polishing.

NaCl   This is perhaps the most commonly used material. The rock salt region is the traditional range (4000–650 cm^{-1}) studied by earlier workers. This material is the least expensive and is easy to polish, but it fractures easily and is only transparent as far as 650 cm^{-1}.

KCl   This is almost as cheap as NaCl and easy to polish, but it fractures when subjected to stress. It is transparent to 500 cm^{-1}.

KBr   This is another traditional material. It costs about 50% more than NaCl or KCl. It is easy to polish but is fragile. KBr is transparent to 400 cm^{-1}.

CsI   This is a very useful material; it does not fracture under mechanical or thermal stress. However, it is very water soluble, very soft, and tricky to polish. Its cost is quite high (about ten times that of NaCl), but it is transparent to 150 cm^{-1}.

CaF$_2$ This is only very slightly soluble in water, so it may be used for aqueous or D_2O solutions. The cost is about five times that of polished NaCl. CaF$_2$ windows should be purchased ready polished since this material is hard and difficult to polish. These windows normally do not need repolishing. Disadvantages of CaF$_2$ are that it is only transparent to 1100 cm^{-1} and it fractures when subjected to thermal or mechanical stress.

TABLE 7-2

Properties of Some Infrared Window Materials

Material	IR Transmission Limit (cm^{-1})	Refractive Index at 4000 cm^{-1}	Solubility in Water (g/100 mL at 20° C)
NaCl	650	1.5	36
KCl	500	1.5	35
KBr	400	1.5	53
CsI	150	1.7	80
CaF$_2$	1100	1.4	0.002
BaF$_2$	850	1.5	0.1
KRS-5	200	2.4	0.02
ZnSe	650	2.4	Insoluble
Polyethylene	50	1.5	Insoluble

BaF₂ This material is somewhat more soluble in water than is CaF_2, but it has an extended transmission range to about 850 cm^{-1}. Like CaF_2, BaF_2 fractures under thermal or mechanical shock.

KRS-5 This material is a mixed thallium bromide iodide compound. It forms bright red crystals, which are sparingly soluble in water, do not cleave, and transmit to 200 cm^{-1}. The main disadvantages of KRS-5 are the high price and the high refractive index that may cause loss of transmitted energy by scattering. KRS-5 also is toxic and may be attacked by some compounds in an alkaline solution.

ZnSe This yellow material (also known as IRTRAN 4) is harder than KRS-5, is insoluble in water, and is not attacked by alkalis. It has a high refractive index and is used as an ATR element (see Section 7-9c). It is very toxic, and when attacked by acids, produces H_2Se gas.

Polyethylene and polytetrafluoroethylene These materials are used in low-cost, disposable cells. They are not usually used as window materials, because two disks made from these materials are not rigid enough to maintain liquid films between them. Polyethylene is used in the far infrared, because it is transparent to below 50 cm^{-1}, with the exception of a sharp band at 70 cm^{-1} and a broader one at 380 cm^{-1}.

Further information on infrared-transmitting optical materials can be found in Chapter 3 of reference 7.5.

7-6a   Liquids and Solutions

Probably the easiest method to obtain a qualitative infrared spectrum of a liquid is to place one drop of the liquid onto a disc of NaCl, KBr, etc., cover the drop with a second disc, and mount the pair in a holder. Teflon spacers may be used to give various pathlengths. Fixed-pathlength, sealed cells are also available. These usually have amalgamated silver or lead spacers. The cells are filled, emptied, or flushed by means of a syringe through conventional Luer ports. Teflon stoppers are used to close the ports.

For the far infrared, polyethylene cells are available with various pathlengths. They are easily contaminated, but are of low enough cost to be disposable.

Recently, the 3M Corporation introduced disposable IR Cards. These consist of microporous polyethylene or polytetrafluoroethylene (PTFE) films mounted in cards. The IR Cards can be used for analysis of liquids or solids soluble in organic solvents. The cards have a circular aperture containing a thin film of the microporous material. A blank card is used to obtain a background spectrum. A small quantity (~50 μl) of a liquid is applied directly to the IR Card film. Solids soluble in organic solvents are applied as solutions and the solvent is allowed to evaporate. The card is then placed in the sample compartment of an FT–IR spectrometer and the spectrum is recorded. The polyethylene card can be used from 4000 to 400 cm^{-1}, except for the region between 3000 and 2800 cm^{-1}. The PTFE card has a more limited range from 4000 to 1300 cm^{-1}, but it can withstand temperatures up to 200° C.

It is often convenient to record the infrared spectrum of a compound in solution. Unfortunately, some of the best solvents have very strong infrared absorption bands that obscure parts of the spectrum of the compound. Water, for example, absorbs strongly throughout the spectrum and is rarely used in routine infrared work. However, H_2O and D_2O can be useful solvents for FT–IR spectroscopy of compounds such as sugars, amino acids, and compounds of biochemical interest, although special window materials such as CaF_2, BaF_2, or KRS-5 must be used. FT–IR instruments have the capability to signal-average over a large number of scans, so that a spectrum can be obtained even when a sample absorbs very strongly and very little radiation passes through it. FT–IR spectrometers and the more sophisticated dispersive IR spectrometers can produce a spectrum from surprisingly little transmission (as low as 0.1% T).

In dispersive instruments, weak to medium solvent absorption can be removed from the spectrum by using a pair of matched cells with the solvent in the reference beam and the solution in the sample beam. With FT–IR instruments, a spectrum of the solvent can be subtracted from the spectrum of the solution. However, it should be noted that, while recording a spectrum in a region in which the solvent absorbs strongly, essentially no energy passes to the detector and the S/N ratio is very poor. Only those regions of the spectrum in which the solvent does not absorb strongly can be used and several solvents may be needed to obtain a complete spectrum. A partial list of solvents and their useful regions is given in Table 7-3.

7-6b   Solids

Solid samples are handled either in the form of mulls or pressed discs. To make a mull, a small amount of the sample is ground in an agate or mullite mortar. Then a drop of mulling material, usually a pure colorless paraffin oil, is added and the grinding continued. The mixture should have the consistency of a thin paste. It is transferred to a window of NaCl, KBr, etc., and covered with a second window. A thin film is produced by gentle pressure with a slight rotating movement. The two plates with the mull between are placed in a cell holder and the spectrum is recorded. There will be strong bands at 2900, 1470, and 1370 cm^{-1} and a weak band at 720 cm^{-1} due to the paraffin oil. If these bands are stronger than the peaks from the sample, then more sample and less oil is required. If the sample peaks are too strong, the two windows can be squeezed together or a small drop of oil can be added. The user will have to experiment with the mull to obtain the best results.

When the region near 2900 cm^{-1} is important, another mulling material, usually a chlorofluorocarbon oil, must be used. This material is completely opaque below 1400 cm^{-1} and it also has a band at 1650 cm^{-1}. Another compound useful for mulls, when the 2900 cm^{-1} region is to be studied, is hexachlorobutadiene. This compound has no absorptions above 1650 cm^{-1}. It also has a useful window between 1500 and 1250 cm^{-1}.

Another sampling method for solid compounds is the pressed pellet technique. In an agate or mullite mortar, a few milligrams of the sample are ground together with about 100 times the quantity of a matrix material that is transparent in the infrared. The usual material is KBr, although other compounds such as CsI, TlBr, and polyethylene are used in special circumstances (see below). The finely ground powder is introduced into a stainless steel die, usually 13 mm in diameter, which is then evacuated for a few minutes with a vacuum pump to remove air from between the particles. The powder is then pressed into a disc between polished stainless steel anvils at a pressure of about 30 tons/in.^{2}. Other devices, such as a hand-held press, are available for making smaller (7, 3, and even 1 mm) KBr pellets.

A well-made KBr pellet will have 80% to 90% transmittance in regions below 3000 cm^{-1} where the sample itself does not absorb. Between 4000 and 3000 cm^{-1} the

---

TABLE 7-3

Useful Solvents for Infrared Solution Spectra

Solvent	Useful Regions (cm^{-1})	Typical Pathlength (mm)
CS$_2$	All except 2200–2100 and 1600–1400	0.5
CCl$_4$	All except 850–700	0.5
CHCl$_3$	All except 1250–1175 and below 820	0.25
CHBr$_3$	All above 700 except 1175–1100 and 3050–3000	0.5
C$_2$Cl$_4$	All except 950–750	0.5
Benzene	All above 750 except 3100–3000	0.1
CH$_2$Cl$_2$	All above 820 except 1300–1200	0.2
Acetone	2800–1850 and below 1100	0.1
Acetonitrile	All except 2300–2200 and 1600–1300	0.1
Cyclohexane	Below 2600	0.1
N,N-Dimethylformamide	2750–1750 and below 1050	0.05
Diethyl ether	All except 3000–2700 and 1200–1050	0.05
Heptane and hexane	All except 3000–2800 and 1500–1400	0.2
Dimethyl sulfoxide	All except 1100–900	0.05

transmission sometimes is low due to scattering effects. The amount of scattering by mulls and pellets depends on the refractive indices of the sample and the matrix or mulling material. When the refractive indices are similar, large particles cause serious scattering at high wavenumbers. At lower wavenumbers, the particle size becomes less important. The particle size should be less than about 20 μm for good pellets. This size can usually be achieved by hand grinding in a hard mortar as discussed above.

Matrices other than KBr may be used. CsI is useful when spectra down to 200 cm^{-1} are required. One disadvantage of CsI is that it is a very hygroscopic material. Lower pressures are required for CsI pellet formation than for KBr. TlBr is used when materials of high refractive index are studied. It has a refractive index of 2.3 in the infrared region and is transparent to 230 cm^{-1}. It is not hygroscopic and can be used in conditions of high humidity. Powdered polyethylene has been used for making pellets for far infrared spectra because, apart from a band at 80 cm^{-1}, its spectrum between 400 and 10 cm^{-1} is free of absorption. In addition to the scattering problems discussed above, the pressed-disc method has other disadvantages. Changes may occur in the sample during the grinding and pressing process and the sample may react with absorbed water or even with the matrix material.

An excellent review of infrared sample handling techniques is given in Chapter 8 of reference 7.5.

7-7 RAMAN SPECTROSCOPY: DISPERSIVE AND FOURIER TRANSFORM

Raman spectrometers give information on vibrational frequencies in the range covered by both mid- and far-infrared spectrometers (4000–10 cm^{-1}). Samples in the form of liquids, solutions, powders, and single crystals can be handled by standard sampling techniques. However, gases, for which the Raman intensity is usually several orders of magnitude weaker, require a more sophisticated arrangement involving a higher power laser and multiple passing of the laser beam through the sample. Part of the low-frequency end of the spectrum is always obscured by Rayleigh scattering of the exciting radiation. Rayleigh scattering is due to elastic scattering of the laser photons by the molecules of the sample. No energy is exchanged and the scattered photons emerge from the sample with the same frequency as that of the laser line. This process is several orders of magnitude more probable than the inelastic (Raman) scattering and gives rise to an extremely strong band centered at 0 cm^{-1} in the Raman spectrum. The width of the Rayleigh line, as it is called, increases from solids to pure liquids to solutions. An example of a Raman spectrum of a liquid organic compound appears in Figure 7-3.

Experienced users of Raman spectroscopy can obtain the same kind of structural information as the infrared spectroscopist. They use a somewhat different set of group frequencies, however, because vibrations that give rise to strong characteristic infrared absorption are often weak in the Raman spectrum. The converse is also true, as can be seen by comparing Figures 7-3 and 7-1. Raman spectra complement infrared spectra, and the two techniques used together provide a important and often unique information for organic structure determination.

Raman spectroscopy has certain advantages over infrared. One is that simpler spectra are more easily observed in the Raman because of the absence of overtone or combination bands, which are an order of magnitude weaker in the Raman than in the infrared and are usually too weak to be observed. A second advantage is the wider choice of solvents for solution spectra—in particular, water can be used—and other solvents have more clear regions in the Raman than in the infrared. Information on the lower frequency region 200–50 cm^{-1}, corresponding to the far infrared, is easily obtained.

There are, however, certain disadvantages of Raman spectroscopy. One is the inherent weakness of Raman spectra, which can often be masked by the background scattering from small suspended particles in liquid or solution samples. Absorption of the laser radiation can cause heating of liquid samples and charring of solid samples. If the compound to be studied fluoresces under visible light, the Raman spectrum is totally obscured. Fluorescence is often a serious problem even when the sample is not fluorescent itself, because a trace of fluorescent impurity can obscure the Raman spectrum. The problem of fluorescence is reduced when the laser excitation is in the red or the near infrared (NIR) (see FT–Raman, Section 7-7b). It is often difficult to get good spectra from solids unless they are crystalline. Thus in practice one often cannot obtain a Raman spectrum of a sample, whereas an infrared spectrum can always be recorded. Another disadvantage of Raman spectroscopy is the high cost of the instruments.

7-7a Dispersive Raman Spectrometers

Some of the instrumentation of a dispersive Raman spectrometer is similar to that of an IR grating spectrometer (see Section 7-5a). Raman scattering is excited by the intense monochromatic radiation from a laser. The scattered light is focused by a lens onto the entrance slit of a monochromator, where it is dispersed into its spectrum. Several basic differences exist between Raman and infrared spectra. The infrared spectrum consists of a continuum with a few parts missing where the sample has absorbed radiation. The Raman spectrum, on the other hand, consists of mostly nothing, with a few narrow regions of radiation emitted from the sample.

The infrared continuum is relatively strong, but the Raman lines in the visible are inherently weak. Infrared detectors, however, have low sensitivity, whereas very sensitive detectors (photomultipliers) can be obtained for radiation in the visible. Slits of several hundred micrometers give reasonable resolution in the infrared, but much smaller slits are needed in the visible. Another very important point is stray light, which if not effectively removed can produce such a high background that the Raman spectrum is lost. To remove as much stray light as possible, a double, or even a triple, monochromator is used. A double monochromator, as the name implies, is simply two monochromators used in series, with a slit between. During a scan the two gratings are rotated simultaneously by a common drive mechanism.

A typical arrangement of a double monochromator is shown in Figure 7-8. Light entering slit S1 is collimated by a concave mirror, M1, onto the first grating, G1, and then diffracted to mirror M2. The radiation then passes through the intermediate slit, S2, into the identical second monochromator. A linear wavenumber counter can be set to zero at any exciting wavenumber. Thus Raman frequencies can be read directly. The radiation is detected by a photomultiplier tube (PMT) mounted at the exit slit S3. The signal from the photomultiplier is amplified and displayed on a recorder. The Raman

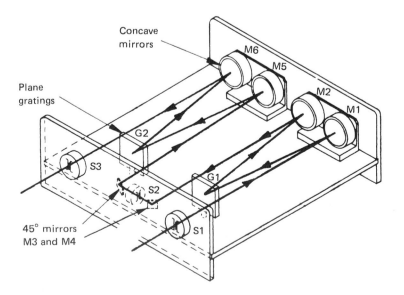

Figure 7-8   A schematic diagram of the SPEX Industries Model 1401 monochromator.

signal is also interfaced through an analog-to-digital converter and sent to a computer where the spectrum can be displayed on the computer screen and stored for future manipulation. Further information on dispersive Raman spectrometers and instrumentation for Raman spectroscopy may be found on pages 148-149 in reference 7.1 and in references 7.5, 7.6, 7.7 and 7.11.

7-7b FT–Raman Spectrometers

It has long been realized that excitation of Raman spectra in the near-infrared region of the spectrum reduces or eliminates fluorescence emission from the sample. Until recently, recording of Raman spectra in the near-infrared was not practical, because the detectors available for this region of the electromagnetic spectrum are not sufficiently sensitive to detect the very weak signals produced by a dispersive Raman spectrometer. Also, Raman spectra are a factor of 20 times weaker in the NIR than in the visible due to the inverse fourth power relationship between Raman intensity and frequency. It was recognized that the advantages of the FT–IR technique should make it feasible to record Raman spectra in the near-infrared region. In the late 1980s the development of very fast, powerful desktop computers made the rapid collection and processing of interferograms possible. This technological advance, together with developments in Nd:YAG lasers, led to the recent introduction of commercial FT–Raman spectrometers.

The rapid display of a Raman spectrum is essential for alignment of the sample, to obtain an indication of the strength of the spectrum or to ascertain if there is any Raman scattering at all from the sample. This procedure only became feasible with the development of the fast processors and large memory capacities of modern computers. Early FT–Raman spectra were obtained from modified near-infrared FT–IR spectrometers in which the infrared source was replaced by a sample irradiated by a Nd:YAG laser beam and some collection optics (see for example reference 7.8). Now the manufacturers of FT–IR spectrometers offer instruments that are either dedicated to FT–Raman or adaptable to record either Raman or infrared spectra. Further information on FT–Raman spectrometers and applications may be found in references 7.9, 7.10 and 7.11.

7-8 RAMAN SAMPLING METHODS

The sampling methods for Raman spectroscopy are usually simpler than those for infrared. In principle, the sample is simply placed in the laser beam in front of the entrance slit of the monochromator of a dispersive Raman instrument, or at the sample position of an FT–Raman spectrometer, and a laser beam is directed to the sample. In either case, the sample itself then becomes the source of radiation passing into the spectrometer. Focusing the laser beam increases the Raman intensity but may also damage the sample. Sample damage can be prevented by reducing the laser power. In the case of liquids, solutions, and gases, Raman scattering may be increased by passing the laser beam through the sample several times by means of mirrors.

The scattered light must be focused into the spectrometer by means of a lens. This lens is usually mounted on an optical bench together with the sample holder, a polarization analyzer, and a polarization scrambler. The laser beam can impinge on the sample from below or from the side (known as 90° illumination), or from the direction of the collection optics (known as 180° illumination).

The schematic diagram of a typical sample compartment of a dispersive Raman instrument is shown in Figure 7-9. Starting from the laser itself, the components have the following uses. The interference filter passes only a narrow band of wavelengths centered on the laser line, thus eliminating non-lasing plasma lines. However, it is often useful to remove this filter and use the scattered plasma lines for frequency calibration of the monochromator. The half-wave plate turns the plane of polarization through 90°. The iris diaphragm also reduces interference from the laser plasma emission. The microscope objective lens focuses the light to a spot in the sample.

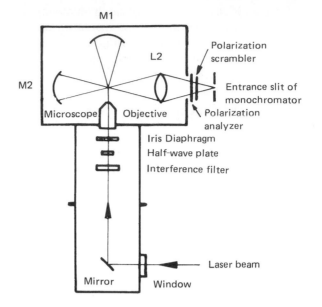

Figure 7-9   The arrangement of the optical components in a typical dispersive Raman spectrometer sample compartment.

The scattered light is focused by means of the lens (L2) onto the entrance slit of the monochromator. A spherical mirror (M1) causes double passing of the beam, and a mirror (M2) reflects 180° Raman scattering back toward the monochromator. The polarization analyzer is made from a piece of polarizing film, and the polarization scrambler is a quartz wedge that ensures that no polarized light gets into the monochromator. The scrambler eliminates any effects produced by the polarizing properties of the monochromator itself. These last two components are used when measurements of depolarization ratios are made (see Section 7-8c).

7-8a   Liquids and Solutions

A small-volume quartz cell (5 mm × 5 mm × 5 cm), such as those supplied for UV absorption work, can be used as a Raman liquid sample container. Although quartz is preferable because it has low fluorescence, cells made from ordinary Pyrex glass are usually satisfactory.

Raman spectroscopy is ideally suited for microsampling. The laser beam can be focused to a very small area, and samples contained in melting-point capillary tubes yield virtually the same spectra as much larger samples. Most instrument manufacturers now offer microscope attachments for their FT–Raman spectrometers. A recent development is a Raman microscope instrument, which incorporates a monochromator and a charge coupled detector (CCD). The Raman scattering is excited in the red by means of a He/Ne laser and in the far red by a diode-pumped laser operating at 780 nm. An application of the use of this instrument is described in reference 7.12.

The techniques for handling solutions are essentially the same as those for pure liquids. However, care must be taken with the solvents. They should be clean, pure, redistilled, and filtered to avoid problems from suspended particles or fluorescence. As we saw in Section 7-6a, most of the common solvents have extensive regions of absorption in the infrared. Although this is not the case for the Raman, there is still the drawback that the intensities obtained from dilute solutions cannot be increased by having a longer pathlength. There is a lower limit to concentrations that can be used, normally 1% to 5% by weight (about 0.1 M).

Of course, every solvent has a Raman spectrum, but Raman lines are usually much narrower than infrared absorption bands. In addition, overtones and combinations are very much weaker in the Raman. Hence there are many completely clear regions in the Raman spectra of all the usual solvents. Water is an excellent solvent for Raman solution studies. This property is extremely important when examining compounds of biological interest, since their natural environment is usually aqueous solution.

Common solvents are listed in Table 7-4. The most important Raman lines are given in the second column. The stronger lines extend for approximately 40–20 cm^{-1} on either side of the frequency quoted, and the weaker lines obscure a region of about 20–10 cm^{-1}. Raman spectra of these and other solvents may be found in reference 7.11. For a Raman spectrum of a dilute solution, the instrument is used at a high sensitivity. Under these conditions even the very weakest lines of the solvent may be observed. These lines are listed in the third column of Table 7-4. Of course, a spectrum of the solvent must be recorded in all cases. As far as possible, the same conditions should be used to observe the spectra of the solution and the solvent. With a computer, the Raman spectrum of the solvent can be subtracted from the spectrum of the solution to give the spectrum of the compound.

7-8b Solids

Polycrystalline powders are best handled in capillary tubes or tamped into a hole in the end of a metal rod. Irregular pieces of solid materials may be glued to a support rod and held in the laser beam. Powders may be pressed into pellets and examined in the same way. High background scatter often is observed, especially from amorphous materials. The interference filter described in Section 7-8 must be used with highly scattering samples to eliminate spurious laser lines from the spectrum.

A crystal 1 or 2 mm long can be mounted on a goniometer head in the same way that crystals are mounted for X-ray examination (see reference 7.13). Single crystals can be carefully positioned so that the laser beam passes along one of the axes of the crystal. In this case, when a polarization analyzer is used, important information concerning the symmetry of the normal vibrations of the molecule can be obtained. Single-crystal Raman spectroscopy is discussed in references 7.6 and 7.7 and Chapter 18 of reference 7.5.

Water-sensitive solid samples can be loaded into capillary tubes in a dry box and sealed. Larger bore tubes (NMR tubes) can be connected by a ground joint to a stopcock and evacuated on a vacuum line. The sensitive compound is transferred to the tube in the glove box and the stopcock is rejoined. After evacuation, the tube can be resealed.

TABLE 7-4
Some Solvents for Raman Spectroscopy

| Solvent | Raman Frequencies* (rounded to nearest 5 cm^{-1}) | |
	Most Important Lines	Other Weak Lines
Carbon disulfide	650 (vs), 795 (m)	400
Carbon tetrachloride	220 (s), 315 (s), 460 (vs), 760–790 (m)	1540
Chloroform	260 (s), 365 (vs), 670 (vs), 760 (m), 1220 (m), 3025 (w)	—
Methanol	1035 (m), 1460 (w), 2840 (m), 2900–2950 (w)	1110, 1165, 3405
Acetonitrile	380 (s), 920 (s), 1375 (w), 2250 (s), 2950 (s)	1440, 3000
Methylene chloride	285 (s), 700 (vs), 740 (w), 2985 (w)	1155, 1425, 3045
Nitromethane	480 (m), 655 (s), 920 (vs), 1370–1410 (m), 2965 (m)	610, 1105, 1560, 2765, 3040, 3060
Acetone	420 (m), 790 (vs), 1070 (m), 1225 (m), 1430 (m), 1710 (m), 2920 (s)	390, 490, 900, 1365, 2845, 2965
Benzene	605 (m), 990 (vs), 1180 (m), 1580–1610 (w), 3040–3070 (m)	405, 780, 825, 850, 2620, 2950
Cyclohexane	385 (w), 430 (w), 800 (s), 1030 (m), 1160 (w), 1265 (m), 1440–1470 (m), 2855 (m), 2920–2950 (m)	1345, 2630, 2670, 2700, 2905
Ethanol	880 (s), 1450–1490 (m)	430, 1050, 1095, 1275, 2875, 2930, 2975
Dimethyl formamide	320 (w), 360 (m), 410 (m), 660 (s), 870 (s), 1095 (m), 1405 (s), 1440 (m), 1660 (m)	1065, 2800, 2860, 2930
Dimethyl sulfoxide	300–350 (s), 385 (m), 650–710 (vs), 955 (w), 1045 (s), 1420 (m), 2915 (m), 3000 (w)	2885
Distilled water	3450 (m, br), 1650 (w)	—

*Key: s = strong; m = medium; w = weak; v = very; br = broad.

Further details of Raman sampling methods may be found in Chapter 15 of reference 7.5 and in Section 5.4 of reference 7.9.

7-8c Depolarization Measurements

Raman spectra can be used to classify molecular vibrations as either totally symmetric or non-totally symmetric, with respect to the elements of symmetry that the molecule possesses. This distinction is accomplished by means of depolarization measurements, which can be made using the polarization analyzer shown in Figure 7-9. Raman scattering is produced by the interaction between the radiation from the plane polarized laser beam and the change in the polarizability of the molecule associated with the molecular vibrations. As a result of this interaction some of the scattered light is no longer plane polarized. Hence there is a difference in intensity between the light transmitted through the analyzer when it is oriented parallel to the plane of polarization of the laser beam ($I_{\parallel}$) and when it is turned through 90° into the perpendicular orientation ($I_{\perp}$). The ratio $I_{\perp}/I_{\parallel}$ is known as the depolarization ratio ρ and it has a theoretical maximum value of 0.75. A band with ρ less than 0.75 is said to be polarized and a band with ρ exactly equal to 0.75 is said to be depolarized. The theory of the depolarization of Raman lines is presented in detail in reference 7.14.

Some molecular vibrations are symmetric with respect to all elements of symmetry that the molecule possesses. These are called totally symmetric modes of vibration, and they give rise to polarized Raman bands (ρ lies between 0 and 0.75). Other vibrations are antisymmetric with respect to some of the symmetry elements and are known as non-totally symmetric vibrations. These modes give rise to depolarized Raman bands ($\rho = 0.75$). Hence depolarization ratios can be used to identify the totally symmetric vibrations of a molecule.

To make depolarization measurements the spectrum is scanned twice with identical instrument settings. The only difference is the orientation of the analyzer (parallel and perpendicular). Peak heights (I) and baselines (I_0) are noted for each orientation, and depolarization ratios are calculated from eq. 7-11. It should be noted that depolarization ratios cannot be measured for powdered solid samples.

$$\rho = \frac{(I - I_0)_{\perp}}{(I - I_0)_{\parallel}} \tag{7-11}$$

Depolarization measurements may be used to separate two overlapping bands, when one is due to a totally symmetric vibration. The intensity of a band due to a totally symmetric vibration is usually drastically reduced on rotating the analyzer through 90°, whereas the intensity of a band due to a non-totally symmetric vibration is reduced only by 25%. Hence the band due to the non-totally symmetric mode predominates in the perpendicular spectrum.

7-9 SPECIAL TECHNIQUES IN INFRARED SPECTROSCOPY

Special techniques have been developed specifically for FT–IR instruments. These are outlined in this section together with some other techniques that can be used with either dispersive or FT spectrometers.

7-9a Gas Phase Infrared Spectroscopy

The gas or vapor phase spectrum of a molecule recorded under low resolution gives essentially the same information as the liquid or solution spectrum. In most cases, rotation of the molecules causes only some broadening of peaks. Gases are usually studied in a 10 cm glass cell with alkali halide windows at pressures of between 10 and 100 mm Hg. A vacuum line is needed to handle a gas or the vapor of a volatile liquid. Less volatile liquids or solids can be studied in the vapor phase if the cell is warmed with a heating tape until sufficient pressure of vapor is produced. If the vapor pressure or concentration of a gas is very low, the radiation can be reflected several times through the sample using special multiple-pass gas cells. These cells have effective pathlengths of up to 40 meters. Details of these cells can be found in reference 7.5.

7-9b   Microsampling

For infrared microsampling of liquids with a standard FT–IR spectrometer, special small-volume cells are available. Very small KBr discs also can be prepared. Two types of microcells are in common use: the cavity cell and a miniaturized version of the standard liquid cell. The former type consists of a small block of material such as NaCl, KBr, etc., with parallel polished faces and a microcavity drilled ultrasonically into the center. Various volumes are available, from a fraction of a microliter up to 0.5 ml. Standard pathlengths in the range 0.05 to 5 mm can be obtained. The second kind of microcell is usually purchased assembled, sealed, and with a fixed pathlength. Because of the small area exposed to the infrared beam, a beam condenser is often used with microcells.

Microgram quantities of solid samples can be pressed into KBr pellets using a special die. The pellets are pressed into openings centered in stainless steel discs. Pellets from 0.5 mm up to 13 mm in diameter can be made. The 0.5 mm pellet requires about 1 μg of sample mixed with 1 mg of KBr.

Infrared microscopes can give spatial resolution as low as 10 μm and very small amounts of sample can be examined. Spectra of fibers and individual components of physical mixtures of solids can be recorded. However, these instruments are sophisticated, very expensive, and not usually available to students.

7-9c   Infrared Reflection Spectroscopy

When a ray of light strikes an interface between two nonabsorbing materials of different refractive index (n_1 and n_2), the light is partially transmitted and partially reflected. This property is used in the manufacture of beam splitters for FT–IR spectrometers. When light enters material 2 from 1 with n_1 less than n_2 (for example air to glass), the result is *external reflection*. For external reflection the reflectivity can never be 100% and is usually much less. In the case when n_1 is greater than n_2, we have *internal reflection,* and in this case the reflection is total when the angle of incidence is between the critical angle and 90° (grazing incidence). Both internal and external reflection can be used to obtain spectra.

Total internal reflection can be observed in a glass of water. When the inside of the glass is viewed through the water surface, it appears to be completely silvered and opaque. However, when the outside of the glass is touched with a finger, details of the ridges and whorls on the skin are clearly seen, but the silvered effect remains between these features. The total reflection is destroyed where the skin actually makes contact with the glass. This result can be explained by penetration of the electromagnetic field of the light into the rarer medium (n_2 smaller refractive index) by a fraction of a wavelength, as illustrated in Figure 7-10.

If the light is in the infrared region and if the rarer medium is a compound that absorbs infrared radiation, then the penetrating radiation field can interact by means of

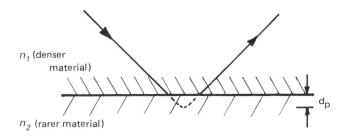

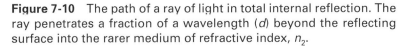

Figure 7-10 The path of a ray of light in total internal reflection. The ray penetrates a fraction of a wavelength (*d*) beyond the reflecting surface into the rarer medium of refractive index, n_2.

PROBLEMS

7-1 Calculate the vibrational frequency (s^{-1}) of a system consisting of two balls of mass 1.66×10^{-27} kg connected by a spring with force constant 510 N m^{-1}.

7-2 A strong infrared absorption band has a peak maximum at 7% T. What is the absorbance of the peak?

7-3 The infrared spectrum of a 0.20 M solution of a compound in CCl_4 solution was recorded using a variable path cell with a pathlength of 0.50 mm. An absorption band was observed at 1730 cm^{-1}, with a peak absorbance of 1.05 absorbance units. The spectrum of a CCl_4 solution of unknown concentration of the same substance was recorded in the same cell but at a pathlength of 0.15 mm. The absorbance of the 1730 cm^{-1} peak was 0.65 absorbance units. Calculate the concentration of the unknown solution.

7-4 Calculate the relative intensity of light scattered at 514.5 nm to that scattered at 1064.8 nm.

7-5 List the advantages of Fourier transform over dispersive infrared spectroscopy. How have these advantages made FT–Raman spectroscopy possible?

7-6 Give reasons why FT–Raman spectroscopy is carried out in the near- infrared (10,000–6000 cm^{-1}) rather than the mid-infrared (4000–400 cm^{-1}) region of the spectrum.

BIBLIOGRAPHY

7.1. J.B. Lambert, H.F. Shurvell, D. Lightner, and R.G. Cooks, *Introduction to Organic Spectroscopy*, Macmillan Publishing, New York, 1987.

7.2. J.R. Durig, ed., *Analytical Applications of FT–IR to Molecular and Biological Systems*, Reidel Publishing Co., Norwell, MA, 1980.

7.3. P.R. Griffiths and J.A. de Haseth, *Fourier Transform Infrared Spectrometry*, John Wiley & Sons, Inc., New York, 1986.

7.4. A.E. Martin, *Infrared Interferometric Spectrometers*, vol. 8 of J.R. Durig, ed., *Vibrational Spectra and Structure*, Elsevier Science Inc., Amsterdam, 1980.

7.5. H.A. Willis, J.H. Van Der Maas, and R.G.J. Miller, eds., *Laboratory Methods in Vibrational Spectroscopy*, 3rd ed., John Wiley & Sons Ltd., Chichester, UK, 1987.

7.6. T.R. Gilson and P.J. Hendra, *Laser Raman Spectroscopy*, Wiley–Interscience, London, 1970.

7.7. D.A. Long, *Raman Spectroscopy*, McGraw-Hill Book Co., London, 1977.

7.8. F.J. Bergin and H.F. Shurvell, *Appl. Spectrosc.* **43,** 516 (1989).

7.9. P. Hendra, C. Jones, and G. Warnes, *Fourier Transform Raman Spectroscopy*, Ellis Horwood, Chichester, UK, 1991.

7.10. D.B. Chase and J.F. Rabolt, eds., *Fourier Transform Raman Spectroscopy*, Academic Press, New York, 1994.

7.11. J.R. Ferraro and K. Nakamoto, *Introductory Raman Spectroscopy*, Academic Press, New York, 1994.

7.12. R.L. Frost, P.M. Fredericks, and H.F. Shurvell, *Can. J. Appl. Spectrosc.*, **41,** 10 (1996).

7.13. M.F.C. Ladd, *Structure Determination by X-Ray Crystallography*, 3rd ed., Plenum Press, New York, 1993.

7.14. L.A. Woodward, *Introduction to the Theory of Molecular Vibrations and Vibrational Spectroscopy*, Oxford University Press, London, 1972.

7.15 N.J. Harrick, *Internal Reflection Spectroscopy*, Interscience, New York, 1967; *Review and Supplement*, Harrick Scientific Corp., Ossining, N.Y., 1985.

7.16 *Optical Spectroscopy: Sampling Techniques Manual*, Harrick Scientific Corp., Ossining, N.Y., 1987.

7.17 P.R. Griffiths and M.P. Fuller, *Advan. Infrared Raman Spectrosc.*, **9,** 63 (1982).

7.18 R. White, *Chromatography/Fourier Transform Infrared and Its Applications*, Marcel Dekker, New York, 1990.

chapter 8

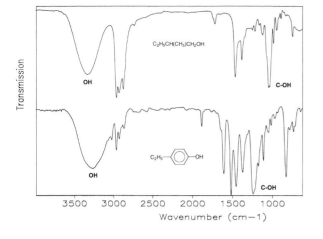

Group Frequencies: Infrared and Raman

8-1 INTRODUCTION TO GROUP FREQUENCIES

Although subject of group frequencies is essentially empirical in nature, it has a sound theoretical basis. Infrared (IR) and Raman spectra of a large number of compounds containing a particular functional group, such as carbonyl, amino, phenyl, nitro, etc., have certain features that appear at generally the same frequency for every compound containing the group. It is reasonable, then, to associate these spectral features with the functional group, provided a sufficiently large number of different compounds containing the group have been studied. For example, the infrared spectrum of any compound that contains a C=O group has a strong band between 1800 and 1650 cm^{-1}. Compounds containing —NH$_2$ groups have two infrared bands between 3400 and 3300 cm^{-1}. The Raman spectrum of a compound containing the C$_6$H$_5$— (phenyl) group has a strong polarized line near 1000 cm^{-1}, and nitro groups are characterized by infrared and Raman bands near 1550 and 1350 cm^{-1}. These are just four examples of the many characteristic frequencies of chemical groups observed in infrared and Raman spectra.

Various pairs of atoms joined by bonds in a molecule can be treated as diatomic molecules. This simple approach gives surprisingly good results when one of the atoms of the pair is a light atom, not bonded to any other atom, for example, C—H and N—H in CH$_3$NH$_2$ or C—H and C=O in (CH$_3$)$_2$CO. The stretching frequencies in cm^{-1} (see Section 7-4 for a discussion of units) of these diatomic groups can be calculated from eq. 8-1, in which k is the force constant (N m^{-1}) and μ is the reduced mass, $m_1 m_2/(m_1 + m_2)$

$$\nu(\text{cm}^{-1}) = 130.3 \sqrt{\frac{k}{\mu}} \tag{8-1}$$

in atomic mass units. The numerical constant $130.3 = 1/2\pi c\sqrt{N} \times 10^{-1}$ (N is Avogadro's number, 6.022×10^{23}, and c is the velocity of light, 2.998×10^8 m s^{-1}). In older texts and papers, force constants have units of mdyn Å^{-1} (10^5 dyn cm^{-1}) and c has units of cm s^{-1}. When these units are used, the numerical constant of eq. 8-1 becomes 1303.

Frequencies of numerous diatomic groups including C=C and C≡C can be calculated from eq. 8-1. A band characteristic of the group is observed in the infrared or

TABLE 8-1

Calculated Frequencies of Some Diatomic Groups

Group	Reduced Mass (amu)	Force Constant ($N\ m^{-1}$)	Frequency (cm^{-1})
O—H	0.94	700	3600
N—H	0.93	600	3300
C—H	0.92	500	3000
C—C	6.00	425	1100
C=C	6.00	960	1650
C=O	6.86	1200	1725
C≡C	6.00	1600	2100
C≡N	6.46	2100	2350

Raman spectrum in the predicted region, provided the vibrational frequency of the group is not close to that of another group in the molecule. Some diatomic group frequencies calculated from eq. 8-1 are given in Table 8-1. These characteristic group frequencies can be used to establish the presence of the functional group in the molecule.

Table 8-1 shows that the values of the force constants of double and triple bonds are approximately twice and four times that of single bonds, respectively. Carbon–carbon single bonds are included in Table 8-1, but C—C stretching does not usually give a well-defined group frequency. Most organic molecules contain several C—C single bonds and other groups that have vibrational frequencies similar to that of the C—C stretching mode. These vibrations interact with each other, and the simple model (eq. 8-1) does not apply. Vibrational interactions can take several forms and are discussed in Sections 8-2b and 8-2c.

When two or more identical diatomic groups are present in a molecule, one might expect to observe a band for each group at a similar frequency in the IR spectrum. These bands are often not resolved. However, if the groups are attached to the same carbon atom or to two adjacent atoms, the frequencies may be spread over a few hundred wavenumbers by strong interactions. For example, in the ethylene molecule (C_2H_4) there are four CH groups and one might expect to find four frequencies near 3000 cm^{-1} due to C—H stretching vibrations. However, the CH groups are attached to the same or adjacent carbon atoms and the four observed CH stretching frequencies are 3270, 3105, 3020, and 2990 cm^{-1}.

The explanation of these characteristic diatomic group frequencies lies in the approximately unchanging values of the stretching force constant of a group in different molecules. Polyatomic groups such as —CH_2, —CH_3, —NH_2, or —C_6H_5 also have characteristic vibrational frequencies that involve both stretching and bending vibrations or combinations of these. Unfortunately, no simple mathematical relationship such as eq. 8-1 can be found to calculate the vibrational frequencies of polyatomic groups. However, examination of the vibrational spectra of a large number of compounds containing polyatomic groups has established frequency ranges for vibrations of these groups.

Raman spectroscopy has not been used routinely to determine the structure of organic compounds. However, the recent introduction of FT–Raman instruments may change this situation, and so some discussion of Raman group frequencies is necessary in a text on organic spectroscopy. Group frequencies in the Raman generally are not the same as those in the infrared, and Raman spectra can provide important additional information on molecular structure and symmetry. An integrated discussion of infrared and Raman group frequencies is presented here to give the reader a more complete perspective on the subject.

8-2   FACTORS AFFECTING GROUP FREQUENCIES

Symmetry, mechanical coupling, Fermi resonance, hydrogen bonding, steric effects, electronic effects, isomerism, physical state, and solvent and temperature effects all contribute to the position, intensity, and appearance of the bands in the infrared and Raman spectra of a compound.

Lowering the temperature usually makes the bands sharper and better resolution can be achieved, especially in solids at very low temperatures. However, there is a possibility of splittings due to crystal effects, which must be considered when examining the spectra of solids under moderately high resolution. Polar solvents can cause significant shifts of group frequencies through solvent-solute interactions, such as molecular association through hydrogen bonding. The effects of the various items listed above on group frequencies are discussed in the next few sections.

8-2a Symmetry

The vast majority of organic molecules have little or no symmetry. Nevertheless, some knowledge of symmetry can be of considerable help in understanding the factors that affect intensities of group frequencies.

For a vibration to give rise to absorption of infrared radiation (to be active), it must be associated with an oscillating electric dipole. A similar statement may be made for Raman active vibrations. In this case the vibration must give rise to a change in the polarizability of the molecule. This change in turn gives rise to an induced dipole through interaction with the electric field of the incident laser radiation. Some vibrations are inactive in the infrared or Raman, usually as consequence of symmetry.

If a molecule possesses a center of symmetry, it has no permanent dipole moment and a vibration that is symmetric with respect to the center of symmetry (a symmetric mode) does not generate an oscillating dipole. This vibration is inactive (does not absorb) in the infrared. However, a vibration that is antisymmetric with respect to the center of symmetry (an antisymmetric mode) produces a transient oscillating dipole moment that interacts with the electric field of the radiation. This vibration is active (absorbs) in the infrared. The symmetric modes, however, are always active (give rise to scattering) in the Raman spectrum, because they produce a change in the polarizability of the molecule. On the other hand the antisymmetric modes are not active in the Raman. This *mutual exclusion rule* between infrared and Raman activity is illustrated by the spectra of *trans*-1,2-dichloroethene (Figure 8-1).

The C=C stretching vibration of *trans*-1,2-dichloroethene is not observed in the infrared, but is seen at 1580 cm^{-1} in the Raman spectrum. The C—Cl stretches give rise to two vibrational modes, a symmetric mode observed in the Raman spectrum at 840

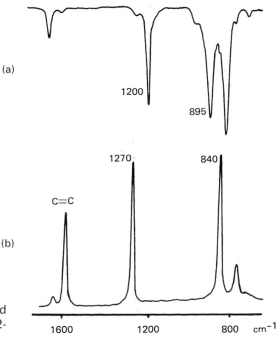

Figure 8-1 Portions of the (a) infrared and (b) Raman spectra of *trans*-1,2-dichloroethene.

cm^{-1} and an antisymmetric mode in the infrared at 895 cm^{-1}. Similarly, the two CH bending modes are observed at 1200 cm^{-1} (infrared) and 1270 cm^{-1} (Raman).

Another example of the effect of a center of symmetry is given by the C≡C stretching mode. For methylacetylene (CH$_3$C≡CH) the vibration is both infrared and Raman active, and infrared and Raman bands are observed at 2150 cm^{-1}. On the other hand, in dimethylacetylene (CH$_3$C≡CCH$_3$), which has a center of symmetry, the exclusion rule applies. A strong C≡C stretching band is found in the Raman spectrum near 2150 cm^{-1}, but no band is observed in the infrared spectrum at this frequency.

Some molecules have threefold or higher axes of symmetry, as well as mirror planes or other symmetry elements (see reference 8.14 for a full discussion of symmetry elements). Such molecules are said to have high symmetry and have simple infrared and Raman spectra. For example, consider the benzene molecule, which has a sixfold axis of symmetry perpendicular to the plane of the ring and many other symmetry elements. It has 12 atoms and therefore has $3N - 6 = 30$ normal modes of vibration. The first effect of the high symmetry is that 10 pairs of these vibrations have identical frequencies (degenerate modes), leaving 20 different normal frequencies. The second effect of the high symmetry is to reduce the number of modes for which there is a change in dipole moment (infrared active) or a change in polarizability (Raman active). In fact, the infrared spectrum of benzene contains only four bands due to fundamentals (defined in Section 7-3), whereas the Raman spectrum contains only six.

The symmetry of the benzene molecule is reduced by substitution, as in 1,3,5-trichlorobenzene, in which the sixfold axis is replaced by a threefold axis. For this molecule the number of infrared and Raman active modes is greater than for benzene, but there are still some degenerate modes and the spectra are relatively simple. When the symmetry is further lowered, as in 1-chloro-2-bromobenzene, which has only the plane of the molecule as a symmetry element, all 30 normal modes are active in both infrared and Raman. However, because of the symmetry of the benzene ring itself, some of these vibrations appear only very weakly and are hard to distinguish from the weak bands due to overtones and combinations.

In larger, more complicated molecules, a local symmetry may exist for a homonuclear diatomic group such as C=C or S—S, so that the infrared absorption from the group vibration may be weak or absent. In such cases a Raman spectrum can confirm the presence (or absence) of the functional group.

The vibrations of a methylene group (CH$_2$) can be described in terms of the local symmetry of the group, which has a twofold axis and two planes of symmetry. Figure 8-2 shows the vibrations associated with a CH$_2$ group when it is attached to a molecule. An isolated CH$_2$ group has three modes, the symmetric and antisymmetric (with respect to the twofold axis), stretching, and the bending (scissors) vibrations. When the group is part of a larger molecule, three additional modes described as twisting, wagging, and rocking are produced. Of these, the twisting mode produces no change in dipole moment and hence is not observed in the infrared. However, it can give rise to a very weak band in the spectrum of an unsymmetrical molecule.

The vibrations of the methyl group (—CH$_3$) in a molecule also can be described in terms of the symmetry of the methyl group itself, which has a threefold axis and three planes of symmetry. An isolated CH$_3$ group would have $3N - 6 = 6$ normal modes of vibration comprising symmetric and degenerate pairs of stretching and bending modes.

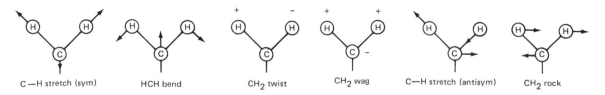

C—H stretch (sym)   HCH bend   CH$_2$ twist   CH$_2$ wag   C—H stretch (antisym)   CH$_2$ rock

Figure 8-2   The vibrations of a CH$_2$ group. The arrows show the direction of motion of atoms in the plane of the CH$_2$ group, while the + and − signs denote motion above and below the plane, respectively.

When the CH_3 group is attached to a molecule, three new modes appear, a torsional mode and a degenerate pair of rocking vibrations. These motions would be rotations in the isolated methyl group. Thus there are four regions of the spectrum in which we expect to find group vibrations of the methyl group. This conclusion is amply documented. The methyl group also contributes three skeletal modes to the vibrations of the molecule. These modes correspond to translations of the free methyl group.

When the methyl group is part of a molecule with low symmetry, the degeneracies are removed, leading to the observation of doublets in some of the regions of the spectrum where the methyl group frequencies are expected. The methyl torsional mode is expected to be inactive in the infrared, because it produces no change in dipole moment. However, due to the low symmetry of the whole molecule, methyl torsions are sometimes observed as weak bands in the far infrared.

The terms twofold, threefold, and sixfold axis, plane of symmetry, and center of symmetry are examples of *symmetry elements*. The collection of all symmetry elements that a molecule possesses is known as a *point group* and provides a way of classifying the symmetry of the molecule. This, in turn, leads to an understanding of the symmetry of the normal vibrations of a molecule and to a prediction of the number of frequencies expected in the infrared and Raman spectra. The student is urged to consult references 8.14 and 8.15 for further information on the applications of symmetry and group theory to vibrational spectroscopy.

8-2b Mechanical Coupling of Vibrations

Two completely free identical diatomic molecules, of course, vibrate with identical frequencies. When the two diatomic groups are part of a molecule, however, they can no longer vibrate independently of each other because the vibration of one group causes displacements of the other atoms in the molecule. These displacements are transmitted through the molecule and interact with the vibration of the second group. The resulting vibrations appear as in-phase and out-of-phase combinations of the two diatomic vibrations. When the groups are widely separated in the molecule, the coupling is very small and the two frequencies may not be resolved. The two C—H stretching modes in acetylene (H—C≡C—H) are observed at 3375 cm^{-1} in the Raman spectrum (in phase) and 3280 cm^{-1} in the infrared (out of phase). In diacetylene (H—C≡C—C≡C—H), the two C—H stretching vibrations have closer frequencies, near 3330 and 3295 cm^{-1}.

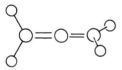

Figure 8-3 The allene molecule.

Figure 8-3 illustrates how mechanical coupling occurs in two C=C groups coupled through a common atom, as in the allene molecule, CH_2=C=CH_2. In the absence of strong coupling one might expect to observe a band in the infrared spectrum near 1600 cm^{-1} from the out-of-phase (unsymmetrical) vibrations of the C=C groups and a line in the Raman spectrum from the in-phase (symmetrical) modes at a similar frequency. For the 1,3-butadiene molecule (CH_2=CH—CH=CH_2), these bands are, in fact, observed at 1640 cm^{-1} in the infrared and 1600 cm^{-1} in the Raman. For allene, however, the observed frequencies are near 1960 and 1070 cm^{-1}. This result can be understood in terms of mechanical coupling of the two C=C group vibrations. When such coupling occurs, it is usually found that the higher frequency mode is the antisymmetric (out-of-phase) vibration and the lower frequency mode is the symmetric (in phase) vibration.

The vibrations of two different diatomic groups are not coupled unless the uncoupled frequencies are similar as the result of a combination of force constant and mass effects. For example, in thioamides and xanthates, the C=S group has a force constant of about 650 N m^{-1} and the reduced mass is 8.72 amu, so that the vibrational frequency calculated from eq. 8-1 is approximately 1120 cm^{-1}. The C—N and C—O groups have force constants of about 480 and 510 N m^{-1}, respectively, and the reduced masses are 6.46 and 6.86 amu. The calculated frequencies are both approximately 1120 cm^{-1}. Consequently, in any compound containing a C=S group adjacent to a C—O or C—N group, there may be an interaction between the stretching vibrations of the groups. In compounds such as thioamides and xanthates, for which the carbon atom is common to both groups, the coupling is large and the two vibrations interact with each other to produce two new frequencies, neither of which is in the expected region of the spectrum.

It is possible for coupling to occur between dissimilar modes such as stretching and bending vibrations when the frequencies of the vibrations are similar and the two groups involved are adjacent in the molecule. An example is found in secondary amides, in which the C—N stretching vibration is of a similar frequency to that of the NH bending mode. Interaction of these two vibrations gives rise to two bands in the spectrum, one at a higher and one at a lower frequency than the uncoupled frequencies. These bands are known as amide II and amide III bands. (The C=O stretching mode is known as the *amide I band*).

Singly bonded carbon atom chains are, of course, not linear, so the simple model used for the allene molecule must be modified. In addition, we ignored the bending of the C=C=C group in allene, which cannot couple with the stretching modes, because it takes place at right angles to the stretching vibrations. Mechanical coupling always occurs between C—C single bonds in an organic molecule, and so there is no simple C—C group stretching frequency. One can expect that there will always be several bands in the infrared and Raman spectra in the 1200–800 cm^{-1} range in compounds containing saturated carbon chains. Certain branched chain structures, such as the *tert*-butyl group, $(CH_3)_3C$—, and the isopropyl group, $(CH_3)_2CH$—, do have characteristic group frequencies involving the coupled C—C stretching vibrations. These systems are discussed further in later sections.

In many molecules, mechanical coupling of the group vibrations is so widespread that few, if any, frequencies can be assigned solely to functional groups. Many such examples are found in aliphatic fluorine compounds, in which the CF and CC stretching modes are coupled with each other and with FCF and CCF bending vibrations. The presence of fluorine can be deduced from several very strong infrared bands in the region between 1400 and 900 cm^{-1}. These vibrations give very weak Raman lines.

8-2c   Fermi Resonance

A special case of mechanical coupling, known as *Fermi resonance,* often occurs. This phenomenon, which results from coupling of a fundamental vibration with an overtone or combination, can shift group frequencies and introduce extra bands. For a polyatomic molecule there are $3N - 6$ energy levels for which only one vibrational quantum number (v_i) is 1 when all the rest are zero. These are called the fundamental levels and a transition from the ground state to one of these levels is known as a *fundamental.* In addition, there are the levels for which one v_i is 2, 3, etc. (*overtones*), or for which more than one v_i is nonzero (*combinations*). There are, therefore, a very large number of vibrational energy levels, and it quite often happens that the energy of an overtone or combination level is very close to that of a fundamental. This situation is termed *accidental degeneracy,* and an interaction known as Fermi resonance can occur between these levels, provided that the symmetries of the levels are the same. Since most organic molecules have no symmetry, all levels have the same symmetry and Fermi resonance effects occur frequently in vibrational spectra.

Normally, an overtone or combination band is very weak in comparison with a fundamental, because these transitions are not allowed. However, when Fermi resonance occurs, there is a sharing of intensity and the overtone can be quite strong. The result is the same as that produced by two identical groups in the molecule. As an example, two peaks are observed in the carbonyl stretching band of benzoyl chloride, near 1760 and 1720 cm^{-1} (Figure 8-4). If this were an unknown compound, one might be tempted to suggest that there were two nonadjacent carbonyl groups in the molecule. However, the lower frequency band is due to the overtone of the CH out-of-plane bending mode at 865 cm^{-1} in Fermi resonance with the C=O stretching fundamental.

Numerous other well-characterized examples of Fermi resonance are known. The N—H stretching mode of the —CO—NH— group in polyamides (nylons), peptides, proteins, etc., appears as two bands near 3300 and 3205 cm^{-1}. The N—H stretching fundamental and the overtone of the N—H deformation mode near 1550 cm^{-1} combine through Fermi resonance to produce the two observed bands. The CH stretching region of the —CHO group in aldehydes provides another example of Fermi resonance. Two

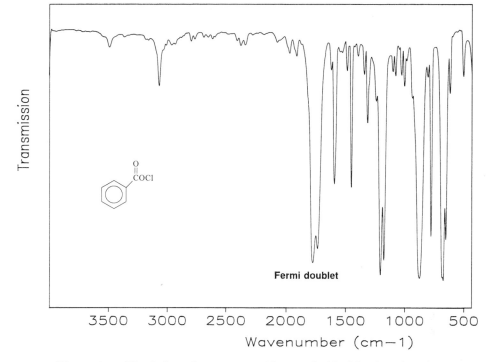

Figure 8-4 The infrared spectrum of benzoyl chloride showing the Fermi doublet at 1760–1720 cm^{-1}.

bands often are observed near 2900 and 2700 cm^{-1} in the infrared spectra of aldehydes (see Figure 9-9). This doubling is attributed to Fermi resonance between the overtone of the C—H deformation mode, which would have a frequency near 2×1400 cm^{-1}, and the C—H stretching mode, which would also occur near 2800 cm^{-1} in the absence of Fermi resonance.

8-2d Hydrogen Bonding

Hydrogen bonding (written X—H$\cdots$Y) occurs between a hydrogen atom bonded to an electronegative element X, as in OH or NH, and an atom Y possessing one or more nonbonding electron pairs, usually O or N. The main effects on the infrared and Raman spectra are broadening of bands in the spectra and shifts of group frequencies. X—H stretching frequencies are lowered by hydrogen bonding, and X—H bending frequencies are raised. Hydrogen bonding also affects the frequencies of the acceptor group, but the frequency shifts are less than those of the X—H group. Solvents such as CCl$_4$, which do not interact with the solute, can reduce the extent of hydrogen bonding and even eliminate the effect in very dilute solutions.

Hydrogen bonding manifests itself in very broad OH and NH stretching bands at frequencies considerably lower than those of the unbonded groups. Changes in the intensity of these bands can be brought about by changes in temperature and concentration, both of which affect the degree of H-bonding. A very broad band centered near 3100 cm^{-1} in the spectrum of a pure carboxylic acid is due mainly to OH stretching of hydrogen-bonded carboxylic acid oligomers. In solutions of carboxylic acids in non-hydrogen bonding solvents such as CCl$_4$, the presence of monomer, dimer, and oligomeric species can be identified by bands in the carbonyl stretching region of the infrared spectrum (see reference 8.16). Monomer-dimer equilibria have been studied for a variety of molecules such as phenols, alcohols, and compounds that self-associate to form hydrogen-bonded species.

In addition to the hydrogen-bonded OH and NH stretching bands between 3500 and 2500 cm^{-1}, the R—OH or R—NH bending modes also can be observed between 1700 and 1000 cm^{-1}. The torsional vibration motion of the R—OH or R—NH bond gives rise to absorption between 900 and 300 cm^{-1}. Stretching of the hydrogen bond itself ($\mathrm{XH}\cdots\mathrm{Y}$) has been observed in the far infrared in many cases between 200 and 50 cm^{-1}, and bending of the hydrogen bond occurs at very low frequencies, usually below 50 cm^{-1}.

Intramolecular hydrogen bonding can occur between OH groups in alcohols or phenols and halogen atoms. In 2-chloroethanol, for example, an intramolecular hydrogen bond stabilizes the gauche rotational isomer (see structure **8-4**, Section 8-2i). The free v_{OH} in the trans conformation absorbs at 3623 cm^{-1}, whereas for the hydrogen-bonded isomer the frequency is 3597 cm^{-1}. Halophenols also show two v_{OH} bands separated by 50–100 cm^{-1} due to bonded and nonbonded conformations. Infrared spectra also indicate that intermolecular OH$\cdots$halogen bonding occurs between alkyl halides and phenols or alcohols.

In this section only a very brief mention has been made of some of the cases in which hydrogen bonding is found. The reader is referred to the books by Bellamy (reference 8.2) and Pimentel and McClellan (reference 8.17) for further discussion and references to the literature of the subject.

8-2e   Ring Strain

The effect of ring strain on group frequencies is quite interesting and useful for diagnostic purposes. As an example, consider the series of alicylic ketones: cyclohexanone, cyclopentanone, and cyclobutanone. The infrared spectra are shown in Figure 8-5. The observed carbonyl stretching frequencies are 1714, 1746, and 1783 cm^{-1}. The increase in $v_{\mathrm{C=O}}$ is attributed to mechanical interaction with the adjacent co-planar C—C single bonds, which changes as the double bond–single bond angle changes (see reference 8.18). A detailed discussion of the effects of bond angle changes in C—CO—C systems is given in Section 5.4(a) of reference 8.2.

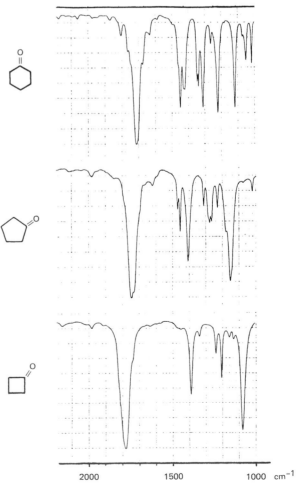

Figure 8-5   Portions of the infrared spectra of three cyclic ketones showing the influence of increasing ring strain on the C=O stretching frequency. (Reproduced with permission of Aldrich Chemical Co., Inc., from *The Aldrich Library of FT–IR Spectra*).

The increase in frequency with increasing angle strain is generally observed for double bonds directly attached (exocyclic) to rings. Frequency changes similar to those observed for cyclic ketones are found in the series of compounds, methylenecyclohexane, methylenecyclopentane, and methylenecyclobutane, in which a C=CH$_2$ group replaces the C=O group. The spectra are shown in Figure 8-6. The observed C=C stretching frequencies are 1649, 1656, and 1677 cm^{-1}, respectively.

In contrast, when the double bond is endocyclic, a decrease in the ring angle causes a *lowering* of the C=C stretching frequency. The observed frequencies for cyclohexene, cyclopentene, and cyclobutene are 1650, 1615, and 1565 cm^{-1}, respectively.

Unlike C=O and C=C groups, the P=O, S=O, and SO$_2$ groups have stretching frequencies that are little affected by being part of a strained ring. This is because the double bond in these groups is not coplanar with the two attached single bonds.

8-2f Electronic Effects

Effects arising from the change in the distribution of electrons in a molecule produced by a substituent atom or group can often be detected in the vibrational spectrum. There are several mechanisms, such as inductive and resonance effects, that can be used to explain observed shifts and intensity changes in a qualitative way. These effects involve changes in electron distribution in a molecule and cause changes in the force constants that are, in turn, responsible for changes in group frequencies. Inductive and resonance effects have been used successfully to explain the shifts observed in C=O stretching frequencies produced by various substituent groups in compounds such as acid chlorides and amides. High C=O stretching frequencies are usually attributed to inductive effects, and low frequencies arise when delocalized structures are possible. For example, in acid chlorides the C=O frequency is near 1800 cm^{-1}, which is high compared

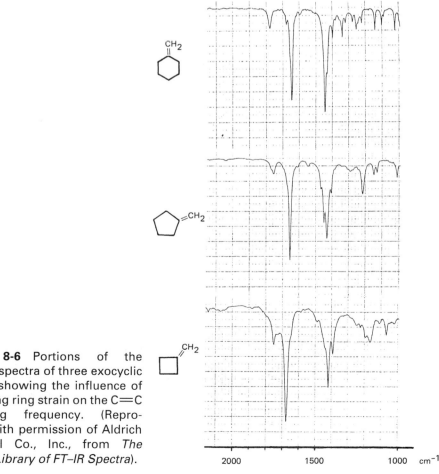

Figure 8-6 Portions of the infrared spectra of three exocyclic alkenes showing the influence of increasing ring strain on the C=C stretching frequency. (Reproduced with permission of Aldrich Chemical Co., Inc., from *The Aldrich Library of FT–IR Spectra*).

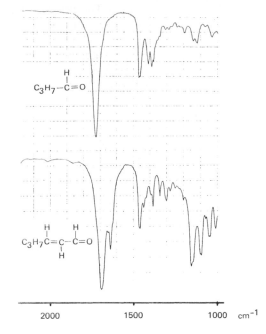

Figure 8-7 Portions of the infrared spectra of two aliphatic aldehydes, showing the influence of conjugation on the C=O stretching frequency. (Reproduced with permission of Aldrich Chemical Co., Inc., from *The Aldrich Library of FT–IR Spectra*).

with that observed for aldehydes or ketones (1730 cm^{-1}). On the other hand, in amides the carbonyl frequency is lower (near 1650 cm^{-1}). An example of this behavior was shown earlier in Figure 7-4. In acid chlorides the electronegative chlorine atom adjacent to the carbonyl group increases electron density in the double bond, raises the C=O stretching force constant, and causes an increase in frequency, whereas in amides the delocalized electronic structure (**8-1** ⟷ **8-2**) lowers the force constant and leads to a decrease of the C=O stretching frequency.

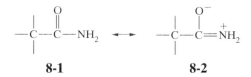

8-1   **8-2**

Conjugation of double bonds tends to lower the double bond character and increase the bond order of the intervening single bond. This result is seen by comparing the two C=C stretching frequencies of isoprene (2-methyl-1,3-butadiene), 1637 and 1604 cm^{-1}, with the C=C stretching frequency of 1,4-pentadiene, 1644 cm^{-1}. For compounds in which a carbonyl group can be conjugated with an alkenic double bond, the C=O stretching frequency is lowered by 30–20 cm^{-1}, as illustrated in Figure 8-7, in which parts of the spectra of two aldehydes are compared.

8-2g Constitutional Isomerism

Constitutional (structural) isomers differ in the way in which their atoms are connected. Different functional groups are present, and the vibrational spectra of the isomers differ considerably. Some examples are the α-amino acid alanine, the urethane ethyl carbamate, and the nitro compound 1-nitropropane. All three compounds have the empirical formula $C_3H_7O_2N$. Infrared spectra of these three compounds are shown in Figure 8-8.

Vibrational spectra are very useful in distinguishing ortho-, meta-, and para-disubstituted benzenes. The CH out-of-plane deformation patterns in the 850–700 cm^{-1} region are different for each isomer (see Section 9-5b for further details). Substituted pyridines, pyrimidines, and other heterocyclic compounds exist as various structural isomers that can be distinguished by their vibrational spectra.

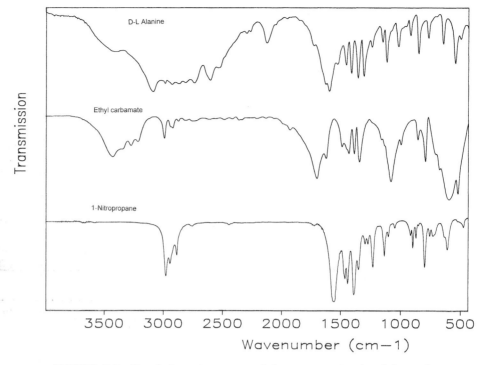

FIGURE 8-8 The infrared spectra of three compounds of formula $C_3H_7O_2N$.

8-2h Stereoisomerism

Stereoisomers can be classified as optical isomers (enantiomers) and geometric or cis-trans isomers (diastereomers). Pure enantiomers are nonsuperimposable mirror images. However, the arrangements of the atoms in the two forms relative to each other are the same, and the molecular vibrations are identical. Therefore, the infrared and Raman spectra of a pair of enantiomers are identical and the same as the spectra of the racemate.

The absence of rotation about carbon–carbon double bonds gives rise to stereoisomers, such as *cis*- and *trans*-1,2-dichloroethene. These isomers are neither superimposable nor mirror images of each other, so they are diastereomers (geometric isomers). Each isomer has a completely different infrared and Raman spectrum, so infrared/Raman spectroscopy is useful in distinguishing between cis and trans isomers. It was stated in Section 8-2a that absorption of infrared radiation by a molecule can only occur if there is a change of dipole moment accompanying a vibration. For cis isomers a dipole moment change occurs for most of the normal vibrations. However, trans isomers usually have higher symmetry, which leads to a zero or very small dipole moment change for some vibrations, and so they are not observed in the infrared spectrum. This situation is illustrated in Figure 8-9 for the C—Cl stretching vibrations of *cis*- and *trans*-1,2-dichloroethene.

Figure 8-9b shows that the symmetric vibration involving stretching of the C—Cl bonds produces no change in dipole moment. If the chlorine atoms were to be replaced by similar, but not identical groups, then a small dipole moment would be produced during the vibration. However, this dipole might be too small to give rise to an observable infrared absorption. Thus, we can conclude that trans compounds often have simpler infrared spectra than the cis isomers.

It was stated in Section 8-2a that for a vibration to be active in the Raman effect there must be a change in polarizability during the vibration. This situation is not so easily visualized, but highly symmetrical modes *always* produce a change in polarizability, whereas less symmetrical vibrations sometimes cause no change and are therefore not seen in the Raman spectrum.

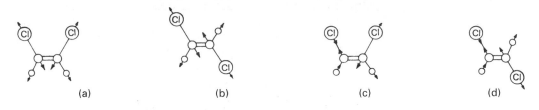

Figure 8-9 The C—Cl stretching modes of *cis*- and *trans*-1,2-dichloroethene: (a) cis (symmetric), the dipole moment changes (observed in both IR and Raman at 710 cm⁻¹); (b) trans (symmetric), no change in dipole moment (not observed in IR but observed in Raman at 840 cm⁻¹; see Figure 8-1); (c) cis (antisymmetric), the dipole moment changes (observed in both IR and Raman at 840 cm⁻¹); (d) trans (antisymmetric), the dipole moment changes, but there is no change in polarizability (observed only in IR at 895 cm⁻¹; see Figure 8-1).

For the C—Cl stretching modes of the two dichloroethenes of Figure 8-9, both vibrations of the cis isomer are seen in the Raman spectrum. However, only the symmetric mode of the trans compound is Raman active. Once again, we conclude that trans isomers have *simpler* spectra than the cis compounds. We also note that for the trans compound some vibrations not seen in the infrared spectrum are Raman active (Figure 8-1b) and, conversely, certain frequencies not seen in the Raman spectrum are infrared active (Figure 8-1a). Trans-substituted alkenes are characterized by a very strong IR band near 970 cm⁻¹ due to the wagging vibration. For the cis isomer a medium to strong band is observed between 730 and 650 cm⁻¹ for this mode. Another point helpful in determining which structural isomer is present is the observation that for a trans isomer the antisymmetric C—X stretching mode in an XC=CX structure is observed at frequencies 40–20 cm⁻¹ higher than for the corresponding cis isomer. Many such examples are found in the spectra of long-chain acids, alcohols, and esters. A similar observation has been made concerning the C=C stretch in unsaturated ketones and unsaturated hydrocarbons. Again, the trans isomer has a slightly higher frequency (10–5 cm⁻¹). However, the infrared absorption may be very weak for the trans compound for the reasons noted above. The lower frequency in the cis compound is probably due to the lower symmetry (or pseudo symmetry), which gives rise to greater coupling of —C—C= or —C=C— vibrations with other lower frequency skeletal modes of the molecule.

8-2i   Conformational Isomerism

In open chain compounds, the barriers to internal rotation about one or more carbon–carbon single bonds may be too high for rapid interconversion between different forms. In such cases, two or more different isomers can exist and their presence may be detected in their infrared or Raman spectra.

In acyclic structures, rotation about a single bond can produce an infinite number of arrangements. Some of these are energetically favored (energy minima). The simplest examples are the substituted ethanes, CH_2XCH_2Y, for which there are several preferred staggered conformations. Two stable conformers, **8-3** and **8-4,** are illustrated at left with the unstable (in this case) eclipsed form, **8-5.** Each stable form will have slightly different infrared and Raman spectra which, in some cases, can be used to distinguish the conformers.

When there is a stabilizing interaction, the eclipsed form may be one of the stable conformations. Many such examples can be found in α-halo ketones, esters, acid halides, and amides. In these compounds, the halogen atom is believed to be either cis (eclipsed), **8-6,** or gauche (staggered), **8-7,** with respect to the carbonyl group. Two C=O stretching bands are observed in such cases. One is at higher frequencies, due to

gauche
(staggered)

8-3

trans
(staggered)

8-4

cis
(eclipsed)

8-5

cis
(eclipsed)

8-6

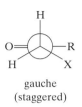

gauche
(staggered)

8-7

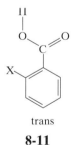

cis

8-10

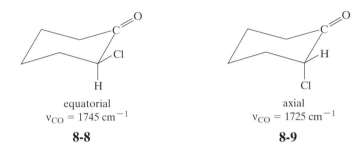

the eclipsed interaction between the halogen atom and the C=O group. The other is found at the normal frequency.

The axial-equatorial conformations of cyclohexane and cyclopentane derivatives are examples of another kind of conformational isomerism. For example, in α-chloro-substituted cyclopentanones or cyclohexanones, two distinct carbonyl stretching frequencies can be observed. For the six-membered ring, one band is found near 1745 cm^{-1}, due to the equatorial conformation, **8-8,** and a second band near 1725 cm^{-1} is attributed to the axial isomer, **8-9.** The relative proportions of axial and equatorial forms change with phase, temperature, and solvent, and such changes can be followed readily in the vibrational spectra. In cyclohexanols, the equatorial C—OH stretching frequency is 1050–1030 cm^{-1}, while in the axial conformation the frequency is 30–10 cm^{-1} lower.

equatorial
$\nu_{CO} = 1745$ cm^{-1}

8-8

axial
$\nu_{CO} = 1725$ cm^{-1}

8-9

Ortho-halogenated benzoic acids also show two carbonyl stretching frequencies, due to the two rotational isomers, **8-10** and **8-11,** which could be described as cis and trans with respect to the halogen and C=O groups.

Vinyl ethers show a doublet for the C=C stretching mode at 1640–1620 and 1620–1610 cm^{-1}. These bands correspond to rotational isomers about the C—O bond. The two bands show variations in intensity that correspond with temperature changes. The CH$_2$ deformation band also is found to be a doublet due to the two different rotational isomers.

8-2j   Tautomerism (Dynamic Isomerism)

Tautomerism is a special case of constitutional isomerism in which isomers are readily interconvertible. Numerous examples of tautomerism can be found in the literature. Infrared spectroscopy offers a useful means of distinguishing between possible tautomeric structures. A simple example is found in β-diketones. The keto form, **8-12,** has two C=O groups, which have separate stretching frequencies. A doublet is often observed in the usual ketone carbonyl stretching region, near 1730 cm^{-1}. The enol form, **8-13,** on the other hand, has only one carbonyl group. The frequency of the carbonyl group is lowered by 100–80 cm^{-1} due to hydrogen bonding and conjugation. This

trans

8-11

keto

8-12

enol

8-13

structure also has an alkenic double bond that should give a band between 1650–1600 cm^{-1}. The C=O and C=C vibrations may then appear as two overlapping bands with closely spaced peaks. An example of such tautomerism is shown in Figure 8-10. The compound ethyl propionylacetate clearly shows both keto and enol forms.

8-3   INFRARED GROUP FREQUENCIES

Group frequencies fall within fairly restricted ranges, regardless of the compound in which the group is found. Mechanical coupling, symmetry, or other effects discussed in the previous section may occasionally cause even a good group frequency to misbe-

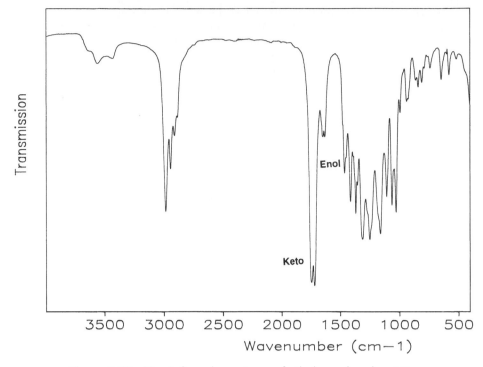

Figure 8-10 The infrared spectrum of ethyl propionylacetate.

have, so one should be aware of this possibility. Some of the vibrations of chemical groups giving rise to characteristic bands in the infrared spectrum between 4000 and 400 cm^{-1} are presented in this section. First, an alphabetical list of groups with frequency ranges and intensities is given in Table 8-2. Table 8-3 presents a list of frequency ranges from 4000 down to 400 cm^{-1} with possible groups that could absorb within a given range. In the next chapter, several tables give further details of frequencies of the vibrations of certain groups in various types of compounds. Tables 8-2 and 8-3 are by no means comprehensive and to make full use of the group frequency method for structure determination, the references cited at the end of this chapter should be consulted.

8-4 *RAMAN GROUP FREQUENCIES*

Depending on the symmetry of a molecule, its vibrational frequencies may give rise to infrared absorption or Raman scattering or both. In the latter case, observed wavenumbers will be numerically the same in Raman and infrared, but the intensities will often be quite different. In most cases, information obtained from the Raman spectrum duplicates that obtained from the infrared, but in some cases the Raman spectrum provides additional information, especially in the low-frequency region where far-infrared spectra may not be available.

Since infrared absorption depends on change of dipole moment, we expect polar bonds or groups to give strong infrared bands. On the other hand, a change in polarizability is necessary for Raman scattering, and so bonds or groups with symmetrical charge distributions are expected to give rise to strong Raman lines. Some of the most important Raman group frequencies are found for the C=C, N=N, C≡C, and S—S stretching modes.

Some Raman group frequencies are given in Table 8-4. A bibliography of sources of Raman group frequencies is listed at the end of this chapter. The book by Dollish, Fateley, and Bentley (reference 8.8) is particularly valuable. A recent book on Fourier transform Raman spectroscopy (reference 8.10) has a useful section on Raman group frequencies.

TABLE 8-2

An Alphabetical Listing of Some Functional Groups and Classes of Compounds
with Their Absorption Frequencies in the Infrared

Groups or Class	Frequency Ranges (cm^{-1}) and Intensities*	Assignment and Remarks
Acid halides $R-\overset{\displaystyle O}{\underset{\displaystyle X}{C}}$		
Aliphatic	1810–1790 (s)	C=O stretch; fluorides 50 cm^{-1} higher
	965–920 (m)	C—C stretch
	440–420 (s)	Cl—C=O in-plane deformation
Aromatic	1785–1765 (s)	C=O stretch; also a weaker band (1750–1735 cm^{-1}) due to Fermi resonance
	890–850 (s)	C—C stretch (Ar—C) or C—Cl stretch
Alcohols		
Primary —CH$_2$OH	3640–3630 (s)	OH stretch, dil CCl$_4$ soln
	1060–1030 (s)	C—OH stretch; lowered by unsaturation
Secondary —CHROH	3630–3620 (s)	OH stretch, dil CCl$_4$ soln
	1120–1080 (s)	C—OH stretch; lower when R is a branched chain or cyclic
Tertiary —CR$_2$OH	3620–3610 (s)	OH stretch, dil CCl$_4$ soln
	1160–1120 (s)	C—OH stretch; lower when R is branched
General —OH	3350–3250 (s)	OH stretch; broad band in pure solids or liquids
	1440–1260 (m–s, br)	C—OH in-plane bend
	700–600 (m–s, br)	C—OH out-of-plane deformation
Aldehydes $R-\overset{\displaystyle O}{\underset{\displaystyle H}{C}}$	2830–2810 (m) ⎱ 2740–2720 (m) ⎰	Fermi doublet; CH stretch with overtone of CH bend
	1725–1695 (vs)	C=O stretch; slightly higher in CCl$_4$ soln
	1440–1320 (s)	H—C=O bend in aliphatic aldehydes
	695–635 (s)	C—C—CHO bend
	565–520 (s)	C—C=O bend
Alkenes		
Monosubst —CH=CH$_2$	—	*See* Vinyl
Disubst —CH=CH—	—	*See* Vinylene
$\overset{\diagdown}{\diagup}C=CH_2$	—	*See* Vinylidene
Trisubst $\overset{\diagdown}{\diagup}C=CH-$	3050–3000 (w)	CH stretch
	1690–1655 (w-m)	C=C stretch
	850–790 (m)	CH out-of-plane bending
Tetrasubst $\overset{\diagdown}{\diagup}C=C\overset{\diagup}{\diagdown}$	1690–1670 (w)	C=C stretch, may be absent for symmetrical compounds
Alkyl	2980–2850 (m)	CH stretch, several bands
	1470–1450 (m)	CH$_2$ deformation
	1400–1360 (m)	CH$_3$ deformation
	740–720 (w)	CH$_2$ rocking
Alkynes RC≡C—H	3300–3250 (m-s)	Terminal ≡C—H stretch
	2250–2100 (w–m)	C≡C, frequency raised by conjugation
	680–580 (s)	—C≡CH bend
Amides		
Primary —CONH$_2$	3540–3520 (m)	NH$_2$ stretch (dil solns); bands shift to
	3400–3380 (m)	3360–3340 and 3200–3180 in solid
	1680–1660 (vs)	C=O stretch (Amide I band)
	1650–1610 (m)	NH$_2$ deformation; sometimes appears as a shoulder (Amide II band)
	1420–1400 (m–s)	C—N stretch (Amide III band)
Secondary —CONHR	3440–3420 (m)	NH stretch (dil soln); shifts to 3300–3280 in pure liquid or solid
	1680–1640 (vs)	C=O stretch (Amide I band)
	1560–1530 (vs)	NH bend (Amide II band)
	1310–1290 (m)	C—N stretch
	710–690 (m)	Assignment uncertain

*Key: s = strong; m = medium; w = weak; v = very; br = broad.

Continued

TABLE 8-2

An Alphabetical Listing of Some Functional Groups and Classes of Compounds
with Their Absorption Frequencies in the Infrared—cont'd

Groups or Class	Frequency Ranges (cm^{-1}) and Intensities*	Assignment and Remarks
Amides—cont'd		
Tertiary —CONR$_2$	1670–1640 (vs)	C=O stretch
General —CONR$_2$	630–570 (s)	N—C=O bend
	615–535 (s)	C=O out-of-plane bend
	520–430 (m–s)	C—C=O bend
Amines		
Primary —NH$_2$	3460–3280 (m)	NH stretch; broad band, may have some structure
	2830–2810 (m)	CH stretch
	1650–1590 (s)	NH$_2$ deformation
Secondary —NHR	3350–3300 (vw)	NH stretch
	1190–1130 (m)	C—N stretch
	740–700 (m)	NH deformation
	450–400 (w, br)	C—N—C bend
Tertiary —NR$_2$	510–480 (s)	C—N—C bend
Amine hydrohalides RNH$_3$$^+X^-$	2800–2300 (m–s)	NH$_3$$^+$ stretch, several peaks
R'NH$_2$R X$^-$	1600–1500 (m)	NH deformation (one or two bands)
Amino acids $\overset{NH_2}{\underset{|}{—C—}}$COOH	3200–3000 (s)	H-bonded NH$_2$ and OH stretch; v broad band in solid state
(or —CNH$_3$COO$^-$)	1600–1590 (s)	COO$^-$ antisym stretch
	1550–1480 (m–s)	—NH$_3$$^+$ deformation
	1425–1390 (w–m)	COO$^-$ sym stretch
	560–500 (s)	COO$^-$ rocking
Ammonium NH$_4$$^+$	3350–3050 (vs)	NH stretch; broad band
	1430–1390 (s)	NH$_2$ deformation; sharp peak
Anhydrides —CO	1850–1780 (variable)	Antisym C=O stretch
O	1770–1710 (m–s)	Sym C=O stretch
—CO	1220–1180 (vs)	C—O—C stretch (higher in cyclic anhydrides)
Aromatic compounds	3100–3000 (m)	CH stretch, several peaks
	2000–1660 (w)	Overtone and combination bands
	1630–1430 (variable)	Aromatic ring stretching (four bands)
	900–650 (s)	Out-of-plane CH deformations (one or two bands depending on substitution)
	580–420 (m–s)	Ring deformations (two bands)
Azides —N=N=N	2160–2080 (s)	N=N=N stretch
Bromo —C—Br	650–500 (m)	C—Br stretch
tert-Butyl (CH$_3$)$_3$C—	2980–2850 (m)	CH stretch; several bands
	1400–1370 (m) and 1380–1360 (s)	CH$_3$ deformations
Carbodiimides —N=C=N—	2150–2100 (vs)	N=C=N antisym stretch
Carbonyl \C=O	1870–1650 (vs, br)	C=O stretch
Carboxylic acids	3550–3500 (s)	OH stretch (monomer, dil soln)
O	3300–2400 (s, v br)	H-bonded OH stretch (solid and liquid states)
R—C	1800–1740 (s)	C=O stretch of monomer (dil soln)
OH	1710–1680 (vs)	C=O stretch of dimer (solid and liquid states)
	960–910 (s)	C—OH deformation
	700–590 (s)	O—C=O bend
	550–465 (s)	C—C=O bend
Chloro —C—Cl	850–550 (m)	C—Cl stretch
Cycloalkanes	580–430 (s)	Ring deformation

TABLE 8-2

An Alphabetical Listing of Some Functional Groups and Classes of Compounds with Their Absorption Frequencies in the Infrared—cont'd

Groups or Class	Frequency Ranges (cm⁻¹) and Intensities*	Assignment and Remarks
Diazonium salts $-\overset{+}{N}\equiv N$	2300–2240 (s)	$N\equiv N$ stretch
Esters $R-C\overset{O}{\underset{OR'}{\rlap{\raise3pt\hbox{$\diagup$}}\diagdown}}$	1765–1720 (vs) 1290–1180 (vs) 645–575 (s)	C=O stretch C—O—C antisym stretch O—C—O bend
Ethers —C—O—C—	1280–1220 (s) 1140–1110 (vs) 1275–1200 (vs) 1250–1170 (s) 1050–1000 (s)	C—O—C stretch in alkyl aryl ethers C—O—C stretch in dialkyl ethers C—O—C stretch in vinyl ethers C—O—C stretch in cyclic ethers R(alkyl)—C—O stretch in alkyl aryl ethers
Fluoroalkyl —CF₃, —CH₂—, etc.	1400–1000 (vs)	C—F stretch
Isocyanates —N=C=O	2280–2260 (vs)	N=C=O stretch
Isothiocyanates —N=C=S	2140–2040 (vs, br)	C=N=S antisym stretch
Ketones $\overset{R}{\underset{R'}{\rlap{\raise3pt\hbox{$\diagup$}}}}C=O$	1725–1705 (vs) 1700–1650 (vs) 1705–1665 (s) and 1650–1580 (m)	C=O stretch in saturated aliphatic ketones C=O stretch in aromatic ketones C=O and C=C stretching in α,β-unsaturated ketones
Lactones	1850–1830 (s) 1780–1770 (s) 1750–1730 (s)	C=O stretch in β-lactones C=O stretch in γ-lactones C=O stretch in δ-lactones
Methyl —CH₃	2970–2850 (s) 2835–2815 (s) 2820–2780 (s) 1470–1440 (m) 1390–1370 (m–s)	CH stretch in C—CH₃ compounds CH stretch in methyl ethers (O—CH₃) CH stretch in N—CH₃ compounds CH₃ antisym deformation CH₃ sym deformation
Methylene —CH₂—	2940–2920 (m) and 2860–2850 (m) 3090–3070 (m) and 3020–2980 (m) 1470–1450 (m)	CH stretches in alkanes CH stretches in alkenes CH₂ deformation
Naphthalenes	645–615 (m–s) and 545–520 (s) 490–465 (variable)	In-plane ring bending Out-of-plane ring bending
Nitriles —C≡N	2260–2240 (w) 2240–2220 (m) 580–530 (m–s)	C≡N stretch in aliphatic nitriles C≡N stretch in aromatic nitriles C—C—CN bend
Nitro —NO₂	1570–1550 (vs) and 1380–1360 (vs) 1480–1460 (vs) and 1360–1320 (vs) 920–830 (m) 650–600 (s) 580–520 (m) 530–470 (m–s)	NO₂ stretches in aliphatic nitro compounds NO₂ stretches in aromatic nitro compounds C—N stretch NO₂ bend in aliphatic compounds NO₂ bend in aromatic compounds NO₂ rocking
Oximes =NOH	3600–3590 (vs) 3260–3240 (vs) 1680–1620 (w)	OH stretch (dil soln) OH stretch (solids) C=N stretch; strong in Raman
Phenols Ar—OH	720–600 (s, br) 450–375 (w)	O—H out-of-plane deformation C—OH deformation
Phenyl C₆H₅—	3100–3000 (w–m) 2000–1700 (w)	CH stretch Four weak bands; overtones and combinations

Continued

TABLE 8-2

An Alphabetical Listing of Some Functional Groups and Classes of Compounds
with Their Absorption Frequencies in the Infrared—cont'd

Groups or Class	Frequency Ranges (cm^{-1}) and Intensities*	Assignment and Remarks
Phenyl—cont'd	1625–1430 (m–s)	Aromatic C=C stretches (four bands)
	1250–1025 (m–s)	CH in-plane bending (five bands)
	770–730 (vs)	CH out-of-plane bending
	710–690 (vs)	Ring deformation
	560–420 (m–s)	Ring deformation
Phosphates $(RO)_3P=O$		
R = alkyl	1285–1255 (vs)	P=O stretch
	1050–990 (vs)	P—O—C stretch
R = aryl	1315–1290 (vs)	P=O stretch
	1240–1190 (vs)	P—O—C stretch
Phosphines $—PH_2$, —PH	2410–2280 (m)	P—H stretch
	1100–1040 (w–m)	P—H deformation
	700–650 (m–s)	P—C stretch
Pyridyl $—C_5H_4N$	3080–3020 (m)	CH stretch
	1620–1580 (vs) and 1590–1560 (vs)	C—C and C—N stretches
	840–720 (s)	CH out-of-plane deformation (one or two bands, depending on substitution)
	635–605 (m–s)	In-plane ring bending
Silanes $—SiH_3$ $—SiH_2—$	2160–2110 (m)	SI—H stretch
	950–800 (s)	Si—H deformation
Silanes (fully substituted)	1280–1250 (m–s)	Si—C stretch
	1110–1050 (vs)	Si—O—C stretch (aliphatic)
	840–800 (m)	Si—O—C deformation
Sulfates $R—O—SO_2—O—R$	1140–1350 (s) and 1230–1150 (s)	S—O stretches in covalent sulfates
$R—O—SO_3^-M^+$ $(M = Na^+, K^+, etc.)$	1260–1210 (vs) and 810–770 (s)	S=O stretches in alkyl sulfate salts C—O—S stretch
Sulfides C—S—	710–570 (m)	C—S stretch
Sulfones $—SO_2—$	1360–1290 (vs)	SO_2 antisym stretch
	1170–1120 (vs)	SO_2 sym stretch
	610–545 (ms)	SO_2 scissor mode
Sulfonic acids $—SO_2OH$	1250–1150 (vs, br)	S=O stretch
Sulfoxides $\backslash S=O /$	1060–1030 (s, br)	S=O stretch
	610–545 (m–s)	SO_2 scissoring
Thiocyanates $—S—C≡N$	2175–2160 (m)	C≡N stretch
	650–600 (w)	S—CN stretch
	405–400 (s)	S—C≡N bend
Thiols —S—H	2590–2560 (w)	S—H stretch; strong in Raman
	700–550 (w)	C—S stretch; strong in Raman
Triazines $C_3N_2Y_3$ 1,3,4,5-trisubst	1600–1500 (vs)	Ring stretching
	1380–1350 (vs)	Ring stretching
	820–800 (s)	CH out-of-plane deformation
Vinyl $—CH=CH_2$	3095–3080 (m) and 3030–2980 (w–m)	$=CH_2$ stretching $=CH$ stretching
	1850–1800 (w–m)	Overtone of CH_2 out-of-plane wagging
	1645–1615 (m–s)	C=C stretch
	1000–950 (s)	CH out-of-plane deformation
	950–900 (vs)	CH_2 out-of-plane wagging
Vinylene $—CH=CH—$	3040–3010 (m)	$=CH_2$ stretching
	1665–1635 (w–m)	C=C stretch (cis isomer)
	1675–1665 (w–m)	C=C stretch (trans isomer)
	980–955 (s)	CH out-of-plane deformation (cis isomer)
	730–665 (s)	CH out-of-plane deformation (trans isomer)
Vinylidene $\backslash C=CH_2 /$	3095–3075 (m)	$=CH_2$ stretching
	1665–1620 (w–m)	C=C stretch
	895–885 (s)	CH_2 out-of-plane wagging

TABLE 8-3

A Numerical Listing of Wavenumber Ranges in Which Some Functional Groups and
Classes of Compounds Absorb in the Infrared

Range (cm⁻¹) and Intensity*	Group and Class	Assignment and Remarks
3700–3600 (s)	—OH in alcohols and phenols	OH stretch (dil soln)
3520–3320 (m–s)	—NH_2 in aromatic amines, primary amines and amides	NH stretch (dil soln)
3420–3250 (s)	—OH in alcohols and phenols	OH stretch (solids and liquids)
3360–3340 (m)	—NH_2 in primary amides	NH_2 antisym stretch (solids)
3320–3250 (m)	—OH in oximes	O—H stretch
3300–3250 (m–s)	≡CH in acetylenes	≡C—H stretch
3300–3280 (s)	—NH in secondary amides	NH stretch (solids); also in polypeptides and proteins
3200–3180 (s)	—NH_2 in primary amides	NH_2 sym stretch (solids)
3200–3000 (v br)	—NH_3^+ in amino acids	NH_3^+ antisym stretch
3100–2400 (v br)	—OH in carboxylic acids	H-bonded OH stretch
3100–3000 (m)	=CH in aromatic and unsaturated hydrocarbons	=C—H stretch
2990–2850 (m–s)	—CH_3 and —CH_2— in aliphatic compounds	CH antisym and sym stretching
2850–2700 (m)	—CH_3 attached to O or N	CH stretching modes
2750–2650 (w–m)	—CHO in aldehydes	Overtone of CH bending (Fermi resonance)
2750–2350 (br)	—NH_3^+ in amine hydrohalides	NH stretching modes
2720–2560 (m)	—OH in phosphorus oxyacids	Associated OH stretching
2600–2540 (w)	—SH in alkyl mercaptans	S—H stretch; strong in Raman
2410–2280 (m)	—PH in phosphines	P—H stretch; sharp peak
2300–2230 (m)	N≡N in diazonium salts	N≡N stretch, aq soln
2285–2250 (s)	N=C=O in isocyanates	N=C=O antisym stretch
2260–2200 (m–s)	C≡N in nitriles	C≡N stretch
2260–2190 (w–m)	C≡C in alkynes (disubstitution)	C≡C stretch; stong in Raman
2190–2130 (m)	C≡N in thiocyanates	C≡N stretch
2175–2115 (s)	N≡C in isonitriles	N≡C stretch
2160–2080 (m)	N=N=N in azides	N=N=N antisym stretch
2140–2100 (w–m)	C≡C in alkynes (monosubstitution)	C≡C stretch
2000–1650 (w)	Substituted benzene rings	Several bands from overtones and combinations
1980–1950 (s)	C=C=C in allenes	C=C=C antisym stretch
1870–1650 (vs)	C=O in carbonyl compounds	C=O stretch
1870–1830 (s)	C=O in β-lactones	C=O stretch
1870–1790 (vs)	C=O in anhydrides	C=O antisym stretch; part of doublet
1820–1800 (s)	C=O in acid halides	C=O stretch; lower for aromatic acid halides
1780–1760 (s)	C=O in γ-lactones	C=O stretch
1765–1725 (vs)	C=O in anhydrides	C=O sym stretch; part of doublet
1760–1740 (vs)	C=O in α-keto esters	C=O stretch; enol form
1750–1730 (s)	C=O in δ-lactones	C=O stretch
1750–1740 (vs)	C=O in esters	C=O stretch; 20 cm⁻¹ lower if unsaturated
1740–1720 (s)	C=O in aldehydes	C=O stretch; 30 cm⁻¹ lower if unsaturated
1720–1700 (s)	C=O in ketones	C=O stretch; 20 cm⁻¹ lower if unsaturated
1710–1690 (s)	C=O in carboxylic acids	C=O stretch; fairly broad
1690–1640 (s)	C=N in oximes	C=N stretch; also imines
1680–1620 (s)	C=O and NH_2 in primary amides	Two bands from C=O stretch and NH_2 deformation
1680–1635 (s)	C=O in ureas	C=O stretch; broad band
1680–1630 (m–s)	C=C in alkenes, etc.	C=C stretch
1680–1630 (vs)	C=O in secondary amides	C=O stretch (Amide I band)
1670–1640 (s–vs)	C=O in benzophenones	C=O stretch
1670–1650 (vs)	C=O in primary amides	C=O stretch (Amide I band)
1670–1630 (vs)	C=O in tertiary amides	C=O stretch

*KEY: s = strong, m = medium, w = weak, v = very, br = broad.

Continued

TABLE 8-3

A Numerical Listing of Wavenumber Ranges in Which Some Functional Groups and Classes of Compounds Absorb in the Infrared—cont'd

Range (cm^{-1}) and Intensity*	Group and Class	Assignment and Remarks
1655–1635 (vs)	C=O in β-ketone esters	C=O stretch; enol form
1650–1620 (w–m)	N—H in primary amides	NH deformation (Amide II band)
1650–1580 (m–s)	NH_2 in primary amines	NH_2 deformation
1640–1580 (s)	NH_3^+ in amino acids	NH_3 deformation
1640–1580 (vs)	C=O in β-diketones	C=O stretch; enol form
1620–1610 (s)	C=C in vinyl ethers	C=C stretch; doublet due to rotational isomerism
1615–1590 (m)	Benzene ring in aromatic compounds	Ring stretch; sharp peak
1615–1565 (s)	Pyridine derivatives	Ring stretch; doublet
1610–1580 (s)	NH_2 in amino acids	NH_2 deformation; broad band
1610–1560 (vs)	COO$^-$ in carboxylic acid salts	—C(=O)O$^-$ antisym stretch
1590–1580 (m)	NH_2 primary alkyl amide	NH_2 deformation (Amide II band)
1575–1545 (vs)	NO_2 in aliphatic nitro compounds	NO_2 antisym stretch
1565–1475 (vs)	NH in secondary amides	NH deformation (Amide II band)
1560–1510 (s)	Triazine compounds	Ring stretch; sharp band
1550–1490 (s)	NO_2 in aromatic nitro compounds	NO_2 antisym stretch
1530–1490 (s)	NH_3^+ in amino acids or hydrochlorides	NH_3^+ deformation
1530–1450 (m–s)	N=N—O in azoxy compounds	N=N—O antisym stretch
1515–1485 (m)	Benzene ring in aromatic compounds	Ring stretch, sharp band
1475–1450 (vs)	CH_2 in aliphatic compounds	CH_2 scissors vibration
1465–1440 (vs)	CH_3 in aliphatic compounds	CH_3 antisym deformation
1440–1400 (m)	OH in carboxylic acids	In-plane OH bending
1420–1400 (m)	C—N in primary amides	C—N stretch (Amide III band)
1400–1370 (m)	*tert*-Butyl group	CH_3 deformations (two bands)
1400–1310 (s)	COO$^-$ group in carboxylic acid salts	—C(=O)O$^-$ sym stretch; broad band
1390–1360 (vs)	SO_2 in sulfonyl chlorides	SO_2 antisym stretch
1380–1370 (s)	CH_3 in aliphatic compounds	CH_3 sym deformation
1380–1360 (m)	Isopropyl group	CH_3 deformations (two bands)
1375–1350 (s)	NO_2 in aliphatic nitro compounds	NO_2 sym stretch
1360–1335 (vs)	SO_2 in sulfonamides	SO_2 antisym stretch
1360–1320 (vs)	NO_2 in aromatic nitro compounds	NO_2 sym stretch
1350–1280 (m–s)	N=N—O in azoxy compounds	N=N—O sym stretch
1335–1295 (vs)	SO_2 in sulfones	SO_2 antisym stretch
1330–1310 (m–s)	CF_3 attached to a benzene ring	CF_3 antisym stretch
1300–1200 (vs)	N—O in pyridine *N*-oxides	N—O stretch
1300–1175 (vs)	P=O in phosphorus oxyacids and phosphates	P=O stretch
1300–1000 (vs)	C—F in aliphatic fluoro compounds	C—F stretch
1285–1240 (vs)	Ar—O in alkyl aryl ethers	C—O stretch
1280–1250 (vs)	Si—CH_3 in silanes	CH_3 sym deformation
1280–1240 (m–s)	C—C (O) in epoxides	C—O stretch
1280–1180 (s)	C—N in aromatic amines	C—N stretch
1280–1150 (vs)	C—O—C in esters, lactones	C—O—C antisym stretch
1255–1240 (m)	*tert*-Butyl in hydrocarbons	Skeletal vibration; second band near 1200 cm^{-1}
1245–1155 (vs)	SO_3H in sulfonic acids	S=O stretch
1240–1070 (s–vs)	C—O—C in ethers	C—O—C stretch; also in esters
1230–1100 (s)	C—C—N in amines	C—C—N bending
1225–1200 (s)	C—O—C in vinyl ethers	C—O—C antisym stretch
1200–1165 (s)	SO_2Cl in sulfonyl chlorides	SO_2 sym stretch
1200–1015 (vs)	C—OH in alcohols	C—O stretch
1170–1145 (s)	SO_2NH_2 in sulfonamides	SO_2 sym stretch
1170–1140 (s)	SO_2— in sulfones	SO_2 sym stretch

TABLE 8-3

A Numerical Listing of Wavenumber Ranges in Which Some Functional Groups and Classes of Compounds Absorb in the Infrared—cont'd

Range (cm^{-1}) and Intensity*	Group and Class	Assignment and Remarks
1160–1100 (m)	$C{=}S$ in thiocarbonyl compounds	$C{=}S$ stretch; strong in Raman
1150–1070 (vs)	$C{-}O{-}C$ in aliphatic ethers	$C{-}O{-}C$ antisym stretch
1120–1080 (s)	$C{-}OH$ in secondary or tertiary alcohols	$C{-}O$ stretch
1120–1030 (s)	$C{-}NH_2$ in primary aliphatic amines	$C{-}N$ stretch
1100–1000 (vs)	$Si{-}O{-}Si$ in siloxanes	$Si{-}O{-}Si$ antisym stretch
1080–1040 (s)	SO_3H in sulfonic acids	SO_3 sym stretch
1065–1015 (s)	$CH{-}OH$ in cyclic alcohols	$C{-}O$ stretch
1060–1025 (vs)	$CH_2{-}OH$ in primary alcohols	$C{-}O$ stretch
1060–1045 (vs)	$S{=}O$ in alkyl sulfoxides	$S{=}O$ stretch
1055–915 (vs)	$P{-}O{-}C$ in organophosphorus compounds	$P{-}O{-}C$ antisym stretch
1030–950 (w)	Carbon ring in cyclic compounds	Ring breathing mode; strong in Raman
1000–950 (s)	$CH{=}CH_2$ in vinyl compounds	${=}CH$ out-of-plane deformation
980–960 (vs)	$CH{=}CH{-}$ in trans disubstituted alkenes	${=}CH$ out-of-plane deformation
950–900 (vs)	$CH{=}CH_2$ in vinyl compounds	CH_2 out-of-plane wag
900–865 (vs)	$CH_2{=}C\begin{smallmatrix}R\\\\R'\end{smallmatrix}$ in vinylidenes	CH_2 out-of-plane wag
890–805 (vs)	1,2,4-trisubstituted benzenes	CH out-of-plane deformation (two bands)
860–760 (vs, br)	$R{-}NH_2$ primary amines	NH_2 wag
860–720 (vs)	$Si{-}C$ in organosilicon compounds	$Si{-}C$ stretch
850–830 (vs)	1,3,5-trisubstituted benzenes	CH out-of-plane deformation
850–810 (vs)	$Si{-}CH_3$ in silanes	$Si{-}CH_3$ rocking
850–790 (m)	$CH{=}C\begin{smallmatrix}R\\\\R'\end{smallmatrix}$ in trisubstituted alkenes	CH out-of-plane deformation
850–550 (m)	$C{-}Cl$ in chloro compounds	$C{-}Cl$ stretch
830–810 (vs)	p-disubstituted benzenes	CH out-of-plane deformation
825–805 (vs)	1,2,4-trisubstituted benzenes	CH out-of-plane deformation
820–800 (s)	Triazines	CH out-of-plane deformation
815–810 (s)	$CH{=}CH_2$ in vinyl ethers	CH_2 out-of-plane wag
810–790 (vs)	1,2,3,4-tetrasubstituted benzenes	CH out-of-plane deformation
800–690 (vs)	m-disubstituted benzenes	CH out-of-plane deformation (two bands)
785–680 (vs)	1,2,3-trisubstituted benzenes	CH out-of-plane deformation (two bands)
775–650 (m)	$C{-}S$ in sulfonyl chlorides	$C{-}S$ stretch; strong in Raman
770–690 (vs)	Monosubstituted benzenes	CH out-of-plane deformation (two bands)
760–740 (s)	o-disubstituted benzenes	CH out-of-plane deformation
760–510 (s)	$C{-}Cl$ alkyl chlorides	$C{-}Cl$ stretch
740–720 (w–m)	${-}(CH_2)_n{-}$ in hydrocarbons	CH_2 rocking in methylene chains; intensity depends on chain length
730–665 (s)	$CH{=}CH$ in cis disubstituted alkenes	CH out-of-plane deformation
720–600 (s, br)	$Ar{-}OH$ in phenols	OH out-of-plane deformation
710–570 (m)	$C{-}S$ in sulfides	$C{-}S$ stretch; strong in Raman
700–590 (s)	$O{-}C{=}O$ in carboxylic acids	$O{-}C{=}O$ bending
695–635 (s)	$C{-}C{-}CHO$ in aldehydes	$C{-}C{-}CHO$ bending
680–620 (s)	$C{-}OH$ in alcohols	$C{-}O{-}H$ bending
680–580 (s)	$C{\equiv}C{-}H$ in alkynes	$C{\equiv}C{-}H$ bending
650–600 (w)	$S{-}C{\equiv}N$ in thiocyanates	$S{-}C$ stretch; strong in Raman
650–600 (s)	NO_2 in aliphatic nitro compounds	NO_2 deformation
650–500 (s)	$Ar{-}CF_3$ in aromatic trifluoro-methyl compounds	CF_3 deformation (two or three bands)
650–500 (s)	$C{-}Br$ in bromo compounds	$C{-}Br$ stretch
645–615 (m–s)	Naphthalenes	In-plane ring deformation
645–575 (s)	$O{-}C{-}O$ in esters	$O{-}C{-}O$ bend
640–630 (s)	${=}CH_2$ in vinyl compounds	${=}CH_2$ twisting
635–605 (m–s)	Pyridines	In-plane ring deformation
630–570 (s)	$N{-}C{=}O$ in amides	$N{-}C{=}O$ bend

Continued

end of this chapter.) Also, the computer attached to a modern FT–IR spectrometer often has a search routine and a library of infrared spectra, which may be searched for spectra that most closely match the spectrum of the unknown compound.

9-2 PRELIMINARY ANALYSIS

9-2a Introduction

The infrared spectrum can be arbitrarily divided into the regions shown in Figure 9-1. The presence of bands in these regions gives immediate information. The *absence* of bands in these regions also is important information, since many groups can be excluded from further analysis, provided the factors discussed in Section 8-2 have been considered. A list of these regions together with absorbing groups and possible types of

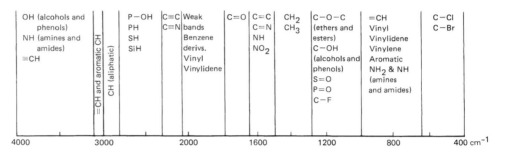

Figure 9-1 Regions of the infrared spectrum for preliminary analysis.

TABLE 9-1

Regions of the Infrared Spectrum for Preliminary Analysis

Region	Group	Possible Compounds Present (or Absent)
3700–3100	—OH	Alcohols, aldehydes, carboxylic acids
	—NH	Amides, amines
	≡C—H	Alkynes
3100–3000	=CH	Aromatic compounds
—	—CH₂ or —CH=CH—	Alkenes or unsaturated rings
3000–2800	—CH, —CH₂—, —CH₃	Aliphatic groups
2800–2600	—CHO	Aldehydes (Fermi doublet)
2700–2400	—POH	Phosphorus compounds
	—SH	Mercaptans and thiols
	—PH	Phosphines
2400–2000	—C≡N	Nitriles
	—N=N=N	Azides
	—C≡C—	Alkynes*
1870–1650	C=O	Acid halides, aldehydes, amides, amino acids, anhydrides, carboxylic acids, esters, ketones, lactams, lactones, quinones
1650–1550	C=C, C=N, NH	Unsaturated aliphatics,* aromatics, unsaturated heterocycles, amides, amines, amino acids
1550–1300	NO₂	Nitro compounds
	CH₃ and CH₂	Alkanes, alkenes, etc.
1300–1000	C—O—C and C—OH	Ethers, alcohols, sugars
	S=O, P=O, C—F	Sulfur, phosphorus, and fluorine compounds
1100–800	Si—O and P—O	Organosilicon and phosphorus compounds
1000–650	=C—H	Alkenes and aromatic compounds
	—NH	Aliphatic amines
800–400	C—halogen	Halogen compounds
	Aromatic rings	Aromatic compounds

*Band may be absent owing to symmetry (see Section 8-2a).

TABLE 9-2
Regions of the Raman Spectrum for Preliminary Analysis

Region (cm^{-1})	Group	Possible Compounds
3100–3050	$=CH_2$	Alkenes
3050–3000	$=CH$ or $—CH=CH—$	Aromatic compounds, alkenes
3000–2900	$—CH, CH_2—, —CH_3$	Aliphatic groups
2600–2500	$—SH$	Thiols, mercaptans
2300–2100	$—C\equiv C—$	Alkynes
2260–2200	$—C\equiv N$	Nitriles
1650	$C=C$	Alkenes
1650–1600	$C=N$	Heterocyclic compounds
1600	$C=C$	Aromatic compounds
1420–1280	$C—N$	Amides, aromatic amines
1390–1330	NO_2	Nitro compounds
1000	Benzene ring	Mono-, m-di-, and 1,3,5-trisubst benzenes
850–650	$C—Cl$	Chloro compounds
750–610	Benzene ring	Aromatic compounds
700–570	$C—S$	Sulfides, mercaptans
650–490	$C—Br$	Bromo compounds
550–430	$S—S$	Disulfides

compounds is given in Table 9-1. This table is, of course, a condensation of Table 8-3. Similar information for Raman spectra is given in Table 9-2.

Certain types of compounds give strong, broad absorptions, which are very prominent in the infrared spectrum. The hydrogen-bonded OH stretching bands of alcohols, phenols, and carboxylic acids are easily recognized at the high-frequency end of the spectrum. The stretching of the NH_3^+ group in amino acids gives a very broad, asymmetric band, which extends over several hundred wavenumbers. Broad bands associated with bending of NH_2 or NH groups of primary or secondary amines are found at the low-frequency end of the spectrum. Amides also give a broad band in this region. Examples of these characteristic absorptions are illustrated in Figure 9-2. These and other broad bands are listed in Table 9-3. It is well worth looking through one of the published collections of infrared spectra (see Section 9-16) to familiarize oneself with these bands. *The Aldrich Library of FT–IR Spectra* is particularly valuable for this purpose, since the spectra are arranged by class of compound.

In the infrared spectra of certain compounds there are weak but characteristic bands that are known to be due to overtones or combinations. Some of these bands are listed in Table 9-4 with assignments and are shown in Figure 9-3.

In the following three sections, suggestions are given for the preliminary analysis of the infrared spectra of hydrocarbons or the hydrocarbon parts of molecules; of compounds containing carbon, hydrogen, and oxygen; and of compounds containing nitrogen. After this preliminary study of the spectrum, the analyst should have some idea of the kind of compound under investigation. The spectrum is then examined in more detail. In later sections, some of the detailed structural information that can be obtained from such an examination is discussed. In many cases, Raman spectra provide important additional information.

9-2b Hydrocarbons or Hydrocarbon Groups

The nature of a hydrocarbon or the hydrocarbon part of a molecule can be identified by first looking in the region between 3100 and 2800 cm^{-1}. If there is no absorption between 3000 and 3100 cm^{-1}, the compound contains no aromatic or unsaturated aliphatic CH groups. Cyclopropanes, which absorb above 3000 cm^{-1}, are an exception. If the absorption is entirely above 3000 cm^{-1}, the compound is probably aromatic or contains $=CH$ or $=CH_2$ groups only. Absorption both above and below 3000 cm^{-1} indicates both saturated and unsaturated or cyclic hydrocarbon moieties (see Figure 9-4 for an example).

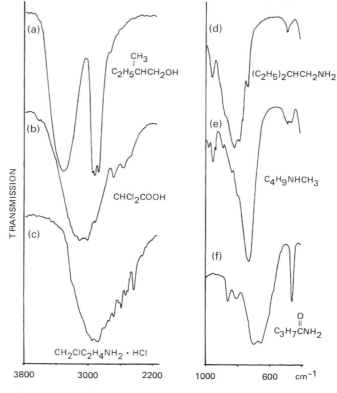

Figure 9-2   Some characteristic broad infrared bands.

The region between 1000 and 650 cm^{-1} should be examined to confirm the conclusions. Strong bands in this region suggest alkenes or aromatic structures. A small, sharp band near 725 cm^{-1} is indicative of a linear chain containing four or more CH$_2$ groups. The presence of CH$_2$ groups is also indicated by a strong band near 1440 cm^{-1} in the infrared and Raman spectra.

9-2c   Compounds Containing Oxygen

If it is known from an elemental analysis that there is an oxygen atom present in the molecule, one should look in three regions for bands due to the oxygen-containing functional group. A strong, broad band between 3500 and 3200 cm^{-1} is likely due to the hydrogen-bonded OH stretching mode of an alcohol or a phenol (see Section 9-9c). Be aware that water also absorbs in this region. When hydrogen bonding is absent, the OH stretching band is sharp and at higher frequencies (3650–3600 cm^{-1}). Carboxylic acids give very broad OH stretching bands between 3200 and 2700 cm^{-1} (see Section 9-9d). Several sharp peaks may be seen on this broad band near 3000 cm^{-1} due to CH stretching vibrations.

One should next look for a very strong band between 1850 and 1650 cm^{-1} due to the C=O stretching of a carbonyl group (see Section 9-4). The third region is between 1300 and 1000 cm^{-1}, where bands due to C—OH stretching of alcohols and carboxylic acids and C—O—C stretching modes of ethers and esters are observed (see Sections 9-9b and 9-9e). Some examples of C—O stretching bands are shown in Figure 9-5. The presence or absence of the bands due to OH, C=O, and C—O stretching gives a good indication of the type of oxygen-containing compound. Infrared spectra of aldehydes and ketones show only the C=O band, whereas spectra of ethers have only the C—O—C band. Esters have bands due to both C=O and C—O—C groups, but none due to the OH group. Alcohols have bands due to both OH and C—O groups, but no

TABLE 9-3

Characteristic Broad Absorption Bands

Range or Band Center and Intensity*	Possible Compounds	Assignment and Remarks
3600–3200 (vs)	Alcohols, phenols, oximes	OH stretch (hydrogen-bonded)
3400–3000 (vs)	Primary amides	NH_2 stretch; usually a doublet
3400–2400 (vs)	Carboxylic acids and other compounds with —OH groups	H-bonded OH stretch of dimers and polymers
3200–2400 (vs)	Amino acids (zwitterion), amine hydrohalides	NH_3^+ stretching; a very broad asymmetric band
3000–2800 (vs)	Hydrocarbons, all compounds containing CH_3 and CH_2 groups	CH stretch Bands due to Nujol obscure this region when spectra are obtained from Nujol mulls
1700–1250 (vs)	Amino acids	C=O stretch; a broad region of absorption with much structure
1650–1500 (vs)	Salts of carboxylic acids	$-\overset{\displaystyle O}{\underset{\displaystyle O}{C}}-$ antisymmetric stretch
ca 1250 (vs)	Perfluoro compounds	CF stretches; may cover the whole region from 1400–1100 cm^{-1} with several bands
ca 1200 (vs)	Esters	C—O—C stretch; ester linkage (not always broad) with much structure
ca 1200 (vs)	Phenols	C—OH stretch
ca 1150 (vs)	Sulfonic acids	S=O stretch; with structure
1150–950 (vs)	Sugars	With structure
ca 1100 (vs)	Ethers	C—O—C stretch
1100–1000 (s)	Alcohols	C—OH stretch
ca 1050 (vs)	Anhydrides	Not always reliable
1050–950 (vs)	Phosphites and phosphates	P=O stretch; often two bands
ca 920 (ms)	Carboxylic acids	H-bonded C—OH deformation
ca 830 (vs)	Primary aliphatic amines	May cover the region 1000–700 cm^{-1}
ca 730 (s)	Secondary aliphatic amines	May cover the region 850–650 cm^{-1}
ca 650 (s)	Amides	May cover the region 750–550 cm^{-1}
800–500 (w)	Alcohols	A weak, broad band

*KEY: s = strong; m = medium; w = weak; v = very.

TABLE 9-4

Some Characteristic Overtone or Combination Bands

Range (cm^{-1})	Classes of Compounds	Assignment and Remarks
ca 3450	Esters	Overtone of C=O stretch
3100–3060	Secondary amides	Overtone of NH deformation
ca 2700	Aldehydes	Part of a Fermi doublet at 2800–2700
2200–2000	Amino acids and amine hydrohalides	Combination of NH_3^+ torsion and NH_3^+ antisym deformation
2000–1650	Aromatic compounds	Overtones and combinations of CH out-of-plane deformations
1990–1960 and 1830–1800	Vinyl compounds	Overtones of CH and CH_2 out-of-plane deformations; high-frequency band is stronger
1800–1780	Vinylidine compounds	Overtone of CH_2 wag

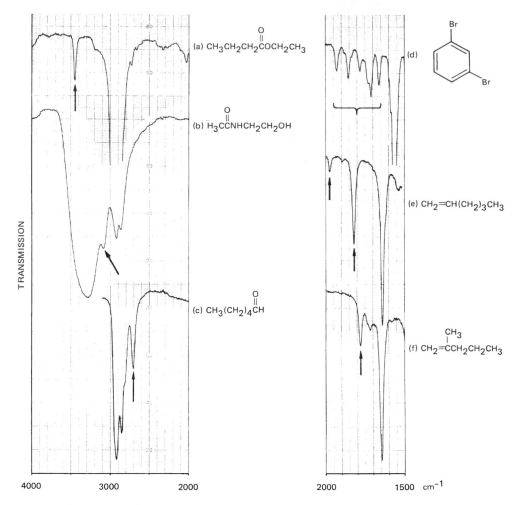

Figure 9-3 Some characteristic weak infrared bands due to overtones or combinations: (a) an ester, (b) a secondary amide, (c) an aldehyde, (d) a substituted benzene, (e) a vinyl compound, (f) a vinylidine compound.

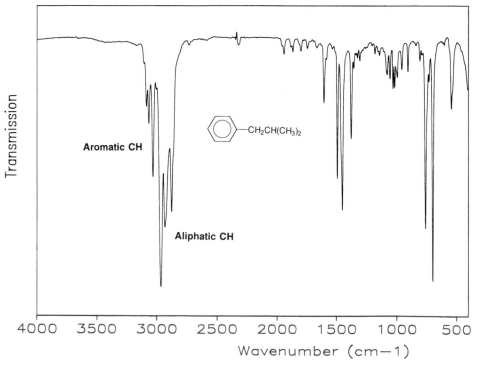

Figure 9-4 The infrared spectrum of *sec*-butylbenzene.

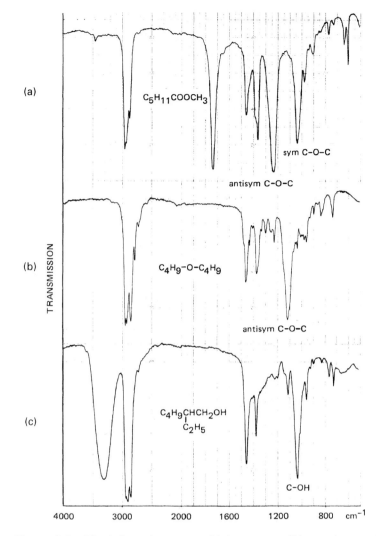

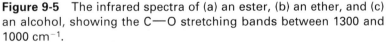

Figure 9-5 The infrared spectra of (a) an ester, (b) an ether, and (c) an alcohol, showing the C—O stretching bands between 1300 and 1000 cm^{-1}.

C=O stretching band. Spectra of carboxylic acids contain bands in all three regions mentioned above.

9-2d   Compounds Containing Nitrogen

In a preliminary examination of the infrared spectrum of a compound known to contain nitrogen, one or two bands in the region 3500–3300 cm^{-1} indicate primary or secondary amines or amides. These bands may be confused with OH stretching bands in the infrared, but are more easily identified in the Raman spectrum, where the OH stretching band is very weak. Amides may be identified by a strong doublet in the infrared spectrum centered near 1640 cm^{-1}. The Raman spectrum of an amide has bands near 1650 and 1400 cm^{-1}. These bands are associated with the stretching of the C=O group and bending of the NH_2 group of the amide. A sharp band near 2200 cm^{-1} in either infrared or Raman spectra is characteristic of a nitrile.

When both nitrogen and oxygen are present, but no NH groups are indicated, two very strong bands near 1560 and 1370 cm^{-1} provide evidence of the presence of nitro groups. An extremely broad band with some structure centered near 3000 cm^{-1} and extending as low as 2200 cm^{-1} is indicative of an amino acid or an amine hydrohalide, whereas a broad band below 1000 cm^{-1} suggests an amine or an amide (see Figure 9-2 for examples of these broad bands).

TABLE 9-5

C—H Stretching Frequencies

Range (cm^{-1}) and Intensity*	Group or Class	Assignment and Remarks
3340–3270 (w–m)	≡CH terminal alkynes	≡CH stretch; sharp band
3100–3000 (w–m)	Aromatic compounds	CH stretch; may be several bands, often weak
3100–3070 (m)	Cyclopropanes (CH$_2$ group)	CH$_2$ antisym stretch
3095–3075 (m–s)	≡CH$_2$ vinyl or vinylidine	≡CH$_2$ stretch
3080–3040 (m)	Epoxides (CH$_2$ group)	CH$_2$ stretch; sharp band
3035–2995 (m)	Cyclopropanes (CH$_2$ group)	CH$_2$ sym stretch
3030–3000 (m–s)	≡CHR aliphatic compounds	≡CH stretch
3000–2970 (m)	Cyclobutanes (CH$_2$ groups)	CH$_2$ antisym stretch
2970–2950 (vs)	—CH$_3$ group (in alkyl groups)	CH$_3$ antisym stretch
2960–2950 (vs)	Cyclopentanes (CH$_2$ groups)	CH$_2$ antisym stretch
2940–2915 (vs)	—CH$_2$— (in alkyl groups)	CH$_2$ antisym stretch
2925–2875 (vs)	Cyclobutanes (CH$_2$ groups)	CH$_2$ sym stretch
2885–2860 (vs)	—CH$_3$ group (in alkyl groups)	CH$_3$ sym stretch
2875–2855 (vs)	Cyclobutanes (CHR groups) and cyclopentanes (CH$_2$ groups)	CH(R) stretch; CH$_2$ sym stretch
2870–2840 (vs)	—CH$_2$— group	CH$_2$ sym stretch
2850–2820 (m) and 2875–2720 (m)	Aromatic aldehydes	Fermi doublet
2840–2815 (m–s)	Methoxy	CH$_3$ stretch
2830–2810 (m) and 2725–2700 (m)	Aliphatic aldehydes	Fermi doublet

*KEY: s = strong; m = medium; w = weak; v = very.

9-3 THE CH STRETCHING REGION (3340–2700 cm^{-1})

9-3a Introduction

Details of CH stretching frequencies in various compounds are summarized in Table 9-5. As indicated in Section 9-2b, this region is usually the first one examined, since certain structural features are immediately revealed by the position of the CH stretching bands. A band at the high-frequency end of the above range indicates the presence of an acetylenic hydrogen atom, whereas a band near 2710 cm^{-1} usually means that there is an aldehyde group in the molecule. It should be noted that, when infrared spectra of solids are recorded from paraffin oil mulls, the CH stretching region is obscured by strong bands from the mulling material. Other mulling materials such as hexachlorobutadiene can be used for this region, or spectra can be recorded from KBr discs (see Section 7-7b).

9-3b Alkynes

If a sharp band is observed near 3300 cm^{-1} in the infrared spectrum, the presence of a terminal ≡CH group is suspected. Confirmation of this group can be made by the observation of a small, sharp peak in the infrared or a strong band in the Raman spectrum near 2100 cm^{-1} due to the C≡C stretch (see Section 9-8c). An example is given in Figure 9-6.

9-3c Aromatic Compounds

Aromatic compounds have one or more sharp peaks of weak or medium intensity between 3100 and 3000 cm^{-1}. However, these bands may be overlooked when they appear only as shoulders on a very strong CH$_3$ or CH$_2$ stretching band. This problem is illustrated by the spectrum of dioctyl phthalate in Figure 9-7.

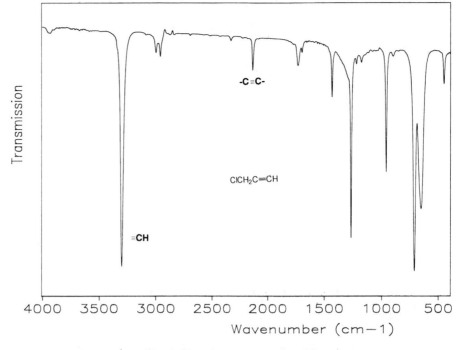

Figure 9-6   The infrared spectrum of 3-chloropropyne.

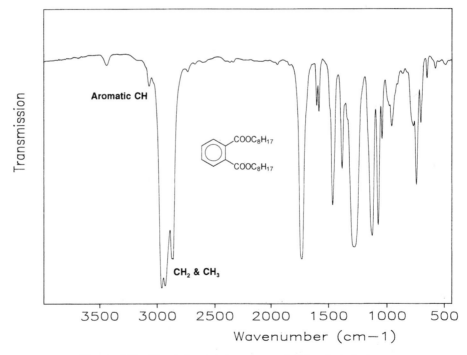

Figure 9-7   The infrared spectrum of dioctyl phthalate.

9-3d   Unsaturated Nonaromatic Compounds

Unsaturated compounds and small aliphatic ring compounds show absorption due to CH stretching in the 3100–3000 cm^{-1} region. Compounds containing the vinylidine (=CH$_2$) group absorb near 3080 cm^{-1}. Di- and trisubstituted ethenes absorb at lower frequencies, nearer to 3000 cm^{-1}, and the band may be overlapped by the stronger CH$_3$ or CH$_2$ absorption (see Section 9-3e). Cyclopropane derivatives have a band between 3100 and 3070 cm^{-1}, whereas epoxides absorb between 3060 and 3040 cm^{-1}.

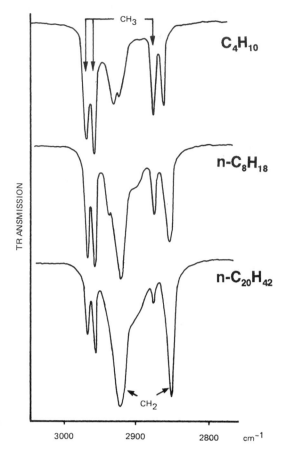

Figure 9-8 The C—H stretching region of n-C_4H_{10}, n-C_8H_{18}, and n-$C_{20}H_{42}$.

9-3e Saturated Hydrocarbon Groups

Saturated compounds can have methyl, methylene, or methine groups, each of which has characteristic CH stretching frequencies. Methyl (CH_3) groups absorb near 2960 and 2870 cm^{-1} and the CH_2 bands are at 2930 and 2850 cm^{-1}. In many cases, only one band in the 2870–2850 cm^{-1} region can be resolved when both CH_3 and CH_2 groups are present in the molecule. Figure 9-8 shows the CH stretching region of three normal saturated molecules. As the carbon chain becomes longer, the CH_2 group bands increase in intensity relative to the CH_3 group absorptions. The doublet at 2960 cm^{-1} is due to the antisymmetric CH_3 stretching mode, which would be degenerate in a free CH_3 group or in a molecule in which the threefold symmetry of the group was maintained, for example, in CH_3Cl. The degeneracy is removed in the saturated molecules, and consequently two individual antisymmetric CH_3 stretching bands are observed. The methine CH stretch can be observed (as a weak peak near 2885 cm^{-1}) only when CH_3 and CH_2 groups are absent. The methoxy group CH_3O— has a characteristic sharp band of medium intensity near 2830 cm^{-1} separate from other CH stretching bands. Saturated cyclic hydrocarbon groups also have characteristic CH stretching bands in both infrared and Raman spectra. The frequencies range from 3100 cm^{-1} in cyclopropanes down to 2900 cm^{-1} in cyclohexanes and larger rings.

9-3f Aldehydes

The CH stretching mode of the aldehyde group appears as a Fermi doublet (see Section 8-2b) near 2820 and 2710 cm^{-1}. The 2710 cm^{-1} absorption is very useful and characteristic of aldehydes, but the higher frequency component often appears as a shoulder on the 2850 cm^{-1} CH_2 stretching band. In the spectrum of a purely aromatic

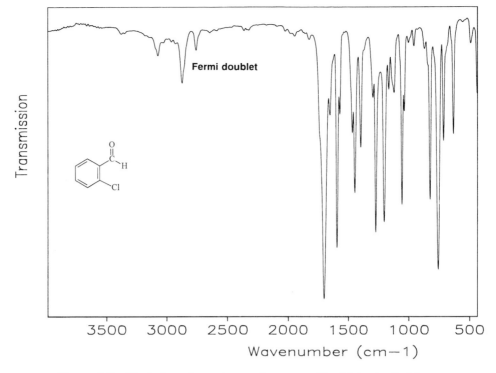

Figure 9-9 The infrared spectrum of an aromatic aldehyde, 2-chloro-benzaldehyde, showing the Fermi doublet at 2750 and 2870 cm^{-1}.

aldehyde there is no overlap of other CH stretching bands and the doublet is clearly resolved. An example is given in Figure 9-9. In ortho-substituted aromatic aldehydes the frequencies of the doublet are about 40 cm^{-1} higher than usual with the high frequency component stronger and broader (Figure 9-9).

9-4 THE CARBONYL STRETCHING REGION (1850–1650 cm^{-1})

9-4a Introduction

A very important region of the spectrum for structural analysis is the carbonyl stretching region. If we were to include metal carbonyls and salts of carboxylic acids, the region of the C=O stretching mode would extend from 2200 down to 1350 cm^{-1}. However, most organic compounds containing the C=O group show very strong infrared absorption in the range of 1850–1650 cm^{-1}. The actual position of the peak (or peaks) within this range is characteristic of the type of compound. At the upper end of the range are found bands due to anhydrides (two bands) and four-membered rings (β lactones), while acyclic amides and substituted ureas absorb at the lower end of the range. Table 9-6 summarizes the ranges in which compounds with a single carbonyl group absorb. Table 9-7 lists compounds that have two interacting carbonyl groups. In these compounds, out-of-phase and in-phase (antisymmetric and symmetric) C=O stretching vibrations can occur (see Section 8-2b). Usually, the out-of-phase mode is found at higher frequencies than the in-phase mode.

9-4b Compounds Containing a Single C=O Group

The type of functional group usually cannot be identified from the C=O stretching band alone, because there may be several carbonyl-containing functional groups that absorb within a given frequency range. However, an initial separation into possible

TABLE 9-6

Carbonyl Stretching Frequencies for Compounds Having One Carbonyl Group

Range (cm^{-1}) and Intensity*	Classes of Compounds	Remarks
1840–1820 (vs)	β lactones	4-membered ring
1810–1790 (vs)	Acid chlorides	Saturated aliphatic compounds
1800–1750 (vs)	Aromatic and unsaturated esters	The C=C stretch is higher than normal (1700–1650 cm^{-1})
1800–1740 (s)	Carboxylic acid monomer	Only observed in dil soln
1790–1740 (vs)	γ lactones	5-membered ring
1790–1760 (vs)	Aromatic or unsaturated acid chlorides	Second weaker combination band near 1740 cm^{-1} (Fermi resonance)
1780–1700 (s)	Lactams	Position depends on ring size
1770–1745 (vs)	α-Halo esters	Higher frequency due to electronegative halogen
1750–1740 (vs)	Cyclopentanones	Unconjugated structure
1750–1730 (vs)	Esters and δ lactones	Aliphatic compounds
1750–1700 (s)	Urethanes	R—O—(C=O)—NHR compounds
1745–1730 (vs)	α-Halo ketones	Noncyclic compounds
1740–1720 (vs)	Aldehydes	Aliphatic compounds
1740–1720 (vs)	α-Halo carboxylic acids	20 cm^{-1} higher frequency if halogen is fluorine
1730–1705 (vs)	Aryl and α,β-unsaturated aliphatic esters	Conjugated carbonyl group
1730–1700 (vs)	Ketones	Aliphatic and large ring alicyclic
1720–1680 (vs)	Aromatic aldehydes	Also α,β-unsaturated aliphatic aldehydes
1720–1680 (vs)	Carboxylic acid dimer	Broader band
1710–1640 (vs)	Thiol esters	Lower than normal esters
1700–1680 (vs)	Aromatic ketones	Position affected by substituents on ring
1700–1680 (vs)	Aromatic carboxylic acids	Dimer band
1700–1670 (s)	Primary and secondary amides	In dil soln
1700–1650 (vs)	Conjugated ketones	Check C=C stretch region
1690–1660 (vs)	Quinones	Position affected by substituents on ring
1680–1630 (vs)	Amides (solid state)	Note second peak due to NH def near 1625 cm^{-1}
1670–1660 (s)	Diaryl ketones	Position affected by substituents on ring
1670–1640 (s)	Ureas	Second peak due to NH def near 1590 cm^{-1}
1670–1630 (vs)	Ortho-OH or -NH$_2$ aromatic ketones	Frequency lowered by chelation with ortho group

*KEY: s = strong; m = medium; w = weak; v = very.

TABLE 9-7

Carbonyl Stretching Frequencies for Compounds Having Two Interacting Carbonyl Groups

Range (cm^{-1}) and Intensity*	Classes of Compounds	Remarks
1870–1840 (m–s) } 1800–1770 (vs) }	Cyclic anhydrides	Low-frequency band is stronger
1825–1815 (vs) } 1755–1745 (s) }	Normal anhydrides	High-frequency band is stronger
1780–1760 (m) } 1720–1700 (vs) }	Imides	Low-frequency band is broad; high-frequency band may be obscured
1760–1740 (vs)	α-Keto esters	Usually only one band
1740–1730 (vs)	β-Keto esters (keto form)	May be a doublet due to two C=O groups
1660–1640 (vs)	β-Keto esters (enol form)	May be a doublet due to a C=O and a C=C group
1710–1690 (vs) } 1640–1540 (vs) }	Diketones	High-frequency band due to keto form; low-frequency band due to enol form
1690–1660 (vs)	Quinones	Frequency depends on substituents
1650–1550 (vs) } 1440–1350 (s) }	Carboxylic acid salts	Two broad bands due to antisym and sym —C stretches

*KEY: s = strong; m = medium; w = weak; v = very.

compounds can be achieved with the information in Table 9-6. For example, a single carbonyl peak in the region 1750–1700 cm^{-1} could indicate an ester, an aldehyde, a ketone (including cyclic ketones), a large ring lactone, a urethane derivative, an α-halo ketone, or an α-halo carboxylic acid. The presence of the halogen could be checked from the elemental analysis, an aldehyde would have a peak near 2700 cm^{-1} (see Section 9-3f), and esters and lactones give a strong band near 1200 cm^{-1}, which is often quite broad in esters (see Section 9-9e). Urethanes would have an NH stretching band near 3300 cm^{-1} if hydrogen bonding is present, or near 3400 cm^{-1} if it is absent.

9-4c   Compounds Containing Two C=O Groups

Compounds with two coupled carbonyl groups are less common, but ambiguities can exist and other parts of the infrared spectrum must be analyzed in conjunction with the positions of the carbonyl bands for a particular structural grouping to be identified. The spectrum in Figure 9-10 is an example of this type of problem. An elemental analysis indicates that only C, H, and O are present and the molecular formula is $C_5H_8O_3$. The CH stretching region (Table 9-5) indicates that there is no unsaturated group present. This conclusion is confirmed by the absence of absorption between 1670 and 1540 cm^{-1}.

A CH$_3$ group or groups (bands at 2980 and 1355 cm^{-1}) and possibly a CH$_2$ group (bands at 2930 and 1415 cm^{-1}) are present. Now, turning to the carbonyl stretching region, we note from Tables 9-6 and 9-7 that the compound could be a cyclic ketone, an ester, a lactone, an aldehyde, or an α-keto ester. We can eliminate aldehyde (no band near 2700 cm^{-1}) and lactones or cyclic ketones (not possible with three oxygens and a methyl group). The bands at 1135 and 1020 cm^{-1} suggest an ester linkage. This assignment is supported by the weak band at 3440 cm^{-1} (see Table 9-4). The compound is then most likely to be an α-keto ester. Two possible structures are **9-1** and **9-2.** Structure **9-2** (ethyl pyruvate) is more likely because an O—CH$_3$ group would give a sharp band near 2830 cm^{-1} (see Section 9-3e).

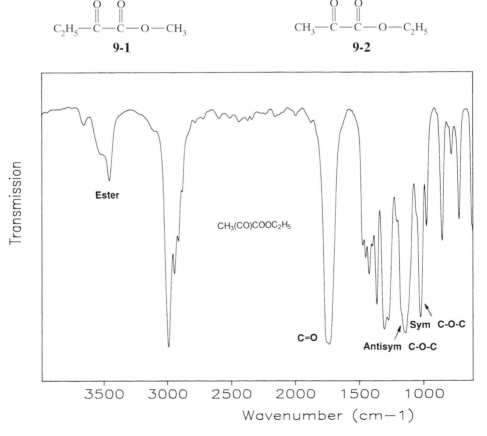

Figure 9-10   The infrared spectrum of an α-keto ester, ethyl pyruvate.

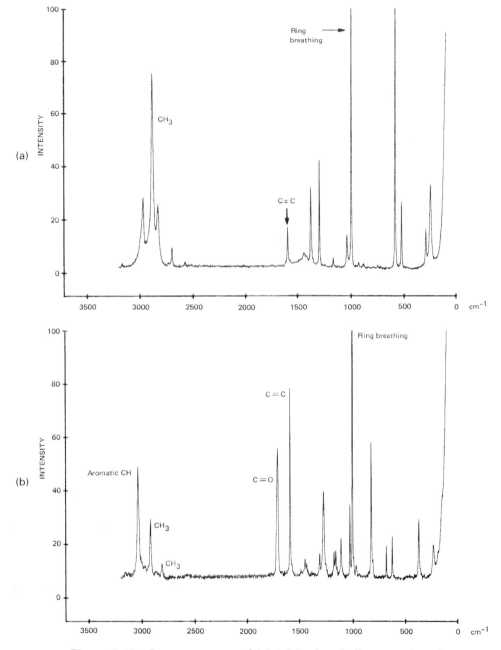

Figure 9-12 Raman spectra of (a) 1,3,5-trimethylbenzene (mesitylene) and (b) methyl benzoate.

9-6 COMPOUNDS CONTAINING METHYL GROUPS

9-6a General

In Section 8-2a, the vibrations of a methyl group were discussed in terms of the symmetry of the group. These vibrations (stretching, bending [deformation], rocking, and torsion) give rise to infrared absorption and Raman scattering in four different regions of the spectrum. The frequencies of CH stretching vibrations of methyl groups have been discussed in Section 9-3e. Antisymmetric deformation of the HCH angles of a CH_3 group gives rise to very strong infrared absorption and Raman scattering in the 1470–1440 cm^{-1} region. Bending of methylene ($-CH_2-$) groups also gives rise to a band in the same region. The symmetric CH_3 deformation gives a strong, sharp infrared band between 1380 and 1360 cm^{-1}. This band appears as a doublet when more than one

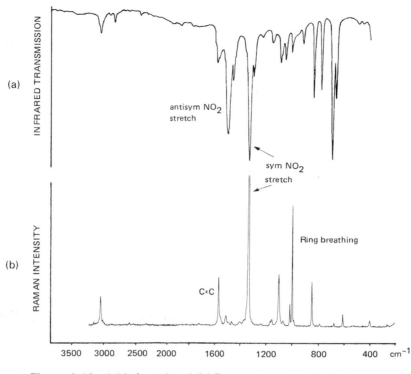

Figure 9-13 (a) Infrared and (b) Raman spectra of nitrobenzene.

TABLE 9-9

Frequencies of the Symmetric CH_3
Deformation in Various Compounds

Compounds	Group	Range (cm⁻¹)
Esters, ethers, etc.	O—CH_3	1460–1430
Amines, amides	N—CH_3	1440–1410
Hydrocarbons	C—CH_3	1380–1360
Sulfoxides, thioethers, etc.	S—CH_3	1330–1290
Phosphines	P—CH_3	1310–1280
Silanes	Si—CH_3	1280–1250
Organomercury compounds	Hg—CH_3	1210–1190

CH_3 group is attached to the same carbon atom and gives a good indication of the presence of isopropyl or *tert*-butyl groups (see Section 9-6b). When the methyl group is attached to an atom other than carbon, there is a significant shift in the symmetric CH_3 deformation frequency. Table 9-9 lists some typical frequency ranges.

The CH_3 rocking vibrations are usually coupled with skeletal modes and may be found anywhere between 1240 and 800 cm⁻¹. Medium to strong bands in both infrared and Raman spectra may be observed, but these are of little use for structure determination. The methyl torsion vibration has a frequency between 250 and 100 cm⁻¹, but often is not observed in either infrared or Raman spectra. In cases where torsional frequencies can be observed or estimated, information on rotational isomerism, conformation, and barriers to internal rotation can be obtained.

9-6b Isopropyl and *tert*-Butyl Groups

Isopropyl and *tert*-butyl groups give characteristic doublets in the symmetric CH_3 deformation region of the infrared spectrum. The isopropyl group gives a strong doublet at 1385/1370 cm⁻¹, whereas the *tert*-butyl group gives a strong band at 1370 cm⁻¹ with a weaker peak at 1395 cm⁻¹. Examples of these doublets are seen in Figure 9-14.

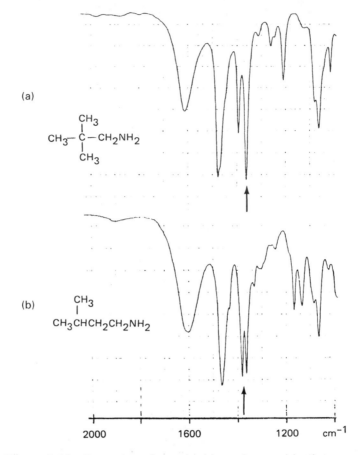

Figure 9-14 Examples of the doublets observed in the symmetric CH$_3$ deformation region for (a) a *tert*-butyl group and (b) an isopropyl group. (Reproduced with permission of Aldrich Chemical Co., Inc., from C.J. Pouchert [ed.], *The Aldrich Library of FT–IR Spectra.*)

9-7 COMPOUNDS CONTAINING METHYLENE GROUPS

9-7a Introduction

There are two kinds of methylene groups, the —CH$_2$— group in a saturated chain and the terminal =CH$_2$ group in vinyl, allyl, or vinylidene compounds. Diagrams of stretching, bending, wagging, twisting, and rocking motions of a CH$_2$ group are given in Figure 8-2. The CH stretching vibrations are discussed in Section 9-3e. However, bending, wagging, and rocking modes of CH$_2$ groups also give rise to important group frequencies.

9-7b CH$_2$ Bending (Scissoring)

The bending (sometimes called scissoring) motion of saturated —CH$_2$— groups gives a band of medium to strong intensity between 1480 and 1440 cm^{-1}. When the —CH$_2$— group is adjacent to a carbonyl or nitro group, the frequency is lowered to 1430–1420 cm^{-1}. A vinyl =CH$_2$ group gives a band of medium intensity between 1420 and 1410 cm^{-1}. This band is sometimes assigned as an in-plane deformation, since the two hydrogen atoms are in the same plane as the C=C group.

9-7c CH$_2$ Wagging and Twisting

The —CH$_2$— wagging and twisting frequencies in saturated groups are observed between 1350 and 1150 cm^{-1}. The infrared bands are weak unless an electronegative atom such as a halogen or sulfur is attached to the same carbon atom. The CH$_2$ twisting modes occur at the lower end of the frequency range and give very weak infrared absorption.

9-7d   CH$_2$ Rocking

A small band is observed near 725 cm^{-1} in the infrared spectrum when there are four or more —CH$_2$— groups in a chain. The intensity increases with the increasing chain length. This band is assigned to the rocking of the CH$_2$ groups in the chain. However, many compounds have bands in this region, so the CH$_2$ rocking band is only useful for aliphatic molecules. The band is always present in spectra recorded from Nujol mulls. An example of this band is seen later (Figure 9-19a).

9-7e   CH$_2$ Wagging in Vinyl and Vinylidene Compounds

The CH$_2$ wagging modes in vinyl and vinylidene compounds are found at much lower frequencies than in the saturated groups. In vinyl compounds, a strong band is observed in the infrared between 910 and 900 cm^{-1}. For vinylidene compounds the frequency range is 10 cm^{-1} lower. The overtone of the CH$_2$ wag often can be clearly seen as a band of medium intensity near 1820 cm^{-1} for vinyl and 1780 cm^{-1} for vinylidene compounds. These frequencies are raised above the normal range by halogens or other functional groups on the carbon atom. More is said about vinyl and vinylidene groups in Section 9-8d.

9-7f   Relative Numbers of CH$_2$ and CH$_3$ Groups

One further useful observation can be made concerning the H—C—H deformation bands in saturated parts of a molecule. When there are more —CH$_2$— groups than CH$_3$ groups present, the 1480–1440 cm^{-1} band is stronger than the 1380–1360 cm^{-1} band (symmetric CH$_3$ deformation). The relative intensities of these two bands, coupled with the 725 cm^{-1} band and the bands in the CH stretching region (see Figure 9-8), can give information on the relative numbers of —CH$_2$— and CH$_3$ groups, as well as on the length of the saturated carbon chain.

9-8   UNSATURATED COMPOUNDS

9-8a   The C=C Stretching Mode

Stretching of a C=C bond gives rise to a strong Raman line near 1650 cm^{-1}. The corresponding infrared band is often weak and sometimes is not observed at all in symmetrical molecules (see Figures 8-1a and 9-15a for examples). Weak absorption due to overtones or combinations may occur in this region and could be mistaken for a C=C stretching mode. In such cases a Raman spectrum will confirm the presence or absence of a C=C bond, because this group always gives a strong Raman line in the 1690–1560 cm^{-1} region (see Figures 8-1b and 9-15b).

The exact frequency of the C=C stretching mode gives some additional information on the environment of the double bond. Tri- or tetraalkyl-substituted groups and trans-disubstituted alkenes have frequencies in the 1690–1660 cm^{-1} range. These bands are weak or absent in the infrared but strong in the Raman. Vinyl and vinylidene compounds as well as cis alkenes absorb between 1660 and 1630 cm^{-1}. Substitution by halogens may shift the C=C stretching band out of the usual range. Fluorinated alkenes have very high C=C stretching frequencies (1800–1730 cm^{-1}). Chlorine and other heavy substituents, on the other hand, usually lower the frequency.

9-8b   Cyclic Compounds

The C=C stretching frequencies of cyclic unsaturated compounds depend on ring size and substitution. Cyclobutene, for example, has its C—C stretching mode at 1565 cm^{-1}, whereas cyclopentene, cyclohexene, and cycloheptene have C=C stretching frequen-

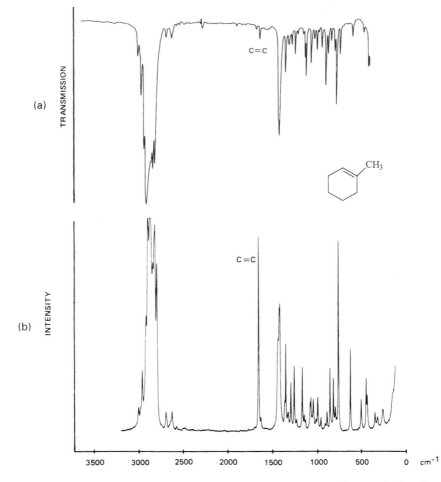

Figure 9-15   (a) Infrared and (b) Raman spectra of 1-methylcyclo-
hexene.

cies of 1610, 1645, and 1655 cm^{-1}, respectively. These frequencies are increased by substitution. An example can be seen in Figure 9-15a, in which the C=C stretching band
of 1-methylcyclohexene is found at 1670 cm^{-1}. The band is weak in the infrared spectrum, but strong in the Raman.

9-8c   The C≡C Stretching Mode

A small, sharp peak due to C≡C stretching is observed in the infrared spectra of terminal alkynes near 2100 cm^{-1} (see Figure 9-7). In substituted alkynes the band is shifted
by 100–150 cm^{-1} to higher frequencies. If the substitution is symmetric, no band is
observed in the infrared because there is no change in dipole moment during the C≡C
stretching vibration. Even when substitution is unsymmetrical, the band may be very
weak in the infrared and could be missed. In such cases, a Raman spectrum is very valuable, since the C≡C stretching mode always gives a strong line. An example is found
in the infrared and Raman spectra of the benzodiazapine drug pinazepam (see reference 9.10). There is a propynyl group attached to nitrogen atom 1 of the diazepine ring
of this compound. In the infrared spectrum only a very weak absorption is observed at
2116 cm^{-1}, whereas the Raman spectrum contains a very strong line at 2118 cm^{-1} due
to the C≡C stretching vibration.

9-8d   CH= and CH₂= Bending Modes

The CH and CH$_2$ wagging or out-of-plane bending modes are very important for structure identification in unsaturated compounds. They occur in the region between 1000

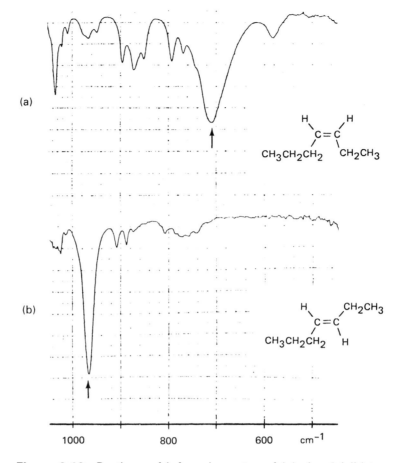

Figure 9-16 Portions of infrared spectra of (a) cis and (b) trans disubstituted alkenes showing the out-of-plane bending modes. (Reproduced with permission of Aldrich Chemical Co., Inc., from C.J. Pouchert (ed.), *The Aldrich Library of FT–IR Spectra.*)

and 650 cm^{-1}. The trans CH bending of a vinyl group gives rise to a strong infrared band between 1000 and 980 cm^{-1}, whereas a trans disubstituted alkene is characterized by a very strong band between 980 and 950 cm^{-1}. Electronegative groups tend to lower this frequency. The cis disubstituted alkenes give a medium to strong, but less reliable, band between 750 and 650 cm^{-1}. Infrared spectra of trans and cis disubstituted alkenes are shown in Figure 9-16.

The CH_2 out-of-plane wagging vibration of vinyl and vinylidene compounds gives a strong band between 910 and 890 cm^{-1}. This band coupled with the trans CH bending mode gives a very characteristic doublet (1000 and 900 cm^{-1}) and distinguishes the vinyl group from the vinylidene group (900 cm^{-1} only). Examples of infrared spectra of vinyl and vinylidene compounds are seen in Figure 9-17. The overtones of the out-of-plane CH bend and the CH_2 wagging modes give weak but characteristic bands near 1950 and 1800 cm^{-1}, respectively. These are clearly seen in Figure 9-17 and provide a useful confirmation of the structural grouping.

Cyclic alkenes usually have a strong band between 750 and 650 cm^{-1} due to out-of-plane bending of the two CH groups in a cis arrangement (Figure 9-18). However, when one of these hydrogens is substituted by another group, such as methyl, the band between 750 and 650 cm^{-1} is absent (Figure 9-15a). There are always several bands of medium intensity between 1200 and 800 cm^{-1} in the infrared spectra of cycloalkyl and cycloalkenyl compounds due to —CH_2— rocking. A summary of the out-of-plane CH bending or wagging modes in alkenic compounds is given in Table 9-10.

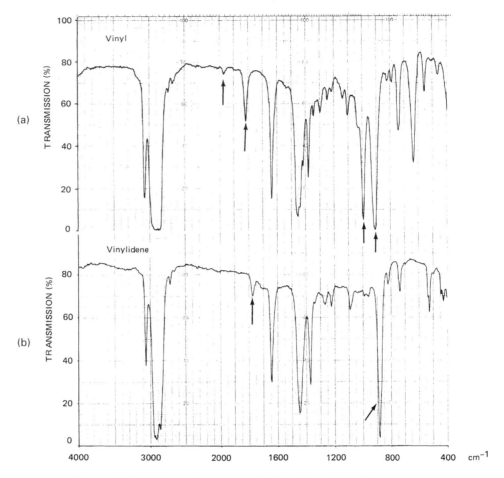

Figure 9-17 Infrared spectra of (a) 3,4-dimethyl-1-hexene, a compound containing a vinyl group, and (b) 2,3-dimethyl-1-pentene, a compound containing a vinylidene group.

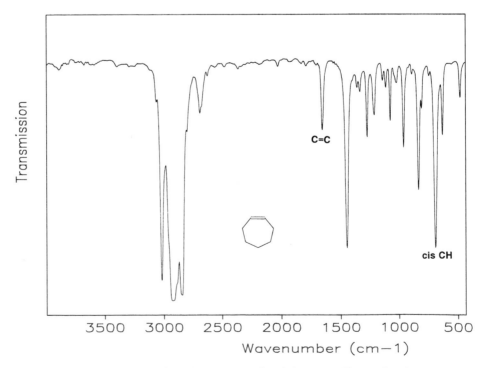

Figure 9-18 The infrared spectrum of cycloheptene, illustrating the out-of-plane bending of the cis CH groups at the double bond (690 cm^{-1}).

TABLE 9-10

CH$_2$ Wagging and CH Out-of-Plane Bending Frequencies of Alkenes in the Infrared

Range (cm^{-1}) and Intensity*	Group or Class	Assignment and Remarks
1000–980 (s)	Vinyl group (—CH=CH$_2$)	trans CH=CH bending, lower in vinyl ethers
980–950 (vs)	trans-Disubstituted alkenes (vinylenes)	CH=CH bending, frequency lowered by halogens
920–900 (s)	Vinyl group (—CH=CH$_2$)	CH$_2$ out-of-plane wagging, higher in vinyl ketones, but lower in vinyl ethers
900–880 (s)	Vinylidene group (C=CH$_2$)	Terminal =CH$_2$ out-of-plane wagging
750–650 (m–s)	cis-Disubstituted and cyclic alkenes	cis CH=CH bending

*Key: s = strong; m = medium; w = weak; v = very.

9-9 COMPOUNDS CONTAINING OXYGEN

9-9a General

Carboxylic acids and anhydrides, alcohols, phenols, and carbohydrates all give strong, often broad infrared absorption bands somewhere between 1400 and 900 cm^{-1}. These bands are associated with stretching of the C—O—C or C—OH bonds or bending of the C—O—H group. The position and multiplicity of the absorption, together with evidence from other regions of the spectrum, can help to distinguish the particular functional group. The usual frequency ranges for these groups in various compounds are summarized in Table 9-11. Aldehydes and ketones have no C—O—C or C—O—H group and are not mentioned in this section. However, they are included in 9-4b.

TABLE 9-11

C—O—C and C—O—H Group Vibrations

Range (cm^{-1}) and Intensity*	Group or Class	Assignment and Remarks
1440–1400 (m)	Aliphatic carboxylic acids	C—O—H deformation; may be obscured by CH$_3$ and CH$_2$ deformation bands
1430–1280 (m)	Alcohols	C—O—H deformation; broad band
1390–1310 (m–s)	Phenols	C—O—H deformation
1340–1160 (vs)	Phenols	C—O stretch; broad band with structure
1310–1250 (vs)	Aromatic esters	C—O—C antisym stretch
1300–1200 (s)	Aromatic carboxylic acids	C—O stretch
1300–1100 (vs)	Aliphatic esters	C—O—C antisym stretch
1160–1000 (s)		C—O—C sym stretch
1280–1220 (vs)	Alkyl aryl ethers	Aryl C—O stretch; a second band near 1030 cm^{-1}
1270–1200 (s)	Vinyl ethers	C—O—C stretch; a second band near 1050 cm^{-1}
1265–1245 (vs)	Acetate esters	C—O—C antisym stretch
1250–900 (s)	Cyclic ethers	C—O—C stretch, position varies with compound
1230–1000 (s)	Alcohols	C—O stretch; see below for more specific frequency ranges
1200–1180 (vs)	Formate and propionate esters	C—O—C stretch
1180–1150 (m)	Alkyl-substituted phenols	C—O stretch
1150–1050 (vs)	Aliphatic ethers	C—O—C stretch; usually centered near 1100 cm^{-1}
1150–1130 (s)	Tertiary alcohols	C—O stretch; lowered by chain branching or adjacent unsaturated groups
1110–1090 (s)	Secondary alcohols	C—O stretch; lowered 10–20 cm^{-1} by chain branching
1060–1040 (s–vs)	Primary alcohols	C—O stretch; often fairly broad
1060–1020 (s)	Saturated cyclic alcohols	C—O stretch; not cyclopropanol or cyclobutanol
1050–1000 (s)	Alkyl aryl ethers	Alkyl C—O stretch
960–900 (m–s)	Carboxylic acids	C—O—H deformation of dimer

*Key: s = strong; m = medium; w = weak; v = very.

9-9b   Ethers

The simplest structure with the C—O—C link is the ether group. Aliphatic ethers absorb near 1100 cm^{-1}, while alkyl aryl ethers have a very strong band between 1280 and 1220 cm^{-1} and another strong band between 1050 and 1000 cm^{-1}. In vinyl ethers the C—O—C stretching mode is found near 1200 cm^{-1}. Vinyl ethers can be further distinguished by a very strong C=C stretching band and the out-of-plane CH bending and CH$_2$ wagging bands, which are observed near 960 and 820 cm^{-1}, respectively. These frequencies are below the usual ranges discussed in Section 9-8d. Examples of these three types of ether are compared in Figure 9-19. Cyclic, saturated ethers such as tetrahydrofuran have a strong antisymmetric C—O—C stretching band in the range 1250–1150 cm^{-1}, whereas unsaturated, cyclic ethers have their C—O—C stretching modes at lower frequencies.

9-9c   Alcohols and Phenols

Alcohols and phenols in the pure liquid or solid state have broad bands due to hydrogen-bonded OH stretching. For alcohols, this band is centered near 3300 cm^{-1}, whereas in phenols the absorption maximum is 50–100 cm^{-1} lower. Phenols absorb near 1350 cm^{-1} due to the OH deformation and give a second broader, stronger band due to C—OH stretching near 1200 cm^{-1}. This second band always has fine structure from underlying aromatic CH in-plane deformation vibrations. Infrared spectra of an alcohol and a phenol are compared in Figure 9-20.

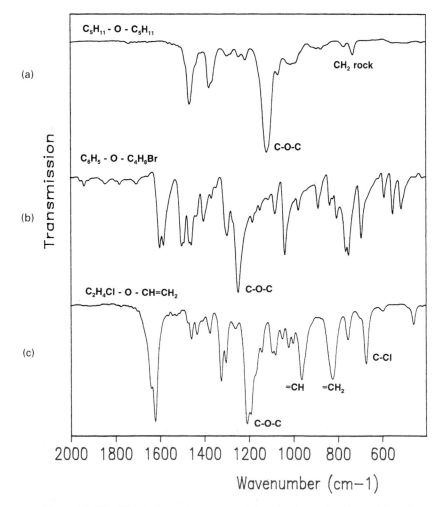

Figure 9-19   The infrared spectra of three types of ethers: (a) a simple aliphatic ether, (b) an alkyl aryl ether, and (c) a vinyl ether.

In simple alcohols the frequency of the C—OH stretch is raised by substitution on the C—OH carbon atom (see Table 9-11). Sugars and carbohydrates give very broad absorption bands centered near 3300 cm^{-1} (OH stretching), 1400 cm^{-1} (OH deformation), and 1000 cm^{-1} (C—OH stretching). The infrared spectrum of a sugar is shown in Figure 9-21.

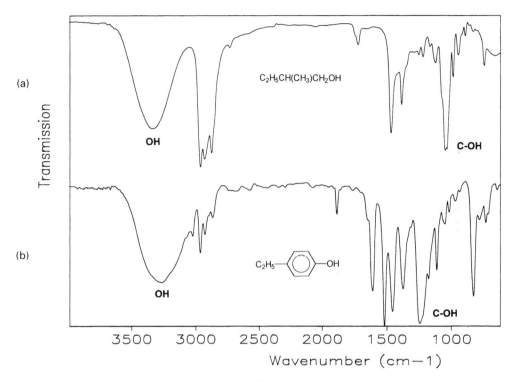

Figure 9-20 Infrared spectra of (a) an alcohol, 2-methyl-1-pentanol, and (b) a phenol, 4-ethylphenol.

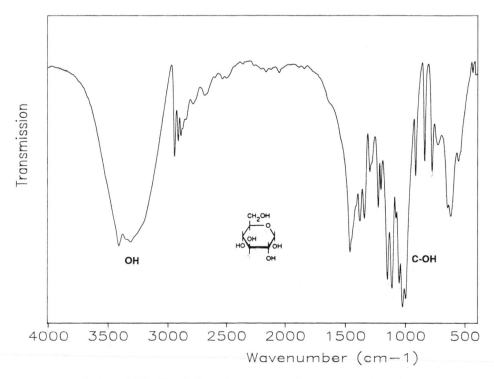

Figure 9-21 The infrared spectrum of a sugar, D-glucose.

9-9d Carboxylic Acids and Anhydrides

Carboxylic acids usually exist as dimers except in dilute solution. The carbonyl stretching band of the dimer is found near 1700 cm^{-1}, whereas in the monomer spectrum the band is located at higher frequencies (1800–1740 cm^{-1}). In addition to the very broad OH stretching band mentioned in Section 9-2a, the following three vibrations are associated with the C—OH group in carboxylic acids: a band of medium intensity near 1430 cm^{-1}, a stronger band near 1240 cm^{-1}, and another band of medium intensity near 930 cm^{-1}. The presence of an anhydride is detected by the characteristic absorption in the C=O stretching region, which consists of a strong sharp doublet with one band at unusually high frequency (1840–1800 cm^{-1}) and a second band at about 60 cm^{-1} lower (1780–1740 cm^{-1}). The C—O—C stretch gives rise to a broad band near 1150 cm^{-1} in open chain anhydrides and at higher frequencies in cyclic structures.

9-9e Esters

The antisymmetric C—O—C stretching mode in esters gives rise to a very strong and often quite broad band near 1200 cm^{-1}. The actual frequency of the maximum of this band can vary from 1290 cm^{-1} in benzoates down to 1100 cm^{-1} in aliphatic esters. There may be structure on this band due to CH deformation vibrations that absorb in the same region. The band may be even stronger than the C=O stretching band near 1750 cm^{-1}. The symmetric C—O—C stretch also gives a strong band at lower frequencies between 1160 and 1000 cm^{-1} in aliphatic esters.

9-10 COMPOUNDS CONTAINING NITROGEN

9-10a General

The presence of primary or secondary amines and amides can be detected by absorption due to stretching of NH$_2$ or NH groups between 3350 and 3200 cm^{-1}. Tertiary amines and amides, on the other hand, are more difficult to identify, because they have no N—H groups. Nitriles and nitro compounds also give characteristic infrared absorption bands near 2250 and 1530 cm^{-1}, respectively. Isocyanates and carbodiimides have very strong infrared bands near 2260 and 2140 cm^{-1}, respectively, where very few absorptions due to other groupings occur. Oximes, imines, and azo compounds give weak infrared bands in the 1700–1600 cm^{-1} region due to stretching vibrations of the —C=N—, or —N=N— group. Raman spectra are useful in these cases, because these groups give rise to strong Raman lines. Some characteristic group frequencies of nonheterocyclic nitrogen-containing compounds are listed in Table 9-12. Much of the information in this table is taken from Tables 8-2 and 8-3, but it is useful to collect the data on nitrogen-containing groups in a single table.

9-10b Amino Acids, Amines, and Amine Hydrohalides

Three classes of nitrogen-containing compounds (amino acids, amines, and amine hydrohalides) give rise to very characteristic broad absorption bands. Some of the most striking of these are found in the infrared spectra of amino acids, which contain an extremely broad band centered near 3000 cm^{-1} and often extending as low as 2200 cm^{-1}, with some structure (Figure 9-22). Amine hydrohalides (ammonium halides) give a similar, very broad band, which has structure on the low-frequency side. The center of the band tends to be lower than in amino acids, especially in the case of tertiary amine hydrohalides, in which the band center may be as low as 2500 cm^{-1}. In fact, this band gives a very useful indication of the presence of a tertiary amine hydrohalide (Figure 9-23).

Both amino acids and primary amine hydrohalides have a weak but characteristic band between 2200 and 2000 cm^{-1} (see Figure 9-23), which is believed to be a combination of the —NH$_3{}^+$ deformation near 1600 cm^{-1} and the —NH$_3{}^+$ torsion near 500 cm^{-1}.

TABLE 9-12

Details of Infrared Frequencies of Some Nitrogen-Containing Groups

Group and Class		Range (cm^{-1}) and Intensity*	Assignment and Remarks
The —NH$_2$ Group			
Primary amides	(dil soln)	3530–3520 (s)	NH$_2$ antisym stretch
		3400–3390 (s)	NH$_2$ sym stretch
	(solid state)	3360–3340 (vs)	NH$_2$ antisym stretch
		3190–3170 (vs)	NH$_2$ sym stretch
		1680–1660 (vs)	C=O stretch; (Amide I band)
		1650–1620 (m)	NH$_2$ deformation (Amide II band)
Primary amines	(dil soln)	3550–3350 (m)	NH$_2$ antisym stretch
		3450–3250 (m)	NH$_2$ sym stretch
	(condensed phase)	3450–3250 (m)	NH$_2$ stretching; br with structure
		1650–1590 (s)	NH$_2$ deformation
		850–750 (s)	NH$_2$ wagging; br
The —NH— Group			
Secondary amides		3450–3400 (m)	NH stretch; dil soln
		3300–3250 (m)	NH stretch; solid state
		3100–3060 (w)	Overtone band
		1680–1640 (vs)	C=O stretch (Amide I band)
		1560–1530 (vs)	Coupled NH deformation and C—N stretch (Amide II band)
		750–650 (s)	NH wag; br
Secondary amines		3500–3300 (m)	NH stretch
		750–650 (s, br)	NH wag
The C≡N Group			
Nitriles			
Saturated aliphatic		2260–2240 (w)	C≡N stretch; strong in Raman
Unsaturated aliphatic adjacent to			C≡N stretch; doublet when the adjacent
a double bond		2230–2220 (m)	double bond is disubstituted
Aromatic		2240–2220 (variable)	C≡N stretch; stronger than in saturated aliphatic nitriles
Isonitriles			
Alkyl		2180–2150 (w–m)	—N≡C stretch; strong in Raman
Aryl		2130–2110 (w–m)	—N≡C stretch; strong in Raman
The C=N Group			
Oximes		1690–1620 (w–m)	C=NOH stretch
Pyridines		1615–1565 (s)	Two bands, due to C=C and C=N stretch in ring
The C—N Group			
Amines and amides			
Primary aliphatic		1140–1070 (m)	C—C—N antisym stretch
Secondary aliphatic		1190–1130 (m–s)	C—N—C antisym stretch
Primary aromatic		1330–1260 (s)	Phenyl—N stretch
Secondary aromatic		1340–1250 (s)	Phenyl—N stretch
The NO$_2$ Group			
Aliphatic nitro compounds		1560–1530 (vs)	NO$_2$ antisym stretch
		1390–1370 (m–s)	NO$_2$ sym stretch
Aromatic nitro compounds		1540–1500 (vs)	NO$_2$ antisym stretch
		1370–1330 (s–vs)	NO$_2$ sym stretch
Nitrates R—O—NO$_2$		1660–1620 (vs)	NO$_2$ antisym stretch
		1300–1270 (s)	NO$_2$ sym stretch
		710–690 (s)	NO$_2$ deformation
The NO Groups (N=O, N—O, $\overset{+}{N}$—$\overset{-}{O}$)			
Nitrites R—ONO		1680–1650 (vs)	N=O stretch; often a weaker band is seen between 1630 and 1600 cm^{-1}
Oximes		965–930 (s)	N—O stretch
Nitrates R—O—NO$_2$		870–840 (s)	N—O stretch
N-Oxides			
Aromatic		1300–1200 (vs)	$\overset{+}{N}$—$\overset{-}{O}$ stretch
Aliphatic		970–950 (vs)	$\overset{+}{N}$—$\overset{-}{O}$ stretch
The N≡N Group			
Azides		2120–2160 (variable)	N≡N stretch; strong in Raman
The N=N Group			
Azo compounds		1450–1400 (vw)	N=N stretch; strong in Raman

*KEY: s = strong; m = medium; w = weak; v = very.

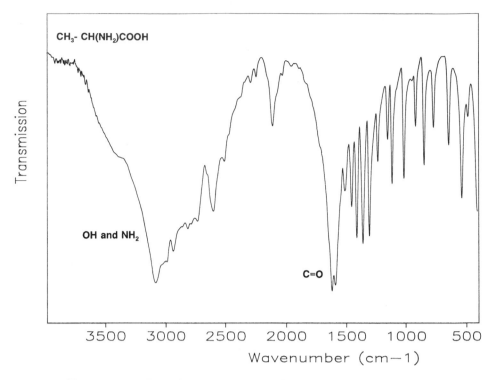

Figure 9-22   The infrared spectrum of an amino acid, L-alanine.

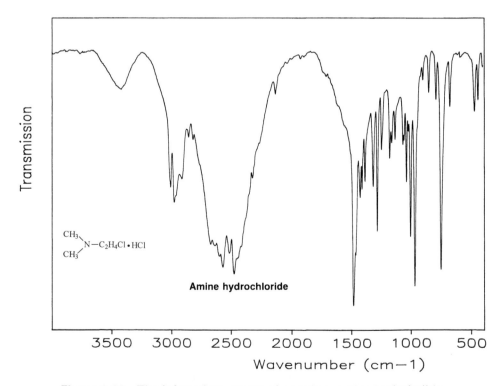

Figure 9-23   The infrared spectrum of a tertiary amine hydrohalide, (2-chloroethyl)dimethylamine hydrochloride.

Primary amines have a fairly broad band in their infrared spectra centered near 830 cm^{-1}, whereas the frequency for secondary amines is about 100 cm^{-1} lower (see Figures 9-2d and 9-2e). This band is not present in the spectra of tertiary amines or amine hydrohalides.

9-10c   Anilines

In anilines, the characteristic broad band shown by aliphatic amines in the 830–730 cm^{-1} region is not present, and so the out-of-plane CH deformations of the benzene ring can be observed. These bands permit the ring substitution pattern to be determined. Of course, when an aliphatic amine is joined to a benzene ring through a carbon chain, both the characteristic amine band and the CH deformation pattern are present.

The infrared spectrum of an aniline derivative is shown in Figure 9-24. Figure 9-25 shows the spectrum of an aliphatic amine joined to a benzene ring. The presence of the benzene ring is identified in both compounds by CH stretching bands between 3100 and 3000 cm^{-1}, out-of-plane CH bending bands between 850 and 700 cm^{-1}, and bands diagnostic of the substitution patterns between 2000 and 1700 cm^{-1}. In the out-of-plane bending region, the single band at 825 cm^{-1} in Figure 9-24 indicates the presence of a para-disubstituted benzene ring, whereas the doublet at 740 and 700 cm^{-1} in Figure 9-25 indicates monosubstitution (see Section 9-5).

9-10d   Nitriles

Saturated nitriles absorb weakly in the infrared near 2250 cm^{-1}, although the band is strong in the Raman spectrum. Unsaturated or aromatic nitriles for which the double bond or ring is adjacent to the C≡N group absorb more strongly in the infrared than saturated compounds, and the band occurs at somewhat lower frequencies near 2230 cm^{-1}.

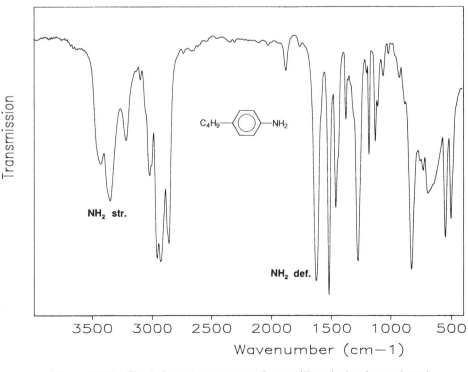

Figure 9-24   The infrared spectrum of an aniline derivative, *p*-butyl-aniline.

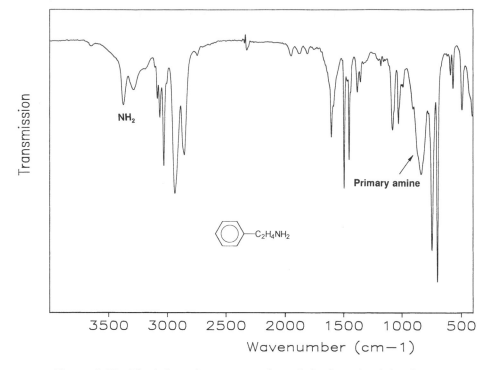

Figure 9-25 The infrared spectrum of an aliphatic amine joined to a benzene ring, phenethylamine.

9-10e Nitro Compounds

Nitro compounds have two very strong absorption bands in the infrared due to the anti-symmetric and symmetric NO_2 stretching vibrations. The symmetric stretch gives rise to a very strong Raman line (see Figure 9-14). In aliphatic compounds, the frequencies are near 1550 and 1380 cm^{-1}, whereas in aromatic compounds the bands are observed near 1520 and 1350 cm^{-1}. These frequencies are somewhat sensitive to nearby substituents. In particular, the 1350 cm^{-1} band in aromatic nitro compounds is intensified by electron-donating substituents in the ring. The out-of-plane CH bending patterns of ortho-, meta-, and para-disubstituted benzene rings are often perturbed in nitro compounds. Other compounds containing N—O bonds have strong characteristic infrared absorption bands (see Table 9-12).

9-10f Amides

9-3

Secondary amides (*N*-monosubstituted amides) usually have their NH and C=O groups trans to each other, as shown in structure **9-3.** The carbonyl stretching mode gives rise to a very strong infrared band between 1680 and 1640 cm^{-1}. This band is known as the Amide I band. A second, very strong absorption that occurs between 1560 and 1530 cm^{-1} is known as the Amide II band. It is believed to be due to coupling of the NH bending and C—N stretching vibrations. The trans amide linkage in structure **9-3** also gives rise to absorption between 1300 and 1250 cm^{-1} and to a broad band centered near 700 cm^{-1} (see Figure 9-2f). Occasionally, the amide linkage is cis in cyclic compounds such as lactams. In such cases, a strong NH stretching band is seen near 3200 cm^{-1} and a weaker combination band occurs near 3100 cm^{-1}, involving simultaneous excitation of C=O stretching and NH bending. The Amide II band is absent, but a cis NH bending mode absorbs between 1500 and 1450 cm^{-1}. This band may be confused with the CH_2 or antisymmetric CH_3 deformation bands.

9-10g Oximes

Oximes are solid derivatives of aldehydes and ketones useful for identification of these compounds. Oximes exhibit a very strong, broad absorption due to the hydrogen-bonded NOH group centered between 3250 and 3150 cm^{-1}. This band is observed between 3650 and 3500 cm^{-1} in CCl$_4$ solution. A weak band due to C=N stretching may be observed between 1685 and 1650 cm^{-1} in aliphatic oximes and about 30 cm^{-1} lower in aromatic oximes. This group vibration usually gives a strong band in the Raman spectrum. There is also a strong absorption between 965 and 930 cm^{-1} due to the N—O stretch.

9-11 COMPOUNDS CONTAINING PHOSPHORUS AND SULFUR

9-11a General

The presence of phosphorus in organic compounds can be detected by the infrared absorption bands arising from the P—H, P—OH, P—O—C, P=O, and P=S groups. A phosphorus atom directly attached to an aromatic ring is also well characterized. The usual frequencies of these groups in various compounds are listed in Table 9-13, and reference 9.10 is a good source of information on infrared spectra of organophosphorus compounds. Most of these groups absorb strongly or very strongly in the infrared, with the exception of P=S. The Raman spectrum is valuable for detecting this group, which has a frequency between 700 and 600 cm^{-1}. There is no characteristic P—C group frequency in aliphatic compounds.

The SO$_2$ and SO groups give rise to very strong infrared bands between 1400 and 1000 cm^{-1}. Other bonds involving sulfur, such as C—S, S—S, and S—H, give very weak infrared absorption, and a Raman spectrum is needed to identify these groups. Characteristic frequencies of some sulfur-containing groups are also listed in Table 9-13. The C=S group has been omitted from the table because the C=S stretching vibration is invariably coupled with vibrations of other groups in the molecule. Frequencies in the 1400–850 cm^{-1} range have been assigned to this group, with thioamides at the low-frequency end of the range. The infrared bands involving C=S groups are usually weak.

9-11b Phosphorus Acids and Esters

Phosphorus acids have P—OH groups that give one or two broad bands of medium intensity between 2700 and 2100 cm^{-1}. Esters and acid salts that have P—OH groups also absorb in this region. The presence of a PH group is indicated by a small, sharp band near 2400 cm^{-1}. In ethoxy and methoxy phosphorus compounds, as well as in other aliphatic compounds with a P—O—C linkage, a very strong and quite broad infrared band is observed between 1050 and 950 cm^{-1}. The presence of a P=O bond is indicated by a strong band close to 1250 cm^{-1}. An example of an aliphatic compound containing P—H, P=O, and P—O—C groups is given in Figure 9-26.

9-11c Aromatic Phosphorus Compounds

Aromatic phosphorus compounds have several characteristic group frequencies. A fairly strong, sharp infrared peak is observed near 1440 cm^{-1} in compounds in which a phosphorus atom is attached directly to a benzene ring. This bond may be seen in the spectrum of chlorodiphenyl-phosphine shown in Figure 9-27. A quaternary phosphorus atom attached to a benzene ring has a characteristic strong, sharp band near 1100 cm^{-1}. The P—O group attached to an aromatic ring gives rise to two strong bands between 1250 and 1160 cm^{-1} and between 1050 and 870 cm^{-1} due to stretching of the Ar—P—O linkage. When the P—O group is attached to an aromatic ring through the oxygen atom, the Ar—O—P group again gives two strong bands, but the higher frequency of these is found between 1350 and 1250 cm^{-1}. This arrangement is illustrated in the spectrum of phenyldichlorophosphate shown in Figure 9-28.

TABLE 9-13

Characteristic Infrared Frequencies of Groups Containing Phosphorus or Sulfur

Group and Class	Range (cm⁻¹) and Intensity*	Assignment and Remarks
The P—H Group		
Phosphorus acids and esters	2425–2325 (m)	P—H stretch
Phosphines	2320–2270 (m)	P—H stretch; sharp band
	1090–1080 (m)	PH_2 deformation
	990–910 (m–s)	P—H wag
The P—OH Group		
Phosphoric or phosphorus acids, esters and salts	2700–2100 (w)	OH stretch; one or two broad and often weak bands
	1040–920 (s)	P—OH stretch
The P—O—C Group		
Aliphatic compounds	1050–950 (vs)	Antisym P—O—C stretch
	830–750 (s)	Sym P—O—C stretch (methoxy and ethoxy phosphorus compounds only)
Aromatic compounds	1250–1160 (vs)	Aromatic C—O stretch
	1050–870 (vs)	P—O stretch
The P—C Group		
Aromatic compounds	1450–1430 (s)	P joined directly to a ring; sharp band
Quaternary aromatic	1110–1090 (s)	P⁺ joined directly to a ring; sharp band
The P=O Group		
Aliphatic compounds R—O—P—	1260–1240 (s)	Strong, sharp band
Aromatic compounds Ar—O—P—	1350–1300 (s)	Lower frequency (1250–1180 cm⁻¹) when OH group is attached to the P atom
Phosphine oxides	1200–1140 (s)	P=O stretch
The S—H Group		
Thiols (mercaptans)	2580–2500 (w)	S—H stretch; strong in Raman
The C—S Group	720–600 (w)	C—S stretch; strong in Raman
The S—S Group		
Disulfides	550–450 (vw or absent)	S—S stretch; strong in Raman
The S=O Group		
Sulfoxides	1060–1020 (vs)	S=O stretch
Dialkyl sulfites	1220–1190 (vs)	S=O stretch
The SO_2 Group		
Sulfones, sulfonamides, sulfonic acids, sulfonates, and sulfonyl chlorides	1390–1290 (vs)	SO_2 antisym stretch
	1190–1120 (vs)	SO_2 sym stretch
Dialkyl sulfates and sulfonyl fluorides	1420–1390 (vs)	SO_2 antisym stretch
	1220–1190 (vs)	SO_2 sym stretch
The S—O—C Group		
Dialkyl sulfites	1050–850 (vs)	S—O—C stretching (two bands)
Sulfates	1050–770 (vs)	Two or more bands

*KEY: s = strong; m = medium; w = weak; v = very.

9-11d   Compounds Containing C—S, S—S, and S—H Groups

Raman spectra of compounds containing C—S and S—S bonds contain very strong lines due to these groups between 700 and 600 and near 500 cm⁻¹, respectively. These group frequencies, especially S—S, are either absent or appear only very weakly in the infrared. The S—H stretching band near 2500 cm⁻¹ is normally quite weak in the infrared but shows a high intensity in the Raman spectrum. The spectrum of a simple aliphatic thiol is shown in Figure 9-29, where the S—H stretch can be seen at 2530 cm⁻¹ and the C—S stretch near 700 cm⁻¹. The compound contains an isopropyl group, and the symmetric deformations of the two methyl groups give rise to a strong characteristic doublet centered at 1375 cm⁻¹ (see Section 9-6b).

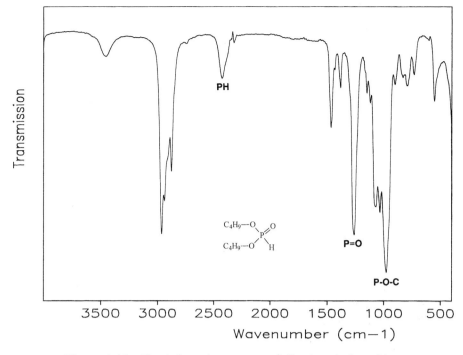

Figure 9-26   The infrared spectrum of di-*n*-butyl phosphite.

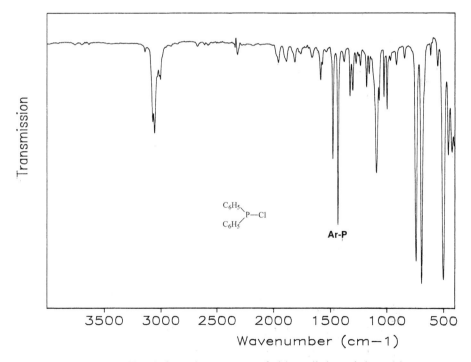

Figure 9-27   The infrared spectrum of chlorodiphenylphosphine.

9-11e Compounds Containing S═O Groups

The stretching vibration of the S═O group in sulfoxides gives rise to strong, broad infrared absorption near 1050 cm^{-1}. In the spectra of alkyl sulfites the S═O stretch is observed near 1200 cm^{-1}. Organic sulfones contain the SO$_2$ group, which gives rise to two very strong bands due to antisymmetric (1369–1290 cm^{-1}) and symmetric (1170–1120 cm^{-1}) stretching modes.

The frequencies of the SO_2 stretching vibrations of sulfonyl halides are about 50 cm^{-1} higher than those of sulfones due to the electronegative halogen atom. In the infrared spectra of sulfonic acids, a very broad absorption is observed between 3000 and 2000 cm^{-1} due to OH stretching. The antisymmetric and symmetric SO_2 stretching modes, respectively, give two strong bands near 1350 cm^{-1} and between 1200 and 1100 cm^{-1} in both alkyl and aryl sulfonic acids and sulfonates.

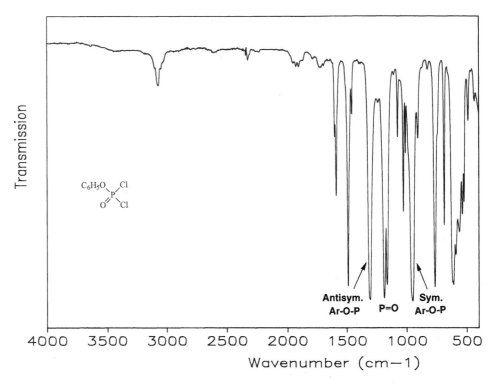

Figure 9-28 The infrared spectrum of phenyl dichlorophosphate.

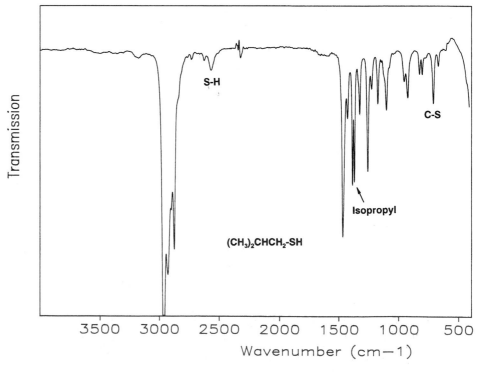

Figure 9-29 The infrared spectrum of an aliphatic thiol.

9-12 HETEROCYCLIC COMPOUNDS

9-12a   General

Heterocyclic compounds containing nitrogen, oxygen, or sulfur may exhibit three kinds of group frequencies: those involving CH or NH vibrations, those involving motion of the ring, and those due to the group frequencies of substituents on the ring. The identification of a heterocyclic compound from its infrared and Raman spectra is a difficult task and beyond the scope of an introductory treatment. However, characteristic frequencies for some heterocyclic compounds are collected in Table 9-14, and the identification of a few types of compounds are discussed in this section. Heterocyclic nitrogen compounds usually have strong Raman spectra, with very strong lines arising from —C≡N— groups when present. Details of infrared and Raman spectra of numerous classes of heterocyclic compounds may be found in references 9.2 and 9.5.

TABLE 9-14

Characteristic Infrared Frequencies for Some Heterocyclic Compounds

Classes of Compounds	Range (cm^{-1}) and Intensity*	Assignment and Remarks
Azoles (imidazoles, isoxazoles, oxazoles, pyrazoles, triazoles, tetrazoles)	3300–2500 (s, br)	H-bonded NH stretch; resembles carboxylic acids
	1650–1380 (m–s)	Three ring-stretching bands
	1040–980 (s)	Ring breathing
Carbazoles	3490–3470 (vs)	NH stretch (dil soln, nonpolar solvents)
1-4-Dioxanes	1460–1440 (vs)	CH_2 deformation
	1400–1150 (s)	CH_2 twist and wag
	1130–1000 (m)	Ring mode; strong in Raman
	850–830 (w)	Very strong in Raman
Furans	3140–3120 (m)	CH stretch; higher than most aromatics
	1600–1400 (m–s)	Ring stretching (three bands)
	770–720 (vs)	Band weakens as number of substituents increases
Indoles	3470–3450 (vs)	NH stretch
	1600–1500 (m–s)	Two bands
	900–600 (vs)	Substitution patterns due to both 6- and 5-membered rings
Pyridines (general)	3080–3020 (w–m)	CH stretch; several bands
	2080–1670 (w)	Combination bands
	1615–1565 (s)	Two bands due to C═C and C═N stretch in ring
	1030–990 (s)	Ring-breathing
2-substituted	780–740 (s)	CH out-of-plane deformation
	630–605 (m–s)	In-plane ring deformation
	420–400 (s)	Out-of-plane ring deformation
3-substituted	820–770 (s)	CH out-of-plane deformation
	730–690 (s)	Ring deformation
	635–610 (m–s)	In-plane ring deformation
	420–380 (s)	Out-of-plane ring deformation
4-substituted	850–790 (s)	CH out-of-plane deformation
disubstituted	830–810 (s)	Two bands due to CH out-of-plane deformations
	740–720 (s)	
trisubstituted	730–720 (s)	CH out-of-plane deformation
Pyrimidines	1590–1370 (m–s)	Ring stretching; four bands
	685–660 (m–vs)	Ring deformation
Pyrroles	3480–3430 (vs)	NH stretch; often a sharp band
	3130–3120 (w)	CH stretch; higher than normal
	1560–1390 (variable)	Ring stretch; usually three bands
	770–720 (s, br)	CH out-of-plane deformation
Thiophenes	1590–1350 (m–vs)	Several bands due to ring stretching modes
	810–680 (vs)	CH out-of-plane deformation; lower than in pyrroles and furans
Triazines	1560–1520 (vs)	Two bands due to ring stretching modes
	1420–1400 (s)	
	820–740 (s)	Out-of-plane ring deformation

*KEY: s = strong; m = medium; w = weak; v = very.

9-12b   Aromatic Heterocycles

Hydrogen atoms attached to carbon atoms in an aromatic heterocyclic ring such as pyridine give rise to CH stretching modes in the usual 3100–3000 cm^{-1} region, or a little higher in furans, pyrroles, and some other compounds. Characteristic ring stretching modes, similar to those of benzene derivatives, are observed between 1600 and 1000 cm^{-1}. The out-of-plane CH deformation vibrations give rise to strong infrared bands in the 1000–650 cm^{-1} region. In some cases, these patterns are characteristic of the type of substitution in the heterocyclic ring, for example, furans, indoles, pyridines, pyrimidines, and quinolines. The in-plane CH bending modes also give several bands in the 1300–1000 cm^{-1} region for aromatic heterocyclic compounds. CH vibrations in benzene derivatives and analogous modes in related heterocyclic compounds can be correlated and may be useful in structure determination.

Overtone and combination bands are observed between 2000 and 1750 cm^{-1} in the infrared spectra. These bands are similar to those observed for benzene derivatives and are characteristic of the position of substitution. In aromatic heterocyclic compounds involving nitrogen, the coupled C=C and C=N stretching modes give rise to several characteristic vibrations. These are similar in frequency to their counterparts in the corresponding nonheterocyclic compounds and give rise to very strong Raman lines. Ring stretching modes are found in the 1600–1300 cm^{-1} region. Other skeletal ring modes include ring breathing modes near 1000 cm^{-1}, in-plane ring deformation between 700 and 600 cm^{-1}, and out-of-plane ring deformation modes, which may be observed between 700 and 300 cm^{-1}.

Nonaromatic heterocyclic compounds usually have one or more CH$_2$ groups present. The stretching and deformation (scissoring) modes give rise to bands in the usual regions (see Section 9-7). However, the wagging, twisting, and rocking modes often interact with skeletal ring modes and may be observed over a wide range of frequencies.

9-12c   Pyrimidines and Purines

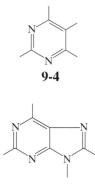

9-4

9-5

Pyrimidines (structure **9-4**) and purines (structure **9-5**) absorb strongly in the infrared between 1640 and 1520 cm^{-1} due to C=C and C=N stretching of the ring. A band near 1630 cm^{-1} is attributed to C=N stretching and a second band between 1580 and 1520 cm^{-1} is assigned to a C=C stretch. The C=N stretch usually gives rise to a very strong Raman line in these and related heterocyclic compounds. Pyrimidines and purines usually have absorption bands between 700 and 600 cm^{-1} due to CH out-of-plane bending. Nitrogen heterocycles can form N-oxides, which have a characteristic very strong infrared band near 1280 cm^{-1}.

9-12d   Five-Membered Ring Compounds

Pyrroles, furans and thiophenes generally have a band in their infrared spectra due to C=C stretching near 1580 cm^{-1}. A strong band also is observed between 800 and 700 cm^{-1} due to an out-of-plane deformation vibration of the CH=CH group, similar to that of cis-disubstituted alkenes. In the spectra of pyrroles, a strong, broad band is observed between 3400 and 3000 cm^{-1} due to the H-bonded N—H stretching mode. Furans have medium to strong infrared bands between 1610 and 1560 cm^{-1}, 1520 and 1470 cm^{-1}, and 1400 and 1390 cm^{-1} due to ring stretching vibrations. All furans have a strong absorption near 595 cm^{-1}, which is attributed to a ring deformation mode.

Thiophenes absorb in the infrared between 3100 and 3000 cm^{-1} (CH stretching), 1550 and 1200 cm^{-1} (ring stretching), and 750 and 650 cm^{-1} (out-of-plane C—H bending). Infrared spectra of thiophenes generally have a band between 530 and 450 cm^{-1} due to an out-of-plane ring deformation. In the Raman spectra of 3-substituted thiophenes, a very strong line near 840 cm^{-1} is attributed to the symmetric C—S stretching mode. A very strong ring stretching band is observed in the Raman spectra of 2-substituted thiophenes between 1440 and 1400 cm^{-1}.

9-12e   NH Stretching Bands

Spectra of heterocyclic nitrogen compounds may contain bands due to a secondary or tertiary amine group. Pyrroles, indoles, and carbazoles in nonpolar solvents have their NH stretching vibrations between 3500 and 3450 cm^{-1}, and the band is very strong in the infrared. In saturated heterocyclics, such as pyrrolidines and piperidines, the band is at lower frequencies. Azoles have a very broad, hydrogen-bonded NH stretching band between 3300 and 2500 cm^{-1}. This band might be confused with the broad OH stretching band of carboxylic acids.

9-13   COMPOUNDS CONTAINING HALOGENS

9-13a   General

A halogen atom attached to a carbon atom adjacent to a functional group often causes a significant shift in the group frequency. Some examples are listed in Table 9-15. Fluorine is particularly important in this regard, and special care must be exercised in conclusions drawn from infrared and Raman spectra when this element is present. Carbon-fluorine stretching bands are very strong in the infrared, usually between 1350 and 1100 cm^{-1}, but they are often weak in the Raman. Other functional groups that absorb in this region of the spectrum may be hidden by the CF stretching band. These groups often can be detected in the Raman spectrum. It should be mentioned that there are many known cases of symmetrical C—F stretching modes at frequencies much lower than the usual 1350–1100 cm^{-1}. The usual regions for the C—X stretching and bending vibrations have been given previously in Tables 8-2, 8-3, and 8-4.

TABLE 9-15
The Effect of Halogen Substituents on Some Group Frequencies

Group or Class		Range (cm^{-1})	Assignment and Remarks
Fluorine			
Fluorocarbons	—FCH— and CF$_2$H	3010–2990	CH stretch; higher frequency than normal
	F$_2$C=CF—	1870–1800	C=C stretch; much higher frequency than normal
	F$_2$C=C$\diagdown^{\diagup}$	1760–1730	C—C stretch; much higher frequency than normal
Acid fluorides	F—C=O	1900–1820	C=O stretch; extremely high carbonyl group frequency
Ketones	—CF$_2$COCH$_2$ and —CD$_2$COCF$_2$	1800–1770	C=O stretch; the normal range for ketones is 1730–1700 cm^{-1}
Carboxylic acids	—CF$_2$—COOH	1780–1740	C=O stretch; the normal range for carboxylic acids (dimers) is 1720–1680 cm^{-1}
Nitriles	—CF$_2$—C≡N	2280–2260	C≡N stretch; 20 cm^{-1} higher than normal
Amides	—CF$_2$—CONH$_2$	1730–1700	C=O stretch; 30 cm^{-1} higher than normal
Chlorine, Bromine, and Iodine			
CH$_2$Cl		1300–1240	CH$_2$ wag; strong IR band
CH$_2$Br		1240–1190	CH$_2$ wag; strong IR band
CH$_2$I		1190–1150	CH$_2$ wag; strong IR band
Acid chlorides		1810–1790	C=O stretch
α-Halo esters		1770–1745	C=O stretch; 20 cm^{-1} higher than the normal range
Noncyclic halo ketones		1745–1730	C=O stretch; 15 cm^{-1} higher than the normal range
α-Halo carboxylic acids		1740–1720	C=O stretch; the normal range is 1720–1680 cm^{-1}
Chloroformates		1800–1760	C=O stretch; near 1720 cm^{-1} in formate esters
α-Chloro aldehydes		1770–1730	Higher than normal aldehydes

9-13b CH₂X Groups

The CH_2 wagging mode in compounds with a CH_2X group gives rise to a strong band whose frequency depends on X. When X is Cl, the range is $1300-1250$ cm^{-1}. For Br, the band is near 1230 cm^{-1}, and for I, a still lower frequency near 1170 cm^{-1} is observed. The infrared absorption regions of halogenomethyl groups in a large number of compounds are discussed in reference 9.1.

9-13c Haloalkyl Groups

In haloalkyl groups, the presence of more than one halogen atom on a single carbon atom shifts the C—X stretching frequency to the high wavenumber end of the range. The antisymmetric stretching frequency of the CCl_3 group gives rise to a band in the infrared spectrum in the $830-700$ cm^{-1} range. The infrared absorption regions of trihalogenomethyl groups in a large number of compounds are discussed in reference 9.1.

9-13d Aromatic Halogen Compounds

Halogen atoms attached to aromatic rings are involved in certain vibrations that are sensitive to the mass of the halogen atom. One of the benzene ring vibrations that involves motion of the substituent atom gives rise to bands between 1250 and 1100 cm^{-1} when the substituent is fluorine, between 1100 and 1040 cm^{-1} for chlorine, and between 1070 and 1020 cm^{-1} for bromine.

9-14 BORON, SILICON, AND ORGANOMETALLIC COMPOUNDS

Boron–carbon and silicon–carbon stretching modes are not usually identifiable, since they are coupled with other skeletal modes. However, the C—B—C antisymmetric stretching mode in phenylboron compounds gives a strong infrared band between 1280 and 1250 cm^{-1}, and a silicon atom attached to an aromatic ring gives two very strong bands near 1430 and 1110 cm^{-1}. Metal–carbon stretching frequencies are found between 600 and 400 cm^{-1} with the lighter metals at the high frequency end of the range, as expected from eq. 7-1. Stretching of these bonds usually gives rise to very strong Raman lines.

TABLE 9-16

Some Infrared Group Frequencies in Boron and Silicon Compounds

	Group	Range (cm⁻¹) and Intensity*	Assignment and Remarks
Boron	—BOH	3300–3200 (s)	Broad band due to H-bonded OH stretch
	—BH and —BH₂	2650–2350 (s)	Doublet for —BH₂ stretch
		1200–1150 (ms)	—BH₂ deformation of B—H bend
		980–920 (m)	—BH₂ wag
		ca 1430 (m–s)	Benzene ring vibration
	B—N	1460–1330 (vs)	B—N stretch; borazines and aminoboranes
	B—O	1380–1310 (vvs)	B—O stretch; boronates, boronic acids
	C—B—C	1280–1250 (vs)	C—B—C antisym stretch
Silicon	—SiOH	3700–3200 (s)	OH stretch, similar to alcohols
		900–820 (s)	Si—O stretch
	—SiH, —SiH₂, and —SiH₃	2150–2100 (m)	Si—H stretch
		950–800 (s)	Si—H deformation and wag
	Si—Ar	ca 1430 (m–s)	Ring mode
		1100 (vs)	Ring mode
	Si—O—C (aliphatic)	1100–1050 (vvs)	Si—O—C antisym stretch
	Si—O—Ar	970–920 (vs)	Si—O stretch
	Si—O—Si	1100–1000 (s)	Si—O—Si antisym stretch

*KEY: s = strong; m = medium; w = weak; v = very.

The B—O and B—N bonds in organoboron compounds give very strong infrared bands between 1430 and 1330 cm^{-1}. The Si—O—C vibration gives a very strong infrared absorption, which is often quite broad in the 1100–1050 cm^{-1} range. Some characteristic frequencies for boron and silicon compounds are listed in Table 9-16. The B—CH$_3$ and Si—CH$_3$ symmetric CH$_3$ deformation modes occur at 1330–1280 cm^{-1} and 1280–1250 cm^{-1}, respectively. The CH$_3$ deformations in metal–CH$_3$ groups give rise to bands between 1210 and 1180 cm^{-1} in organomercury and organotin compounds and between 1170–1150 cm^{-1} in organolead compounds.

The infrared spectra of aromatic organometallic compounds usually contain a fairly strong, sharp band near 1430 cm^{-1} due to a benzene ring vibration. This band has been observed for compounds in which As, Sb, Sn, Pb, B, Si, and P atoms are attached directly to the ring.

The mercury–carbon bond in aliphatic organomercury compounds is characterized by a very strong Raman band between 550 and 500 cm^{-1}. For aromatic mercury compounds, a band between 250 and 200 cm^{-1} is assigned to the phenyl–Hg stretch. These bands are very strong and they can be seen in Raman spectra of dilute aqueous solutions.

9-15 ISOTOPICALLY LABELED COMPOUNDS

9-15a The Effect of ^{2}H and ^{13}C Isotopic Substitution on Stretching Modes

In Section 8-1 the vibrational frequency of a diatomic group was given by the equation $v(\text{cm}^{-1}) = 130.3\sqrt{k/\mu}$, in which k is the force constant (N m^{-1}) and μ is the reduced mass $(m_1 m_2 / m_1 + m_2)$. Isotopic substitution does not change the force constant, so the frequency of the isotopically substituted group is given by:

$$v(\text{isotopic}) = v(\text{normal})\sqrt{\mu(\text{normal})/\mu(\text{isotopic})}$$

For example, consider the CH group for which $\mu(\text{normal}) = 0.923$ amu. The reduced mass for the CD group is $\mu(\text{isotopic}) = 1.714$ amu. If the CH stretch is observed at 3000 cm^{-1}, the CD stretch is expected at: $3000 \times 0.923/1.714 = 2200$ cm^{-1}. In general, bands due to CD stretching modes are observed in the range 2300–2100 cm^{-1}. Similar calculations may be made for stretching modes of other X—D groups. The results are given in Table 9-17. Much smaller changes in the Raman and infrared spectra (band shifts) are predicted for ^{13}C isotopic substitution. Calculated frequencies for ^{13}C—H, ^{13}C=O, and ^{13}C≡N groups are included in Table 9-17.

9-15b The Effect of Deuterium Substitution on Bending Modes

All types of C—H bending vibrations are displaced by a factor of approximately 0.7 ($\sqrt{1/2}$) to lower frequencies when hydrogen atoms are replaced by deuterium. The disappearance of a band due to a C—H bending mode is usually of more analytical

TABLE 9-17

Calculated Wavenumbers of Isotopically Substituted Groups

Group	Normal Frequency (cm^{-1})	Isotopic Frequency (cm^{-1})	Isotope Shift (cm^{-1})
O—D	3500	2550	−950
N—D	3300	2400	−900
C—D	3000	2200	−800
^{13}C—H	3000	2990	−10
^{13}C=O	1720	1680	−40
^{13}C≡N	2350	2300	−50
X—C—D bend	1400	990	−500

value than the appearance of the corresponding C—D band. The new CD group vibration is likely to couple with C—C skeletal stretching vibrations and often cannot be identified. The most important CH group vibrations are the methylene bending (scissoring) vibration found between 1470 and 1400 cm^{-1} and the symmetric methyl deformation mode observed between 1385 and 1360 cm^{-1}.

For example, selective deuteration at the active α-positions of cyclopentanone can be observed in the infrared spectrum. In the spectrum of cyclopentanone, bands due to CH bending are observed at 1455 and 1406 cm^{-1}. In the cyclopentanone-$\alpha,\alpha,\alpha,\alpha$-$d_4$ spectrum, only the 1455 cm^{-1} band remains in this region. This observation shows that the 1406 cm^{-1} band is due to the active α-methylenes, whereas the 1455 cm^{-1} band is due to the β-methylene groups.

The infrared spectrum of 3-pentanone ($CH_3CH_2COCH_2CH_3$) contains bands at 1461, 1414, 1379, and 1365 cm^{-1}. Selective deuteration of this molecule is demonstrated by the disappearance of bands in the infrared spectrum at 1461 and 1379 cm^{-1} when the methyl groups are deuterated ($CD_3CH_2COCH_2CD_3$) and the disappearance of bands at 1414 and 1365 cm^{-1} when the CH_2 groups are deuterated ($CH_3CD_2COCD_2CH_3$). The bands at 1461 and 1379 cm^{-1} are due to the antisymmetric and symmetric methyl deformations, the band at 1414 cm^{-1} is attributed to the CH_2 scissors vibration, and the 1365 cm^{-1} band is due to a methylene vibration coupled with the C—CO stretch.

Infrared bands arising from the —CH_2—CO— group in some natural products can be identified by deuteration. For example, in the infrared spectrum of methyl laurate ($CH_3(CH_2)_9CH_2COOCH_3$), a shoulder at 1420 cm^{-1} due to the CH_2 "scissors" vibration disappears when the α-CH_2 group is deuterated. At the same time a second band at 1365 cm^{-1} due to a methylene vibration coupled with the C—CO stretch also disappears. Similarly, deuteration of the α-methylene group in the 17-ketosteroid **9-6** results in the disappearance of a band in the infrared spectrum at 1408 cm^{-1}. There are two α-methylene groups in the 3-ketosteroid **9-7**, and bands at 1432 and 1422 cm^{-1} both disappear on deuteration at carbon atoms 2 and 4.

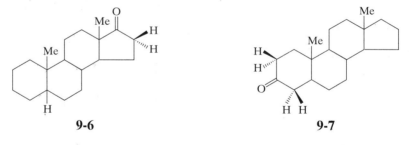

9-6   9-7

9-16 USING THE LITERATURE ON VIBRATIONAL SPECTRA

We have seen that vibrational spectra are extremely useful for checking for the presence of a functional group in a compound. Positive identification of a compound is obtained if its infrared or Raman spectrum exactly matches the spectrum of a known compound. To make use of this method, collections of spectra or references to spectra in the literature are needed. Published spectra, either in the literature or in collections, are currently available for several hundred thousand compounds.

In this Section, sources of collections of spectra, sources of literature, and references to spectra are listed.

9-16a Collections of Spectra

In addition to the very extensive *Sadtler Standard Infrared Spectra* collection and other large collections of infrared spectra, there are several smaller collections of infrared, FT–IR, and Raman spectra. Perhaps the most useful of these are *The Aldrich Library of FT–IR Spectra* and *The Coblentz Society Desk Book of Infrared Spectra*.

9-16b Infrared Spectra

1. *Sadtler Standard Infrared Spectra,* Sadtler Research Laboratories, Inc., 3314 Spring Garden Street, Philadelphia, PA 19104.

    The main collection consists of prism or grating spectra in loose-leaf volumes, containing 1000 spectra per volume. The format of the earlier spectra is linear in wavelength (2–15 µm). The index to this collection consists of the following four sections: Chemical Classes, Alphabetical, Molecular Formula, and Numerical.

2. *The Aldrich Library of FT–IR Spectra,* C.J. Pouchert (ed.), Aldrich Chemical Co., Inc., P.O. Box 355, Milwaukee, WI 53201.

    A three-volume set containing FT–IR spectra arranged 8 to a page, with alphabetic and molecular formula indices.

3. *Selected Infrared Spectral Data,* American Petroleum Institute (API), Research Project 44, Department of Chemistry, Texas A&M University, College Station, TX 77843.

    A large collection of spectra that is continually updated. The presentation is usually linear in wavelength for the older entries in the collection, but linear in wavenumber for the more recent spectra. This is the most extensive collection of spectra of high-purity petroleum hydrocarbons; also included are nitrogen and sulfur compounds found in petroleum.

4. *Coblentz Society Spectra,* P.O. Box 9952, Kirkwood, MO 63122.

    Ten thousand spectra in volumes of 1000 spectra each, in notebook format.

5. *The Coblentz Society Desk Book of Infrared Spectra,* 2nd ed. C.D. Craver (ed.), The Coblentz Society Inc., P.O. Box 9952, Kirkwood, MO 63122 (1982).

    Eight hundred and seventy grating spectra grouped by chemical classes, with text; designed as a reference and teaching aid.

9-16c Raman Spectra

1. *Sadtler Standard Raman Spectra,* Sadtler Research Laboratories, Inc., 3314 Spring Garden Street, Philadelphia, PA 19104.

    A collection of Raman spectra. Both parallel and perpendicular polarized spectra are presented together with the corresponding infrared spectrum.

2. *Selected Raman Spectra Data,* American Petroleum Institute (API), Research Project 44, Department of Chemistry, Texas A&M University, College Station, TX 77843.

    This compilation is produced in the same format as the API infrared spectra (see Section 9-16b, reference 3). There are 500 Raman spectra obtained using mercury vapor lamp excitation and 200 laser-excited spectra.

3. *Thermodynamic Research Center Data Project,* Chemistry Department, Texas A&M University, College Station, TX, 77843.

    Similar to the API publication described above, but with the emphasis on spectra of petrochemicals and other major industrial chemicals.

4. *Characteristic Raman Frequencies of Organic Compounds,* F.E. Dollish, W.G. Fateley, and F.F. Bentley, Wiley–Interscience, New York, 1973.

    This work includes 108 representative Raman spectra.

5. *Introductory Raman spectroscopy,* J.R. Ferraro and K. Nakamoto, Academic Press, San Diego, 1994.

    This book contains Raman spectra of 30 solvents.

6. *Analytical Chemistry,* **19,** 700-765 (1947) and **22,** 1074-1114 (1950).

    These two articles contain Raman spectra with tables of frequencies and relative intensities of 291 hydrocarbons and oxygenated compounds. The spectra were obtained using mercury vapor lamp excitation.

7. *Ramanspektren,* K.F.W. Kohlrausch, Heyden and Sons Ltd., London, 1972.

    Reprinted from the original German edition. This work contains data on Raman spectra obtained using mercury vapor lamp excitation.

9-16d   Sources of References to Published Spectra

To find a spectrum in a collection or in the original literature it is necessary to have a reference. The following list of indices is quite useful. The ultimate source is, of course, the *Chemical Abstracts Index*. To use this index, first look for the reference to a spectrum of a compound in the 5- or 10-year cumulative indices under Spectra, Infrared (or Raman), then look up the abstract. The abstract gives the reference to the paper in which the spectrum was published. However, this can be a tedious procedure, leading to many papers in which the complete spectrum may not be included.

1. *American Society for Testing and Materials (ASTM),* distributed by Sadtler Research Laboratories, Inc., 3314 Spring Garden Street, Philadelphia, PA 19104.

    This source contains comprehensive indices for the infrared spectra in the general collections listed above in Section 9-16b, plus infrared spectra abstracted from technical journals through 1972. There is a molecular formula list and a serial number list, each with names and references to published infrared spectra. There also is an alphabetical list of compound names, formulas, and references.

2. *Atlas of Spectral Data & Physical Constants for Organic Compounds,* 2nd ed., J.G. Grasselli and W.M. Ritchey (eds.), CRC Press, Inc., 2000 Corporated Blvd. NW, Boca Raton, FL 33431, 1975.

    This index contains coded infrared spectra for 22,000 compounds. It lists strong bands in the infrared and includes Raman, UV, NMR, and mass spectra data when available.

3. H.M. Hershenson, *Infrared Absorption Spectra,* (Indices for 1945–1957 and 1958–1962), Academic Press, New York, 1959 and 1964.

    A total of 36,000 references to infrared absorption spectra. The indices are alphabetic and references are made to 66 journals and one collection of spectra.

PROBLEMS

Deducing the structure of a compound from its infrared and Raman spectra is not easy; in fact, for large, complicated molecules it is not possible. The best way to learn how to obtain structural information from spectra is by practice. A list of sources of interpreted spectra and problems is given in references 9.11 through 9.17.

It is useful to have a checklist of questions giving basic information on the structure. For example, which elements are present, what is the molecular formula, what other information is available from other instrumental or chemical methods? Then, from an infrared survey spectrum, further clues can be gathered. Again, a checklist is useful. Are there any broad absorption bands? (see Table 9-3) Is there an aromatic ring present? (see Section 9-5) What kind of X—H bonds are present (OH, NH, CH, SH, etc.)? Is there a carbonyl group in the molelcule? Then a systematic analysis of the bands in the infrared spectrum can be made, first using Table 9-1, then Table 8-3. Tentative assignments can be made to each band, cross-checking where possible in other regions of the spectrum and referring to Tables 8-2 and 9-5 through 9-16. A Raman spectrum may be useful in some cases to confirm or eliminate certain assignments. When a tentative identification has been made to a class of compounds, the infrared spectrum of the unknown compound should be compared with spectra of similar compounds in a library of infrared spectra (see Section 9-15).

9-1 The compounds represented in the following spectra contain only carbon and hydrogen. Try to identify the type of compound and suggest possible structures.

(a) A volatile liquid hydrocarbon. The molecular weight determined by mass spectrometry is 84.

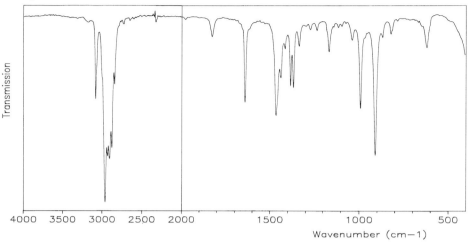

Wavenumber (cm−1)

(b) A liquid hydrocarbon with the molecular formula C_7H_{14}.

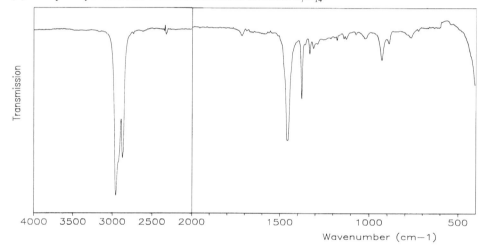

Wavenumber (cm−1)

(c) A liquid hydrocarbon. The boiling point is 159° C and the molecular weight is 120.

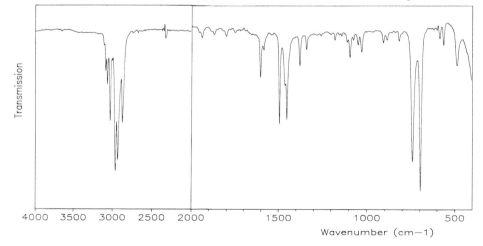

Wavenumber (cm−1)

(f) A solid with a low melting point (50–52° C) and molecular weight of 164.

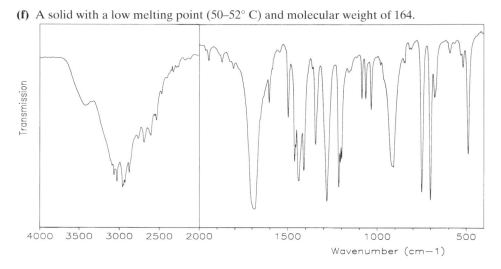

(g) An aromatic compound containing only C, H, and O.

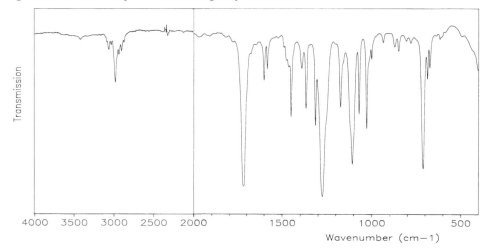

9-3 Compounds containing only C, H, and N yielded the following infrared spectra. In each compound, there is only one kind of nitrogen-containing functional group. Consult Section 9-10, as well as the sections on carbon–hydrogen vibrations to deduce the structures.

(a) A compound with a molecular weight of 103.

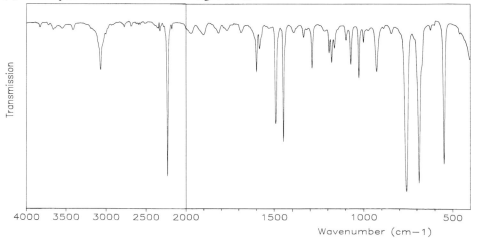

(b) A liquid compound with a molecular weight of 101.

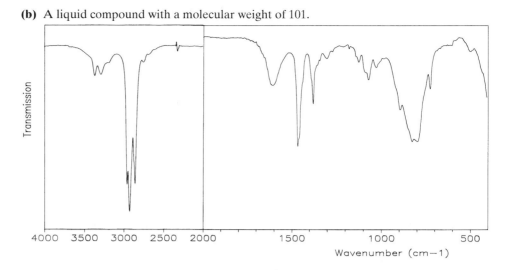

(c) A compound with a molecular formula $C_{10}H_{15}N$. Try first to deduce the functional groups present.

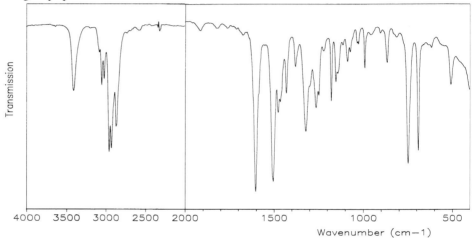

(d) A compound with a molecular weight of 79.

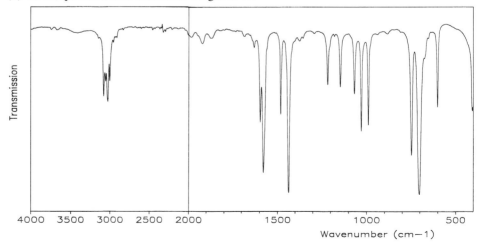

BIBLIOGRAPHY

Interpretation of Spectra

9.1.   N.P.G. Roeges, *A Guide to the Complete Interpretation of Infrared Spectra of Organic Structures,* John Wiley & Sons Ltd., Chichester, UK 1994.

9.2.   G. Socrates, *Infrared Characteristic Group Frequencies,* 2nd ed., John Wiley & Sons Ltd., Chichester, UK 1994.

9.3.   D. Lin-Vien, N.B. Colthup, W.G. Fateley, and J.G. Grasselli, *The Handbook of Infrared and Raman Characteristic Frequencies of Organic Molecules,* Academic Press, San Diego, 1991.

9.4.   L.J. Bellamy, *The Infrared Spectra of Complex Molecules,* Vols. 1 and 2, Chapman and Hall Ltd., London, 1975/80.

9.5.   R.R. Hill and D.A.E. Rendell, *The Interpretation of Infrared Spectra: A Programmed Introduction,* Heyden & Son Ltd., London, 1975.

9.6.   F.R. Dollish, W.G. Fateley, and F.F. Bentley, *Characteristic Raman Frequencies of Organic Compounds,* John Wiley & Sons, Inc., New York, 1974.

9.7.   J.R. Ferraro and K. Nakamoto, *Introduction to Raman Spectroscopy,* Boston Academic Press, Boston, 1994.

9.8.   N.B. Colthup, L.H. Daly, and S.E. Wiberly, *Introduction to Infrared and Raman Spectroscopy,* 3rd ed., Academic Press, San Diego, 1990.

9.9.   K. Nakamoto, *Infrared Spectra of Inorganic and Coordination Compounds,* 4th ed., John Wiley & Sons, Inc., New York, 1986.

9.10.   L.C. Thomas, *Interpretation of Infrared Spectra of Organophosphorus Compounds,* Heyden, London, 1975.

Sources of Interpreted Spectra and Problems

9.11.   K. Nakanishi and P.H. Solomon, *Infrared Absorption Spectroscopy, 2nd ed.,* Holden Day, San Francisco, 1977. (Contains 100 problems with detailed solutions.)

9.12.   H.A. Szymanski, *Interpreted Infrared Spectra,* 3 vols., Plenum Press, New York, 1964, 1966, 1967.

9.13.   T. Cairns, *Spectroscopic Problems in Organic Chemistry,* Heyden & Son Ltd., London, 1964.

9.14.   A.J. Baker, T. Cairns, G. Eglinton, and F.J. Preston, *More Spectroscopic Problems in Organic Chemistry,* Heyden & Son Ltd., London, 1967.

9.15.   D. Steele, *The Interpretation of Vibrational Spectra,* Chapman and Hall, London, 1971. (Contains 26 infrared [and other] spectra of organic molecules with interpretation.)

9.16.   R.K. Smalley and B.J. Wakefield, "Infrared Spectroscopic Problems and Answers," in *An Introduction to Spectroscopic Methods for the Identification of Organic Compounds,* vol. 1, *Nuclear Magnetic Resonance and Infrared Spectroscopy,* F. Scheinmann (ed.), Pergamon Press, Oxford, UK 1970. (The chapter contains 14 problems followed by detailed answers.)

9.17.   R. Davis and C.H.J. Wells, *Spectral Problems in Organic Chemistry,* Chapman & Hall, New York, 1984. (Contains 56 problems based on infrared, ^{1}H and ^{13}C NMR, and mass spectra together with either analytical data or a molecular formula. Solutions are given in the form of references to the catalog of The Aldrich Chemical Co., Inc., and other commonly available listings of organic compounds.)

part III

ELECTRONIC ABSORPTION AND CHIROPTICAL SPECTROSCOPY

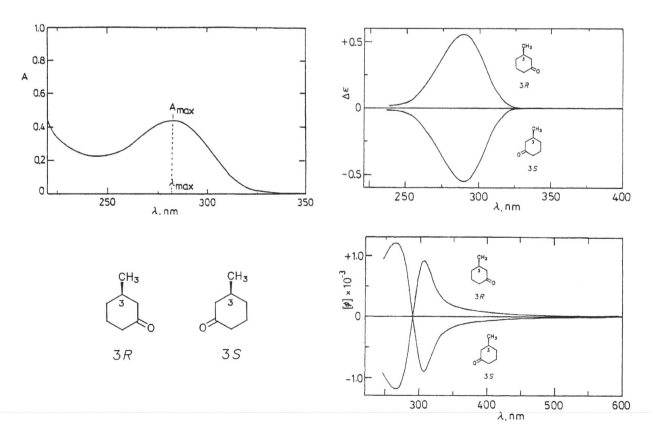

chapter 10

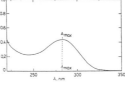

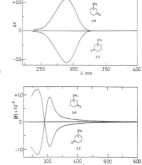

Introduction and Experimental Methods

10-1 INTRODUCTION

Ultraviolet (UV) and visible (vis) spectroscopies were among the earliest spectroscopic techniques used in organic structure determination and were predated only by refractive index and optical rotation measurements. The latter evolved into the chiroptical spectroscopic methods termed optical rotatory dispersion (ORD) and circular dichroism (CD), which are powerful tools for determining of molecular absolute configuration and conformation. Optical rotations are still measured to detect and quantify optical activity. In structure determination UV–vis spectroscopy is used to detect the presence of chromophores, especially conjugated chromophores such as aromatics, dienes and polyenes, α,β-unsaturated ketones, etc. Both UV–vis and ORD–CD are useful and sensitive probes in quantitative measurements where knowledge of concentration or concentration changes is important.

   UV–vis, ORD, and CD spectroscopy fall into the broad classification called electronic absorption spectroscopy. UV–vis spectroscopy is the most common of these, and it is used to measure energy and intensity of ordinary light absorption for any organic compound. The results are typically displayed as a plot of absorbance (A) intensity on the vertical axis and the wavelength of light (λ, in nanometers) on the horizontal axis giving UV–vis absorption curves such as that shown in Figure 10-1a for 3-methylcyclohexanone. ORD and CD spectroscopies are somewhat less common and are generally limited to optically active compounds. These spectroscopic methods differ from UV–vis in that they involve *difference* measurements of the absorption or refraction of light: the difference in absorption of left and right circularly polarized light or the difference in their refractive indices. Thus, although the horizontal scale is still displayed as λ, the vertical scale for CD and ORD curves is always signed positive or negative, as illustrated in Figures 10-1b and 10-1c respectively, for 3(R)- and 3(S)-methylcyclohexanone, which have ORD or CD curves of identical magnitudes but opposite signs. CD curves look very much like signed UV curves, but ORD curves look very different. There is a sense of handedness in CD and ORD spectra to match the handedness of nonsuperimposable mirror-image molecules, in this example to distinguish the R and S enantiomers. How-

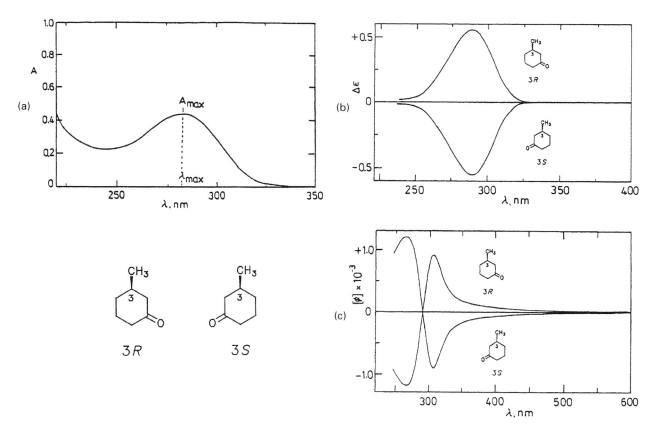

FIGURE 10-1 (a) UV–vis spectrum of 3(R)-methylcyclohexanone, 0.0245 M in methanol, run in a 1.0 cm cuvette. The spectrum is identical for the 3(S) isomer. UV–vis curves are typically displayed as absorbance (A) on the ordinate and wavelength (λ, in nanometers) on the abscissa. The wavelength at maximum absorbance (A_{max}) is called λ_{max}. UV–vis spectra may also be plotted as ε vs. λ, where ε is the molar absorptivity (molar extinction coefficient). (b) CD curves of 3(R)- and 3(S)-methylcyclohexanone in methanol. CD curves are typically displayed as $\Delta\varepsilon$ versus λ, or molar ellipticity [θ] versus λ. (c) ORD curves of 3(R)- and 3(S)-methylcyclohexanone in methanol. ORD curves are typically displayed as molecular rotation [ϕ] versus λ. (UV–vis spectrum from Portia Mahal Sabido; ORD–CD curves from Professor J.E. Gurst.)

ever, no CD or ORD can be detected for an equimolar mixture of enantiomers, such as a racemic mixture of 3(R)- and 3(S)-methylcyclohexanones. Ordinary UV–vis spectroscopy cannot distinguish between enantiomers; 3(R)- and 3(S)-methylcyclohexanone have identical UV–vis curves.

Electronic spectra provide information about the electronic properties of molecules, especially unsaturated compounds. All organic compounds absorb light in the ultraviolet region of the electromagnetic spectrum, and some absorb light in the visible region as well. Absorption of UV or visible light occurs only when the energy of incident radiation is the same as that of a possible electronic transition in the molecules studied (quantization of energy). Such absorption of energy is termed *electronic excitation* and is typically associated with moving a single electron from an occupied to an unoccupied molecular orbital (Figure 10-2), whereby the molecule is promoted from the molecular ground state to a higher energy, electronically excited state. In a given molecule many different electronic transitions are possible; and those important in organic chemistry often involve promoting of an electron from the **H**ighest **O**ccupied bonding or nonbonding **M**olecular **O**rbital (HOMO) to the **L**owest **U**noccupied **M**olecular **O**rbital (LUMO).

Figure 10-2 Idealized representation on a potential energy scale of occupied and unoccupied molecular orbitals in the electronic ground state (left) and electronic configuration of an excited state arising by promotion of an electron from the highest occupied molecular orbital to the lowest unoccupied molecular orbital (right). The electrons and their relative spin orientations are represented by small arrows.

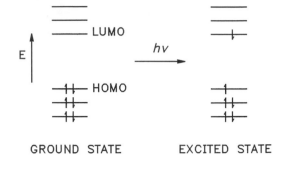

TABLE 10-1

Electronic Absorption Data for Isolated Chromophores*

Chromophore	Example	Solvent	λ_{max} (nm)†	ε (liter mol^{-1} cm^{-1})
C=C	1-Hexene	Heptane	180	12,500
—C≡C—	1-Butyne	Vapor	172	4,500
(benzene ring)	Benzene	Water	254	205
			203.5	7,400
	Toluene	Water	261	225
			206.5	7,000
C=O	Acetaldehyde	Vapor	298	12.5
			182	10,000
	Acetone	Cyclohexane	275	22
			190	1,000
	Camphor	Hexane	295	14
—COOH	Acetic acid	Ethanol	204	41
—COCl	Acetyl chloride	Heptane	240	34
—COOR	Ethyl acetate	Water	204	60
—CONH$_2$	Acetamide	Methanol	205	160
—NO$_2$	Nitromethane	Hexane	279	15.8
			202	4,400
$\overset{+}{=}$N=$\overset{-}{N}$	Diazomethane	Diethyl ether	417	7
—N=N—	*trans*-Azomethane	Water	343	25
C=N—	C$_2$H$_5$CH—NC$_4$H$_9$	Isooctane	238	200

* From J.B. Lambert, H.F. Shurvell, L. Verbit, R.G. Cooks, and G.H. Stout, *Organic Structural Analysis,* Macmillan Publishing, New York, 1976.
† Chromophores often have more than one absorption band.

Although electronic transitions arise between ground and excited states of the *entire* molecule, in UV–vis absorption most of the action can usually be assigned to parts of the molecule (*chromophores*) where valence electrons are found, such as the nonbonding (*n*) or π electrons. One speaks thus of an electronic transition in a chromophore, which in organic molecules is typically a functional group, such as a carbonyl group, a carbon–carbon double bond, or an aromatic ring. Representative chromophores are shown in Table 10-1, along with λ_{max}, the wavelength at which light absorption reaches a maximum, and ε_{max}, the molar absorptivity constant at which light absorbance reaches a maximum (Figure 10-1a).

10-2 *MEASUREMENT OF ULTRAVIOLET–VISIBLE LIGHT ABSORPTION*

The *UV–vis spectrum* typically represents the absorption of light as a plot (Figure 10-1a) of energy (usually reported in organic chemistry as wavelength, λ, in nanometers, from $E = hc/\lambda$) vs. the intensity of absorption (as absorbance, A, or molar extinction coefficient, ε, where ε is a rough measure of the transition probability). The wavelength at maximum absorbance (A_{max}) for each electronic transition is termed λ_{max}.

TABLE 10-2

Definitions of Terms and Equations

Quantity	Unit of measure	Dimensions
Wavelength (λ)	Nanometer (nm)	$1\ nm = 10^{-9}\ m$
	Ångstrom (Å)	$1\ \text{Å} = 10^{-10}\ m$
Frequency (ν)	Hertz, Hz, or s^{-1}	Cycles per second
Energy	Depends on the units of h	6.626×10^{-27} erg s
		6.626×10^{-34} J s
		9.534×10^{-14} kcal–s/mol
		1.583×10^{-34} cal–s/molec

TABLE 10-3

Useful Conversion Factors

cm^{-1}	Hz	kcal mol^{-1}	kJ mol^{-1}
1	3.00×10^{10}	2.86×10^{-3}	1.20×10^{-2}
3.33×10^{-11}	1	9.53×10^{-14}	3.99×10^{-13}
3.50×10^2	1.05×10^{13}	1	4.18

10-2a   Wavelength and λ_{max}

Electromagnetic radiation (see Section 10-3) may be described by the wavelength λ between its waves, by the frequency ν (s^{-1}), or by the wavenumber $\bar{\nu}$ (cm^{-1}). ($\lambda\nu = c$, the velocity of light; $\bar{\nu} = 1/\lambda$). Commonly used wavelength units in ultraviolet and visible light regions are nanometers (nm) and Ångstroms (Å). According to Planck's equation (eq. 10-1), frequency is directly proportional to energy. Table 10-2 lists the units

$$\Delta E = h\nu \qquad (10\text{-}1)$$

commonly used for λ and ν, and Table 10-3 gives some useful conversion factors. Eq. 10-2 is convenient for calculation of energies in the familiar units of kcal mol^{-1}. Hence

$$\Delta E = \frac{hc}{\lambda} = \frac{28,636}{\lambda}\ \text{kcal mol}^{-1} = \frac{119,809}{\lambda}\ \text{kJ mol}^{-1}\ \text{(for } \lambda \text{ in nm)} \qquad (10\text{-}2)$$

light of 300 nm wavelength corresponds to an energy of 95.4 kcal mol^{-1} or 399 kJ mol^{-1}, depending on the units of h.

Wavenumbers are directly proportional to energy, so that a given number of reciprocal centimeters (cm^{-1}) represent the same energy anywhere in the electromagnetic spectrum. For example, a shift of $\bar{\nu}$ of 700 cm^{-1} anywhere in the spectrum corresponds to 1.95 kcal mol^{-1}. On the other hand, wavelength is inversely proportional to energy and thus the relationship is not linear. For example, an energy change of 1.95 kcal mol^{-1} at 200 nm corresponds to a shift of 2.7 nm, but the same energy change at 800 nm corresponds to a shift of approximately 4.4 nm.

At the lower end of the visible spectrum, below 400 nm, is the UV region. It is convenient to divide the UV into two parts: the near UV, 190–400 nm (53,000–25,000 cm^{-1}), and the far or vacuum UV, below 190 nm (>53,000 cm^{-1}). This division is because atmospheric oxygen begins to absorb around 190 nm. Oxygen must be removed from the spectrophotometer, either by using a vacuum instrument or by vigorous purging with nitrogen.

10-2b   The Beer–Bouger–Lambert Law and ε_{max}

The laws of Lambert, Bouger, and Beer (or more simply, Beer's law) state that at a given wavelength the proportion of light absorbed by a transparent medium is independent of the intensity of the incident light and is proportional to the number of absorbing

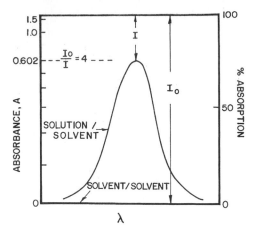

Figure 10-3 Measurement of solute absorbance A by a double-beam spectrophotometer.

molecules through which the light passes. According to eq. 10-3, I_0 is the intensity of incident light, I the intensity of transmitted light, k the absorption coefficient, and l the

$$I = I_0 e^{-kl} \quad \text{or} \quad I = I_0 10^{-\varepsilon cl} \quad \text{or} \quad \log (I_0/I) = \varepsilon cl \tag{10-3}$$

pathlength (cm). The absorption coefficient k is related to the more common molar extinction coefficient (ε), or molar absorptivity, by the equation $k \approx 2.303 \, \varepsilon c$, where c is the concentration in mol liter^{-1}. Since $\log (I_0/I)$ is defined as the absorbance (A) and A is the quantity actually measured, eq. 10-3 is rewritten as eq. 10-4, for i absorbing

$$A = \varepsilon lc \quad \text{or} \quad A = l \sum_i \varepsilon_i c_i \tag{10-4}$$

species. The units of ε are cm^2 mol^{-1} (liter mol^{-1} cm^{-1}) but are usually omitted. (For a derivation of Beer's law, see reference 10.2.)

In practice, the quantities actually measured are the relative intensities of the light beams transmitted by a reference cell containing pure solvent and by an identical cell containing the solution. When the intensities are taken as I_0 and I, respectively, the resulting absorption is that of the dissolved solute only (Figure 10-3). One also can see from Figure 10-3 that ε is different at different λ, and one refers to ε_{max} at λ_{max} for a given absorption.

10-2c Shape of Absorption Curves: The Franck–Condon Principle

Absorption of UV–vis light is typically recorded as broad absorption maxima (see Figure 10-1) and not as single, sharp lines representing the absorption in an extremely narrow energy range. The absorption curves are broadened because the electronic levels have vibrational levels superimposed on them.

For simplicity, let us look at the ground and excited electronic states of a diatomic molecule. In the more common case, the bond strength in the excited electronic state is less than that in the ground state, and the equilibrium internuclear distance is longer than in the ground state. A typical potential energy diagram is shown in Figure 10-4.

Most molecules exist mainly in their ground vibrational state at room temperature. Excitation can occur to any of the excited state vibrational levels, so that the absorption due to the electronic transition consists of a large number of lines. In practice, the lines overlap so that a continuous band is observed. Hence the shape of an absorption band is determined by the spacing of the vibrational levels and by the distribution of the total band intensity over the vibrational subbands. The intensity distribution is determined by the *Franck–Condon principle,* which states that *nuclear motion is negligible during the time required for an electronic excitation.* For example, the time required for an electron to circle a hydrogen nucleus can be calculated from Bohr's

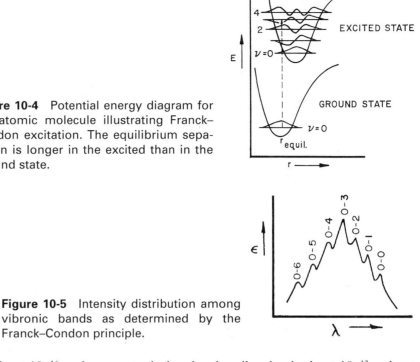

Figure 10-4 Potential energy diagram for a diatomic molecule illustrating Franck–Condon excitation. The equilibrium separation is longer in the excited than in the ground state.

Figure 10-5 Intensity distribution among vibronic bands as determined by the Franck–Condon principle.

model to be about 10^{-16} s, whereas a typical molecular vibration is about 10^{-13} s, about a thousand times longer. Another statement of the Franck–Condon principle based on classical mechanics is that the most probable vibrational component of an electronic transition is one that involves no change in the position of the nuclei, a so-called *vertical transition*. As represented by the vertical arrow in Figure 10-4, the most probable transition is to the excited $v = 3$ vibrational state. This state has a maximum at the same internuclear distance (r) as that corresponding to the starting point of the transition.

Figure 10-5 shows the vibrational-electronic (vibronic) spectrum corresponding to Figure 10-4, with the 0–3 band (from $v = 0$ in the ground state to $v = 3$ in the excited state) the most intense spike. Note that the other transitions, including the 0–0 band, have significant probabilities. This result is because even in the ground electronic state (zeroth vibrational level), the internuclear distance is described by a probability distribution (Figure 10-4). Therefore, transitions may originate over a range of r values so that more than one band originating from $v = 0$ may be observed.

Sometimes, on raising the temperature, the vibrational structure of a band is lost. This band broadening is due to the population of several ground vibrational states at higher temperature so that a larger number of possible vibrational transitions can occur on electronic excitation. Featureless or broad bands are also observed at ambient temperatures, usually in solution spectra, such as Figure 10-1, where solute-solvent vibrational interactions become important.

10-2d   Solvent Effects and λ_{max} Shifts

Moving an electron from the ground state to an excited state configuration typically leads to an excited state that is more polar than the ground state and more sensitive to solvation effects. The influence of solvent on UV–vis spectra is related directly to the degree of interaction between the solute and the solvent, which becomes greater as the polarity of the solvent increases. The least polar solvents (for example, hydrocarbons) have the least effect on UV–vis spectra and are least effective in inhibiting vibrational fine structure (Figure 10-5). Polar solvents, on the other hand, interact more strongly with solutes. They tend to smooth out vibrational structure on the absorption bands and to cause spectral shifts.

In polar solute molecules, the excited state usually receives predominant contributions from highly polar structures, such as $R_2C^+\!-\!O^-$ in the $\pi \longrightarrow \pi^*$ state of ketones,

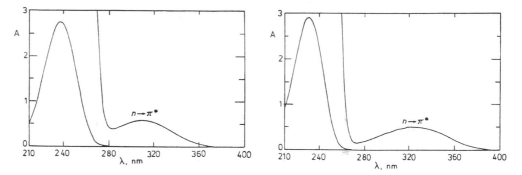

Figure 10-6 UV–vis absorption spectra of 4-methyl-3-penten-2-one (mesityl oxide) in (left) methanol and (right) heptane. Spectra in methanol were run in a 1 cm cuvette at 0.0105 M and 2.63×10^{-4} M concentrations. Spectra in heptane were run at 0.014 M and 2.8×10^{-4} M concentrations. The vertical axis is in absorbance units. The ~320 nm absorption is the $n \longrightarrow \pi^*$ transition, the ~240 nm is mainly $\pi \longrightarrow \pi^*$. (Spectra run by Portia Mahal Sabido.)

and is more polar than the ground state. In accord with simple electrostatic theory, such polar excited states are expected to be stabilized by polar solvents, thereby facilitating electronic excitation and resulting in a red (bathochromic) shift of the absorption band. Only when the ground state is more polar than the excited state can one expect to find a blue (hypsochromic) shift with increasing solvent polarity.

Consider the case of the α,β-unsaturated ketone, 4-methyl-3-penten-2-one (mesityl oxide). Its UV–vis spectrum (Figure 10-6) shows two readily accessible UV transitions: (1) a weak absorption near 320 nm ($\varepsilon \sim 50$) associated with the promotion of a nonbonding (*n*) oxygen electron to an antibonding π^* orbital ($n \longrightarrow \pi^*$ transition), and (2) a strong absorption near 230 nm ($\varepsilon \sim 12{,}000$) associated with the promotion of a bonding π electron to an antibonding π^* orbital ($\pi \longrightarrow \pi^*$ transition).

For $\pi \longrightarrow \pi^*$ excited states (see Section 11-1), dipole–dipole interactions and hydrogen bonding with solvent molecules tend to lower the energy of the excited state more than the ground state, with the result that the λ_{max} *increases* (red shift) about 10 nm in going from a fairly noninteractive solvent like heptane to methanol (Figure 10-6). For $n \longrightarrow \pi^*$ excited states (see Section 11-1), both the ground and the excited states are lowered in energy by dipole–dipole and hydrogen-bonding interaction with solvent. In hydrogen-bonding solvents, the ground state *n* electrons coordinate with the solvent more strongly than excited state *n* electrons, with the result that the λ_{max} *decreases* about 15 nm in going from heptane to methanol solvent (Figure 10-7). Solvent effects on $n \longrightarrow \pi^*$ and $\pi \longrightarrow \pi^*$ transitions are summarized for the example of mesityl oxide in Table 10-4.

10-3 MEASUREMENTS USING POLARIZED LIGHT

The light beam used in UV–vis spectroscopy is essentially unpolarized (or isotropic). Use of linearly polarized (anisotropic) light (sometimes less rigorously referred to as plane-polarized light) to investigate optically active (chiral) molecules is a powerful

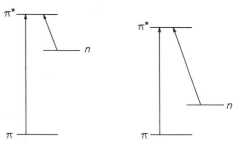

Figure 10-7 Influence of solvent on the orbitals involved in $\pi \longrightarrow \pi^*$ and $n \longrightarrow \pi^*$ electronic transitions.

/ Introduction and Experimental Methods

TABLE 10-4

Influence of Solvent on the UV λ_{max} and ε_{max} of the $n \longrightarrow \pi^*$ and $\pi \longrightarrow \pi^*$ Excitations of 4-Methyl-3-penten-2-one (Mesityl Oxide)*

$$(CH_3)_2C{=}CH{-}\overset{\displaystyle \overset{O}{\|}}{C}{-}CH_3$$

	$\pi \longrightarrow \pi^*$ Transition		$n \longrightarrow \pi^*$ Transition	
Solvent	λ_{max} (nm)	ε_{max} (liter mol^{-1} cm^{-1})	λ_{max} (nm)	ε_{max} (liter mol^{-1} cm^{-1})
Hexane	229.5	12,600	327	97.5
Diethyl ether	230	12,600	326	96
Ethanol	237	12,600	325	78
Methanol	238	10,700	312	74
Water	244.5	10,000	305	60

*From H.H. Jaffé and M. Orchin, *Theory and Applications of Ultraviolet Spectroscopy*, John Wiley & Sons, Inc., New York, 1962.

technique for obtaining structural and stereochemical information. How is linearly (or plane-polarized) light generated?

Light is electromagnetic radiation, which can be considered to behave as two wave motions (electric and magnetic) propagated in time and generated by oscillating electric and magnetic dipoles. Although the energies associated with electric and magnetic waves are equal, most optical measurements (UV–vis and chiroptical) are concerned only with the electric field (Figure 10-8). Figure 10-8 considers light in the context of a wave phenomenon caused by transverse vibrations of the electric field vector (vertical arrows). There is an associated magnetic field vector perpendicular to the oscillating electric field vector, but we can ignore it for purposes of the present discussion. Note that the electric field vector vibrates perpendicular to the direction of travel of the light wave and that there are an infinite number of planes that can pass through the line oz in Figure 10-8.

Ordinary light thus consists of different wavelengths vibrating in many different planes. If we could place ourselves at the point z and look toward o, we would see the cross section of the light wave depicted in Figure 10-9a, a schematic representation of unpolarized light. The radial electric field vectors (arrows) are meant to indicate that no single direction predominates in completely unpolarized light. However, if the light is filtered (through a polarizing filter) to remove all waves other than in one direction (as in polaroid sunglasses, for example) that lying in the xz plane, the light is linearly polarized (as represented in Figure 10-9b).

Even if we were to use unpolarized light of a single wavelength, it would still consist of waves vibrating in many planes at right angles to the direction of propagation. With polarized light, since several directions of propagation are possible within a plane, it is correct to refer to light traveling in a specified direction as linearly polarized rather than plane polarized.

The linearly polarized light of Figure 10-9b is vertically polarized, as represented by the wave in Figure 10-10a. A perpendicular (horizontal) plane of polarization

Figure 10-8 Electric field wave motion propagated in the z direction by transverse vibration; λ is the wavelength. The arrows denote the electric field vector (E) at a given instant as the light wave progresses along the z (or time) axis.

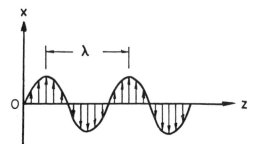

Figure 10-9 Schematic representation of unpolarized and linearly polarized light. (a) Cross section of a narrow beam of ordinary light traveling directly toward the observer. Vibration of the light may be in any direction that is perpendicular to the direction of travel, as indicated by the numerous arrows. (b) A beam of polarized light has vibration in only one direction. This direction is the plane of polarization, in this example the *xz* plane of Figure 10-8.

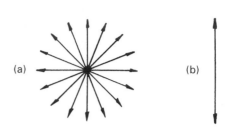

(Figure 10-10b) could just as easily have been selected by rotating the polarizing filter about the axis of propagation. These two orthogonal polarization states are represented in Figures 10-10a and b, and any radiation in any arbitrary state may be projected onto them (or any orthogonal polarization states).

Linearly polarized light may be decomposed into another pair of orthogonal polarization states that are important in understanding chiroptical phenomena: left and right circularly polarized light. In the case of circularly polarized light, the transverse vibrations trace out a helix as a function of time. The helix may be either left-handed (Figure 10-10c) or right-handed (Figure 10-10d). Viewed in cross section, that is, if an observer were situated on the *z* axis looking toward the light source, the transverse vibrations would trace out a circle. Light whose electric field vector traces out a right-handed helical pattern is termed *right circularly polarized light*. The cross-sectional appearance of clockwise rotation of the electric field vector is obtained by pushing the helix forward through a perpendicular plane without rotating it. In other words, the helix is moved forward, but it is not turned like a mechanical screw.

Another type of polarized light, which we have not pictured, resembles a flattened helix and has a cross section that is an ellipse. Elliptically polarized light may also be right- or left-handed.

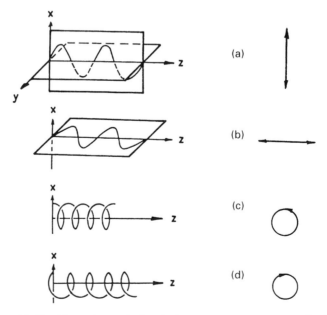

Figure 10-10 Linearly and circularly polarized radiation. (Left) The light wave as a function of time. (Right) Cross-section of the light wave. Light polarized (a) linearly and vertically, (b) linearly and horizontally, (c) left circularly, (d) right circularly. Light having a right-handed helical pattern is termed right circularly polarized. The cross-sectional clockwise rotation of the electric field vector is obtained as the helix is moved forward without rotation through a perpendicular plane.

The French physicist Biot discovered early in the nineteenth century that certain naturally occurring organic compounds possess the unusual property of rotating the plane of polarization of a linearly polarized incident light beam. A few years later, in 1817, Biot and his countryman Fresnel independently found that the extent of optical rotation of a compound increases as one uses light of increasingly shorter wavelength for the measurement. The change in optical rotation with wavelength is termed *optical rotatory dispersion (ORD)*.

Thirty years later, Haidenger reported his observations on the differences in the absorption of the left- and right-handed components of circularly polarized light by crystals of amethyst quartz. Such differential absorption of left- and right-handed circularly polarized light is termed *circular dichroism (CD)*.

Since both CD and ORD involve optical measurements on chiral molecules, they have been termed chiroptical methods.

10-3a Ordinary Absorption and Chiroptical Spectroscopy

Ordinary absorption (UV–vis) spectroscopy and chiroptical (ORD and CD) spectroscopy arise from the same photophysical process: typically the promotion of an electron from a ground state orbital to an excited state orbital. As such they are intimately related. They differ only in that the former measures the absorption of ordinary light associated with the promotion of an electron, whereas the latter measures the absorption of left and right circularly polarized light and displays the difference in their absorption. (The incident light can be separated into its left [L] and right [R] circularly polarized components, giving light with a handedness.) The difference in absorption, $\Delta A = A_L - A_R$, or difference in molar extinction coefficient, $\Delta \varepsilon = \varepsilon_L - \varepsilon_R$, is recorded at each λ of the absorption band (see Figure 10-1b). Since A_L and A_R typically have very nearly the same magnitudes, ΔA and $\Delta \varepsilon$ are quite small and may take on either positive or negative values, as in the case of optically active compounds, or they may be zero, as in the case of achiral compounds and racemic mixtures.

10-3b Circular Birefringence and Circular Dichroism

As a useful model for conceptualizing the rotation of linearly polarized light, consider the light as composed of two oppositely rotating coherent beams of circularly polarized light. The linearly polarized light is then the vector sum of the left and right circularly rotating components as shown in Figure 10-11a. The vector sums are indicated at points 1 to 5 with the resultant vectors having the properties of a linearly polarized light wave, as in the *xz* plane of Figure 10-11b.

When left and right circularly polarized light beams pass through an achiral sample or racemic mixture (isotropic medium), they travel with the same velocity and thus enter and exit the medium in phase. Hence there is no rotation of the plane of linearly polarized light as it passes through and exits the medium.

In 1825, Fresnel postulated that when the circularly polarized light passes through an optically active medium, which may be a solid, liquid, or gas, the refractive index (n) for one circularly polarized component is different from that for the other. The medium is said to be *circularly birefringent* and to have the property given by eq. 10-5, in which

$$\Delta n = n_L - n_R \neq 0 \qquad (10\text{-}5)$$

n_L and n_R are the refractive indices for left and right circularly polarized light, respectively. (Ordinary indices of refraction are of the order of unity, and typically, Δn is only about one millionth of the absolute values of n_L or n_R.) Differences in refractive indices correspond to differences in light velocities. Consequently, one of the two circularly polarized components of the linearly polarized light becomes retarded with respect to the other (Figure 10-11c). Upon emerging from the optically active medium, left circularly polarized component (E_L) lags behind the right (E_R) in the example of

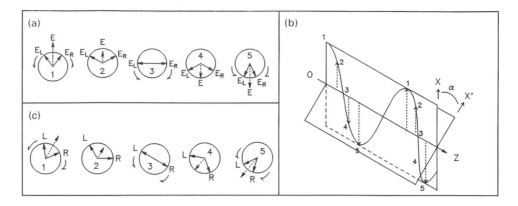

Figure 10-11   (a) and (c) Representations of linearly polarized radiation as the vector sum (E, dotted arrows) of two oppositely rotating beams of circularly polarized radiation E_L and E_R. In (c), the vector for left circularly polarized light (E_L) lags behind that for right circularly polarized light (E_R) relative to the picture in (a). (b) Linearly polarized light waves. The *xz* plane of polarization corresponds to the situation in (a). The *x'z* plane, rotated from the *xz* plane by the angle of rotation α, corresponds to the situation in (c).

Figure 10-11. Since the two circularly polarized components did not travel with the same velocity through the anisotropic medium, the two components are no longer in phase, and the resultant (linear polarization) vectors (in plane *x'z*) have been rotated by the angle α to the original plane of polarization (Figures 10-11c and 10-12). Optical rotations at a single wavelength, such as 589 nm (sodium D-line), have been used to detect and quantitate optical activity. Optical rotations measured over a range of wavelengths, as in ORD spectroscopy, have been used to determine absolute configuration.

In the region of an absorption band, the two circularly polarized components, in addition to sustaining a differential retardation because of the circular birefringence of the medium, also are absorbed to different extents. In other words, the optically active medium has an unequal molar absorptivity coefficient ε for left and right circularly polarized light. This difference in molar absorptivity (eq. 10-6) is termed *circular dichroism*.

$$\Delta\varepsilon \ = \ \varepsilon_L - \varepsilon_R \ \neq \ 0 \tag{10-6}$$

Upon emerging from the optically active medium, the two circularly polarized components are not only out of phase but also of unequal amplitude. The resultant vector no longer oscillates along a single line but now traces out an ellipse, as shown in Figure 10-13. The linearly polarized light beam has been converted to elliptically polarized light by the unequal absorption of its two circularly polarized components. Circular dichroism is used to determine the absolute stereochemistry of compounds.

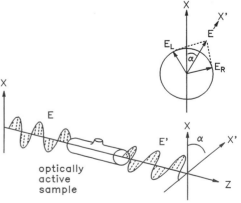

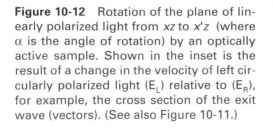

Figure 10-12   Rotation of the plane of linearly polarized light from *xz* to *x'z* (where α is the angle of rotation) by an optically active sample. Shown in the inset is the result of a change in the velocity of left circularly polarized light (E_L) relative to (E_R), for example, the cross section of the exit wave (vectors). (See also Figure 10-11.)

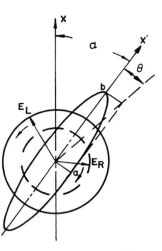

Figure 10-13 Elliptically polarized light caused by the unequal speed and unequal absorption of left and right circularly polarized light by a chiral medium. The tangent of the ratio of the minor axis *a* to the major axis *b* is θ, the angle of ellipticity. The major axis of the ellipse lies along the *x'* axis and forms the angle of rotation, α, to the original plane of polarization, the *xz* plane of Figures 10-11 and 10-12.

10-3c Optical Rotatory Dispersion (ORD) and Circular Dichroism (CD) Quantities

The difference in refractive indices for left and right circularly polarized light $(n_L - n_R)$ is related to the angle of rotation per unit length (ϕ) of plane (linearly) polarized light (Figure 10-12) or elliptically polarized light (Figure 10-13) by eq. 10-7. In this equation,

$$\phi = (\pi/\lambda)\,(n_L - n_R) \tag{10-7}$$

λ is the wavelength of the light used in the measurement and the units of ϕ are radians per cm. Multiplying eq. 10-7 by $1800/\pi$ converts it to the more familiar units of degrees per decimeter as defined by the angle of rotation α in eq. 10-8. We may note that for an

$$\alpha = \phi\,\frac{1800}{\pi} = \frac{1800}{\lambda\,(\text{cm})}\,(n_L - n_R) \tag{10-8}$$

observed optical rotation of $\alpha = 1°$ at 360 nm in a 1 decimeter (10 cm) cell, $n_L - n_R$ is 2 $\times 10^{-8}$. Typical indices of refraction are of the order of unity. Hence the difference in refractive indices is extremely small, on the order of one millionth of 1%.

Spectropolarimeters are polarimeters used to make measurements at a variety of wavelengths. They record the angle of rotation α as a function of wavelength (λ) and at temperature *T*. Eq. 10-9 is used to calculate the specific rotation $[\alpha]$ from the observed

$$[\alpha]_\lambda^T = \frac{a}{c'l} = \frac{\text{observed rotation (degrees)}}{\text{concn (g ml}^{-1}) \times \text{length of sample tube (dm = 0.1 m)}} \tag{10-9}$$

rotation, the concentration c' is in g ml^{-1}, and the tube length in dm. For ORD work, it is more common to use the molar rotation $[\phi]$, which is the specific rotation multiplied by the molecular weight *M* over 100 (eq. 10-10). The units of $[\phi]$ are degrees cm^2 dmol^{-1}.

$$[\phi] = \left(\frac{\alpha}{c'l}\right)\left(\frac{M}{100}\right) = \frac{[\alpha]\,M}{100} \tag{10-10}$$

The quantities used in CD spectroscopy are $\Delta\varepsilon$ (eq. 10-6) and θ, the angle of ellipticity per unit length (Figure 10-13), which is defined in eq. 10-11 by the difference in

$$\tan\theta = \tanh\,[(\pi/\lambda)\,(\kappa_L - \kappa_R)l] \tag{10-11}$$

absorption indices (κ) for left and right circularly polarized light $(\kappa_L - \kappa_R)$, with *l* the pathlength in cm, λ the wavelength of light used, and κ related to the absorption coefficient (*k* of eq. 10-3) by $k = (4\pi/\lambda)\kappa$. Since in practice $(\kappa_L - \kappa_R)$ represents only a small fraction (rarely more than a few hundredths) of their absolute value, the ellipse is usually extremely elongate, and $(\kappa_L - \kappa_R)l$ is typically much less than unity. Thus, θ may be approximated by eq. 10-12, which has the same form as eq. 10-7.

$$\theta = (\pi/\lambda)\,(\kappa_L - \kappa_R) \tag{10-12}$$

As with the angle of rotation ϕ, the ellipticity angle θ may be converted from units of radians per cm to degrees per decimeter (eq. 10-13). In CD work, the molar ellipticity

$$\theta = \frac{1800}{\lambda(\text{cm})}\,(\kappa_L - \kappa_R) = \frac{1800}{4\pi}\,(k_L - k_R) \tag{10-13}$$

$[\theta]$ and $\Delta\varepsilon$ are commonly used. The molar ellipticity, with units of degrees $cm^2\,dmol^{-1}$, has a form similar to that of eq. 10-10 and is given by eq. 10-14, or, in terms of molar

$$[\theta] = \left(\frac{\theta}{c'l}\right)\left(\frac{M}{100}\right) \tag{10-14}$$

extinction coefficients, by substituting eq. 10-13 and 10-6 into 10-14 and recognizing that $k \approx 2.303\,\varepsilon c$ (eq. 10-3), to give eq. 10-15. This equation thus relates molar ellipticity to

$$[\theta] \approx 2.303\left(\frac{4500}{\pi}\right)(\varepsilon_L - \varepsilon_R) = 3300\,(\Delta\varepsilon) \tag{10-15}$$

the difference in molar extinction coefficients between left and right circularly polarized light ($\Delta\varepsilon$), and hence via eqs. 10-4 and 10-14 it relates angle of ellipticity in CD to absorbance difference. Just as $(n_L - n_R)$ is small in magnitude compared with the mean index of refraction, so is the difference $(\varepsilon_L - \varepsilon_R)$ between the molar absorptivity coefficients for left and right circularly polarized light compared to ε, on the order of 10^{-2} to $10^{-4} \times \varepsilon$.

Most CD instruments measure angle of ellipticity vs. λ, which can be adjusted to read the differential absorbance, $\Delta A = A_L - A_R$. This quantity is related to the difference in molar absorptivity, $\Delta\varepsilon = (\varepsilon_L - \varepsilon_R)$, by eq. 10-16, in which c is in mol liter^{-1} and

$$\Delta A = \Delta\varepsilon c l \tag{10-16}$$

l is the pathlength in cm — a direct analogy to eq. 10-4. It should be kept in mind that any medium that exhibits circular birefringence also exhibits circular dichroism, because the effects occur simultaneously. Strictly speaking, one should refer to the rotation of the major axis of the ellipse rather than of a plane of polarization. However, since the ellipticities encountered in practice are typically small, the distinction becomes meaningful only in the immediate vicinity of an absorption band.

The hand-in-hand relationship between birefringence and dichroism may be expressed simultaneously by defining the complex rotatory power Φ through eq. 10-17,

$$\Phi = (\phi - i\theta) = (\pi/\lambda)\,(N_L - N_R) \tag{10-17}$$

in which $i = \sqrt{-1}$ and N indicates the complex index of refraction, $N = n - i\kappa$. The real and imaginary parts of eq. 10-17 are related by a pair of integral transforms, called Kronig–Kramers relationships, described in detail for optical activity in reference 10.4. For the ~300 nm electronic transition of ketones, one derives the relationship expressed in eq. 10-18 between the amplitude (a) of an ORD curve and $\Delta\varepsilon$ or $[\theta]$ of a CD curve,

$$a \approx 40.28\,(\Delta\varepsilon_{max}) \approx 0.0122\,([\theta]_{max}) \tag{10-18}$$

in which a is the difference divided by 100 between $[\phi]$ at the peak and trough of an ORD Cotton effect, as shown in the following section.

10-3d   ORD and CD Spectra and Cotton Effects

Unlike UV–vis and CD spectroscopy, which detect and record *absorption* only for discrete wavelengths of incident-light corresponding to quantization energies for electron promotion, ORD may be detected over all wavelengths. This is because ORD is based on the unequal refraction of light and on the difference in refractive indices for left and right circularly polarized light. All optically clear substances exhibit a molecular refraction at almost any wavelength of incident irradiation. As with CD, ORD is linked to chiral substances and is not observed for achiral compounds or racemic mixtures. In ORD, one generally measures the sign and magnitude rotation of plane polarized light (usually $[\phi]$) vs. wavelength of incident light. An optically active substance such as $3(R)$-methylcyclohexanone has a measurable optical rotation at 589 nm, but its nearest absorption band lies near 290 nm (see Figure 10-1a). Its rotation varies with wavelength, and the ORD spectrum assumes the general shape of a plot of refractive index versus vs. λ, with the most rapid changes occurring in the vicinity of the absorption bands (see Figure 10-1c). Note that ORD involves measurement of a rotation, whereas CD involves measurement of an absorption, namely the differential absorption of left- and right-handed circularly polarized radiation. Hence CD occurs only in the vicinity of an absorption band, whereas ORD is theoretically finite everywhere.

The shape and appearance of a CD curve closely resembles that of the ordinary (UV–vis) absorption curve of the electronic transition to which it corresponds. Unlike ordinary absorption curves, such as Figure 10-1a, however, CD curves may be positive or negative, as in Figure 10-1b. CD curves plot $\Delta\varepsilon$ (or $[\theta]$) vs. wavelength. They are *difference spectra* representing the difference in absorption of left and right circularly polarized light, hence the *signed* nature of the curve. Each CD curve for each electronic transition also represents a positive or negative *Cotton effect* for the transition. For every CD Cotton effect there exists a corresponding ORD Cotton effect of the same sign (Figure 10-14). (The S-shaped ORD curve is known as a Cotton effect, in honor of the French physicist Aimé Cotton, who observed both ORD and CD phenomena, beginning in 1896.)

Cotton effects also are seen in ORD spectra (Figure 10-14) because, in the region of an absorption band, left and right circularly polarized light is not only absorbed to different extents but also propagated with different velocities. ORD curves, especially those of close-lying electronic transitions, are generally more difficult to read than the bell-shaped CD curves, but information can be derived beyond the immediate region (λ) of the electronic transitions. In ORD curves, the angle of rotation α, or more often the molar rotation $[\phi]$, of a chiral compound is plotted against wavelength. A typical ORD curve for a dextrorotatory ketone having an absorption maximum at 295 nm is shown in Figure 10-15. The part of the ORD curve above about 325 nm, labeled the plain ORD region, is characteristic of compounds that have no optically active absorption bands in the spectral region being measured. The smoothly rising part of the curve is referred to as a plain dispersion curve; in the example shown it is a plain, *positive* dispersion curve. A plain, *negative* curve would be one whose rotational values fall or become increasingly less negative on going toward longer wavelength. The so-named anomalous part of the ORD curve falls in the region of an electronic transition and corresponds to the Cotton effect.

As measurements of optical rotation in Figure 10-15 are made to shorter wavelengths, the rotation increases. It is found to increase rapidly as the absorption maximum is approached. Somewhat before this maximum, rotational values reach a maximum (termed a *peak*), then drop drastically, *going through zero rotation,* until another inflection point (termed a *trough*) is reached. The rotation then tends to increase again. In the ideal case for which the molecule possesses no other absorption bands near the one measured, λ_0 closely corresponds to λ_{max}, the maximum of the absorption band. The

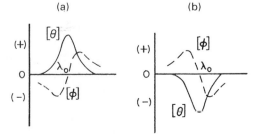

Figure 10-14 Positive (a) and negative (b) CD Cotton effects of an isolated absorption band with their corresponding ORD curves (dashed lines) superimposed. The wavelength (λ_0) at the sign change cross-over point of the ORD curve corresponds to λ_{max} of the CD curve.

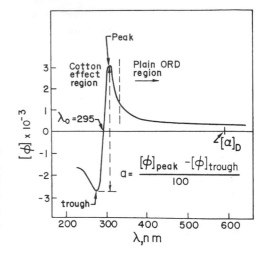

Figure 10-15 Optical rotatory dispersion curve of a typical saturated ketone. The amplitude of the Cotton effect *a* is defined as shown. The crossover point from positive to negative rotational values, $\lambda_0 = 295$ nm, corresponds closely to the UV absorption maximum of the ketone.

vertical distance between the peak and trough divided by 100 is termed the amplitude *(a)*, as defined in Figure 10-15. If the peak precedes the trough on measuring from longer to shorter wavelength, the Cotton effect is termed positive. Conversely, if the trough precedes the peak, it is a negative Cotton effect.

10-3e Relative Advantages of ORD and CD

The most significant difference between ORD and CD is that CD gives a signal only in the vicinity of an optically active absorption band. In contrast, the change in optical rotation with wavelength is everywhere finite. Thus even chromophores in the far UV contribute to the sign and magnitude of the ORD signal. Overlapping ORD curves sum up to often very complex looking spectra. The rather broad, S-shaped form of the ORD curve often causes difficulties in separating the contributions of neighboring electronic transitions and in detecting vibrational fine structure. This problem generally is not present with CD curves, which do not overlap unless the electronic transitions lie very close; when two CD curves do overlap, they are usually easily resolved—especially if they have opposite signs. On the other hand, the dispersive nature of ORD can constitute an advantage in cases such as obtaining spectroscopic-structural information just beyond the wavelength limits of the instrument by determining the sign and magnitude of the ORD curve at these limits. Compounds devoid of chromophores in the accessible spectral region are transparent by CD. Occasionally, the capability to measure rotations at various wavelengths can be useful for comparison purposes with ORD data in the literature. Generally speaking, however, it seems fair to state that if a choice must be made, then CD is preferred over ORD, as the inherent simplicity of CD spectra recommends them over ORD for most of applications.

10-4 QUANTITATIVE MEASUREMENTS

As explained in Section 10-2b, the proportion of light absorbed by a transparent medium is independent of the incident-light intensity but proportional to the number of absorbing molecules through which the light passes. When the absorptivity constants are large, such as $\varepsilon \sim 10{,}000$, UV–vis and chiroptical spectroscopy become ideal for determining concentration using Beer's law (eq. 10-8).

10-4a Difference Spectroscopy

Difference spectroscopy is a sensitive method for detecting and recording changes in a UV–vis spectrum associated with a change in the chromophore, as in solvent perturbation or chemical reaction. In this technique, if a double-beam spectrophotometer is

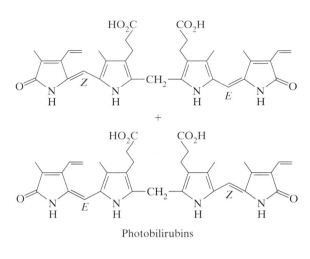

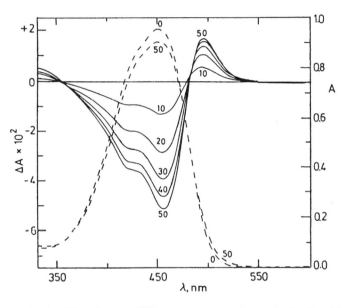

Figure 10-16 Absorbance difference spectra (———) obtained from irradiation of a 1.5×10^{-5} M solution of bilirubin in $CHCl_3$ 1% ethanol 10% triethylamine at 450 nm (10 nm bandpass). The cumulative irradiation time (sec) is indicated on each scan. Absorption curves (– – –) of the sample solution before and after irradiation are superimposed on the absorbance difference spectra. (From the data of D.A. Lightner, T.A. Wooldridge, and A.F. McDonagh, *Biochem. Biophys. Res. Comm.,* **86**, 235 [1979].)

used, the reference and sample compartments each contain identical samples at identical concentrations. The net UV–vis spectrum is a flat line. Then only one, for instance, the sample in the sample compartment, is allowed to change. As it does, a difference spectrum emerges. Alternatively, in a single-beam instrument, the UV–vis spectrum of the sample is recorded, stored electronically, and subtracted from UV–vis spectrum of the sample as it undergoes change.

For example, the bile pigment bilirubin has an intense UV–vis absorption near 450 nm (Figure 10-16). Bilirubin is known to undergo photoisomerization to give a mixture of two photobilirubins that have UV–vis spectra only slightly different from that of bilirubin. After 50 seconds of photoirradiation, a slight drop and slight broadening of the UV–vis spectrum is observed (Figure 10-16). However, spectral changes associated with the photoisomerization may be observed much more easily by sensitive absorbance

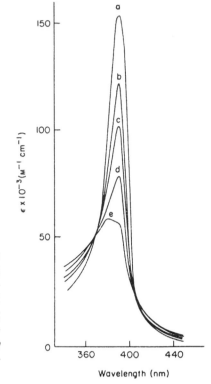

Figure 10-17 Spectra of coproferriheme in phosphate buffer (pH 6.97) showing variation due to aggregation. Concentrations were (a) 3.97×10^{-8} M; (b) 6.36×10^{-6} M; (c) 2.78×10^{-5} M; (d) 9.93×10^{-5} M; (e) 3.18×10^{-4} M. (Reproduced with permission from S.B. Brown, ed. *Introduction to Spectroscopy for Biochemists,* Academic Press, New York, 1980.)

difference spectroscopy, which shows the emergence of negative peaks near 450 nm and a positive peak near 500 nm. The most rapid spectral changes occur early, and the absorbance difference lessens over time, indicating that photoequilibrium is being approached.

10-4b Deviations From Beer's Law

In practice, Beer's law, which states that the absorbance is proportional to the number of absorbing molecules, has been found to hold for a large number of compounds over a considerable range of concentrations. However, since the molar absorption coefficient ε depends on wavelength, Beer's law can be strictly true only for pure monochromatic light. True deviations may be expected when the concentration of absorbing molecules is so high that they interact with each other. Effects such as association and dissociation are common causes of deviations from the Lambert–Bouger–Beer law. For example, compounds that tend to form dimers, such as aqueous dye solutions, seldom follow Beer's law over any extended concentration range. A test of Beer's law can be made by dilution of the sample solution to a different volume, which should then show the correct absorbance corresponding to the dilution.

For example, porphyrins are known to aggregate in aqueous solutions. As illustrated in Figure 10-17, coproferriheme solutions in pH 6.97 aqueous phosphate buffer and at constant ionic strength deviate from Beer's law. Plots (ε vs. λ) of spectra over a wide range of coproferriheme concentrations, 3.18×10^{-4} to 3.97×10^{-8} M, should be identical and thus coincident if Beer's law were obeyed but clearly are not because the solute is forming soluble aggregates.

10-4c Isosbestic Points

Isosbestic points are shared cross-over points where ε remains invariant in a series of overlapping UV–vis curves. Note in Figure 10-17 that there are two wavelengths at which the ε values are the same for all spectra. These isosbestic points correspond to wavelengths where the ε values of two absorbing species are identical. The presence of a third absorbing species is highly improbable, because it too would be required to have identical ε values at the isosbestic points. Isosbestic points therefore are usually a diagnostic for the presence of only two absorbing species—in the example shown, most likely a monomer and dimer. Isosbestic points are also shown in Figures 10-16 and 11-11.

10-5 EXPERIMENTAL ASPECTS

10-5a Solvents

The methods and procedures for obtaining UV–vis and CD measurements are virtually the same. Most measurements are carried out on fairly dilute solutions (10^{-2} to 10^{-6} M) of the sample in an appropriate solvent. Such a solvent should not interact with the solute, should not absorb in the spectral region of interest, and, in the case of chiroptical measurements, should not be optically active—unless one is looking for solvent-induced CD. It is well to point out that, in both ordinary absorption and chiroptical techniques, measurements may be made on pure liquids, gases, and solids.

An important difference between UV–vis spectrophotometers and CD instruments is that the former may be double-beam instruments, whereas the latter invariably are single-beam instruments. Nonetheless, most newer UV–vis instruments are computer controlled and often single beam. In such instruments and in CD and ORD instruments, measurements are taken of a solution of the desired compound followed by rescanning the spectrum with all parameters held the same and using pure solvent in the sample cell to obtain the baseline.

Some useful solvents and their short wavelength cutoff limits are given in Table 10-5. Note the significant advantage to be gained by the use of short pathlength cells (1 mm or less) where there is less solvent in the pathlength.

It is of the utmost importance to use solvents or other reagents with known purity. Commercially available solvents with such designations as "spectral grade" and "for spectroscopy" are not necessarily pure but have been specially prepared to ensure the absence of impurities *absorbing* in the UV–vis region. Nonabsorbing contaminants may well be present.

Commonly used polar solvents are 95% ethanol, water, and methanol. Aliphatic hydrocarbons (such as hexane, heptane, and cyclohexane) are examples of nonpolar solvents that have good spectral transparency (low λ cut-off) (Table 10-5) and have boiling points high enough so that solvent evaporation does not become a problem. However, they must be rigorously purified because these hydrocarbons may contain alkenic impurities or traces of aromatic compounds. It has been observed that fluoroalkanes have enhanced transparency relative to the alkanes, and a similar finding has been made for the fluorinated alcohols such as 2,2,2-trifluoroethanol.

Organic cyanides such as acetonitrile and propionitrile are polar, nonhydroxylic solvents with excellent spectral transparency. A widely used polar, nonhydroxylic solvent is 1,4-dioxane, which is transparent to about 205 nm.

Several mixed solvents have found use in spectroscopic studies, particularly at very low temperatures, usually down to liquid nitrogen temperatures, about $-190°$ C. These solvent systems do not crystallize when cooled but instead become viscous and glassy. Low temperature solvents include (1) EPA, a 5/5/2 mixture by volume of diethyl ether, isopentane, and ethanol, (2) methanol and glycerol, 9/1 v/v, (3) tetrahydrofuran and diglyme, 4/1 v/v, and (4) methylcyclohexane and isopentane, 1/3 v/v. The degree of contraction of these solvents over the range 25° to $-190°$ C is 29.4% in methylcyclohexane/isopentane (1/3), 29% in EPA, 24.4% in methanol/glycerol (9/1), and 26.1% in tetrahydrofuran/diglyme (4/1). Tabulated values for intermediate temperatures may be found in reference 10.5.

10-5b Cells (Cuvettes) and Sample Preparation

Cells for use in the visible region are usually made of Pyrex or similar glass that is transparent to about 380 nm. In the UV region, quartz cells are necessary, and those made of high purity fused silica (Ultrasil, Spectrosil, Supersil) are recommended. Rectangular

TABLE 10-5

Short Wavelength Cut-off Limits of Various Solvents*

Solvent	Cut-off Point, λ (nm)†		Boiling Point (°C)
	10 mm Cell	0.1 mm Cell	
Acetonitrile	190	180	81.6
2,2,2-Trifluorethanol	190	170	79
Pentane	190	170	36.1
2-Methylbutane	192	170	28
Hexane	195	173	68.8
Heptane	197	173	98.4
2,2,4-Trimethylpentane (isooctane)	197	180	99.2
Cyclopentane	198	173	49.3
Ethanol (95%)	204	187	78.1
Water	205	172	100.0
Cyclohexane	205	180	80.8
2-Propanol	205	187	82.4
Methanol	205	186	64.7
Methylcyclohexane	209	180	100.8
Dibutyl ether	210	195	142
EPA‡	212	190	—
Diethyl ether	215	197	34.6
1,4-Dioxane	215	205	101.4
Bis(2-methoxyethyl) ether (glyme)	220	199	162
1,1,2-Trichlorotrifluorethane	231	220	47.6
Dichloromethane	232	220	41.6
Chloroform	245	235	62
Carbon tetrachloride	265	255	76.9
N,N-Dimethylformamide	270	258	153
Benzene	280	265	80.1
Toluene	285	268	110.8
Tetrachloroethylene	290	278	121.2
Pyridine	305	292	116
Acetone	330	325	56
Nitromethane	380	360	101.2
Carbon disulfide	380	360	46.5

* From J.B. Lambert, H.F. Shurvell, L. Verbit, R.G. Cooks, and G.H. Stout, *Organic Structural Analysis,* Macmillan Publishing, New York, 1976.
† The cutoff point is taken as the wavelength at which the absorbance in the indicated cell is about one.
‡ 5/5/2 mixture by volume of ethyl ether, isopentane, and ethanol.

cuvettes are routinely used in isotropic absorption work. The preferred pathlength is 10 mm, since *l* in eq. 10-8 then is equal to unity.

Short pathlength cells are essential when solvent absorption must be minimized. Cells in the range 0.01–2 mm are used for work in the far-UV region. Fused cylindrical cells with fixed windows are commonly used for chiroptical work. Cells to be used in ORD work should be checked for stress birefringence by observing whether the empty cell exhibits optical rotation. Cells are calibrated by the manufacturer but may be checked by determining the absorbance or optical activity of known substances (see Section 10-5d).

For most purposes cells may be cleaned by rinsing *at least ten times* with the solvent used in the measurement. A final rinse with methanol or acetone is suitable if the cells are to be air dried. In general, oven drying is recommended, particularly for short pathlength cells. In this case the final two rinses should be with distilled water so that no trace of a flammable organic liquid remains. Although a drying oven set at a high temperature will not damage the cells (remember that they have been fused during manufacture at temperatures above 1000° C), removing a fragile quartz cell from a hot oven can be troublesome. A fast and convenient drying method is to place the cell in a vacuum desiccator, heated to about 50° C if possible. In this manner, cells may be dried in less than 5 minutes.

TABLE 10-6

Values for CD and ORD Reference Standards: (+)-Camphorsulfonic Acid (CSA), Isoandrosterone, Androsterone, D-Pantolactone, and $\Lambda\text{-}[Co(en)_3]I_3 \cdot H_2O$

Standard	CD		ORD		
	$\Delta\varepsilon_{max}$	$[\theta]_{max} \, (\lambda_{max})$	$[\phi]_{max} \, (\lambda_{max})$	$[\phi]_{min} \, (\lambda_{min})$	a
CSA·H$_2$O*	+2.36	+7,780 (290)	+4,464 (305)	−5,586 (270)	+100.5
Isoandrosterone[†]	+3.30	+10,900 (304)	—	—	—
Androsterone[†]	+3.39	+11,200 (304)	—	—	—
D-Pantolactone[‡]	−3.97	−13,100 (220)	—	—	—
$\Lambda\text{-}[Co(en)_3]I_3 \cdot H_2O$[§]	+1.94	+6,410 (490)	+2,570 (520)	−5,310 (461)	+78.8

* 1.94×10^{-3} M in H$_2$O, $[\alpha]_D^{23} +21.4°$ (c 4.6, H$_2$O).
† In dioxane solvent.
‡ 9.22×10^{-4} M in CH$_3$OH, $[\alpha]_D^{23} -51.0°$ (c 2, H$_2$O).
§ 2.20×10^{-3} M in H$_2$O, $[\alpha]_D^{23} +91.1°$ (c 0.606, H$_2$O).

Above all, cell windows should never be touched with the fingers. It is good practice when cleaning a cell to rinse the outside a few times. Any material that still remains on the outside optical faces should be wiped off with a lintless wiper, such as Kimwipes, *soaked in solvent.* Never use a dry cloth or tissue.

Quantitative analytical techniques are applied to sample preparation. Volumetric glassware and cells must be clean and dry. Solid samples should be dried to constant weight in a desiccator to remove adhering water or solvent. A typical 1–10 mm cell holds anywhere from 0.2–3 ml of solution, so an appropriate amount of stock sample solution should be prepared. Since the measurement is nondestructive, the sample may be recovered by evaporation of the solvent. If the amount of sample available permits, 10–25 ml of solution is a convenient size to prepare. The amount of sample required for this volume is sufficient to minimize errors associated with weighing small quantities. Naturally, one must use an analytical balance capable of weighing directly to at least 0.1 mg. Because most compounds being measured probably will not have been run previously by the operator, an initial sample concentration should be approximately 0.05% (about 0.5 mg ml^{-1}) for small molecules and about 0.005% for polypeptides and large macromolecules. A peak absorbance in the range 0.7–1.2 absorbance units is desirable for most instruments, since it gives a good pen deflection and the electronics are usually most sensitive in this range.

10-5c Calibration of ORD and CD Instruments

Spectropolarimeters may be calibrated with aqueous sucrose solutions or with a standardized quartz plate. The data for sucrose solutions are readily available in the reprint of Lowry's classic book (see reference 10.6). CD and ORD instruments are usually calibrated against a carefully prepared 1.94×10^{-3} M aqueous solution of (+)-camphorsulfonic acid monohydrate (CSA) standard, which absorbs at 290 nm. The standard values, taken from reference 10.7, are reported in Table 10-6.

10-5d Possible Sources of Error

Errors may arise from the nature of the sample being examined, from the instrument, and, last but not least, from the operator. The simplest error may be one in which Beer's law (eq. 10-8) is not obeyed (see Section 10-4b).

The problem of stray light occurs in most double monochromator instruments at high sample absorbances. Stray light can be defined as the ratio of spurious light to the desired wavelength. It can also arise from scattering of the light beam from any of the surfaces that it encounters: lenses, prisms, slit edges, etc., as well as dirty cell windows. Turbid solutions scatter light, and the scattering becomes more important at shorter wavelengths. Sometimes dust particles are responsible for the scattering. Such solutions should be passed through a fine filter, such as Millipore, or centrifuged.

In ORD and CD instruments, stray light may be produced by nonpolarized radiation reaching the detector. A loose sample compartment cover or other light leak may allow room light to enter the instrument and cause stray light effects. It is a common occurrence for stray light effects to occur when solvent absorption becomes severe toward the short wavelength end of a scan. The slits are then wide open to pass the maximum amount of energy, and the absorbance appears to drop sharply. It is important to recognize this effect in order to know the absorbance limits to which valid measurements may be made. Rescanning the spectrum is of no help since stray light effects are usually highly reproducible!

Oxygen has a series of absorption bands that begin at about 195 nm and extend to shorter wavelength. Hence, to work in this spectral region, air must be excluded from the instrument. Otherwise much of the light is lost in both the sample and the reference beam because of oxygen absorption. Oxygen is most easily removed by flushing the entire optical path of the instrument with pure, dry nitrogen. A liquid nitrogen cylinder having a gaseous take-off valve provides a convenient source of high purity nitrogen. Optimum flow rates can be determined by observing the disappearance of the oxygen absorption spectrum.

PROBLEMS

10-1 Calculate the molar extinction coefficient of 3-methylcyclohexanone at 280 nm and 320 nm from the data of Figure 10-1.

10-2 A compound $C_5H_8O_2$ has a UV λ_{max} at 270 nm ($\varepsilon = 32$) in methanol; in hexane, the band shifts to 290 nm ($\varepsilon_{max} = 40$). What functional groups could be present? Draw a possible structure.

10-3 Calculate the absorbance (A) of a 0.005 M solution of 3-methylcyclohexanone in isooctane in a 10 cm pathlength quartz cuvette at $\lambda = 280$ nm.

10-4 A compound, 0.0002 M in methanol, shows $\lambda_{max} = 235$ nm with an absorbance (A) of 1.05 when measured in a 0.5 cm pathlength quartz cuvette. Calculate its ε_{max} at 235 nm.

10-5 A molar absorptivity of benzoic acid (mol. wt. = 122.1) in ethanol at 273 nm is about 2000. If an absorbance not exceeding 1.35 is desired, what is the maximum allowable concentration in g L^{-1} that can be used in a 2.00 cm cell?

10-6 A 250 mg sample containing a colored component X is dissolved and diluted to 250 ml. The absorbance of an aliquot of this solution, measured at 500 nm in a 1.00 cm cell, is 0.900. Pure X (10.0 mg) is dissolved in 1 L of the same solvent. The absorbance measured in a 0.100 cm cell at the same wavelength is 0.300. What is the percent of X in the first sample?

10-7 Substances X and Y, which are colorless, form the colored compound XY: X + Y $\rightleftharpoons$ XY. When 2.00×10^{-3} mol of X is mixed with a large excess of Y and diluted to 1 L, the solution has an absorbance that is twice as great as when 2.00×10^{-3} mol of X is mixed with 2.00×10^{-3} mol of Y and treated similarly. What is the equilibrium constant for the formation of XY?

10-8 Absorbances were measured for three solutions containing A and B separately and in a mixture, all in the same cell. Calculate the concentrations of A and B in the mixture.

		Absorbance	
		475 nm	670 nm
0.001 M	A	0.90	0.20
0.01 M	B	0.15	0.65
Mixture		1.65	1.65

10-9 The CD spectrum of a chiral compound, 0.0005 M in isooctane, shows $\lambda_{max} = 285$ nm with $\Delta A = -0.0003$ when measured in a 1.0 cm pathlength quartz cuvette. Calculate $\Delta\varepsilon_{max}$ at 285 nm.

10-10 The amplitude of the ORD $n \longrightarrow \pi^*$ Cotton effect of 3-methylcyclohexanone was determined to be $[a] = -25°$. Calculate $\Delta\varepsilon_{max}$.

BIBLIOGRAPHY

10.1. J.B. Lambert, H.F. Shurvell, L. Verbit, R.G. Cooks, and G.H. Stout, *Organic Structural Analysis*, Macmillan Publishing, New York, 1976.

10.2. R.L. Pecsok, L.D. Shields, T. Cairns, and I.B. McWilliam, *Modern Methods of Chemical Analysis*, 2nd ed., John Wiley & Sons, Inc., New York, 1976.

10.3. H.H. Jaffé and M. Orchin, *Theory and Applications of Ultraviolet Spectroscopy*, John Wiley & Sons, Inc., New York, 1962.

10.4. A. Moscowitz, in *Adv. Chem. Phys.*, **4,** 67, (1962).

10.5. O. Korver and J. Bosma, *Anal. Chem.*, **43,** 1119 (1971).

10.6. T.M. Lowry, *Optical Rotatory Power*, Dover Publications, Mineola, NY, 1964 (reprint of 1935 first publication).

10.7. K. Timura, T. Konno, H. Meguro, M. Hatano, T. Murakami, K. Kashiwabara, K. Saito, Y. Kondo, and T.M. Suzuki, *Anal. Biochem.*, **81,** 167 (1977).

chapter 11

UV–Vis, CD, and ORD

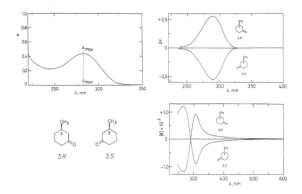

UV–vis and CD spectroscopies measure the probability and energy of exciting a molecule from its ground electronic state to an electronically excited state (promoting an electron from an occupied to an unoccupied molecular orbital). This process is illustrated in Figure 11-1 for the aldehyde or ketone carbonyl chromophore, as in 3-methyl-cyclohexanone (see Figure 10-1a). Note that both singlet (S) and triplet (T) states are possible. If the spins of pairs of electrons are antiparallel, the state is a singlet; if the spins of two electrons are parallel, three states are possible (jointly called a triplet state). Selection rules permit S $\longrightarrow$ S and T $\longrightarrow$ T processes but not S $\longrightarrow$ T and T $\longrightarrow$ S. Ground states are usually singlets; thus most excitations are to singlet excited states. Triplet excited states are usually formed by intersystem crossing from an excited singlet state, such as S_1, rather than by direct excitation from the S_0 ground state. The electronically excited state may decay unimolecularly back to the ground state *photophysically* by emitting energy of fluorescence (from an excited singlet state) or of phosphorescence (from an excited triplet state). Alternatively, it might decay *photochemically* to a different ground state (hence a different structure). One can thus measure absorption and emission from molecules. In this part we shall be concerned mainly with absorption processes.

11-1 ELECTRONIC TRANSITIONS AND CHROMOPHORES

A UV–vis absorption spectrum of a given compound may exhibit many absorption bands, with λ_{max} of each band corresponding roughly to the energy associated with the formation of a particular excited state. This situation is illustrated in Figure 10-1a and (especially) in Figure 10-6, where λ_{max} corresponds to the excitation energy and ε_{max} to the intensity of the electronic transition—a measure of the probability of promoting an electron, given the requisite excitation energy.

As indicated earlier, electronic excitations are typically assigned to chromophores in a molecule. Although the entire molecule is in the excited energy state, the excited state energy is mainly localized within the chromophore for simple transitions.

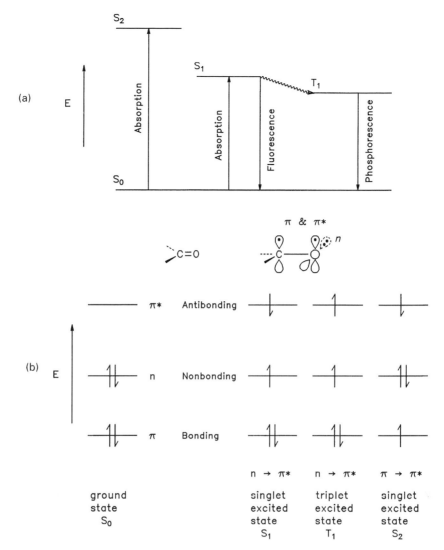

Figure 11-1 (a) Energy diagram for electronic excitation and decay processes in the (ketone) carbonyl chromophore. (b) Diagram of selected electronic molecular orbital energies for the isolated carbonyl chromophore of 3-methylcyclohexanone, showing the ground state and excited state configurations. Singlet states (S) all have electron spins paired, triplet states (T) have two spins parallel. Note that the n orbital containing two electrons is orthogonal to the π and π^* orbitals. Subscript 0 refers to ground state, 1 to the first excited state, 2 to the second excited state.

11-1a Classification of Electronic Transitions

The wavelength of an electronic transition depends on the energy difference between the ground state and the excited state. It is a useful *approximation* to consider the wavelength of an electronic transition to be determined by the energy difference between the molecular orbital originally occupied by the electron and the higher orbital to which it is excited. Saturated hydrocarbons contain only strongly bound σ electrons. Their excitation to antibonding σ^* orbitals ($\sigma \longrightarrow \sigma^*$) or to molecular Rydberg orbitals (involving higher valence shell orbitals, 3s, 3p, 4s, etc.) requires relatively large energies, corresponding to absorption in the far-UV region. One exception is cyclopropane, which has λ_{max} at 190 nm. Contrast this cycloalkane to propane, which has λ_{max} about 135 nm.

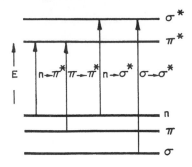

Figure 11-2 Relative electronic orbital energies and selected transitions in order of increasing energy.

TABLE 11-1

Effect of Extended Conjugation in Alkenes on Position of Maximum Absorption[*]

n in H(CH=CH)$_n$H	λ_{max} (nm)	ε_{max} (liter mol^{-1}cm^{-1})	Color
1	162	10,000	Colorless
2	217	21,000	Colorless
3	258	35,000	Colorless
4	296	52,000	Colorless
5	335	118,000	Pale yellow
8	415	210,000	Orange
11	470	185,000	Red
15	547[†]	150,000	Violet

[*]From J.B. Lambert, H.F. Shurvell, L. Verbit, R.G. Cooks, and G.H. Stout, *Organic Structural Analysis,* Macmillan Publishing, New York, 1976.
[†]Not a maximum.

Electronic transitions commonly observed in the readily accessible UV (above ~190 nm) and visible regions have been grouped into several main classes* (Figure 11-2).

n ⟶ π Transitions.* These transitions involve the excitation of an electron in a nonbonding atomic orbital, such as unshared electrons on O, N, S, or halogen atom, to an antibonding π* orbital associated with an unsaturated center in the molecule. The transitions occur with compounds that possess double bonds involving heteroatoms, for example, C=O, C=S, N=O. A familiar example is the low intensity absorption in the 285–300 nm region of saturated aldehydes and ketones (see Figure 10-1).

π ⟶ π Transitions.* Molecules that contain double or triple bonds or aromatic rings can undergo transitions in which a π electron is excited to an antibonding π* orbital. Although ethylene itself does not absorb strongly above about 185 nm, conjugated π electron systems are generally of lower energy and absorb in the accessible spectral region. An important application of UV–vis spectroscopy is to define the presence, nature, and extent of conjugation. Increasing conjugation generally moves the absorption to longer wavelengths and finally into the visible region; this principle is illustrated in Table 11-1.

n ⟶ σ Transitions.* These transitions, which are of less importance than the first two classes, involve excitation of an electron from a nonbonding orbital to an antibonding σ* orbital. Since *n* electrons do not form bonds, there are no antibonding

* In addition to that given here, several other systems of classification exist. For example, the designation N-V is used to describe transitions from a bonding to an antibonding orbital (σ ⟶ σ*, π ⟶ π*). The term N-Q designates transitions from a nonbonding atomic orbital to a higher energy molecular orbital (*n* ⟶ σ*, *n* ⟶ π*). Burawoy has termed π ⟶ π* transitions K-bands (from the German *Konjugation*) and *n* ⟶ π* transitions R-bands (from an early theory that the excited state was a radical). Numerous terms also exist in the literature for the classification of ground and excited states on the basis of symmetry (see reference 11.1).

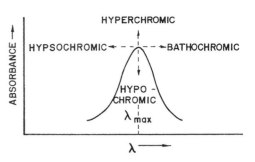

Figure 11-3 Terminology of shifts in the position of an absorption band.

orbitals associated with them. Some examples of $n \longrightarrow \sigma^*$ transitions are CH_3OH (vapor), λ_{max}, 183 nm, ε 150; trimethylamine (vapor), λ_{max} 227 nm, ε 900; and CH_3I (hexane), λ_{max} 258 nm, ε 380.

Rydberg transitions are mainly to higher excited states. For most organic molecules, they occur at wavelengths below about 200 nm. A Rydberg transition is often part of a series of molecular electronic excitations that occurs with systematically narrowing spacings toward the short wavelength side of the UV range and terminates at a limit representing the ionization potential of the molecule.

Groups that give rise to electronic absorption are known as *chromophores* (color bearer, from an early theory of color). The term *auxochrome* (color increaser) is used for substituents containing unshared electrons (OH, NH, SH, halogens, etc.). When attached to π electron chromophores, auxochromes generally move the absorption maximum to longer wavelengths (lower energies). Such a movement is described as a *bathochromic* or *red shift*. The term *hypsochromic* denotes a shift to shorter wavelength (*blue shift*). Increased conjugation usually results in increased intensity, termed *hyperchromism*. A decrease in intensity of an absorption band is termed *hypochromism*. These terms are summarized in Figure 11-3.

11-1b Singlet and Triplet States

Most molecules have ground electronic states with all electron spins paired, and most excited electronic states also have electron spins all paired, even though in the excited state there may be two orbitals that each possess only one electron (see Figure 11-1). Such states are known as *singlet states* ($S_1, S_2, \ldots$) and have no net spin angular momentum. Imposition of a reference direction by an applied magnetic or electric field can produce only the single component of zero angular momentum in the field direction.

For molecules having an even number of electrons, regardless of whether or not the ground state is a singlet, there are excited states in which a pair of electrons have their spins parallel, giving the molecule a net spin angular momentum. Angular components along a given direction can now have values of $+1, 0$, and -1 times the angular momentum. Such an electronic configuration is known as a *triplet state* ($T_1, T_2, \ldots$). As indicated earlier, $S \longrightarrow S$ and $T \longrightarrow T$ are spin-allowed excitations, but $S \longrightarrow T$ and $T \longrightarrow S$ are spin forbidden and typically have small to vanishing ε_{max} values.

11-1c Allowed and Forbidden Transitions

Electronic transitions may be classed as intense or weak according to their magnitude, roughly measured by ε_{max}. These correspond to *allowed* or *forbidden* transitions. Allowed transitions are those for which (1) there is no change in the orientation of electron spin, (2) the change in angular momentum is 0 or ± 1, and (3) the product of the electric dipole vector and the group theoretical representations (see reference 11.1) of the two states is totally symmetric.

The first rule is the *spin selection rule* and may be stated as follows: transitions between states of different spin multiplicities (such as, S and T) are invariably forbidden since electrons cannot undergo spin inversion during the change of electronic state. The second rule usually presents no problem since most states are within one unit of angular momentum of each other. The last rule is the *symmetry selection rule*. If the direct product of the group theoretical representations (see reference 11.1) to which the initial and final state functions belong is different from all the representations to which the coordinate axes belong, the transition moment of that transition is zero. A good example of this rule is the $n \longrightarrow \pi^*$ transition of saturated alkyl ketones, where a carbonyl n electron is promoted to an *orthogonal* π^* orbital (90° movement of charge) (Figure 11-1). Such a transition is said to be symmetry forbidden. For most organic molecules, such forbidden transitions are usually observable but of weak intensity. They arise because the intensity of the electronic absorption band really depends on the average of the electronic transition moments over all the nuclear orientations of the vibrating molecule and this average is not necessarily zero. When the symmetry of a molecule is periodically changed by some vibration that is not totally symmetric, the symmetry of the electronic wave functions is also periodically changed since the electrons adapt instantaneously to the motion of the nuclei. Hence a symmetry forbidden transition may become allowed. The intensity of a transition that is symmetry forbidden but has become vibrationally allowed is much less than that of an ordinarily allowed transition. Such vibrational contributions are temperature dependent.

11-1d   Absorption Intensity: Oscillator and Rotatory Strengths

As we have seen in Section 10-2c, the shape of an electronic absorption band is due primarily to the vibrational sublevels of the electronic states. It is a widespread and convenient practice among many chemists to describe the intensities of absorption bands in terms of the molar absorptivity ε_{max} (for UV–vis spectra) and $\Delta\varepsilon_{max}$ (for CD spectra). Unfortunately, the values of ε_{max} and $\Delta\varepsilon_{max}$ *are not directly related to any quantity obtainable from theory*.

Although intensity of the interaction causing the electronic transition is typically approximated nicely by ε_{max}, a more accurate description is determined by the wavelength-weighted *area* under the particular band rather than to any particular value of the absorbance. A convenient unit characterizing the intensity of this excitation for an electronic transition, suggested by Milliken and termed the dipole strength D (eq. 11-1),

$$D = 9.184 \times 10^{-39} \int_0^\infty \frac{\varepsilon}{\lambda} d\lambda \qquad (11\text{-}1)$$

represents the electronic transition probability and hence the intensity of the UV–vis absorption band. Application of the definition of D to typical experimental curves reveals that the dipole strength varies from 10^{-34} to 10^{-38} erg · cm³. The quantity D may be approximated roughly from ε_{max}, λ_{max} and $\Delta\lambda$ (the bandwidth at $\varepsilon_{max}/2$) by eq.11-2.

$$D \approx 9.188 \times 10^{-39} (\varepsilon_{max}) \Delta\lambda/\lambda_{max} \qquad (11\text{-}2)$$

In an electronic transition, an electron is promoted from one molecular orbital to a higher-lying molecular orbital while the molecule goes from its ground state to an excited state. This migration of charge creates a momentary dipole, called the electric transition dipole moment **(μ),** which has both direction and intensity and is therefore a vector quantity. The intensity of **μ** is the square root of D (of eq. 11-1 or 11-2).

In CD spectroscopy, the integrated area under the CD curve is called the rotatory strength (R), and the equation (11-3) relating R to $\Delta\varepsilon$ takes on a form quite analogous

$$R = 2.297 \times 10^{-39} \int_0^\infty \frac{\Delta\varepsilon}{\lambda} d\lambda \qquad (11\text{-}3)$$

to that of the dipole strength for a UV–vis transition (see reference 11.3). Values of R are typically 10^{-38} to 10^{-42} cm · erg · gauss^{-1}. The quantity R may be approximated roughly in terms of $\Delta\varepsilon_{max}$, λ_{max} and $\Delta\lambda$ by eq. 11-4, or more simply as eq. 11-5 in which

$$R \approx 2.297 \times 10^{-39} \, (\Delta\varepsilon_{max}) \; \Delta\lambda/\lambda_{max} \tag{11-4}$$

$$[\theta]_{max} \approx 994 \, [R] \quad \text{or} \quad \Delta\varepsilon_{max} \approx 0.30 \, [R] \tag{11-5}$$

$[R]$ is the reduced rotatory strength, $[R] \approx 1.078 \times 10^{40} \, R$.

A CD transition involves both the electric dipole transition moment (μ) and a magnetic transition dipole moment (**m**). Just as the dipole strength (D) is related to μ by $D = \mu \cdot \mu \, (= \mu^2)$, so the rotatory strength (R) is related to both μ and **m** by the scalar product, $R = \mu \cdot \mathbf{m} = |\mu| |\mathbf{m}| \cos\beta$, where β is the angle between the transition moments μ and **m**. When β is acute ($0 < \beta < 90°$), the Cotton effect is positive; when β is obtuse ($90 < \beta < 180°$), the Cotton effect is negative. When $\beta = 90°$, there is no CD, as the scalar product $\mu \cdot \mathbf{m} = 0$.

11-2 ISOLATED CHROMOPHORES

11-2a Carbonyl: Ketone and Aldehyde Absorption

The longest wavelength transition in aliphatic aldehydes and ketones, the $n \longrightarrow \pi^*$ band, is probably the best studied of any electronic transition (see references 11.1, 11.3, and 11.4 and Figure 11-1). It is a weak ($\varepsilon \sim 10\text{--}20$) and rather broad band, occurring in the neighborhood of 270–300 nm. As noted in Section 10-2d, its position is quite solvent sensitive. The $n \longrightarrow \pi^*$ transition involves the promotion of an electron from a nonbonding p orbital on oxygen to the antibonding π^* orbital associated with the entire carbonyl group. The transition is symmetry forbidden (see Section 11-1c), hence the low intensity (Figure 10-1a).

A second carbonyl band, attributed to a $\pi \longrightarrow \pi^*$ transition, occurs near 190 nm in ketones. The $\pi \longrightarrow \pi^*$ transition is allowed and is considerably more intense than the $n \longrightarrow \pi^*$ transition. This wavelength region is near the practical wavelength cut-off of most UV instruments, so that often only the beginning of the band is observed as so-called end absorption. Transitions at wavelengths shorter than about 190 nm are most likely due to excitations from the carbonyl sigma (σ) bond ($\sigma_{CO} \longrightarrow \pi^*$) and from Rydberg transitions.

The symmetries of the n, π and π^* orbitals involved in the transitions are important. The bonding π orbital and the antibonding π^* orbital lie in the same plane, whereas the nonbonding n orbital is in an orthogonal plane. Hence promotion of an electron from the nonbonding orbital is not possible without a significant change in the geometry of the molecule. The weak $n \longrightarrow \pi^*$ intensity is due to nonsymmetrical vibrations that slightly deform the molecule and lower its symmetry, allowing the transition to acquire a finite probability.

Cyclic ketones absorb at longer wavelength than the corresponding open-chain analogues. In addition, there is a variation in the position of the absorption band with ring size in nonpolar solvents, as illustrated in Table 11-2.

The $n \longrightarrow \pi^*$ transition of ketones and aldehydes is sensitive to the presence of nonalkyl substituents, which when located α to the carbonyl group affect the position (λ_{max}) and intensity (ε_{max}) of the absorption band. For example, the presence of an α-bromine in the cyclohexanone series causes a bathochromic shift of λ_{max} of about 23 nm when the bromine is axial, but a 5 nm shift when it is equatorial. Equatorially substituted 2-chloro-4-*tert*-butylcyclohexanone has its $n \longrightarrow \pi^*$ maximum at a slightly shorter wavelength than that of the parent ketone, whereas the axial chlorine isomer has a more intense absorption band at a considerably longer wavelength (Table 11-2). The strong bathochromic and hyperchromic effect of an α-halo substituent also is observed in steroidal ketones.

In general, $n \longrightarrow \pi^*$ transitions are easily recognizable by their low intensities, by the spectral shifts caused by substitution, and by the sensitivity of the position of the band to solvent effects. Shifts in the position of absorption bands on going from the

TABLE 11-2

Absorption Data for Aliphatic Aldehydes and Ketones*

Compound	Solvent	$n \longrightarrow \pi^*$ Transition		$\pi \longrightarrow \pi^*$ Transition	
		λ_{max} (nm)	ε_{max} (liter mol^{-1}cm^{-1})	λ_{max} (nm)	ε_{max} (liter mol^{-1}cm^{-1})
Formaldehyde	Vapor	304	18	175	18,000
	Isopentane	310	5		
Acetaldehyde	Vapor	289	12.5	182	10,000
Acetone	Vapor	274	13.6	195	9,000
	Cyclohexane	275	22	190	1,000
Butanone	Isooctane	278	17		
2-Pentanone	Hexane	278	15		
4-Methyl-2-pentanone	Isooctane	283	20		
Cyclobutanone	Isooctane	281	20		
Cyclopentanone	Isooctane	300	18		
Cyclohexanone	Isooctane	291	15		
Cycloheptanone	Isooctane	292	17		
Cyclooctanone	Isooctane	291	15		
Cyclononanone	Isooctane	293	17		
Cyclodecanone	Isooctane	288	15		
2-Chloro-4-*tert*-butyl-cyclohexanone					
Equatorial Cl	Isooctane	286	17		
Axial Cl	Isooctane	306	49		

*From J.B. Lambert, H.F. Shurvell, L. Verbit, R.G. Cooks, and G.H. Stout, *Organic Structural Analysis,* Macmillan Publishing, New York, 1976.

vapor phase to solution or from one solvent to another are caused mainly by differences in the solvation energies of the solute in the ground and excited electronic states. The effect of solvent on the position of the $n \longrightarrow \pi^*$ absorption has served as an important diagnostic tool (see Section 10-2d). Indeed, the fundamental role of the unshared electron pair in this transition can be demonstrated by the disappearance of the $n \longrightarrow \pi^*$ band in acid solution in which the unshared pair is protonated.

Changing from a nonpolar to a polar solvent results in a significant hypsochromic shift in the position of the $n \longrightarrow \pi^*$ transition. It has been shown that hydroxylic solvents of comparable dielectric constant cause a larger blue shift than do nonhydroxylic, polar solvents. The larger shifts occasioned by hydroxylic solvents are attributable in part to greater hydrogen bonding to the carbonyl oxygen lone pairs than to the π^* electrons, thus lowering the energy of the ground state relative to that of the excited state. An example of solvent effects on the $n \longrightarrow \pi^*$ band of acetone is shown in Figure 11-4, and of an α,β-unsaturated ketone in Table 10-4.

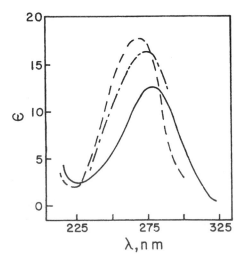

Figure 11-4 Solvent effects on the $n \longrightarrow \pi^*$ transition of acetone in hexane (———), in 95% ethanol (— · —), and in water (– – –).

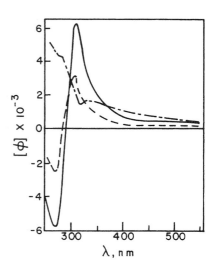

Figure 11-5 The ORD curves in methanol solution of 5α-cholestan-1-one **(11-1)** (— · —), -2-one **(11-2)** (———), and -3-one **(11-3)** (– – –). (Adapted with permission from C. Djerassi, *Optical Rotatory Dispersion,* McGraw-Hill Book Co., New York, 1960.)

11-2b Carbonyl: Ketone ORD and CD

The renaissance of the chiroptical techniques in the mid-1950s began with Djerassi's studies of the ketone $n \longrightarrow \pi^*$ transition (see references 11.3 and 11.5). Optical rotatory dispersion was first applied to stereochemical and configurational investigations of compounds such as the 5α-cholestanones **(11-1, 11-2,** and **11-3).**

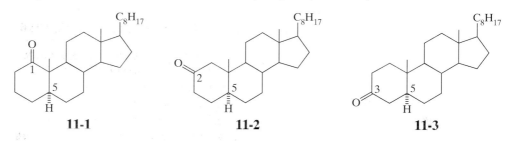

11-1   **11-2**   **11-3**

The ORD curves of the isomeric 1-, 2-, and 3-keto steroids are shown in Figure 11-5. Although **11-2** and **11-3** exhibit positive Cotton effects, the 1-keto isomer **(11-1)** shows only a weak, negative $n \longrightarrow \pi^*$ Cotton effect ($a = -25$) superimposed on a positive background rotation due to more intense Cotton effects at shorter wavelengths. The sign and magnitude of the Cotton effects are due to the chiral environment in the vicinity of the carbonyl chromophore. In principle, then, optically active chromophores in a molecule can be used as stereochemical probes.

Distinguishing between a 2- and a 3-keto steroid by means of their UV spectra is practically impossible; even the IR spectra of such compounds show only slight differences. However, ORD–CD allows a clear distinction. As seen in Figure 11-5, the magnitudes of the two positive Cotton effects are quite different: $a = +121$ for the 2-keto isomer **(11-2)** and +55 for the 3-keto isomer **(11-3).** Since these compounds are derived from natural products, they are optically pure and the amplitudes of their 290 nm Cotton effects can serve to differentiate them.

The corresponding CD spectra of **11-1** and **11-3** are shown in Figure 11-6. The negative Cotton band of the 1-keto steroid **(11-1)** is seen much more clearly than by ORD, owing to the absence of background effects.

11-2c Carbonyl: Acid, Ester, and Amide Absorption

Introduction of a heteroatom at the carbon atom of the carbonyl chromophore causes a large blue shift of the $n \longrightarrow \pi^*$ band, although the intensity remains about the same as for aldehydes and ketones. The effect of such substituents on the $n \longrightarrow \pi^*$ transition is given in Table 11-3.

The heteroatom at the carbonyl group in carboxylic acids, esters, acid chlorides, amides, etc., can donate electron density by conjugation to the carbonyl function. The

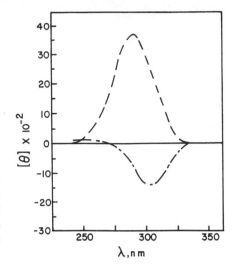

Figure 11-6 The CD spectra in methanol solution of 5α-cholestan-1-one (**11-1**) (— · —) and -3-one (**11-3**) (– – –). (Reproduced with permission from C. Djerassi, H. Wolf, and E. Bunnenberg, *J. Am. Chem. Soc.*, **84**, 4552 [1962]. Copyright 1962 American Chemical Society.)

TABLE 11-3

Effect of Heteroatom Substituents on the Carbonyl $n \longrightarrow \pi^*$ Transition*

X in CH$_3$$\overset{\overset{\displaystyle O}{|}}{C}$—X	Solvent	λ_{max} (nm)	ε_{max} (liters mole^{-1} cm^{-1})
—H	Vapor	290	10
—CH$_3$	Hexane	279	15
	95% ethanol	272.5	19
—OH	95% ethanol	204	41
—SH	Cyclohexane	219	2200
—OCH$_3$	Isooctane	210	57
—OC$_2$H$_5$	95% ethanol	208	58
	Isooctane	211	58
—O—$\overset{\overset{\displaystyle O}{\|}}{C}$—CH$_3$	Isooctane	225	47
—Cl	Heptane	240	40
—Br	Heptane	250	90
—NH$_2$	Methanol	205	160

*From J.B. Lambert, H.F. Shurvell, L. Verbit, R.G. Cooks, and G.H. Stout, *Organic Structural Analysis,* Macmillan Publishing, New York, 1976.

energy of the antibonding π^* orbital is raised by interaction with the unshared pair of the substituent, whereas the p orbital occupied by the n electrons in the ground state is not affected. The $n \longrightarrow \pi^*$ transition energy thus is raised, and the absorption shifts to shorter wavelength.

Two bands have been identified in the UV spectra of aliphatic carboxylic acids and esters. The first is an $n \longrightarrow \pi^*$ transition in the vicinity of 210 nm ($\varepsilon = 40–60$). The second band corresponds to a $\pi \longrightarrow \pi^*$ transition at about 165 nm ($\varepsilon = 2500–4000$).

Carboxylic acids are extensively dimerized in the liquid state, in nonpolar solvents, and even to some extent in the vapor phase (eq. 11-6). Hence, depending on solvent and

$$2\ \text{RCOOH} \rightleftharpoons R—C\overset{\overset{\textstyle O \cdots H—O}{}}{\underset{\underset{\textstyle O—H \cdots O}{}}{}}C—R \tag{11-6}$$

concentration, one may be dealing with pure dimer, pure monomer, or a mixture of the two. In the case of carboxylic acids, it is recommended that spectra, either isotropic absorption or chiroptical, be determined at more than one concentration. Protic solvents such as alcohols shift the above equilibrium toward monomer. Conversion of the acid to an ester obviates the possibility of hydrogen bonding according to eq. 11-6. Essentially identical spectra from a carboxylic acid and, for example, its methyl or ethyl ester are indicative that dimerization is absent or is not spectroscopically important.

UV and chiroptical spectra of amides have received considerable attention because of the importance of the amide chromophore in polypeptides and proteins. At least five transitions of the amide group have been identified (see reference 11.6). The two lowest energy transitions are a weak $n \longrightarrow \pi^*$ band ($\varepsilon \sim 100$) near 220 nm in nonpolar solvents and a relatively strong $\pi \longrightarrow \pi^*$ transition in the 173–200 nm region ($\varepsilon \sim 8000$). Both of these transitions are strongly perturbed by the nitrogen 2p orbital of the conjugated π system, which extends over the N, C, and O atoms. The $n \longrightarrow \pi^*$ amide transition exhibits the usual solvent effect, being blue-shifted on going from nonpolar to hydroxylic solvent.

The question of whether aliphatic amides protonate on oxygen or nitrogen was examined by UV spectroscopy. It was reasoned (see reference 11.7) that if protonation occurs on nitrogen, the nitrogen atom would be removed from the amide system and the spectral properties of the simple carbonyl group should result, for instance, a large red shift of the $n \longrightarrow \pi^*$ band and a large blue shift of the $\pi \longrightarrow \pi^*$ band. Studies of the acidity dependence of the $\pi \longrightarrow \pi^*$ absorption of N,N-dimethylacetamide near 195 nm showed no significant shifts, indicating that the oxygen-protonated amide is the dominant species in dilute acid solutions.

11-2d Alkenes

Most alkenes absorb in the far-UV region, near 200 nm. Substitution of the double bond influences λ_{max} somewhat, with more highly substituted double bonds having λ_{max} bathochromically shifted relative to the less highly substituted (Table 11-4). Ethylene itself absorbs well outside of the generally accessible UV region with a broad absorption maximum at about 162 nm. The rather intense absorption ($\varepsilon \sim 10,000$) is attributed to a $\pi \longrightarrow \pi^*$ absorption maximum. Geometrical isomers of disubstituted alkenes often can be differentiated, although not necessarily systematically. *trans*-2-Butene has a longer wavelength $\pi \longrightarrow \pi^*$ absorption than *cis*-2-butene, but *trans*-2-octene has a shorter wavelength $\pi \longrightarrow \pi^*$ transition than *cis*-2-octene compounds (Table 11-4). Solution spectra are displaced to longer wavelength than the corresponding vapor spectra. When two alkenic chromophores in the same molecule are insulated from each other by saturated carbons, their spectrum approximates the sum of the two chromophores, such as 1,5-hexadiene in Table 11-4.

The double bond, when rendered optically active by a chiral environment, gives rise to Cotton effects. Although the UV spectra of the isomeric steroidal alkenes

TABLE 11-4

Absorption Data for Alkenes*

Compound	Solvent	λ_{max} (nm)	ε_{max} (liter mol^{-1} cm^{-1})
Ethylene	Vapor	162	10,000
cis-2-Butene	Vapor	174	—
trans-2-Butene	Vapor	178	13,000
1-Hexene	Vapor	177	12,000
	Hexane	179	—
cis-2-Octene[†]	Vapor	183	13,000
trans-2-Octene[†]	Vapor	179	15,000
Allyl alcohol	Hexane	189	7,600
Cyclohexene	Vapor	176	8,000
	Cyclohexane	183.5	6,800
Cholest-4-ene	Cyclohexane	193	10,000
1,5-Hexadiene	Vapor	178	26,000

* From A.E. Hansen and T.D. Bouman, *Adv. Chem. Phys.,* **44,** 545 (1980).
† Data from J.R. Platt, H.B. Klevens and W.C. Price, *J. Chem. Phys.,* **17,** 466 (1949).

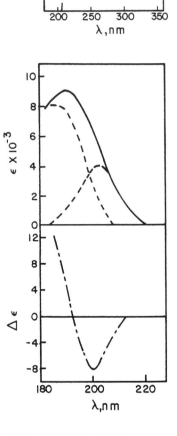

Figure 11-7 The ORD curves in cyclohexane of cholest-4-ene **(11-4)** (———) and cholest-5-ene **(11-5)** (– – –). (Reproduced with permission from A. Yogev, D. Amar, and Y. Mazur, *Chem. Commun.*, 339 [1967].)

Figure 11-8 Isotropic absorption (———), deconvoluted isotropic absorption (– – –), and CD (— · —) spectra of cholest-4-ene **(11-4)** in cyclohexane. (Reproduced with permission from A. Yogev, J. Sogiv, and Y. Mazur, *Chem. Commun.*, 411 [1972].)

cholest-4-ene **(11-4)** and cholest-5-ene **(11-5)** are almost identical, the ORD curves are quite different. The data are presented in Figure 11-7. The shape of the ORD curve

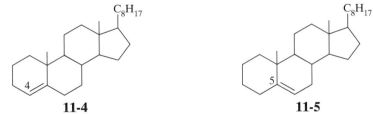

11-4   **11-5**

in the Cotton effect region is characteristic of the position of the double bond for various steroidal alkenes. Hence, once the ORD curves are known for these model compounds, they can be used to locate the position of double bonds in unknown compounds, so long as the latter do not contain other chromophores that absorb in the same spectral region as the alkenic chromophore or interact with it.

The UV spectra of unconjugated alkenes have been shown to exhibit more than just the high intensity $\pi \longrightarrow \pi^*$ transition in the accessible spectral region. Within the past decade, other bands have been observed in the vapor phase spectra of alkyl-substituted alkenes: higher energy Rydberg transitions and a weaker band somewhat above 200 nm, which has been assigned to a $\pi \longrightarrow \sigma^*$ transition (see references 11.8, 11.9, 11.10). Studies by Yogev, Sagiv, and Mazur (reference 11.11) based on linear dichroism (the measurement of $\varepsilon_{\parallel}$ and $\varepsilon_{\perp}$ of an oriented molecule) have shown that the broad UV band of **11-4** and **11-5** near 190 nm is actually composed of two transitions, a weaker one at about 203 nm (ε 4000) and a more intense band at approximately 185 nm (ε 8000). The first band is polarized at an angle of about 16.5° to the double bond axis, whereas the second transition is polarized along the direction of the double bond. The observed and deconvoluted UV spectra, together with the CD curve of cholest-4-ene **(11-4),** are shown in Figure 11-8. This example illustrates the usefulness of CD in detecting transitions that are not resolved in the isotropic absorption spectrum. The CD spectrum of **11-4** reveals a negative Cotton effect associated with the longer wavelength transition and, overlapping it on the shorter wavelength side, a more intense positive Cotton effect whose maximum was not reached but which presumably is associated with the second transition. A caveat should be given here. Oppositely signed, overlapping Cotton effects appear essentially as the algebraic sum of the component Cotton effects. Hence, depending upon their relative magnitudes and proximity to each other, the observed bands can be shifted in position and diminished in intensity relative to the isolated Cotton effects.

11-3 CONJUGATED CHROMOPHORES

UV–vis spectroscopy has long been used as a structural diagnostic to detect conjugation when two or more chromophores are present, as in dienes and polyenes, α,β-unsaturated carbonyl compounds, or aromatic chromophores.

11-3a Dienes and Polyenes

Conjugation of π electron systems of double bonds results in dramatic bathochromic shifts and increased intensities. Spectral data illustrating some of the above effects are given in Table 11-5.

The bathochromic shifts resulting from increased double bond conjugation may be illustrated in terms of Hückel molecular orbital theory (more sophisticated theories lead to the same result). The molecular orbitals and their relative energies for ethylene, 1,3-butadiene, and two higher conjugated homologues are depicted in Figure 11-9. Note

TABLE 11-5

Absorption Data for Alkenes*

Compound	Solvent	λ_{max} (nm)	ε_{max} (liter mol^{-1} cm^{-1})
1-Hexene	Vapor	177	12,000
	Hexane	179	—
1,3-Hexadiene	Hexane	224	26,400
1,3,5-Hexatriene	Isooctane	268	43,000
1,3,5,7-Octatetraene	Cyclohexane	304	—
1,3,5,7,9-Decapentaene	Isooctane	334	121,000
1,3,5,7,9,11-Dodecahexaene	Isooctane	364	138,000

*From A.J. Merer and R.S. Mulliken, *Chem. Rev.,* **63**, 639 (1969).

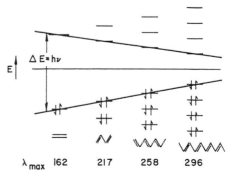

Figure 11-9 Schematic Hückel molecular orbital diagram illustrating the effect of conjugation on the $\pi \longrightarrow \pi^*$ absorption maximum.

TABLE 11-6

Absorption Data for Substituted Dienes in Ethanol*

Compound	λ_{max} (nm)	ε_{max} (liter mol^{-1} cm^{-1})
$CH_2{=}CH{-}CH{=}CH_2$	217.5	22,400
$CH_3CH{=}CH{-}CH{=}CH_2$	223	25,000
$CH_2{=}C(CH_3){-}CH{=}CH_2$	222.5	22,800
$CH_3CH{=}CH{-}CH{=}CHCH_3$	226	23,800
$CH_2{=}C(CH_3){-}C(CH_3){=}CH_2$	226.5	20,300
$CH_3CH{=}C(CH_3){-}C(CH_3){=}CH_2$	231.5	19,200

*From W.F. Forbes, R. Shilton, and A. Balasubramanian, *J. Org. Chem.*, **29**, 3527 (1964).

that the energies of the highest occupied molecular orbitals (HOMO) increase, while those of the lowest unfilled molecular orbitals (LUMO) decrease with increasing conjugation. The observed transition involves promotion of an electron from the HOMO to the LUMO. Figure 11-9 shows that as the conjugated π system increases in length, the energy required for the transition becomes less, that is, a bathochromic shift results.

The diene chromophore $\pi \longrightarrow \pi^*$ absorption band is shifted in a systematic way by substitution, as illustrated in Table 11-6. This pattern was recognized more than 40 years ago by Woodward and by the Fiesers and led to the formulation of the Woodward–Fieser rules (see Section 12-1), which have enjoyed wide use in structure determination.

11-3b α,β-Unsaturated Carbonyls

Many organic molecules of interest contain both carbonyl and alkene chromophores. If the groups are separated by two or more sigma bonds, there is generally (but with important exceptions, see below) little electronic interaction and the effect of the two chromophores on the observed spectrum is essentially additive. Compounds in which the double bond is conjugated with a carbonyl group exhibit spectra in which both the alkenic $\pi \longrightarrow \pi^*$ and the carbonyl $n \longrightarrow \pi^*$ absorption maxima of the isolated chromophores have undergone bathochromic shifts of 15–45 nm, although each band is not necessarily displaced by an equal amount. Photoionization data indicate that the n orbital energy is relatively constant, so that the red shift is most probably caused by a lowering of the energy of the π^* orbital.

Crotonaldehyde ($CH_3CH{=}CH{-}CH{=}O$) in ethanol solution has an intense band at 220 nm ($\varepsilon = 15{,}000$) and a weak band at 322 nm ($\varepsilon = 28$). The low intensity and hypsochromic shift in hydroxylic solvents relative to hydrocarbon solvents suggest that the 322 band is the $n \longrightarrow \pi^*$ transition. The bathochromic shift relative to a saturated carbonyl indicates that the excited π^* orbital now extends over all the atoms of the conjugated carbonyl group. The phenomenon is general and may be seen in α,β-unsaturated ketones as well as aldehydes, in conjugated acids, and in conjugated esters (Table 11-7).

The solvent effect on the $\pi \longrightarrow \pi^*$ transition is opposite to that on the $n \longrightarrow \pi^*$ peak; the $\pi \longrightarrow \pi^*$ absorption shifts to longer wavelength with increasing solvent polarity. The effect of solvent on the $n \longrightarrow \pi^*$ and $\pi \longrightarrow \pi^*$ transitions of mesityl oxide, $(CH_3)_2C{=}CH{-}CO{-}CH_3$, is given in Table 10-4.

As the number of double bonds conjugated with the carbonyl increases, the $\pi \longrightarrow \pi^*$ transition shifts to longer wavelength and its intensity increases, causing the much weaker $n \longrightarrow \pi^*$ absorption to appear as a shoulder or to become completely obscured by the more intense, overlapping $\pi \longrightarrow \pi^*$ band.

Alkyl substitution shifts the $\pi \longrightarrow \pi^*$ and $n \longrightarrow \pi^*$ maxima in opposite directions, the $\pi \longrightarrow \pi^*$ being displaced to longer wavelength. Such effects of substitution on the position of the $\pi \longrightarrow \pi^*$ transition can be predicted through the use of empirical rules first formulated by R.B. Woodward, then modified by the Fiesers. These rules, which have played an important role in assigning the structures of steroids and other natural products, are discussed Section 12-1b.

Table 11-7 illustrates the effects of conjugation for aldehydes, ketones, acids, esters, and amides. The data indicate a considerable similarity in the locations of the $\pi \longrightarrow \pi^*$

TABLE 11-7

UV–Vis Absorption Bands of α,β-Unsaturated Carbonyl Compounds in Ethanol

Compound	$\pi \longrightarrow \pi^*$		$n \longrightarrow \pi^*$	
	λ_{max} (nm)	ε_{max} (liter mol^{-1}cm^{-1})	λ_{max} (nm)	ε_{max} (liter mol^{-1}cm^{-1})
CH_2=CH—CHO	207	11,200	322	28
	203*	12,000*	345*	20*
CH_3CH=CH—CHO	217	17,900	327	
CH_2=C(CH$_3$)CHO	216	11,000		
CH_3CH=C(CH$_3$)CHO	226	16,100		
CH_2=CHCOCH$_3$	215	3,600		
	203*	9,600*	331*	25*
CH_3CH=CHCOCH$_3$	221	12,300		
CH_2=C(CH$_3$)COCH$_3$	214	7,550		
CH_3CH=C(CH$_3$)COCH$_3$	223	13,600		
CH_2=CH—CO$_2$H	200	10,000		
CH_3CH=CHCO$_2$H	206	14,000		
CH_2=C(CH$_3$)CO$_2$H	210			
CH_3CH=CHCO$_2$Et	210	12,600		
CH_3CH=C(CH$_3$)CO$_2$H	213	12,500		
CH_3CH=C(CH$_3$)CONH$_2$	214	12,100		

*Cyclohexane.

transitions and a similar sensitivity toward substitution of the carbon–carbon double bond. In general, the $\pi \longrightarrow \pi^*$ transition lies between 200 and 220 nm for α,β-unsaturated carbonyls, and alkyl-substituted systems have their $\pi \longrightarrow \pi^*$ transition shifted toward the higher end of the range. As with dienes, extended conjugation causes bathochromic shifts of the $\pi \longrightarrow \pi^*$ band and a set of Woodward's rules has been developed for α,β-unsaturated carbonyls (see Section 12-1).

11-4 AROMATIC COMPOUNDS

The benzene ring ranks with the carbonyl group and alkenes as one of the most widely studied chromophores. The spectrum of benzene above 180 nm consists of three well-defined absorption bands due to $\pi \longrightarrow \pi^*$ transitions (Figure 11-10). An intense, structureless band occurs at about 185 nm, with a somewhat weaker band (λ_{max} ~200 nm) of poorly resolved vibrational structure overlapping the 185 nm absorption. The longest wavelength transition is a low-intensity system centered near 255 nm that exhibits a characteristic vibrational structure. Data on the benzene absorption bands and some of the various nomenclature systems used to describe them are given in Table 11-8.

The benzene absorptions at 254 and 204, termed ^1L_b and ^1L_a in the Platt notation (Table 11-8), are both forbidden, but the ^1L_a band is able to borrow intensity from the allowed ^1B transition, which overlaps it at shorter wavelength. The different transition

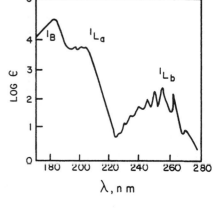

Figure 11-10 The UV spectrum of benzene in hexane.

TABLE 11-8

Notation Systems for Benzene Absorption Bands

$\lambda_{max}(\varepsilon_{max})$ in Hexane			Origin of the Spectral Notation*
184 nm (68,000)	204 nm (8800)	254 nm (250)	
^1B	^1L_a	^1L_b	a
β	p (para)	α	b
$^1E_{2u}$	$^1B_{1u}$	$^1B_{2u}$	c
Second primary	Primary	Secondary	d
	K (conjugation)	B (benzenoid)	e

*KEY: a = Platt free-electron method notation: J.R. Platt, *J. Chem. Phys.,* **17**, 484 (1949); b = Empirical notation of Clar based on behavior of bands with temperature: E.P. Clar, *Aromatische Kohlenwasserstoffe,* Springer Verlag, Berlin, 1952; c = Molecular orbital approach based on the group theoretical notation of the transitions; d = Empirical notation: L. Doub and J.M. Vandenbelt, *J. Am. Chem. Soc,* **69**, 2714 (1947), **71**, 2414 (1949); e = Early empirical notation: A. Burawoy, *J. Chem. Soc.,* 1177 (1939); 20 (1941).

probabilities relate to configuration interaction because of the degeneracy of the highest occupied and lowest vacant orbitals in benzene. The superscript 1 indicates that the transition is to a singlet excited state. Benzene belongs to the D_{6h} point group, and the intensity of the symmetry forbidden 254 nm ^1L_b transition should be zero. However, vibrational distortions from hexagonal symmetry result in a small net transition dipole moment and the observed low intensity.

The ^1L_b absorption, sometimes called the benzenoid band, is usually easily identifiable; it has about the same intensity in benzene and its simple derivatives, $\varepsilon = 250-300$. It also has similar well-defined vibrational structure with up to six vibrational bands, as in benzene itself. The vibrational structure is less evident in polar solvents and more sharply defined in vapor spectra or in nonpolar solvents.

Alkyl substitution shifts the benzene absorption to longer wavelengths and tends to reduce the amount of vibrational structure. Increases in band intensities are observed commonly. In general, substitution can perturb the benzene ring by both inductive and resonance effects. A methyl substituent causes the largest wavelength shift, hyperchromism, and the greatest change in vibrational intensities. The effect decreases as the methyl hydrogens are replaced by alkyl groups. This result is often cited as evidence for the importance of C—H hyperconjugation ($\sigma \longrightarrow \pi$ electron interaction) (see reference 11.1).

The absorption data for the xylenes given in Table 11-9 illustrate that bathochromic shifts caused by alkyl disubstitution are usually in the order para > meta > ortho. Alkylbenzenes, like alkyl-substituted alkenes, normally do not undergo any significant spectral changes when the solvent is varied. On the other hand, polar substituents such as —NH_2, —OH, —OCH_3, —CHO, —COOH, and —NO_2 cause marked spectral changes. With these groups, the intensity of the ^1L_b band is enhanced. Much of the fine structure is lost in polar solvents, although it may be observed to some extent in nonpolar solvents. In addition, the ^1L_a band is shifted bathochromically; for example, in aniline, thiophenol, and benzoic acid it occurs in the 230 nm region (Table 11-9).

Carbonyl substituents possess nonbonding and π-electrons. Their π system can conjugate with the π system of an aromatic ring. Since the energy of the π^* state is lowered by delocalization over the entire conjugated system, both the $\pi \longrightarrow \pi^*$ and the $n \longrightarrow \pi^*$ absorptions occur at longer wavelength than in the corresponding unconjugated chromophoric substituent. For example, acetophenone, C_6H_5—CO—CH_3, exhibits an $n \longrightarrow \pi^*$ absorption at 320 nm and the ^1L_b aromatic transition at 276 nm. Both bands are bathochromically shifted and considerably increased in intensity partly as the result of conjugation of the benzene π system with the π system of the carbonyl group. A similar effect on the $\pi \longrightarrow \pi^*$ benzene transition is seen in styrene, where a C=C bond is conjugated with the benzene ring.

Substitution of benzene by auxochromes (nonchromophoric groups, usually containing unshared electrons such as —Cl, —NR_2, —OH), chromophores, or fused rings has varying effects on the absorption spectrum. Because of their importance, we shall consider these effects in some detail.

wavelength and intensified by co
group of the carboxylic acid. The
more than twice that for the cis and
length in the trans compound. A si
A generally applicable rule for m:
$\pi \longrightarrow \pi^*$ transition occurs at longer

The spectral data for bipheny
of adjacent benzene chromophore:
and intense bands at 202 nm (ε 44.
the 248 nm band indicates that it co
by conjugation between the two i
broad envelope of the intense ^1L_a l

The deviation from coplanari
stituents increase the deviation of
conjugation. This effect can be see
closely resembles the sum of two ii

The presence of a saturated
molecule may result in complete lc
for diphenylmethane (C_6H_5—CH_2
ε of 500, almost exactly the sum of

Like the carbonyl group, a r
inherently achiral but chirally pert:
chromophores are surprisingly rec:
reported, and disagreement existec
chiral molecule could exhibit opti
caused by conflicting reports on c
investigators reported that this alc
Cotton effect in the 260 nm region.
reported no Cotton effect, only
phenylethanol (Figure 11-12) clea
containing vibrational fine structur
Cotton band near 210 nm. Both C
only the benzene ring absorbs in th
the symmetry-forbidden ^1L_b transi
why previous investigators someti
superimposed on a steeply falling b
caused by the more intense negativ
the sign of this compound in the
zenoid Cotton effects were a quest
sic difference in the nature of the b

The benzenoid transitions are
and co-workers (see reference 11.1

CH₃ structure (11-7): H—C—OH with CH₃ above and C₆H₅ below

11-7

TABLE 11-9
Absorption Data for Benzene and Derivatives*†

Compound	Solvent	λ_{max} (nm)	ε_{max}	λ_{max} (nm)	ε_{max}	λ_{max} (nm)	ε_{max}	λ_{max} (nm)	ε_{max}
Benzene	Hexane	184	68,000	204	8,800	254	250		
	Water	180	55,000	203.5	7,000	254	205		
Toluene	Hexane	189	55,000	208	7,900	262	260		
	Water			206	7,000	261	225		
Ethylbenzene	Ethanol‡			208	7,800	260	220		
tert-Butylben-zene	Ethanol			207.5	7,800	257	170		
o-Xylene	25% methanol			210	8,300	262	300		
m-Xylene	25% methanol			212	7,300	264	300		
p-Xylene	Ethanol			216	7,600	274	620		
1,3,5-Trimethyl-benzene	Ethanol			215	7,500	265	220		
Fluorobenzene	Ethanol			204	6,200	254	900		
Chlorobenzene	Ethanol			210	7,500	257	170		
Bromobenzene	Ethanol			210	7,500	257	170		
Iodobenzene	Ethanol			226	13,000	256	800		
	Hexane			207	7,000	258	610	285(sh)	180
Phenol	Water			211	6,200	270	1,450		
Phenoxide ion	aq NaOH			236	9,400	287	2,600		
Aniline	Water			230	8,600	280	1,400		
	Methanol			230	7,000	280	1,300		
Anilinium ion	aq acid			203	7,500	254	160		
N,N-Dimethyl-aniline	Ethanol			251	14,000	299	2,100		
Thiophenol	Hexane			236	10,000	269	700		
Anisole	Water			217	6,400	269	1,500		
Benzonitrile	Water			224	13,000	271	1,000		
Benzoic acid	Water			230	10,000	270	800		
	Ethanol			226	9,800	272	850		
Nitrobenzene	Hexane			252	10,000	280(sh)	1,000	330(sh)	140
Benzaldehyde	Hexane			242	14,000	280	1,400	328	55
	Ethanol			240	16,000	280	1,700	328	20
Acetophenone	Hexane			238	13,000	276	800	320	40
	Ethanol			243	13,000	279	1,200	315	55
Styrene	Hexane			248	15,000	282	740		
	Ethanol			248	14,000	282	760		
Cinnamic acid									
cis-	Hexane	200	31,000	215	17,000	280	25,000		
trans-	Hexane	204	36,000	215	35,000	283	56,000		
	Ethanol			215	19,000	268	20,000		
Stilbene									
cis-	Ethanol			225	24,000	274	10,000		
trans-	Heptane	202	24,000	228	16,000	294	28,000		
Phenylacetylene	Hexane	202	44,000	248	17,000	hidden			
2,2'-Dimethyl-biphenyl	Hexane	198	43,000	228(sh)	6,000	264	800		
Diphenyl-methane	Ethanol			220	10,000	262	500		

* From J.B. Lambert, H.F. Shurvell, L. Verbit, R.G. Cooks, and G.H. Stout, *Organic Structural Analysis*, Macmillan Publishing, New York, 1976.
† If vibrational structure is present, λ_{max} refers to the subband of highest intensity.
‡ "Ethanol" should be taken to mean 95% ethanol.

11-6

The spectral changes found when phenol is converted to the phenoxide (pheny-late) anion and when the anilinium cation is converted to aniline are of considerable interest and practical importance. In the case of aniline, the 280 nm band is most probably due to the benzenoid ^1L_b transition, red-shifted and enhanced by electron donation from the amino group to the ring. Resonance structures involving intramolecular charge transfer (**11-6**) make a substantial contribution to the ground electronic state,

Figure 11-12 The CD in he
(−)-α-phenylethanol (11-7). (
permission from L. Verbit, .
Soc., **87**, 1617 [1965]. Co
American Chemical Society.)

but their predominant contrib
benzenes for which this type o
shifted spectra relative to benz

Conversion of aniline to
the nonbonding electron pair,
ring (eq. 11-7). The absorption

zene. The blue shift observed i
of the spectral changes due to
in structure elucidation.

Conversion of phenol to
bonding electrons available to
intensities of the absorption b
mation obtainable in the anili
a phenolic group may be deter
a neutral and in an alkaline (p

The aniline–anilinium o
demonstrate the presence of
isosbestic point in the UV s
which obeys Beer's law, are ir
constant total concentration
which the absorbances of the
11-11 for 4-methoxy-2-nitropl

In the UV spectra of tl
the band at about 280 nm rep

Figure 11-11 The spectra of
nitrophenol as a function of p
has the shorter waveleng
maximum; the phenoxide t
curves meet at a common p
bestic point where the absc
phenol and phenolate are e
with permission from H.H.
Orchin, *Theory and Applicat*
olet Spectroscopy, John V
Inc., New York, 1962, p. 562.

TABLE 11-10
Important Chromophores of Proteins*

Residues	Chromophore	Location (nm)	$\log \varepsilon_{max}$	Assignment
Peptide bond	CONH	162	3.8	$\pi^+ \longrightarrow \pi^-$
		188	3.9	$\pi^0 \longrightarrow \pi^-$
		225	2.6	$n \longrightarrow \pi^-$
Aspartic, glutamic	COOH	175	3.4	$n \longrightarrow \pi^*$
		205	1.6	$n \longrightarrow \pi^*(?)$
Aspartate, glutamate	COO$^-$	200	2	$n \longrightarrow \pi^*$
Lysine, arginine	N—H	173	3.4	$\sigma \longrightarrow \sigma^*$
		213	2.8	$n \longrightarrow \sigma^*$
Phenylalanine	Phenyl	188	4.8	
		206	3.9	$\pi \longrightarrow \pi^*$
		261	2.35	
Tyrosine	Phenolic	193	4.7	
		222	3.9	$\pi \longrightarrow \pi^*$
		270	3.16	
Tyrosine (ionized)	Phenolate ion	200?	5	
		235	3.97	$\pi \longrightarrow \pi^*$
		287	3.41	
Tryptophan	Indole	195	4.3	
		220	4.53	$\pi \longrightarrow \pi^*$
		280	3.7	
		286	3.3	
Histidine	Imidazole	211	3.78	$\pi \longrightarrow \pi^*(?)$
CysSH	S—H	195	3.3	$n \longrightarrow \sigma^*$
CysS—	S—	235	3.5	$n \longrightarrow \sigma^*$
Cystine	—S—S—	210	3	$n \longrightarrow \sigma^*$
		250	2.5	

*Modified from J. Donovan, *Physical Principles and Techniques of Protein Chemistry*, Part A, S. Leach, ed., Academic Press, New York, 1960.

TABLE 11-11
Circular Dichroism Properties of Aromatic Amino Acids and of Cystine

Amino Acid (as the Hydrochloride)	Chromophore	λ (nm)	$[\theta]$ (deg $\cdot$ cm^2 dmole^{-1})	References*
α-Phenylglycine	Benzenoid	267, 260, 254	−1,000	a
		218	34,000	
Phenylalanine	Benzenoid	266, 263, 257	50-75	a, b
		217	14,000	
Tyrosine	Phenolic	274	1,200	b, c
		225	8,000	
Tyrosine	Phenolate ion	293	1,000	b, d
		230	−2,000	
		210	7,000	
Tryptophan	Indole	286, 276, 269	1,500–2,500	c
		225	20,000	
		209	−15,000	
Histidine	Imidazole	216	8,000	b, c
Cystine	Disulfide	250	−2,000	b, e, f
		220	20,000	
		196	−44,000	

*KEY: a = L. Verbit and P.J. Heffron, *Tetrahedron*, **24**, 1231 (1968); b = M. Legrand and R. Viennet, *Bull. Soc. Chim. France*, 479 (1965); c = L. Verbit and P.J. Heffron, *Tetrahedron*, **23**, 3865 (1967); d = T.M. Hooker, Jr., and J.A. Schellman, *Biopolymers*, **9**, 1319 (1970); e = P.C. Kahn and S. Beychok, *J. Am. Chem. Soc.*, **90**, 4168 (1968); f = J.P. Casey and R.B. Martin, *J. Am. Chem. Soc.*, **94**, 6141 (1972).

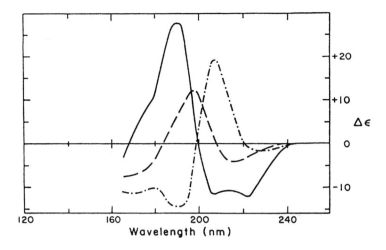

Figure 11-16 The CD in aqueous solution of poly-L-glutamic acid at pH 4.5 representing the α-helix (———), poly(Lys-Leu) in 0.1 M NaF at pH 7 representing the antiparallel β sheet (– – –), and poly(Ala₂-Gly₂) representing the β turn (– · –). (Reproduced with permission from the *Annual Review of Physical Chemistry* in W.C. Johnson, ed., *Ann. Rev. Phys. Chem.*, **29**, 93 [1978].)

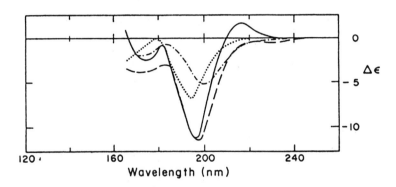

Figure 11-17 The CD of a number of peptides in aqueous solution thought to represent the unordered amide chromophore. Poly-L-glutamic acid at pH 8 (———), poly(Pro-Lys-Leu-Lys-Leu) in a salt-free solution (– – –), collagen at 45° C in a 0.01 M sodium phosphate buffer pH 3.5 (– · –), and *N*-acetyl-L-alanine-*N'*-methylamide (·······). (Reproduced with permission from the *Annual Review of Physical Chemistry*. W.C. Johnson, ed., *Ann. Rev. Phys. Chem.*, **29**, 93 [1978].)

periodic structural modes obtained from X-ray data of the same proteins. The results indicate good agreement with experiment for the α-helical spectra, fair agreement for the β sheet conformation, and only qualitative similarities for the random coil form.

Poly-L-glutamic acid is a polypeptide that is almost entirely in the α-helical form at pH 4. Three transitions are observed in the CD spectrum above 185 nm: a positive band near 192 nm and two strongly overlapped negative Cotton effects at 208 and 222 nm. The 192 and 208 nm bands arise from the split $\pi \longrightarrow \pi^*$ transition of the amide chromophore and are polarized, respectively, perpendicular and parallel to the helix axis. The negative 222 nm Cotton effect is due to the $n \longrightarrow \pi^*$ transition.

Above pH 6, poly-L-glutamic acid is in the random coil conformation with a weak $n \longrightarrow \pi^*$ Cotton effect near 216 nm (Figure 11-17), a still weaker band at about 235 nm, and a stronger, negative band at 198 nm, probably from the $\pi \longrightarrow \pi^*$ transition, although there is a significant wavelength difference with the corresponding absorption maximum (191 nm).

TABLE 11-12

Absorption Data for Nucleosides and Their Bases*

		λ_{max}	ε_{max}	pH
Adenine structure	Adenosine (R=X)†	260	14,900	6.4
	Adenine (R=H)	261	13,300	7.0
Guanine structure	Guanosine (R=X)†	253	13,700	6.0
	Guanine (R=H)	245	10,700	7.0
Uracil structure	Uridine (R=X)†	262	10,100	7.2
	Uracil (R=H)	276	8,150	7.2
		259	8,100	
Thymine structure	Thymidine (R=X)†	267	9,650	7.2
	Thymine (R=H)	265	7,890	7.2
Cytosine structure	Cytidine (R=X)†	271	9,100	7.2
	Cytosine (R=H)	267	6,130	7.2

*From G.H. Beaven, E.R. Holiday and E.A. Johnson, *The Nucleic Acids,* vol. 1, E. Chargaff and J.N. Davidson, eds., Academic Press, New York, 1965, pp 493-545.

†R=X=Ribose

11-5b   Nucleic Acids and Polynucleotides

Nucleic acids, the building blocks of DNA, contain purine (adenine and guanine) and pyrimidine (uracil, thymine, and cytosine) bases, which are rich in π electrons and exhibit π ⟶ π* UV–vis transitions in the 240–275 nm region (Table 11-12). Nucleosides (which are purine and pyrimidine bases containing a sugar unit) and nucleotides (which are nucleoside 5'-phosphates) have almost identical UV–vis spectra that are very similar to those of their component bases. They are also the component chromophores of polynucleotides DNA and RNA. Although the CD of DNA (and RNA) differ somewhat according to source, the spectra are thought to arise from exciton interactions involving the chromophore bases (see reference 11.14) and have intense CD absorption near 180 nm and much weaker multiple bands between 220 and 300 nm (Figure 11-18).

11-5c   Porphyrins and Metalloporphyrins

Porphyrins are common natural products found as metal complexes in the heme **(11-9)** of blood and in the chlorophyll-*a* **(11-10)** of green leaves. Chlorophyll-*a* is a dihydroporphyrin (chlorin), but the macrocyclic conjugation of porphyrins is still main-

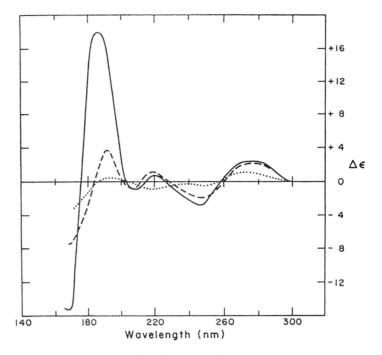

Figure 11-18 The CD of native (———) and heat denatured (– – –) *Escherichia coli* DNA in 0.001 M NaF, pH 6.8; and the average CD (·······) for the four deoxynucleotide monomers in 0.01 M sodium phosphate buffer, pH 7.0. (Reproduced with permission from the *Annual Review of Physical Chemistry.* W.C. Johnson, ed., *Ann. Rev. Phys. Chem.,* **29,** 93 [1978].)

tained. The UV-vis spectra of porphyrins and metalloporphyrin are characterized by an intense absorption near 400 nm (the Soret band), as illustrated for protoporphyrin-IX

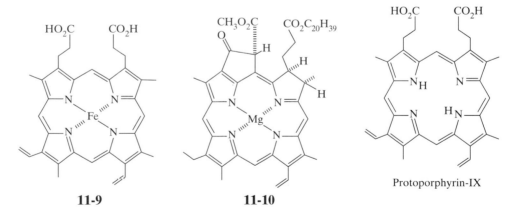

11-9   **11-10**   Protoporphyrin-IX

in Figure 11-19. In addition to the Soret band, there are also characteristic longer wavelength bands that are weak. These bands are very sensitive to changes in the porphyrin structure, especially to the presence of a metal atom and its ligand, as may be seen in comparing the protoporphyrin-IX UV–vis spectrum to that of the chlorin **11-10.** Note also the sensitivity of the various human hemoglobins to different ligands on the iron of the porphyrin (Figure 11-20), and the change spectrum of chlorophyll-*a* relative to that of hemin (Figure 11-19). In chlorophyll-*a* the fused cyclopentanone ring and phytyl ester make little or no contribution to the UV–vis spectrum, but the reduction of one of the pyrrole rings leads to a decrease in the Soret band intensity and a significant increase in the 650 nm absorption.

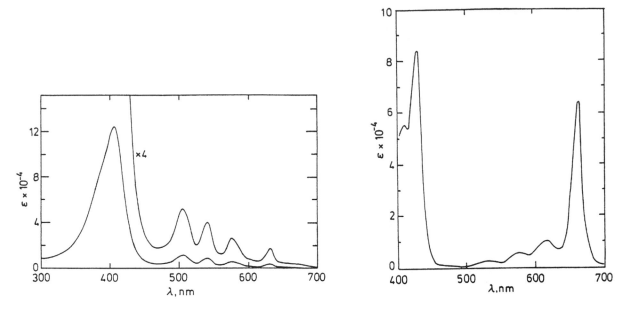

Figure 11-19 UV–vis spectra of (left) protoporphyrin-IX in dimethyl-sulfoxide, and (right) chlorophyll-*a* in chloroform. (Right spectrum from Portia Mahal Sabido; left spectrum from data of A.S. Holt, *Chemistry and Physiology of Plant Pigments,* vol. 14, T.W. Goodwin, ed., Academic Press, New York, 1976.)

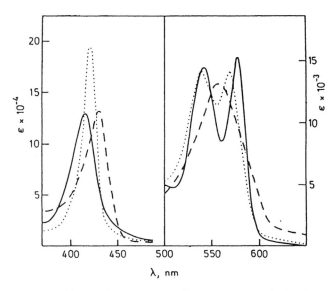

Figure 11-20 Absorption spectra of hemoglobin derivatives. Oxyhemoglobin (———), deoxyhemoglobin (– – –), carbonmonoxyhemoglobin (······). (From the data of M.R. Waterman, *Methods in Enzymology,* vol. LII, S. Fleischer and L. Packer, eds., Academic Press, New York, 1978.)

11-6 WORKED PROBLEMS

Problem 11-1

Explain the following data for azobenzene 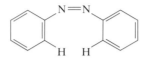 in isooctane.

trans (anti) λ_{max} = 318 nm ε_{max} = 22,600
cis (syn) λ_{max} = 282 nm ε_{max} = 5,200

Answer. As in the stilbenes (Chapter 12-3), steric hindrance in the syn or cis isomer inhibits mutual coplanarity of the two phenyl rings with the connecting —N=N—

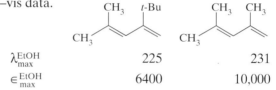

linkage thereby, limiting conjugation and raising the excitation energy (decreasing λ_{max}). The reduced ε_{max} in the cis isomer follows from a shorter distance between the ends of the conjugated system. As a rough rule of thumb, the greater the distance between the ends of a conjugated chromophore, the greater is ε_{max}.

Problem 11-2

Explain the following UV–vis data.

	225	231
λ_{max}^{EtOH}	225	231
ε_{max}^{EtOH}	6400	10,000

Answer. A nonbonded steric interaction between CH_3 and *tert*-butyl is more effective than a nonbonded CH_3 and CH_3 interaction in destabilizing the s-trans conformation shown. In the *tert*-butyl analogue, conjugation is inhibited by rotation about the sp^2-sp^2 C—C bond, leading to a twisted diene.

Problem 11-3

Explain the following UV–vis data for

n	λ_{max}(nm)	ε_{max}
2	258	10,000
3	248	7,500
4	228	5,600
6	< 215	end absorp.

Answer. With increasing ring size, the diene increasingly deviates from coplanarity. As the diene twists out of coplanarity, the $\pi \longrightarrow \pi^*$ transition energy increases (decreasing λ_{max}) and the probability of the excitation (ε) decreases.

Problem 11-4

How would you distinguish the following pairs of compounds using UV–vis spectroscopy?
(a) $CH_3CH=CH—CH=CH—CH_3$ (b)   (c)
and
$CH_3CH=CH—CH_2—CH=CH_2$

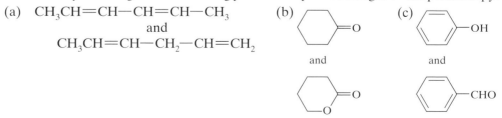

Answer. In (a) the upper diene is conjugated and should therefore absorb longer wavelength light than the lower diene, which is not conjugated (see Table 11-5).

11-7 Account for the following observations.

(a)

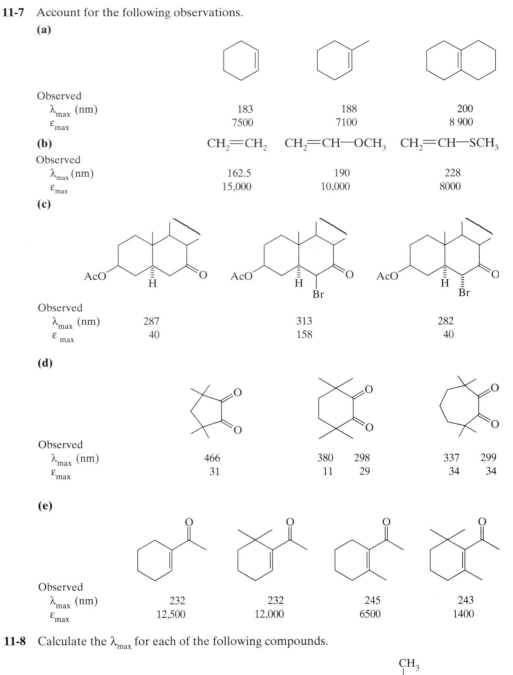

Observed		
λ_{max} (nm)	183	188	200
ε_{max}	7500	7100	8 900

(b)

$CH_2\!\!=\!\!CH_2$ $CH_2\!\!=\!\!CH\!\!-\!\!OCH_3$ $CH_2\!\!=\!\!CH\!\!-\!\!SCH_3$

Observed		
λ_{max} (nm)	162.5	190	228
ε_{max}	15,000	10,000	8000

(c)

Observed			
λ_{max} (nm)	287	313	282
ε_{max}	40	158	40

(d)

Observed				
λ_{max} (nm)	466	380	298	337	299
ε_{max}	31	11	29	34	34

(e)

Observed				
λ_{max} (nm)	232	232	245	243
ε_{max}	12,500	12,000	6500	1400

11-8 Calculate the λ_{max} for each of the following compounds.

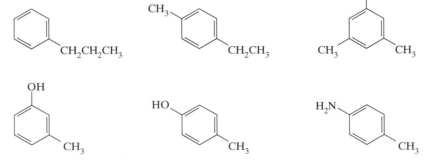

BIBLIOGRAPHY

11.1. H.H. Jaffé and M. Orchin, *Theory and Applications of Ultraviolet Spectroscopy,* John Wiley & Sons, Inc., New York, 1962.

11.2. J.B. Lambert, H.F. Shurvell, L. Verbit, R.G. Cooks, and G.H. Stout, *Organic Structural Analysis,* Macmillan Publishing, New York, 1976.

11.3. C. Djerassi, *Optical Rotatory Dispersion,* McGraw-Hill Book Co., New York, 1960.

11.4. D.A. Lightner, T.D. Bouman, W.M.D. Wijekoon, and A.E. Hansen, *J. Am. Chem. Soc.,* **108,** 4484 (1986).

11.5. W. Moffitt, R.B. Woodward, A. Moscowitz, W. Klyne, and C. Djerassi, *J. Am. Chem. Soc.,* **83,** 4013 (1961).

11.6. M.B. Robin, F.A. Bovey, and H. Basch, in *The Chemistry of Amides,* J. Zabichy, ed., Wiley–Interscience, New York, 1970.

11.7. H. Benderly and K. Rosenheck, *Chem. Commun.,* 179 (1972).

11.8. A.J. Merer and R.S. Mulliken, *Chem. Rev.,* **63,** 639 (1969).

11.9. A.E. Hansen and T.D. Bouman, *Adv. Chem. Phys.,* **44,** 545 (1980).

11.10. T.D. Bouman, A.E. Hansen, B. Voigt, and S. Ruttrup, *Int. J. Quantum Chem.,* **23,** 595 (1983).

11.11. A. Yogev, J. Sagiv, and Y. Mazur, *Chem. Commun.,* 411 (1972).

11.12. P. Salvadori, L. Lardicci, R. Menicagli, and C. Bertucci, *J. Am. Chem. Soc.,* **94,** 8598 (1972).

11.13. D.A. Lightner, D.T. Hefelfinger, T.W. Powers, G.W. Frank, and K.N. Trueblood, *J. Am. Chem. Soc.,* **94,** 3492 (1972).

11.14. C.W. Deutsche, D.A. Lightner, R.W. Woody, and A. Moscowitz, *Ann. Rev. Phys. Chem.,* **20,** 407 (1969).

11.15. W.C. Johnson, Jr., *Ann. Rev. Phys. Chem.,* **29,** 93 (1978).

chapter 12

Structural Analysis

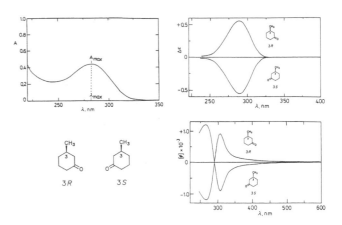

As indicated earlier in part III, UV–vis spectroscopy is used in structure determination to detect and identify conjugated chromophores and the substitution patterns on them. The Woodward–Fieser rules played an important early role in structure determination by allowing the detection and identification of substitution patterns of dienes, polyenes, and conjugated carbonyl compounds. More recently, ORD and CD spectroscopy have become very important tools for assigning absolute stereochemistry and analyzing the conformation of optically active compounds. In this connection, the octant rule is doubtless the most prominent and successful of the various sector rules derived for chromophores. Most recently, the exciton chirality rule has been formulated and applied to the determination of absolute configuration for a multitude of different compounds. The octant rule and the exciton chirality rule are discussed after the Woodward–Fieser rules. This chapter then ends with applications of UV–vis and chiroptical spectroscopy to conformational analysis.

12-1 APPLICATIONS OF THE WOODWARD–FIESER RULES

12-1a Conjugated Dienes and Polyenes

Extensive studies of the UV spectra of alkenes, particularly those of terpenes and steroids, led Woodward and the Fiesers to formulate empirical rules for the prediction of the $\pi \longrightarrow \pi^*$ absorption maxima of various dienes and polyenes. These rules have proved quite useful for solving organic structure problems, especially among natural products. The rules assign a base-level absorption maximum for the *parent* chromophore, then add specified increments to the parent value for each substituent attached to the parent π electron system. The values used in the Woodward–Fieser rules for diene absorption are given in Table 12-1.

Application of the rules to the dienes and trienes below clearly reveals the importance of the large contributions due to a homoannular diene and to extended conjugation.

TABLE 12-1

Woodward–Fieser Rules for the Calculation of Absorption
Maximum of Dienes and Polyenes (good to about ± 3 nm)*

	λ (nm)
Parent chromophore	214

Each alkyl substituent (at any position) add	5
Each exocyclic double bond add	5

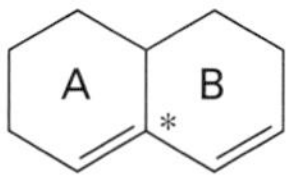

*exocyclic to ring B only

Each additional conjugated double bond (one end only) add	30
Each homoannular (rather than acyclic or heteroannular) add	39

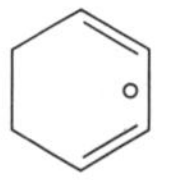

°homoannular (same ring)

NOTE: In cases for which both types of diene systems are present, the one with the
longer wavelength is designated as the parent system.
Do *not* count the double bond as a substituent, since this effect is included.

Each polar group	
—O—acyl	0
—OR	6
—SR	30
—Cl, —Br	5
—NR$_2$	60
Solvent correction	0

* From R.B. Woodward, *J. Am. Chem. Soc.,* **63,** 1123 (1941); **64,** 72, 76 (1942); L.F. Fieser and M. Fieser, *Natural Products Related to Phenanthrene,* Reinhold, New York, 1949.

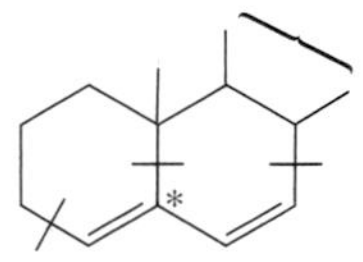

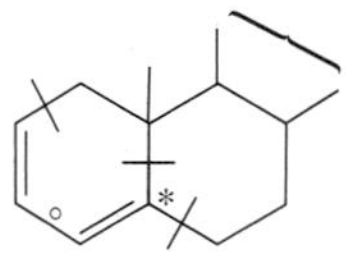

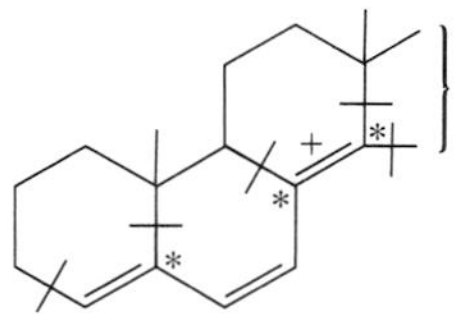

Symbol				
—┼—	alkyl substituent (5)	$\times 3 = 15$	$\times 3 = 15$	$\times 5 = 25$
*	exocyclic (5)	$\times 1 = 5$	$\times 1 = 5$	$\times 3 = 15$
+	extra conjugation (30)	$\times 0$	$\times 0$	$\times 1 = 30$
°	homoannular (39)	$\times 0$	$\times 1 = 39$	$\times 0$
	Parent	214	214	214
	Predicted λ_{max}(nm)	234	273	284
	Observed* λ_{max}(nm)	235	275	283

*Steroids. Data from A.I. Scott, *Interpretation of the Ultraviolet Spectra of Natural Products,* Pergamon Press, New York, 1964.

Examples of their application are illustrated in the following.

1. Abietic acid and levopimaric acid are both found in nature and each is known
to have two carbon–carbon double bonds. Their UV–vis spectra clearly indi-
cate that each is a conjugated diene, one with $\lambda_{max} = 237.5$ (ε 10,000) and the
other with λ_{max} 272.5 (ε 7000). Thus they can be distinguished readily. Application of the Woodward–Fieser rules predicts that abietic acid has a UV–vis band
near 239 nm and levopimaric acid has a UV–vis band near 278 nm.

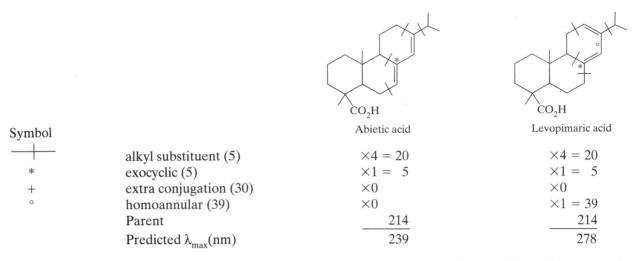

Abietic acid   Levopimaric acid

Symbol		Abietic acid	Levopimaric acid
—	alkyl substituent (5)	×4 = 20	×4 = 20
*	exocyclic (5)	×1 = 5	×1 = 5
+	extra conjugation (30)	×0	×0
∘	homoannular (39)	×0	×1 = 39
	Parent	214	214
	Predicted λ_{max}(nm)	239	278

2. Levopimaric acid is a homoannular diene; abietic acid is a heteroannular diene. It is usually easy to distinguish homoannular from heteroannular by UV–vis spectroscopy because the former is bathochromically shifted by 39 nm. Could levopimaric acid be distinguished from its isomeric homoannular diene, palustric acid, by UV–vis spectroscopy? Could abietic acid be distinguished from its isomeric diene, neoabietic acid?

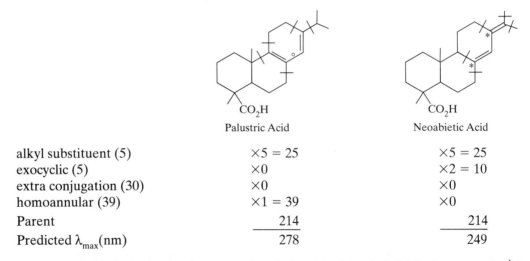

Palustric Acid   Neoabietic Acid

Symbol		Palustric Acid	Neoabietic Acid
—	alkyl substituent (5)	×5 = 25	×5 = 25
*	exocyclic (5)	×0	×2 = 10
+	extra conjugation (30)	×0	×0
∘	homoannular (39)	×1 = 39	×0
	Parent	214	214
	Predicted λ_{max}(nm)	278	249

3. Similarly, isomeric trienes may be distinguished by their UV–vis spectra (λ_{max}^{obs}) and the application of the Woodward–Fieser rules. $\Delta^{2,4,6}$-Cholestatriene, $\Delta^{3,5,7}$-cholestatriene, and $\Delta^{5,7,9(11)}$-androstan-3β,17β-diol have UV–vis λ_{max}^{obs} at 324, 306, and 315 nm respectively. The predicted λ_{max} are as follows:

Symbol				
—	alkyl substituent (5)	×3 = 15	×4 = 20	×5 = 25
*	exocyclic (5)	×1 = 5	×2 = 10	×3 = 15
+	extra conjugation (30)	×1 = 30	×1 = 30	×1 = 30
∘	homoannular (39)	×1 = 39	×1 = 39	×1 = 39
	Parent	214	214	214
	Predicted λ_{max}(nm)	303	313	323
	Observed* λ_{max}(nm)	306	315	324

*Data from A.I. Scott, *Interpretation of the Ultraviolet Spectra of Natural Products,* Pergamon Press, New York, 1964.

12-1b Conjugated Ketones, Aldehydes, Acids, and Esters

As with dienes and polyenes, extensive studies of terpene and steroid enones led Woodward and the Fiesers to formulate empirical rules (Table 12-2) for predicting the wavelength (λ_{max}) of maximum UV–vis absorption for the $\pi \to \pi^*$ transition. Application of the rules to structure proof has been quite successful and useful. To apply the rules, the base absorption of the parent unit is supplemented according to substituent position and type, extended conjugation, etc. The rules predict λ_{max} for ethanol solutions, but adjustments in λ_{max} can be made for solvents, as noted. Application of the Woodward–Fieser rules is illustrated in the unsaturated steroid ketones below.

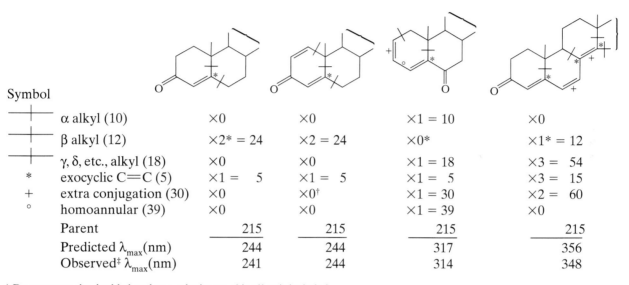

Symbol					
	α alkyl (10)	$\times 0$	$\times 0$	$\times 1 = 10$	$\times 0$
	β alkyl (12)	$\times 2^* = 24$	$\times 2 = 24$	$\times 0^*$	$\times 1^* = 12$
	γ, δ, etc., alkyl (18)	$\times 0$	$\times 0$	$\times 1 = 18$	$\times 3 = 54$
*	exocyclic C=C (5)	$\times 1 = 5$	$\times 1 = 5$	$\times 1 = 5$	$\times 3 = 15$
+	extra conjugation (30)	$\times 0$	$\times 0^\dagger$	$\times 1 = 30$	$\times 2 = 60$
°	homoannular (39)	$\times 0$	$\times 0$	$\times 1 = 39$	$\times 0$
	Parent	215	215	215	215
	Predicted λ_{max}(nm)	244	244	317	356
	Observed‡ λ_{max}(nm)	241	244	314	348

* Do *not* count the double bond as a substituent; this effect is included.
† Cross conjugated, count only the more substituted double bond.
‡ Data from A.I. Scott, *Interpretation of the Ultraviolet Spectra of Natural Products,* Pergamon Press, New York, 1964.

The rule does very well in predicting the values of the simple enone, the homoannular dienone, and the cross-conjugated dienone, but less well for the trienone. Examples of applications are illustrated as follows.

1. $\Delta^{3,5}$-Cholestan-2-one and $\Delta^{3,5}$-cholestan-7-one are regioisomeric ketones. Can they be distinguished by UV–vis spectroscopy? Could they be distinguished from $\Delta^{2,6}$-cholestan-3-one?

Symbol			
	α alkyl (10)	$\times 0$	$\times 0$
	β alkyl (12)	$\times 1 = 12$	$\times 0$
	γ, δ, etc., alkyl (18)	$\times 1 = 18$	$\times 2 = 36$
*	exocyclic C=C (5)	$\times 1 = 5$	$\times 1 = 5$
+	extra conjugation (30)	$\times 1 = 30$	$\times 1 = 30$
	Parent	215	215
	Predicted λ_{max}(nm)	280	286
	Observed* λ_{max}(nm)	277	290

*Data from A.I. Scott, *Interpretation of the Ultraviolet Spectra of Natural Products,* Pergamon Press, New York, 1964.

TABLE 12-2

Rules for the Calculation of the Position of $\pi \longrightarrow \pi^*$
Absorption Maximum of Unsaturated Carbonyl Compounds

$$\overset{\beta}{\underset{|}{C}} = \overset{\alpha}{\underset{|}{C}} - \overset{R}{\underset{|}{C}} = O \quad \text{and} \quad \overset{\delta}{\underset{|}{C}} = \overset{\gamma}{\underset{|}{C}} - \overset{\beta}{\underset{|}{C}} = \overset{\alpha}{\underset{|}{C}} - \overset{R}{\underset{|}{C}} = O$$

		λ (nm)
Parent α,β-unsaturated carbonyl compound		
(acyclic, six-membered, or larger ring ketone) R=alkyl		215
(5-membered ring ketone)		202
(aldehyde) R=H		207
(acid or ester) R=OH or OR		193
Each alkyl substituent:		
α add		10
β add		12
If other double bonds, for each γ, δ, etc., add		18
Each exocyclic carbon–carbon double bond add		5
Each extra conjugation add		30
(do not count double bond as substituent, as this effect is included)		
Each homoannular add		39
Each polar group		
—OH	α	35
	β	30
	δ	50
—O—Ac	α, β, or δ	6
—OR	α	35
	β	30
	γ	17
	δ	31
—SR	β	85
—Cl	α	15
	β	12
—Br	α	25
	β	30
—NR$_2$	β	95
Solvent correction		0
Ethanol, methanol		0
Chloroform		1
Dioxane		5
Diethyl ether		7
Hexane, cyclohexane		11
Water		-8

*From R.B. Woodward, *J. Am. Chem. Soc.*, **63**, 1123 (1941); **64**, 72, 76 (1942); L.F. Fieser and M. Fieser, *Natural Products Related to Phenanthrene*, Reinhold, New York, 1949; A.I. Scott, *Interpretation of the UV Spectra of Natural Products*, Pergamon Press, New York, 1964.

2. Eremophilone and *allo*-eremophilone are isomers. Can you distinguish between them by UV–vis spectroscopy?

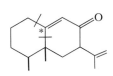

	Eremophilone	*allo*-Eremophilone
Symbol		
α-alkyl (10)	$\times 1 = 10$	$\times 0$
β-alkyl (12)	$\times 1 = 12$	$\times 2 = 24$
exocyclic C=C (5)	$\times 1 = 5$	$\times 1 = 5$
Parent	215	215
Predicted λ_{max}(nm)	242	244

3. The trienone steroids shown below have UV–vis λ_{max} at 348 and 388 nm. Use the Woodward–Fieser rules to distinguish them.

Symbol			
──┼──	α alkyl (10)	×0	×0
──┼──	β alkyl (12)	×1 = 12	×1 = 12
──┼──	γ, δ, etc., alkyl (18)	×1 = 18	×3 = 54
*	exocyclic C=C (5)	×1 = 5	×1 = 5
+	extra conjugation (30)	×2 = 60	×2 = 60
○	homoannular diene (39)	×1 = 39	×1 = 39
	Parent	215	215
	Predicted λ_{max}(nm)	349	385

12-2 DETERMINATION OF ABSOLUTE CONFIGURATION

The following discussion describes two well-known chirality rules used to determine absolute configuration of organic compounds: the octant rule and the exciton chirality rule.

12-2a The Octant Rule

The octant rule for saturated ketones and aldehydes is one of the earliest sector rules for determining absolute configuration (see references 12.1 and 12.2). As long ago as 1961, a considerable amount of chiroptical data for the carbonyl chromophore had accumulated, and a semiempirical generalization, the *octant rule,* was proposed (see references 12.3 and 12.4). The rule predicts the sign of the $n \longrightarrow \pi^*$ Cotton effect (see Section 10-3d) of a ketone from its structure through contributions of perturbing groups lying in each of eight octant sectors surrounding the carbonyl group. Thus the octant rule allows one to determine the absolute configuration and also a likely conformation of the molecule from the sign of the Cotton effect.

To understand the octant rule, let us consider the symmetry properties of the orbitals involved in the $n \longrightarrow \pi^*$ transition near 290 nm. As depicted in Figure 12-1a, the n orbital on oxygen has a nodal plane that bisects the R—C—R' bond angle and is perpendicular to the plane defined by the carbonyl group (C=O) and its attached atoms from R and R'. For the π^* orbital, the plane containing the carbonyl group is a nodal plane. There is another nodal surface, not necessarily a plane, that lies perpendicular to the C=O bond axis and intersects it approximately midway between the C and O. The three nodal surfaces combine together to divide the entire space into the octants having the signs shown in Figure 12-2 for cyclohexanone.

Figure 12-1 Nodal surfaces for a saturated ketone, R—CO—R'. (a) The nodal plane of the *n* orbital. It bisects the R—C—R' angle and is perpendicular to the plane of the ketone. (b) The nodal surfaces of the *n** orbital. The plane of the carbonyl group is a nodal plane and there is another nodal surface, not necessarily a plane, perpendicular to the C—O axis and intersecting it between the carbon and oxygen atoms.

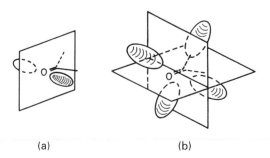

(a)   (b)

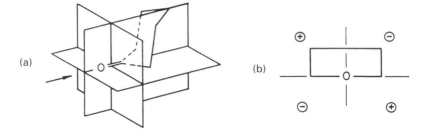

Figure 12-2 (a) Octants for saturated cyclohexanones. (b) Signs of the four rear octants viewed along the carbonyl bond axis from oxygen to carbon. (The front octants have opposite signs.)

The octant rule states that substituents lying in the nodal planes make no contributions to the $n \longrightarrow \pi^*$ Cotton effect. Substituents within an octant contribute the sign of that octant to the overall sign of the Cotton effect. Since most substituents are usually on the same side of the nodal surface as the carbonyl carbon, the octant diagram is simplified by considering only the four rear octants. Relative intensities are determined qualitatively. For example, when both negative rear octants are occupied, the magnitude of the Cotton effect is enhanced.

A few substituents such as $-F$, $-\overset{+}{N}Me_3$, and cyclopropyl exhibit *antioctant* behavior, and several breakdowns of the octant rule for certain aliphatic ketones have been reported (see references 12.1, 12.5 and 12.6). Ketones having an axial α-chloro or α-bromo substituent display Cotton effects whose signs are determined only by the octant location of the halogen, even if most of the molecule occupies an oppositely signed octant; this is the *axial halo ketone rule* of Djerassi and Klyne (see references 12.3 and 12.7).

As an illustration of the use of the octant rule, consider (+)-3-methylcyclohexanone, which exhibits a positive $n \longrightarrow \pi^*$ Cotton effect (Figure 10-1). We would predict that the chair form with the equatorial methyl group (Figure 12-3a) would be energetically favored over the conformer with the axial methyl group (Figure 12-3b). In the octant projection of (+)-3-methylcyclohexanone (Figure 12-3d), carbon atoms 2, 4, and 6 lie in nodal planes and thus make no contribution to the sign of the Cotton effect. The contribution of C-3 is canceled by the equal but opposite contribution of C-5. However, the methyl group at C-3 lies in a positive octant and is responsible for the sign of the positive Cotton effect. The absolute configuration of (+)-3-methylcyclohexanone is deduced to be R, since the enantiomeric molecule would lead to an octant prediction of a negative $n \longrightarrow \pi^*$ Cotton effect.

The octant rule is also valuable in the field of natural products. Figure 11-6 illustrates that the Cotton effect of cholestan-2-one **(11-2)** is about twice as large as that of the 3-isomer **(11-3).** The octant diagrams for the two steroids are shown in Figure 12-4. In the octant diagram for **11-3,** the contributions of rings A and C are seen to cancel each

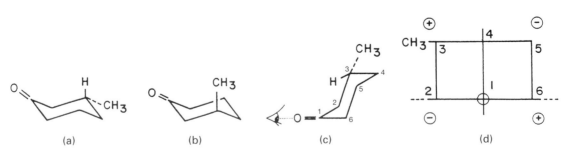

Figure 12-3 Equatorial (a) and axial (b) conformations of (+)-3-methylcyclohexanone and the octant rule, projection of the equatorial conformer (c, d).

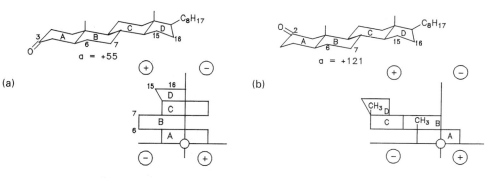

Figure 12-4 Octant diagrams for (a) cholestan-3-one and (b) cholestan-2-one.

other. The two angular methyl groups lie in the vertical nodal plane, so that only the methylene groups C-6 and C-7 (and C-15 and C-16, these two being very remote from the chromophore) contribute to the Cotton effect (Figure 12-4a). According to the octant diagram, these contributions should be positive. In the case of cholestan-2-one (Figure 12-4b), the positive contributions are very strong and give rise to the relatively intense Cotton effect. The contribution of a substituent to the total Cotton effect generally depends on its octant for the sign and upon the distance from the chromophore for its magnitude, that is, the more distant from the chromophore, the smaller the contribution.

An example of the use of the octant rule to deduce stereochemical information is provided for cafestol, a diterpene isolated from the coffee bean. Degradation of cafestol gave ketone **A**, whose ORD curve was found to be almost the mirror image of that of 4α-ethylcholestan-3-one (Figure 12-5). Since the observed Cotton effect curves are

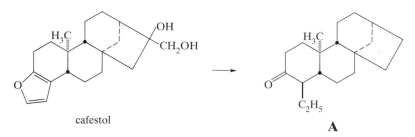

cafestol   **A**

essentially mirror images and the two ketones **A** and 4α-ethylcholestan-3-one are structurally the same in the vicinity of the carbonyl group, the conclusion was reached that **A,** and hence cafestol, from which it was obtained, possess enantiomeric stereochemistry at the A/B ring junction, as depicted in Figure 12-6.

Both the $\pi \longrightarrow \pi^*$ and $n \longrightarrow \pi^*$ transitions of α,β-unsaturated aldehydes and ketones give rise to Cotton effects. The $\pi \longrightarrow \pi^*$ transition near 240 nm usually exhibits

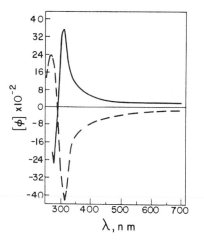

Figure 12-5 The ORD curves of the ketone from cafestol (– – –) and of 4α-ethyl-cholestan-3-one (———). (Adapted with permission from C. Djerassi, M. Cais, and L.A. Mitscher, *J. Am. Chem. Soc.*, **81**, 2386 (1959). Copyright 1959 American Chemical Society.)

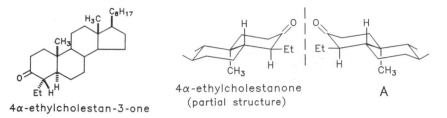

4α-ethylcholestan-3-one

4α-ethylcholestanone
(partial structure)

A

Figure 12-6 Conformation of the A-B ring portion of the cafestol-derived ketone (A) and of the model compound 4α-ethylcholestan-3-one showing the mirror image relationship. See Figure 12-5 for the ORD curves.

a Cotton effect of opposite sign to that of the $n \longrightarrow \pi^*$ band. Recently several authors have reported a new Cotton effect of relatively high intensity near 215 nm, which overlaps the $\pi \longrightarrow \pi^*$ transition and sometimes obscures its sign (see reference 12.8). Thus caution should be used in the analysis of chiroptical spectra for α,β-unsaturated ketones. When the conjugated chromophore is not planar, electronic interactions of two *achiral* chromophores, the alkene and the carbonyl, can result in an inherently chiral α,β-unsaturated ketone chromophore. In these cases, the octant rule, which is based on the concept of the inherently achiral carbonyl group, generally cannot be applied.

12-2b   Exciton Coupling and the Exciton Chirality Rule

A more recent chirality rule, the exciton chirality rule, was advanced by Harada and Nakanishi (reference 12.9) and has proved to be a useful and powerful method for predicting absolute configuration. This rule depends on the phenomenon called exciton coupling. Absorption of UV–vis light causes changes in the distribution of electron density in a molecule (or in a chromophore in the molecule). The movement of electron density due to promoting the molecule from its electronic ground state to an excited state creates a momentary dipole, called an electric transition dipole (μ of Section 11-1d). Consequently, with each electronic transition there is a polarization (electric transition dipole) that has both a direction and an intensity that vary according to the chromophore and the particular excitation. Suppose two chromophores (represented by X and Y in Figure 12-7) are brought into close proximity, but orbital overlap and electron exchange are negligible. The chromophores may interact through dipole–dipole coupling of their locally excited states to produce a delocalized excitation (called an exciton) and a splitting (called an exciton splitting) of the locally excited states (Figure 12-7).

The excited state dipole–dipole interactions, which lie at the heart of exciton coupling, are most effective when the electric dipole transitions are strongly allowed (as in $\pi \longrightarrow \pi^*$ UV–vis transitions, such as the 310 nm transition of the *p*-dimethylaminobenzoate chromophore with $\varepsilon \approx 30{,}000$ liter mol^{-1} cm^{-1}). Exciton coupling leads to shifted and broadened, if not split, UV–vis spectra of the composite molecule. For a chiral orientation of two chromophores, exciton coupling can be seen very clearly in two oppositely-signed CD Cotton effects flanking the relevant UV–vis absorption band(s) (Figure 12-8). The signed order of the CD transitions is correlated with the relative orientation of the relevant electric dipole transition moment from each chromophore and thus the absolute configuration of the composite molecule (exciton chirality rule). Applications of the exciton chirality rule have become numerous during recent years and claim an extraordinarily high degree of success. In most applications of exciton chirality to the determination of absolute configuration, the molecule under study is derivatized with a suitable chromophore, such as *p*-dimethylaminobenzoate in the case of diols, and its circular dichroism spectra are measured and analyzed.

It is the orientation-dependence of the exciton coupling that forms the basis for the exciton chirality rule. The UV–vis and CD spectra of the composite system originate not only from the electronic excitation spectral properties of the component chro-

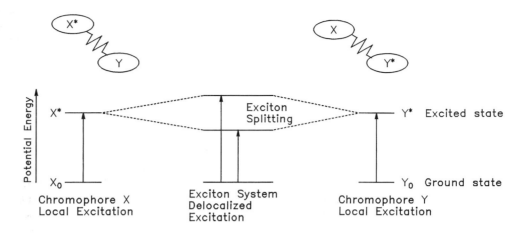

Figure 12-7 Diagrammatic representation for a system with two chromophores (X and Y) held together by covalent bonding or weak intermolecular forces. Local excitations are shown (left and right) for the chromophores in their locally excited (X* or Y*) monomer states. In the composite molecule or system (center), excitation is delocalized between the two chromophores and the excited state (exciton) is split by resonance interaction of the local excitations. Exciton coupling may take place between identical chromophores (X = Y) or nonidentical chromophores (X ≠ Y), but is less effective when the excitation energies are very different, such as when the relevant UV–visible bands do not overlap.

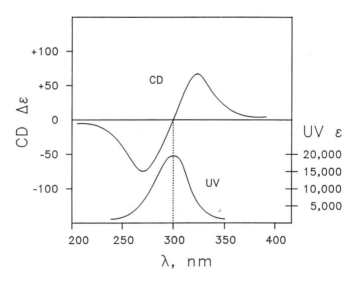

Figure 12-8 Typical exciton coupling CD (upper) and UV (lower) spectra for chromophores with electronic transitions near 300 nm. The bisignate shape of the observed CD curve is due to overlapping, oppositely-signed (+) and (−) CD curves associated with electronic excitation into the two exciton states (Figure 12-7).

mophores, but they also depend on interchromophoric distance, mutual orientation, and geometry. Evidence of exciton interaction in bichromophoric or polychromophoric systems often consists of spectral shifts (which can be to the blue or red) and splittings. Such spectral shifts and splitting magnitudes can be correlated with the intensity and the relative orientation of the electric transition dipole moments associated with the particular UV–vis absorption band of each chromophore (Figure 12-9). Since excitons arise from electric transition dipole coupling, strongly allowed electronic transitions with large dipoles are intrinsically more effective than are weakly allowed transitions.

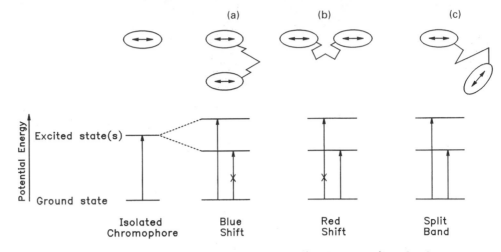

Figure 12-9 Diagram for the origin of UV–vis spectral shifts and splittings due to exciton interaction of two chromophores. In two limiting orientations (a) only excitation into the higher lying exciton state is allowed for the parallel orientation (resulting in a blue shift), and (b) only excitation into the lower lying exciton state is allowed for the in-line orientation (resulting in a red-shift)—relative to the isolated chromophore. For orientations lying between these two limiting cases, (c), a split or broadened band is typically observed. Chromophores are represented by ellipses and electric transition dipoles by (⟵⟶).

This result arises since dipole–dipole interaction falls off with the inverse cube of the separation distance. Consequently, the most useful chromophores for examining exciton coupling phenomena over short and especially over long distances are those with accessible, electric dipole allowed UV–vis transitions, typically $\pi \longrightarrow \pi^*$, such as aromatic chromophores and polyenes. From a practical standpoint, the stronger the absorption of the parent chromophore, the more sensitive it becomes for detecting exciton coupling and exciton chirality at extremely low concentrations (see reference 12.2).

Anthracene is a good example of a chromophore useful for exciton coupling. It exhibits several $\pi \longrightarrow \pi^*$ transitions, including ^1L_a, a short axis–polarized UV absorption near 360 nm ($\varepsilon \sim 7{,}500$) and ^1B_b, a very intense long axis–polarized intense UV absorption near 250 nm ($\varepsilon \sim 200{,}000$) (Figure 12-10). In a composite system, where two anthracenes are fused to bicyclo[2.2.2]octane (Figure 12-10), the intense long axis–polarized transition dipoles are oriented neither parallel nor in-line, but intersect at an obtuse angle and lie one in each of two intersecting planes (dihedral angle of ~120°). The UV–vis spectrum of the composite molecule (Figure 12-10) clearly resembles that of the component anthracene chromophores, yet it differs in two significant ways. The spectrum is not simply the sum of two *independent* anthracene transitions, as it would be if the chromophores did not interact; rather, it is due to two overlapping *exciton* transitions (Figure 12-9c), which can be seen clearly by a broadening and bathochromic shift of the ^1B_b transition(s) from 252 nm band to 267 nm. The broadening is due to unresolved exciton splitting, and the bathochromic shift is due to alignment of the ^1B_b transition dipoles at a 151° intersection angle.

Exciton coupling can be detected by shifted and broadened, if not split, UV–vis spectra. However, when the chromophores are held in a chiral orientation, exciton coupling can be detected even more clearly in the CD spectrum (Figure 12-11) as two oppositely-signed Cotton effects typically corresponding to the relevant UV–vis absorption band(s). In Figure 12-11, the CD spectrum shows an exciton splitting very clearly as two oppositely-signed, very intense Cotton effects near 267 nm. (One key to recognizing an exciton is a very intense bisignate CD Cotton effect.) The signed order of the Cotton

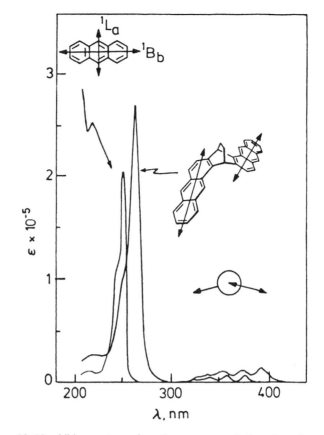

Figure 12-10 UV spectra of anthracene and the ring fused bis-anthracene. On the anthracene structure are shown the in-plane orientations of the electric dipole transition moment vectors associated with the intense short wavelength UV bands (^1B_b) near 250 nm and the weak long wavelength bands ($1L_a$) near 360 nm. The λ_{max} for the exciton bands of the bis-anthracene are red shifted from the corresponding bands in the parent, as the vectors intersect at >90°. (Reproduced with permission from D.A. Lightner in *Analytical Applications of Circular Dichroism,* N. Purdie and H.G. Brittain, eds., Elsevier Science Inc., Amsterdam, 1994.)

effects can be correlated with the relative orientation of the ^1B_b transition moments from the two anthracenes. In the example shown, the extraordinarily intense exciton couplet has a positive band at 268 nm and a negative band at 250 nm, corresponding to a (+) helical orientation of the transition dipoles and hence a positive (+) exciton chirality. In fact, a (+) chirality is predicted for this example by the exciton chirality rule: When the relevant transition moments are oriented in a (+) chirality, the long wavelength component of the associated exciton couplet can be expected to exhibit a (+) Cotton effect (Figure 12-12). When they are oriented in a (−) chirality, the long wavelength Cotton effect is negative. Thus, from the CD, one can determine the helical orientation of the transition moments and therefore the absolute configuration of the molecule.

In most applications of the exciton chirality rule, a chiral molecule is derivatized to introduce suitable chromophores. In many of these studies, hydroxyl or amino is the typical resident functional group, which is derivatized with appropriate acids containing chromophores suitable for exciton coupling. The ideal chromophore has a very intense UV–visible transition, located in a convenient spectral window, and with the orientation of its electric transition moment being well-defined relative to the alcohol R—OH bond. One of the most useful is *p*-dimethylaminobenzoate, which has an intense (ε ~30,000) transition in an easily accessible, generally noninterfering region (near 310 nm). The associated electric dipole transition moment is oriented along the long axis of the molecule, from nitrogen to carboxyl. Although the chromophore might

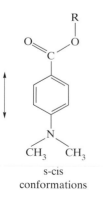

s-cis
conformations

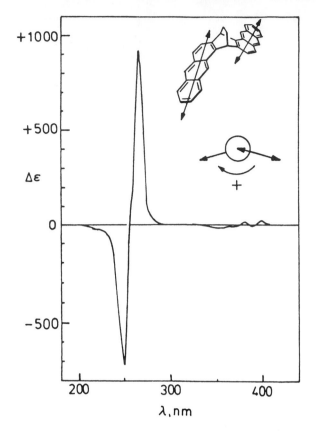

Figure 12-11 CD spectrum of the chiral ring fused bis-anthracene. The ~250 nm electric transition dipoles (◄──►) of each chromophore intersect at a (+) torsion angle of ~151°. The signed order of the Cotton effects of the 250 nm exciton couplet confirms a (+) helicity and thus the absolute configuration shown. (Reproduced with permission from D.A. Lightner in *Analytical Applications of Circular Dichroism,* N. Purdie and H.G. Brittain, eds., Elsevier Science Inc., Amsterdam, 1994.)

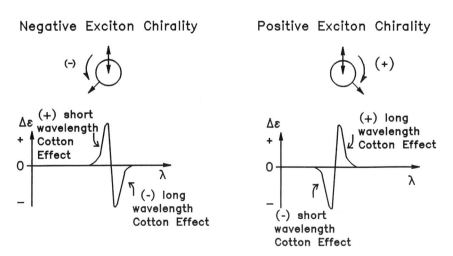

Figure 12-12 Basic elements of the exciton chirality rule relating the torsion angle or helicity of two electric dipole transition moments (◄──►) (upper) to the signed order of the circular dichroism Cotton effects (lower). (Reproduced with permission from D.A. Lightner in *Analytical Applications of Circular Dichroism,* N. Purdie and H.G. Brittain, eds., Elsevier Science Inc., Amsterdam, 1994.)

TABLE 12-3

Application of the Exciton Chirality Rule to *trans*-1,2-Cyclohexanediol[*][†]

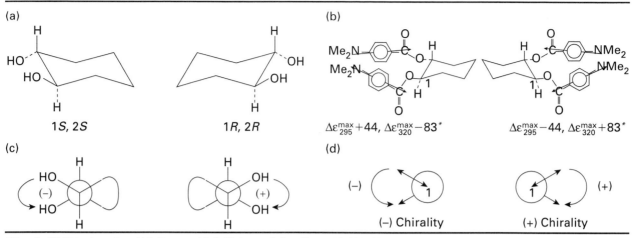

(a)

1S, 2S   1R, 2R

(b)

$\Delta\varepsilon_{295}^{max}+44,\ \Delta\varepsilon_{320}^{max}-83^{*}$ $\Delta\varepsilon_{295}^{max}-44,\ \Delta\varepsilon_{320}^{max}+83^{*}$

(c)

(–)   (+)

(d)

(–)   (+)

(–) Chirality   (+) Chirality

[*] Data in CH_3OH solvent from Y.S. Byun and D.A. Lightner, *J. Org. Chem.,* **56,** 6027-6033 (1991).

[†] Conformational structures (a and b) and Newman projection diagrams (c and d) of (1S,2S) and (1R,2R)-cyclohexane diol (a) and their bis-*p*-dimethylaminobenzoate derivatives (b). CD data for the bisignate Cotton effects of the latter are shown.

adopt various conformations (relative to its point of attachment on the chiral molecule) by rotation about the ester bonds, one conformation (*s*-cis) predominates, and the relevant transition dipoles are thus aligned parallel to the ester R—O bond. With this in mind, one can determine the (+) or (−) relative helicity of the transition dipoles by inspection and thus the assignment of absolute configuration from the CD spectrum. Chromophores with electric dipole transitions not aligned parallel to the ester R—O ester bond are less satisfactory. A good derivatizing agent is thus a symmetric acid in which the alignment of strongly allowed electric dipole transition moment(s) is known with a high degree of certainty (see reference 12.2).

In *trans*-1,2-cyclohexanediol (Table 12-3), the diequatorial configuration is preferred when the cyclohexane ring adopts the stable chair conformation. The two enantiomers exhibit oppositely signed O—C—C—O torsion angles. And for the reasons cited above, when the diols are derivatized as their bis-*p*-dimethylaminobenzoate, (or other suitable chromophores), the electric transition moments lie parallel to the alcohol C—O bonds. Thus the relative orientation (helicity) of the two transition dipoles correlates with the signs of the torsion angles. According to the exciton chirality rule, a (+) exciton chirality is predicted for the (1R,2R) enantiomer and a (−) exciton chirality for the (1S,2S) enantiomer. This conclusion is in complete agreement with the observed bisignate Cotton effects for the intense ~310 nm electronic transition(s) of the bis-*p*-dimethylaminobenzoates. The Cotton effects of the corresponding monoderivatives are monosignate and very weak because there is only one (inherently symmetric) chromophore being perturbed by dissymmetric vicinal action. The contrasting CD spectra of the mono and diesters are fully evident. The diester CD is not simply the sum of two mono ester CD curves; its marked difference implies a different origin consistent with exciton coupling. The signed order of the Cotton effects is consistent with the predictions of the exciton chirality rule.

Examples of applications of the octant rule and the exciton chirality rule to organic stereochemistry are illustrated in the worked structure problems of this chapter.

12-3   STERIC EFFECTS

Stilbene can be isolated as two structural isomers: trans (**12-1**) and cis (**12-2**). Only the trans isomer can easily adopt a conformation with maximum coplanarity of the aromatic ring π systems with the central C=C. Consequently, one expects a greater λ_{max} and ε_{max} for the long wavelength electronic transition.

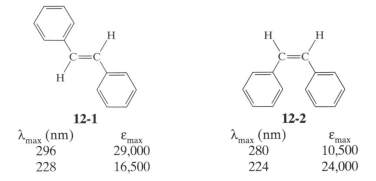

12-1		**12-2**	
λ_{max} (nm)	ε_{max}	λ_{max} (nm)	ε_{max}
296	29,000	280	10,500
228	16,500	224	24,000

As explained earlier (Section 11-4), the coupling of two benzene chromophores in biphenyl lowers the energy of the allowed benzene ^1L_a transition from 204 to 248 nm. However, when coplanarity, and hence maximum π overlap of the two rings of biphenyl, is severely sterically inhibited, the biphenyl derivative behaves more like the sum of the two independent aromatic chromophores. Compare the data for benzene and biphenyl with data for *m*-xylene and 2,2',6,6'-tetramethylbiphenyl.

Benzene		Biphenyl		*m*-Xylene		2,2',6,6'-Tetramethylbiphenyl	
λ_{max} (nm)	ε_{max}	λ_{max} (nm)	ε_{max}	λ_{max} (nm)	ε_{max}	λ_{max} (nm)	ε_{max}
254 (^1L_b)	204			265 (^1L_b)	285	269 (^1L_b)	450
204 (^1L_a)	7900	246 (^1L_a)	16,300	211 (^1L_a)	8400	218 (^1L_a)	23,500

Examples of steric hindrance to conjugation are not limited to biphenyls. Other examples of steric hindrance UV–vis spectra are not hard to find. *p*-Nitroaniline **(12-3)** shows ε_{max} 16,000 at λ_{max} 375 nm, but its *o,o'*-dimethyl analogue **(12-4)** shows ε_{max} 4800 at λ_{max} 385 nm. The methyl groups of **12-4** inhibit coplanarity (hence maximum *p* orbital overlap) of the nitro π system with the aromatic ring π system, leading to a greatly diminished ε_{max} and a red shift of the absorption (see reference 12.11). Acetophenone **(12-5)** has benzenoid bands at 199 nm (ε_{max} 20,000) and 278 nm (ε_{max} 1000), an $n \longrightarrow \pi^*$ conjugated carbonyl absorption at 320 nm (ε_{max} 45), and a 243 nm absorption (ε_{max} 12,600, ethanol), which has been ascribed (see reference 12.12) to an electron transfer (ET) band. Substitution at the ortho or meta position by a methyl group is known to cause a 3 nm bathochromic shift of the ET band, and a para methyl causes a 10 nm bathochromic shift (ε_{max}^{252} 15,100). Yet, in *o,o',p*-trimethylacetophenone **(12-6)**, the ET band is not shifted (λ_{max}= 242 nm) and the ε_{max} is reduced (to 3600). Two ortho methyls effectively inhibit coplanarity of the carbonyl and aromatic π-systems.

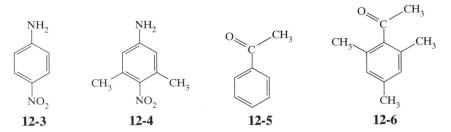

12-4 SOLVENT EFFECTS AND TAUTOMERISM

Cyclohexane-1,3-dione in cyclohexane solvent exhibits weak absorption near 295 nm (ε_{max} ~50), but in ethanol the UV–vis spectrum changes: λ_{max} 255 nm (ε_{max} 12,500). In basified ethanol the spectrum changes again: λ_{max} 280 nm (ε_{max} 20,000). The diketone can exhibit the equilibrium shown in eq. 12-1. In the hydrocarbon solvent, the equilibrium lies largely in favor of the diketo tautomer, which exhibits the weak $\lambda_{max} = 295$ $n \longrightarrow \pi^*$ absorption. In the polar, protic solvent, ethanol, the β-hydroxyenone tautomer is favored, and this form shows the λ_{max} at 255 nm, as predicted by the Woodward–Fieser rules (215 [parent] + 12 [β-alkyl] + 30 [β-OH] = 257 nm) for the $\pi \longrightarrow \pi^*$ absorption. Deprotonation of this enol gives the enolate anion, which absorbs more intensely and at longer wavelengths than the enol form.

$$+ \quad H^+ \qquad (12\text{-}1)$$

5-Amino-4-hexyl-2-mercaptothiazole **(12-7)** may exist as the thione **(12-7a)** or the thiol **(12-7b)** tautomer. In neutral solution **12-7** exhibits λ_{max} 334 (ε_{max} 11,200), but in alkaline solution **12-7** exhibits λ_{max} 304 (ε_{max} 5500). The latter band is also present in **12-8**, which suggests that **12-7a** is favored in neutral solution, whereas in alkaline solution **12-7b** predominates (as its thiolate). However, the 4-phenyl analogue **(12-9)** exhibits only the 305 nm band in both alkaline and neutral solution. Apparently the conjugating influence of the phenyl ring forces the hetero-ring into the aromatic form (of the thiol).

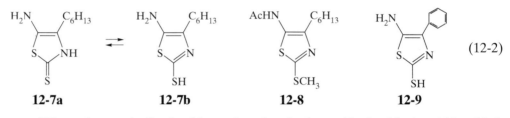

$$(12\text{-}2)$$

| **12-7a** | **12-7b** | **12-8** | **12-9** |

When a ketone is dissolved in methanol and a drop of hydrochloric acid is added, an equilibrium is established in which the ketone is converted to a dimethyl ketal, as shown in eq. 12-3. (An analogous reaction can be written for conversion of an aldehyde to an acetal.)

$$RR'C{=}O \quad \xrightleftharpoons{CH_3OH/H^+} \quad R-\underset{\underset{OCH_3}{|}}{\overset{\overset{OCH_3}{|}}{C}}-R' \quad + \quad H_2O \qquad (12\text{-}3)$$

Although the ketal can be isolated only after removal of the acid catalyst, its formation in solution is readily monitored by UV or chiroptical spectroscopy, since ketal formation causes the carbonyl group and its associated $n \longrightarrow \pi^*$ transition to disappear. The reaction of cholestan-3-one with acidified methanol was investigated by Zalkow and coworkers, who found that the ketone-ketal equilibrium is strongly dependent on the amount of water present. This result is illustrated in Figure 12-13,

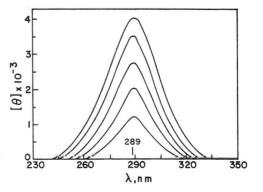

Figure 12-13 A CD investigation of the ketone-ketal equilibrium for cholestan-3-one in acidified methanol. The Cotton effect due to free ketone increases with successive additions of water to the ketal. (Plotted from data of L. H. Zalkow, R. Hale, K. French, and P. Crabbe, *Tetrahedron*, **26**, 4947 [1970].)

which also shows the use of CD as a kinetic tool. The Cotton effect at 289 nm is due to the concentration of free ketone present; the ketal absorbs at much shorter wavelength. Successive additions of small amounts of water shift the ketone-ketal equilibrium toward the free ketone. The same workers also found that ketal formation depends upon the structure of the alcohol, as well as on stereochemical factors, such as the size of groups in the molecule near the carbonyl function. Thus, cholestan-3-one gives 96% of the dimethyl ketal, 84% of the diethyl ketal, and only 25% of the diisopropyl ketal.

12-5 CONFORMATIONAL ANALYSIS

Solvent-induced conformational changes have been detected for $(+)$-*trans*-6-chloro-3-methylcyclohexanone (eq. 12-4). Thus, in isooctane solvent, a negative $n \longrightarrow \pi^*$ Cotton effect is observed, corresponding to a predominance of the conformer **(12-10a)** with the

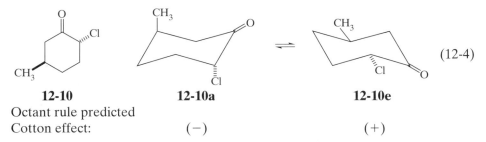

	12-10	**12-10a**		**12-10e**	(12-4)
Octant rule predicted					
Cotton effect:		$(-)$		$(+)$	

methyl and chloro groups oriented diaxially (Figure 12-14). In a more polar solvent, EPA (5/5/2 by vol. diethyl ether–isopentane–ethanol), the Cotton effect sign begins to become reversed with the bisignate form at 25° C. This observation suggests that more of the conformer **(12-10e)** with the equatorial methyl and chloro groups is present in the equilibrium. At −192° C in EPA, very little of the diaxial form is present. Apparently **12-10** attempts to minimize its net dipole moment in the more hydrocarbon (low dielectric) solvents by orienting the C=O and C—Cl bond dipoles as nearly opposite to each other as can be achieved, that is, in the conformation with an axial Cl.

Solvent effects on conformational equilibria have also been detected by CD spectroscopy for $(-)$-menthone **(12-11)** (Figure 12-15). The octant rule predicts a $(+) n \longrightarrow \pi^*$ Cotton effect for the diequatorial conformer **(12-11e)** and a $(-)$ Cotton effect for the diaxial conformer **(12-11a)**. However, the twist boat conformer **(12-11tb)**, which can be expected to relieve the serious axial–alkyl interactions of **12-11a**, is also predicted to have a $(-)$ Cotton effect (eq. 12-5). From the data of Figure 12-15, one might tend to think in terms of the increasing presence of **12-11a** – **12-11tb** as the solvents change from methanol to isooctane. The bisignate nature of the CD curves is explained by the overlapping of a shorter wavelength negative $n \longrightarrow \pi^*$ Cotton effect (due to **12-11a** or **12-11tb**) with a longer wavelength positive $n \longrightarrow \pi^*$ Cotton effect (due to **12-11e**).

12-11

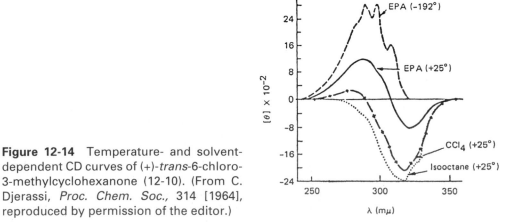

Figure 12-14 Temperature- and solvent-dependent CD curves of (+)-*trans*-6-chloro-3-methylcyclohexanone (12-10). (From C. Djerassi, *Proc. Chem. Soc.*, 314 [1964], reproduced by permission of the editor.)

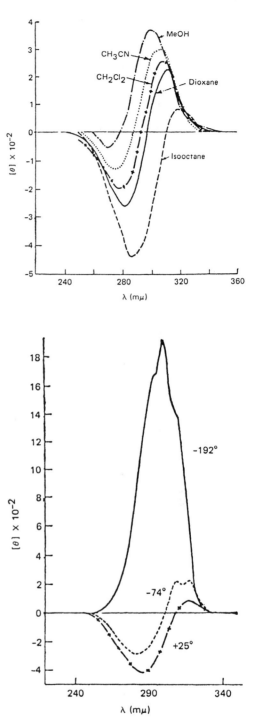

Figure 12-15 CD curves of (−)-menthone **(12-11)** in methanol, acetonitrile, methylene chloride, dioxane, and isooctane. (Reproduced with permission from C. Djerassi, *Proc. Chem. Soc.,* 314 [1964].)

Figure 12-16 CD curves of (−)-menthone **(12-11)** in isopentane-methylcyclohexane at −192°C and in decalin at −74° and +25°. (From C. Djerassi, *Proc. Chem. Soc.,* 314 [1964], reproduced by permission of the editor.)

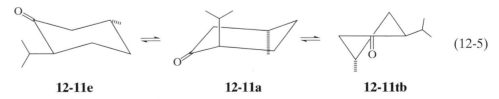

$$\textbf{12-11e} \qquad\qquad \textbf{12-11a} \qquad\qquad \textbf{12-11tb} \qquad (12\text{-}5)$$

Interestingly, the CD curves of (−)-menthone **(12-11)** measured in the glass-forming hydrocarbon solvent, isopentane-methylcyclohexane, change sign between +25° and −192° C (Figure 12-16). These data suggest that the conformational equilibrium (eq. 12-5) is sensitive to temperature and that at very low temperature, high energy conformers with (+) Cotton effects (such as **12-11a** and **12-11tb**), are less prevalent and the equilibrium lies largely in favor of the (expectedly more stable) diequatorial conformer, **12-11e.**

12-6 HYDROGEN BONDING STUDIES

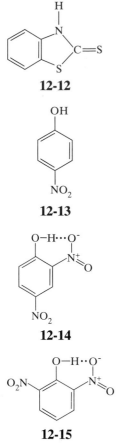

12-12

12-13

12-14

12-15

Hydrogen bonding between molecules is an important factor that may result in relatively large spectral shifts. These spectral shifts are used to deduce information about the strengths of hydrogen bonds. Experimentally it is found that absorption bands are blue-shifted when the chromophore under investigation functions as a hydrogen bond *acceptor* and are red-shifted when it serves as a *donor*. For example, benzthiazoline-2-thione **(12-12)** can function as a hydrogen bond acceptor at the thione group or as a hydrogen bond donor at the NH group. The compound exhibits blue shifts in hydroxylic solvents $(-O-H\cdots S=C\langle\ )$, but undergoes bathochromic shifts in the presence of acceptor molecules such as acetone, with which it functions as a hydrogen bond donor $(-N-H\cdots O=C\langle\ )$. In indifferent solvents such as CCl_4, where there is extensive self-association and the thione acts as both donor and acceptor, no shifts in the absorption spectra are observed. The *N*-methyl derivative of benzthiazoline-2-thione, which cannot function as a hydrogen bond donor, exhibits only a blue shift in donor solvents.

The absorption spectra of compounds engaged in *intramolecular* hydrogen bonding are generally solvent insensitive when studied in donor or acceptor solvents. Compare, for example, the behavior of *p*-nitrophenol **(12-13)** with that of 2,4- or 2,6-dinitrophenol **(12-14** and **12-15)** relative to cyclohexane solution. A substantial bathochromic shift is observed when *p*-nitrophenol is placed in the presence of a proton acceptor (from 286 to 297 nm with dioxane and from 286 to 307 nm with triethylamine). On the other hand, no spectral shifts are found for 2,4- and 2,6-dinitrophenols when these proton acceptors are added. A vast literature exists on spectral aspects of hydrogen bonding, and several monographs on the topic have been published (see reference 12.13).

Hydrogen bonding is also important in ORD–CD studies. For example, the reddish bile pigments *d*-urobilin **(12-16)** and *l*-stercobilin **(12-17)** hydrochloride exhibit

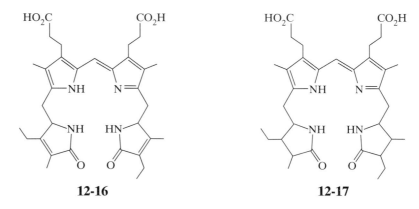

12-16   **12-17**

very strong ORD bands centered near 490 nm in chloroform solvent, but when trichloroacetic acid is added, the intensity drops precipitously (Figure 12-17). The very intense ORD observed for *d*-urobilin indicates that the central dipyrrylmethene chromophore adopts a dissymmetric (helical) shape. In principle, either a left- or a right-handed helix is possible; yet from the ORD it appears that only one helix is preferred. This result seems odd, since all asymmetric centers in the pigment lie removed from the chromophore and in the lactam rings. However, strong intramolecular hydrogen bonding occurs between the opposing lactam and pyrrole units, which forces the pigment to adopt a lock-washer shape. The choice of helicity is determined by the absolute config-

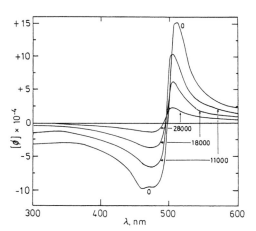

Figure 12-17 ORD curves for the titration of *d*-urobilin HCl in CHCl₃ with trichloracetic acid. Numbers near curves give mole ratio of trichloroacetic acid to *d*-urobilin HCl. (From A. Moscowitz, W.C. Krueger, I.T. Kay, G. Skewes, and S. Bruckenstein, *Proc. Natl. Acad, Sci.* (U.S.), 52, 1190 [1964].)

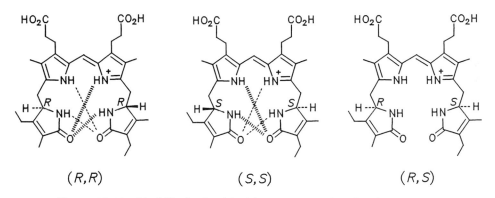

(R,R) (S,S) (R,S)

Figure 12-18 Urobilin hydrochloride structures showing intramolecular hydrogen bonding. The (R,R) enantiomer is forced to adopt a left-handed (M) helix; the (S,S) enantiomer is forced to adopt a right-handed (P) helix. In the (R,S) diastereomer either helix is formed with equal facility, since the full complement of intramolecular hydrogen bonds cannot be formed in either helix.

uration of the asymmetric center near the lactam nitrogen (Figure 12-18). Thus, R,R-urobilin can form intramolecular hydrogen bonds most effectively when the chromophore helicity is left-handed (M), whereas S,S-urobilin forms intramolecular hydrogen bonds most effectively in the right-handed helix (P). (R,S)-Urobilin cannot form intramolecular hydrogen bonds as effectively and exhibits no marked helical preference.

d-Urobilin **(12-16)**, with a strong $(+)$ ORD Cotton effect (Figure 12-17), is thought to adopt the M or left-handed helix, consistent with the R,R-configuration. When trichloroacetic acid is added, the ORD drops to very low values. This seemingly odd behavior in the presence of trichloroacetic acid is due to breaking of intramolecular hydrogen bonds and concomitant loss of the helical preference dictated by the asymmetric centers.

12-7 HOMOCONJUGATION

Homoconjugation, such as that exemplified in β,γ-unsaturated ketones can have a profound effect on the UV–vis spectrum. In norbornenone, for example, the $n \longrightarrow \pi^*$ transition is bathochromically shifted and intensified, as compared with norbornanone. In addition, a new moderately intense UV–vis band appears near 215 nm. The shifts are due to coupling or interaction of locally excited states from the component chromophores.

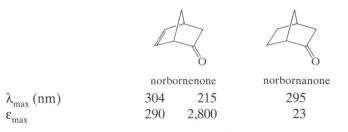

	norbornenone		norbornanone
λ_{max} (nm)	304	215	295
ε_{max}	290	2,800	23

The phenomenon is general, but the magnitude of the wavelength shifts and intensity (ε) changes depends very much on the relative orientation of the two chromophores, as demonstrated by the following examples.

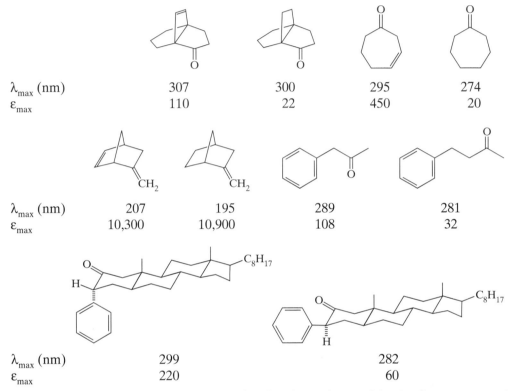

λ_{max} (nm)	307	300	295	274
ε_{max}	110	22	450	20

λ_{max} (nm)	207	195	289	281
ε_{max}	10,300	10,900	108	32

λ_{max} (nm)	299	282
ε_{max}	220	60

Even more remote homoconjugation has been detected in the δ,ε-unsaturated ketone *trans*-5-cyclodecenone. The transannular interaction between the two chromophores produces a blue-shifted $n \longrightarrow \pi^*$ band and a new (charge-transfer) band near 215 nm. Here again, the alignment of the two chromophores is important, as the transannular interaction appears to be absent in the cis isomer.

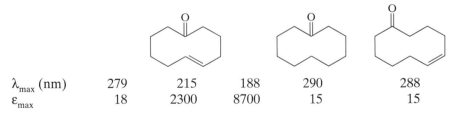

λ_{max} (nm)	279	215	188	290	288
ε_{max}	18	2300	8700	15	15

Unsaturated carbonyl compounds such as β,γ-unsaturated ketones also can form an inherently chiral chromophoric system. Spectroscopically, in these compounds there is a chromophoric interaction, a coupling, in which the forbidden $n \longrightarrow \pi^*$ transition borrows intensity from the $\pi \longrightarrow \pi^*$ band, resulting in a substantial increase in the $n \longrightarrow \pi^*$ Cotton effect near 310 nm.

A chirality rule, depicted in Figure 12-19, has been proposed (see reference 12.14) for this chromophore to correlate the sign of the $n \longrightarrow \pi^*$ Cotton band. The plus and minus signs refer to the sign of the $n \longrightarrow \pi^*$ Cotton effect. An application of this chirality

Figure 12-19 Correlation between chirality and sign of Cotton effect for inherently chiral β,γ–unsaturated ketones. The (+) and (−) signs refer to the $n \longrightarrow \pi^*$ Cotton effect near 300 nm. (See A. Moscowitz, K. Mislow, M.A.W. Glass, and C. Djerassi, *J. Am. Chem. Soc.*, **84**, 1945 [1962].)

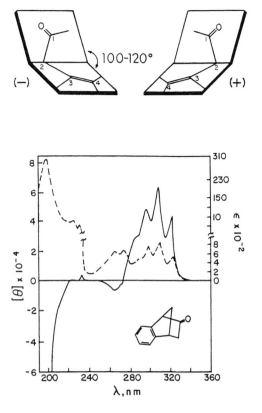

Figure 12-20 The UV (– – –) and CD (——) spectra in isooctane of (1*R*)-(+)-2-benzonorbornenone, a homoconjugated system. (Adapted with permission from D.J. Sandman, K. Mislow, W.P. Giddings, J. Dirlam, and G.C. Hanson, *J. Am. Chem. Soc.*, **90**, 4877 [1968]. Copyright 1968 American Chemical Society.)

rule for β,γ-unsaturated ketones is given in Figure 12-20, which shows the UV and CD spectra of (+)-2-benzonorbornenone. The high rotational strength expected of such inherently chiral chromophores is reflected in the large molar ellipticity, $[\theta]_{307.5}$ +62,000 ($\Delta\varepsilon$ +18.8). To have the orientation corresponding to that shown in the chirality rule (Figure 12-19), the structure shown on the UV and CD spectra would need to be turned over. The positive $n \longrightarrow \pi^*$ Cotton effect centered at 307.5 nm then corresponds to the positive geometry in Figure 12-19. (The three fingers and a shoulder of this Cotton effect are due to vibrational transitions within the $n \longrightarrow \pi^*$ electronic transition.) On the basis of the chiroptical correlation, which was in agreement with chemical evidence, Mislow and coworkers (see reference 12.16) were able to assign the 1*R* configuration to (+)-2-benzonorbornenone.

12-8 CHARGE TRANSFER BANDS

Dissolving a sample in a solvent or mixing two compounds in an indifferent solvent for UV or chiroptical measurements may lead to the formation of a new band due to a charge transfer (ct) complex (see reference 12.16). For example, iodine in hexane has a violet color but in benzene it is brown. Tetracyanoethylene (TCNE) and aniline each form colorless solutions in chloroform, but when mixed, the solution becomes deep blue. Iodine and benzene form a ct complex; so do TCNE and aniline. The observed ct bands arise when a donor molecule, such as iodine or aniline, which has a filled orbital, forms a complex with an acceptor molecule such as benzene or TCNE, which possesses an unoccupied orbital of appropriate symmetry at a slightly higher energy. The ct band is not present in either the isolated donor or acceptor molecule but is found, usually as a new absorption band or Cotton effect, in the ct complex. Charge transfer bands are usually broad and quite strong, an important asset since often the equilibrium constants for formation of the complexes are small.

Almost any type of orbital (such as σ, π, d), can function as a donor or an acceptor, but the most common examples involve π orbitals. Some π donors, in increasing order of basicity, are benzene < mesitylene < naphthalene < anthracene < *N,N,N',N'*-tetramethyl-*p*-phenylenediamine. Some examples of π acceptors or π acids, in increasing order of acidity, are *p*-benzoquinone < 1,3,5-trinitrobenzene < chloranil < tetracyanoethylene. The donor-acceptor complex is analogous to the combination of a Lewis acid with a Lewis base. In contrast to the latter complex, however, ct complexes are characterized by very weak interactions.

UV spectroscopy has also been used in an investigation of the detrimental effects of chlorinated hydrocarbons on living organisms. It is known that these substances attack the central nervous system and block the transport of ions across nerve membranes. The absorption spectrum of the insecticide DDT **(12-18)** has a band at 240 nm,

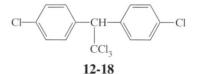

12-18

which in the presence of cockroach nerve axons shifts to 245 nm, with a shoulder appearing at 270 nm. DDT may function as a charge transfer acceptor with the nerve as the donor. The resultant ct complex is responsible for the deactivation of the nerve function.

The UV spectrum of (*S*)-(+)-*N*-[1-(*p*-anisyl)-2-propyl]-4-cyanopyridinium chloride **(12-19)** shows an absorption band at about 350 nm, which is attributed to an

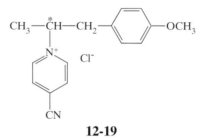

12-19

intramolecular ct transition. The band was shown to be optically active and to exhibit a positive Cotton effect.

Tetracyanoethylene (TCNE), a strong electron-deficient compound, forms intermolecular charge transfer complexes with a variety of electron-donor compounds. This marriage between electron donor and acceptor may lead to new absorption in the UV–vis, typically at longer wavelengths than found in the separate components. Such long wavelength charge-transfer bands have been seen in the colored complexes between colorless components: TCNE and *p*-xylene (460 nm), and TCNE and paracyclophane (521 nm).

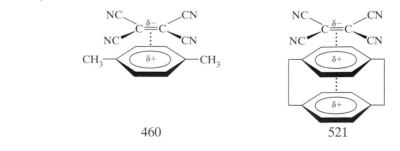

$\lambda_{max}^{CH_2Cl_2}$ 460   521

Intramolecular charge transfer is thought to be responsible for the weak "mystery band" corresponding to the inflection near 220 nm in 7-norbornenone **(12-20)** in heptane. In 2,2,3,3-tetrafluorpropanol the $n \longrightarrow \pi^*$ band shifts to 226 nm (ε_{max} 49) and the mystery band is resolved ($\varepsilon_{max}^{235} = 58$) (see reference 12.17). No such band is found in the saturated analogue, 7-norbornanone **(12-21)** whose $n \longrightarrow \pi^*$ transition lies some 20 nm red shifted relative to 7-norbornenone. This band, which is typically difficult to resolve in the UV–vis spectrum, is fully evident in the CD spectrum of the optically active deuterio analogue **(12-22)** in heptane. It is a mixed transition and thought to involve ~15% $\pi_{C=C} \longrightarrow \pi^*_{C=O}$ component (see reference 12.17). The remainder is very approximately ~30% $n(\sigma)_{C=O} \longrightarrow \pi^*_{C=O}$, ~30% $\sigma_{C=O} \longrightarrow \pi^*_{C=O}$ and ~15% $\sigma_{C-C\alpha} \longrightarrow \pi^*_{C=O}$ (components originating in three different carbonyl in-plane $\sigma \longrightarrow \pi^*$ excitations).

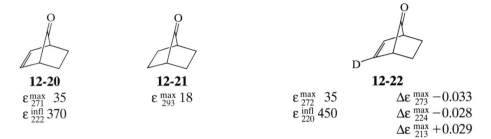

12-20	**12-21**	**12-22**
ε^{max}_{271} 35	ε^{max}_{293} 18	ε^{max}_{272} 35 $\quad \Delta\varepsilon^{max}_{273}$ -0.033
ε^{infl}_{222} 370		ε^{infl}_{220} 450 $\quad \Delta\varepsilon^{max}_{224}$ -0.028
		$\Delta\varepsilon^{max}_{213}$ $+0.029$

12-9 WORKED PROBLEMS

Problem 12-1

Calculate λ_{max} for the following dienes.

alkyl substituent (5)	$0 \times 5 = 0$	$2 \times 5 = 10$	$3 \times 5 = 15$
Parent	214	214	214
Answer	214	224	229
Observed λ^{EtOH}_{max} (nm)	217	226	231
ε^{EtOH}_{max}	22,400	23,800	10,000

Problem 12-2

The compounds below have UV–vis absorption λ_{max} at 234, 244, and 273 nm. Which compound has which λ_{max}?

Symbol

alkyl substituent (5)	$3 \times 5 = 15$	$4 \times 5 = 20$	$4 \times 5 = 20$
exocyclic (5)	$1 \times 5 = 5$	$2 \times 5 = 10$	$0 \times 5 = 0$
homoannular (39)	0	0	39
Parent	214	214	214
Answer	234	244	273

Problem 12-3

A diene, $C_{11}H_{16}$, was thought to have the structure 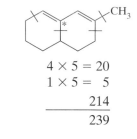 . Its UV–vis spectrum showed λ_{max} = 263 nm, with A_{max} = 0.85.

(a) Can the structure be correct?

(b) Using the same carbocyclic skeleton, draw a structure that satisfies the spectrum.

(c) If the recorded spectrum was determined in a 1 cm cell with 3 mg of compound in 250 mL of ethanol, determine ε_{max}.

(d) From your knowledge of transition allowedness, is the ε_{max} in (iii) reasonable?

Answers

(a) Use the Woodward–Fieser rules to calculate λ_{max}

alkyl substituent (5)	$4 \times 5 = 20$
exocyclic (5)	$1 \times 5 = 5$
Parent	214
	239

The calculated λ_{max} is far away from that observed.

(b) With such a large λ_{max}, it is appropriate to assume a homoannular diene. Subtracting 39 for the homoannular contribution and 214 for the parent leaves 10, which is due to a total of two alkyl substitutions and exocyclic double bonds combined. Only [structure] CH_3 fits the data.

(c) Using Beer's law, $A_{max} = \varepsilon \cdot l \cdot c$ and a MW = 148

$$0.85 = \varepsilon \cdot (1) \frac{3}{148(250)}$$

$$\varepsilon \approx 10,500$$

(d) An ε_{max} of ~10,000 for a homoannular diene is about right.

Problem 12-4

The ketones below have λ_{max} at 236, 244, and 256 nm. Which ketone has which absorption?

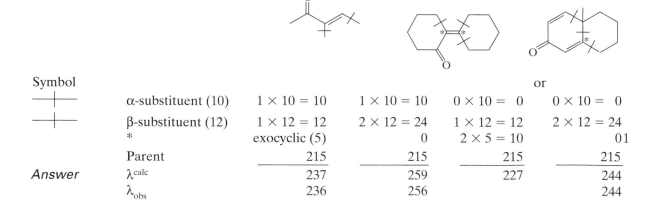

Symbol					or	
╪	α-substituent (10)	$1 \times 10 = 10$	$1 \times 10 = 10$	$0 \times 10 = 0$	$0 \times 10 = 0$
╫	β-substituent (12)	$1 \times 12 = 12$	$2 \times 12 = 24$	$1 \times 12 = 12$	$2 \times 12 = 24$
*		exocyclic (5)	0	$2 \times 5 = 10$	01
	Parent	215	215	215	215
Answer	λ^{calc}	237	259	227	244
	λ_{obs}	236	256		244

Problem 12-5

Two isomeric enamines are formed from 10-methyl-$\Delta^{1(9)}$-octalin-2-one. Can they be distinguished by UV–vis spectroscopy? Application of Woodward's rules leads to the following predictions:

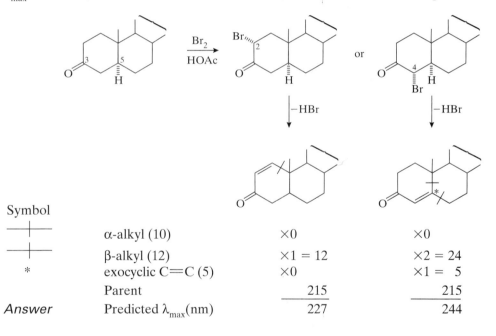

Symbol			
—+—	alkyl substituent (5)	×3 = 15	×3 = 15
*	exocyclic (5)	×1 = 5	×1 = 5
+	extra conjugation (30)	×0	×0
o	homoannular (39)	×1 = 39	×0
	polar group (60)	×1 = 60	×1 = 60
	Parent	214	214
Answer	Predicted λ_{max}(nm)	333	294

Problem 12-6

Bromination of 5α-cholestan-3-one might occur at either C(2) or C(4), the positions α to the ketone group to give two regio isomers. How might you distinguish between them? Dehydrobromination affords two different α,β-unsaturated ketones, one with λ_{max} at 230 nm, the other at 241 nm. From which bromoketone did each originate?

Symbol			
—+— —+—	α-alkyl (10)	×0	×0
	β-alkyl (12)	×1 = 12	×2 = 24
*	exocyclic C=C (5)	×0	×1 = 5
	Parent	215	215
Answer	Predicted λ_{max}(nm)	227	244

Problem 12-7

Conversion of cholesta-4,6-dien-3-dione (right) to its enol acetate gave a product with $\lambda_{max}^{EtOH} = 302$ nm, $\varepsilon_{max}^{EtOH} = 12{,}600$. Which enol acetate was formed?

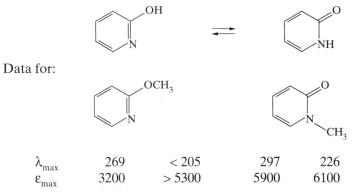

Symbol				
$+$	alkyl substituent	(5)	$3 \times 5 = 15$	$4 \times 5 = 20$
	acetate	(0)		
$*$	exocyclic	(5)	$1 \times 5 = 10$	$2 \times 5 = 10$
$+$	extended conjugation	(30)	$1 \times 30 = 30$	$1 \times 30 = 30$
$\circ$	homoannular	(39)	$1 \times 39 = 39$	$1 \times 39 = 39$
	Parent		214	214
Answer	λ_{max}^{calc}		303	313
	λ_{max}^{obs}		302	316

Problem 12-8

Determine the position of the tautomeric equilibrium between hydroxypyridine and pyridinone from the observed UV–vis spectrum (λ_{max} 224, ε_{max} 7200 and λ_{max} 293, ε_{max} 5900).

Data for:

λ_{max}	269	< 205	297	226
ε_{max}	3200	> 5300	5900	6100

Answer. The equilibrium lies largely to the pyridinone side.

Problem 12-9

Explain the following observations.

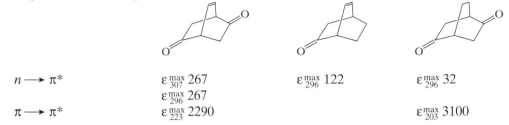

$n \longrightarrow \pi^*$ ε_{307}^{max} 267 ε_{296}^{max} 122 ε_{296}^{max} 32

                ε_{296}^{max} 267

$\pi \longrightarrow \pi^*$ ε_{223}^{max} 2290 ε_{203}^{max} 3100

Answer. The ene-dione exhibits a more intense ε_{max} with a longer wavelength λ_{max} than in either the dione or the enone, suggesting that the C=C acts to link the two C=O groups electronically. The ε_{max} value of the ene-dione is slightly more than twice that of the enone, and λ_{max} is shifted to longer wavelength, suggesting that the former is something more than two independent enones. The band near 223 nm in the ene-dione suggests charge transfer.

Problem 12-10

Two optically active isomeric 2-methyl-4-*tert*-butylcyclohexanones were prepared, and their CD spectra were run. One ketone had an $n \longrightarrow \pi^*$ Cotton effect $\Delta\varepsilon_{max} = +1.4$; the other had $\Delta\varepsilon_{max} = -0.3$. Both ketones had the same optical purity (enantiomeric excess). Assign the absolute configuration of each ketone.

Answer. The *tert*-butyl group is expected to occupy an equatorial site on the cyclohexanone ring, leading to two possible stereoisomers and their mirror images. The octant diagrams are shown below each isomer. According to the octant rule, a strong positive Cotton effect would be predicted for the axial 2(*S*)-methyl isomer and a weak negative Cotton effect for the equatorial 2(*R*)-isomer. So the ketone with $\Delta\varepsilon_{max} = +1.4$ is assigned to the 2(*R*)-axial methyl isomer, and that with $\Delta\varepsilon_{max} = -0.3$ corresponds to the 2(*S*)-equatorial isomer.

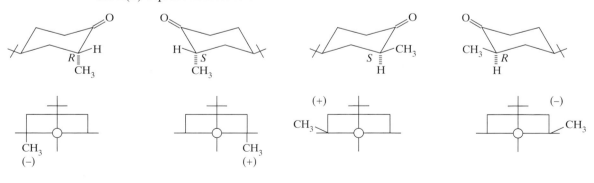

Problem 12-11

4(e)-methyladamantanone has a negative $n \longrightarrow \pi^*$ Cotton effect. Assign the absolute configuration.

Answer. According to the octant rule, the 4(s) isomer shown is predicted to have a negative Cotton effect.

Problem 12-12

2-Oxo-*p*-menthol exhibits a positive CD Cotton effect($\Delta\varepsilon_{max}$ +1.76) in methanol and a negative Cotton effect ($\Delta\varepsilon_{max}$ −1.36) in isopentane. Explain.

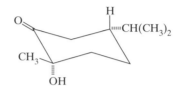

Answer. The conformation shown is predicted to have a positive $n \longrightarrow \pi^*$ Cotton effect by the octant rule and is thus the favored conformation in methanol. In hydrocarbon solvent, intramolecular hydrogen bonding stabilizes a new conformation, which according to the octant rule exhibits a negative Cotton effect.

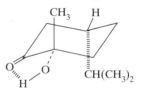

Problem 12-13

Explain the following.

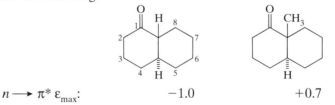

$$n \longrightarrow \pi^* \, \varepsilon_{max}: \qquad\qquad -1.0 \qquad\qquad +0.7$$

Answer. The conformations and octant diagrams shown below indicate a negative Cotton effect for the parent decalone, due principally to the octant contributions of ring CH_2 groups at C-5 and C-6. Introduction of an axial angular methyl α to the ketone inverts the Cotton effect. The methyl makes a large positive contribution that outweighs the negative contributions from ring atoms 5 and 6.

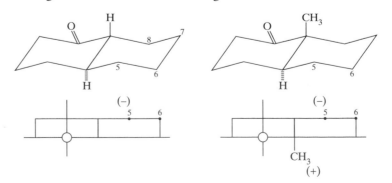

Problem 12-14

3(R)-methylcyclohexanone exhibits an $n \longrightarrow \pi^*$ $\Delta\varepsilon$ +0.6, but for 3(R)-methylcyclopentanone it is $\Delta\varepsilon$ +2.1. Explain.

Answer. 3(R)-Methylcyclohexanone adopts a chair conformation in which the methyl group occupies an equatorial position. In the chair conformation, ring atoms do not contribute to the $n \longrightarrow \pi^*$ Cotton effect. 3(R)-Methylcyclopentanone adopts a half-chair conformation in which ring atoms 3 and 4 lie off the octant symmetry planes and thus contribute to the Cotton effect. The more stable twist-chair has an equatorial methyl, and in this conformation 3(R)-methylcyclopentanone has more octant contributions than does 3(R)-methylcyclohexanone.

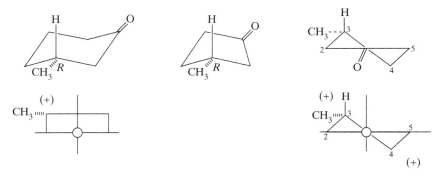

Problem 12-15

How can one correlate absolute and relative configuration in prostaglandins? Recent widespread research efforts directed toward the total synthesis of prostaglandins have made important use of chiroptical techniques. The ORD and CD curves of natural prostaglandin E_1 (PGE) of established configuration have been reported by Korver (see reference 12.18). Miyano and Dorn used chiroptical methods to correlate the absolute configuration of PGE with several intermediates involved in prostaglandin synthesis. They resolved 7-(2-*trans*-styryl-3-hydroxy-5-oxocyclopentenyl)heptanoic acid **(12-23)**, the key intermediate in their total synthesis of racemic prostaglandins, and converted the resolved acids into (8S,12S,15S)-dihydro PGE **(12-24)** and its diastereomer (8S,12S,15R)-dihydro PGE **(12-25)** by an unambiguous series of reactions. Chiroptical studies in the region of the $n \to \pi^*$ transition (Table 12-4) showed that the ORD and CD curves of **12-24** and **12-25** are both mirror images of those of natural PGE. In agreement with octant rule projections, the signs of the Cotton effects indicate that the stereochemistry about the cyclopentanone ring must be the same for both diastereomers and must be enantiomeric to that of natural PGE.

TABLE 12-4

Chiroptical Data for the $n \longrightarrow \pi^*$ Absorption Region of PGE, 12-24, and 12-25

	ORD				CD	
	Peak		Trough			
	[φ]	λ (nm)	[φ]	λ (nm)	[θ]	λ (nm)
PGE	+7200	272	−6200	314	−11,000	296
12-24	+3800	315	−4100	273	—	—
12-25	+3700	315	−5600	274	+7,600	295

Data for methanol solution.
See M. Miyano and C.R. Dorn, *J. Am. Chem. Soc.,* **95,** 2664 (1973).

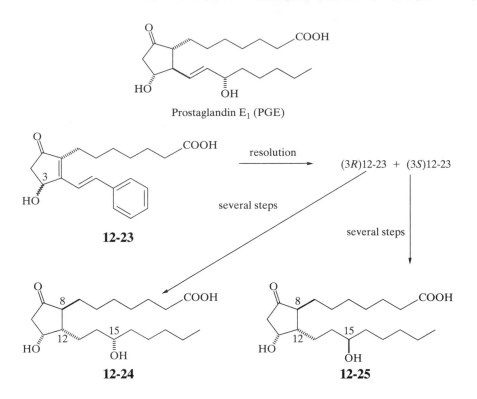

Prostaglandin E$_1$ (PGE)

12-23

resolution

$(3R)$12-23 + $(3S)$12-23

several steps

several steps

12-24   12-25

Problem 12-16

Racemic bicyclo[2.2.1]octen-2-ol was resolved into its enantiomers. Oxidation of one enantiomer gave a ketone with a positive $n \longrightarrow \pi^*$ Cotton effect ($\Delta\varepsilon$ +12.0). Which enantiomer is it?

Answer.   According to the extended octant rule (Figure 12-19), it is

Problem 12-17

Enzymatic dihydroxylation of toluene by *Pseudomonas putida* affords (+)-*cis*-1,2-dihydroxy-3-methyl-3,5-cyclohexadiene. The relative stereochemistry was determined to be cis, but the absolute configuration was unknown. How would you determine the absolute configuration?

(a)   **(b)**

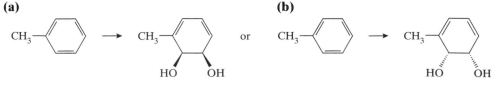

Answer. Catalytic reduction of the diene gave a stereochemically well-resolved cyclohexane diol suitable for introducing the benzoate chromophore. The dibenzoate

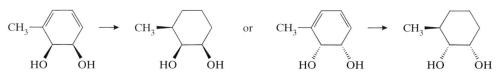

gave a (+) exciton chirality CD, and thus the organism produces the diene-diol shown in reaction (a), as reasoned in the following. There are two enantiomeric cis-diene-diols coming from reactions (a) and (b). The saturated diols are expected to adopt chair cyclohexane conformations eae, where only one substituent is axial. The eae conformers are expected to be more stable than the aea, and the predicted exciton chirality for the eae conformers is given: (+) in (c); (−) in (d).

(c)   **(d)**

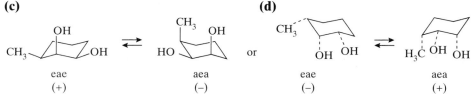

eae   aea   eae   aea
(+)   (−)   (−)   (+)

Problem 12-18

How might the absolute configuration of *trans*-7,8-dihydroxy-7,8-dihydrobenzo[a]pyrene, a carcinogenic metabolite of benzo[a]pyrene, be determined?

Answer. Hydroxylation was determined to yield the trans stereochemistry (Figure 12-21), but there are two mirror image trans-diols. The stereochemistry of the flexible dihydroxylated ring was determined from the vicinal coupling constant of the hydrogens at C-7 and C-8 ($J_{7,8}$ = 8 Hz). With the conformation of the ring known, and the CD spectrum of the diol bis-*p*-dimethylaminobenzoate showing a (−) exciton chirality, the absolute configuration was assigned (7R,8R).

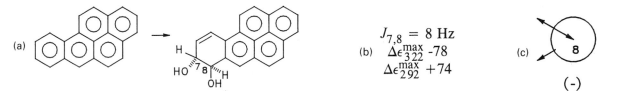

Figure 12-21 (a) Dihydroxylation of benzo[a]pyrene. (b) Coupling constant and exciton chirality CD CEs of the bis-*p*-dimethyl-aminobenzoate. (c) (−) Exciton chirality and relative orientation of the transition moment vectors, hence the absolute configuration of the diol.

Problem 12-19

The absolute configuration of (−)-spiro[4.4]nonane-1,6-dione was determined by application of Horeau's method of optical rotations. How might it be determined using exciton chirality?

Answer. In support of the assignment, and with greater certainty, the exciton chirality rule was applied to the bis-*p*-dimethylaminobenzoate of the cis,trans-diol obtained following reduction of the diketone. The cis,trans-diol is readily distinguished by NMR from the C_2-symmetry cis,cis and trans,trans-diols. Since the cis,trans-diol dibenzoate from the (−)-dione exhibits a (−) exciton chirality CD, it follows that the absolute configuration of the dione is that shown below. The absolute stereochemistry is easily determined by CD.

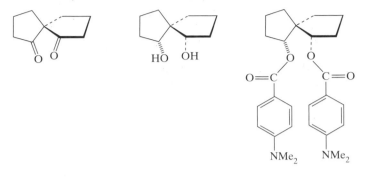

Problem 12-20

How might one determine the absolute configuration of binaphthyls and bianthryls?

Answer. In the 1,1'-bianthryl shown in Figure 12-22, the ethano bridge fixes the interplanar angle of the two anthracene planes at ~30°. The CD corresponding to the ~250 nm ^1B_b transition is extraordinarily strong. When a (−) CD chirality is observed, the molecule has the *P*-helicity or (*R*) absolute configuration. The absolute configuration of (−)-2,2'-bis(bromomethyl)-1,1'-binaphthyl shown was determined by the anomalous diffraction X-ray method on a crystal of the resolved (−)-binaphthyl. The exciton coupling CD for the ~225 nm ^1B_b transition correlates with a (+) chirality, in accord with the X-ray results. The exciton chirality rule is thus confirmed by an independent method, but the CD determination is usually faster.

$\Delta\epsilon_{267}^{max}$ -1100, $\Delta\epsilon_{248}^{max}$ +1100 $\Delta\epsilon_{231}^{max}$ +342, $\Delta\epsilon_{224}^{max}$ -329

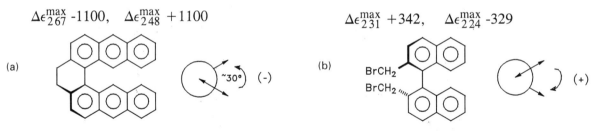

Figure 12-22 (a) Absolute configuration and exciton chirality of an ethano-bridged 1,1'-bianthryl. (b) Absolute configuration and exciton chirality of (−)-2,2'-bis-(bromomethyl)-1,1'-binaphthyl.

Problem 12-21

Tryptycenes can be chiral, but how can one determine the absolute configuration?

Answer. The CD of (+)-dimethyl 5,12-dihydro-5,12[1',2']benzonaphthacene-1,15-dicarboxylate shows an intense (+) chirality exciton couplet ($\Delta\varepsilon_{243}^{max}$ +151, $\Delta\varepsilon_{220}^{max}$ −178) for the intense (ε_{max} ~84,000) 233 nm transition. As with the pyranosides, the binary couplets must be examined, then summed. In the case of the benzotryptycene, the coupling between the two methyl benzoate chromophores is expected to be small because the relevant electric transition moments are parallel. However, the couplings between each methyl benzoate chromophore and the naphthalene chromophore are expected to be large, given the ~90° angle between the transition moment vectors (Figure 12-23). An examination of the structure would predict a (+) chirality for each of the two couplings and a net (+) exciton chirality for the enantiomer shown (Figure 12-23). Determination of absolute configuration by other means would be much more difficult for this substance than by using CD spectroscopy and the exciton chirality rule.

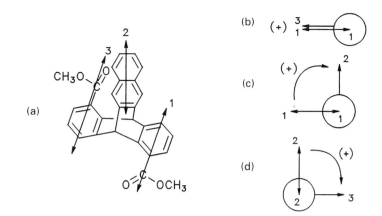

Figure 12-23 (a) Absolute configuration of (5*S*,12*S*)-(+)-dimethyl-5,12-dihydro-5,12[1′,2′]benzonaphthacene-1,15-dicarboxylate (b), (c), and (d) orientations of pairs of electric transition dipole moments from the chromophores: 1 and 3 from the methylbenzoate chromophores; 2 from the naphthalene. The helicities are shown to the right. The 1,3 couplet is predicted to be zero since the dipoles are parallel. The 1,2 and 2,3 couplets have (+) chirality. The net chirality is predicted to be (+).

PROBLEMS

12-1 Calculate λ_{max} for the following steroids.*

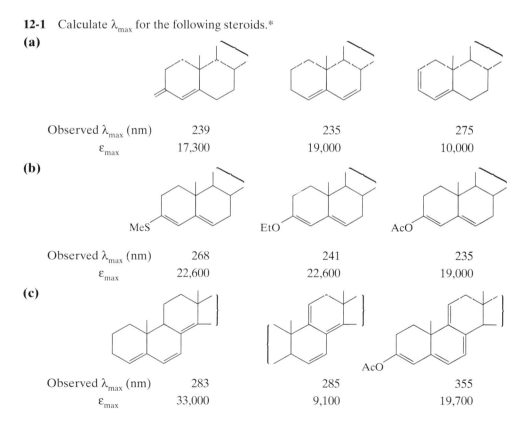

(a)

Observed λ_{max} (nm)	239	235	275
ε_{max}	17,300	19,000	10,000

(b)

Observed λ_{max} (nm)	268	241	235
ε_{max}	22,600	22,600	19,000

(c)

Observed λ_{max} (nm)	283	285	355
ε_{max}	33,000	9,100	19,700

* Observed data from A.I. Scott, *Interpretation of the Ultraviolet Spectra of Natural Products,* Pergamon Press, New York, 1964.

12-12 Spiroenones were prepared of structures **(a)** and **(b)**. One showed an intense λ_{max} at 247 nm, the other at 241 nm. Assign the structures.

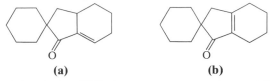

(a)   (b)

12-13 Predict and explain whether UV–vis spectroscopy can be used for distinguishing members of the isomeric pairs.

(a)

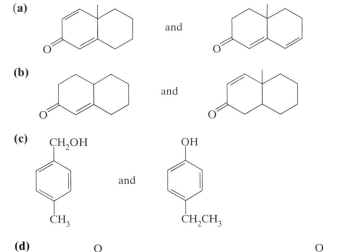

and

(b)

and

(c) CH₂OH   and   OH

CH₃   CH₂CH₃

(d)

 and

(e) $CH_3CH_2COOCH_3$ and $CH_3COOCH_2CH_3$

(f) CH_3—CH=CH—CH_2—CH=CH—CH_3 and
$$CH_3CH_2—CH=CH—CH=CH—CH_3$$

12-14 Compounds X (C_6H_8) and Y (C_6H_8) each take up 2 mole equivalents of H_2 in the presence of Pd/C to give cyclohexane. What are the structures of X and Y when X has λ_{max} 267 nm and Y has λ_{max} 190 nm?

12-15 A compound (A), $C_{11}H_{16}$, has λ_{max} 288 nm. On treatment with Pd/C (which dehydrogenates cyclic compounds completely to aromatic compounds without rearrangement), α-methylnaphthalene (below) is produced. What is the structure of A?

12-16 An unknown monocyclic hydrocarbon A, C_8H_{14}, has a λ_{max} at 234 nm and could be selectively ozonized to yield B, $C_7H_{12}O$, which has λ_{max} at exactly 239 nm. Reduction of B with LiAlH₄ and careful elimination of water (dehydration) gave C, C_7H_{12}, which has a λ_{max} at 267 nm. Give structures for A, B, and C.

12-17 An optically active allylic alcohol (A), $C_8H_{14}O$, was treated with hot H_3PO_4, and three new products (B, C, and D) were distilled. B and C were optically active, and their formation did not involve rearrangement. B, C, and D could be further dehydrogenated (Pd/C + heat) to give p-xylene (1,4-dimethylbenzene). The λ_{max} for B was 229, for C 268, and for D 273 nm. Provide the structures for B to D and deduce the structure of A from the information given.

12-18   Explain the following observations.

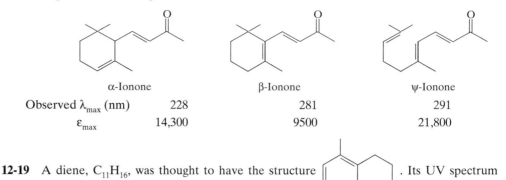

	α-Ionone	β-Ionone	ψ-Ionone
Observed λ_{max} (nm)	228	281	291
ε_{max}	14,300	9500	21,800

12-19   A diene, $C_{11}H_{16}$, was thought to have the structure 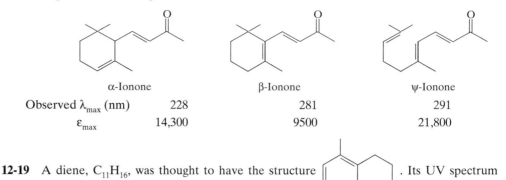 . Its UV spectrum showed λ_{max} 245 nm with $A = 1.47$.

(a) Can the suggested structure be correct? (Apply the Woodward–Fieser rules to it.)
(b) Write a structure that satisfies the UV spectrum and λ_{max} and has the same carbon skeleton.
(c) If the recorded spectrum was determined in a 1 cm cell with 3 mg of compound in 250 ml of solvent, determine the molecular extinction coefficient ε.
(d) From your knowledge of transition allowedness, is the ε determined in (c) reasonable?
(e) What would the value of ε be if the measurement in (c) had been determined in a 1.0 m cell and gave the recorded spectrum? Is this a reasonable value?

12-20   A diene, $C_{11}H_{16}$, was thought to have the structure . Its UV spectrum showed λ_{max} 261 nm with $A = 1.92$.

(a) Can the suggested structure be corrected? (Apply the Woodward–Fieser rules to it.)
(b) Write a structure that satisfies the UV spectrum and λ_{max} and has the same carbon skeleton.
(c) If the recorded spectrum was determined in a 1 cm cell with 3 mg of compound in 250 ml of solvent, determine the molecular extinction coefficient ε.
(d) From your knowledge of transition allowedness, is the ε determined in (c) reasonable?
(e) What would the value of ε be if the measurement in (c) had been determined in a 1.0 m cell and gave the recorded spectrum? Is this a reasonable value?

12-21   An unsaturated ketone, $C_9H_{12}O$, was thought to have the structure . Its UV spectrum showed λ_{max} 300 nm with $A = 1.34$.

(a) Can the suggested structure be corrected? (Apply the Woodward–Fieser rules to it.)
(b) Write a structure that satisfies the UV spectrum and λ_{max} and has the same carbon skeleton.
(c) If the recorded spectrum was determined in a 1 cm cell with 4 mg of compound in 200 ml of solvent, determine the molecular extinction coefficient ε.

12-22   Optically pure 3-methylcyclohexanone shows a CD Cotton effect $\Delta\varepsilon = +2.0$ at $\lambda = 295$ nm. To what electronic transition does this value correspond? Using the octant rule, determine the absolute configuration.

12-23   Predict the sign and approximate $\Delta\varepsilon$ for the $n \longrightarrow \pi^*$ CD Cotton effects of (S)-9-methyl-*trans*-decalone-2 and cholestan-2-one.

12-24   The compounds below have negative (−) Cotton effects. Draw the absolute configuration and conformation of each.

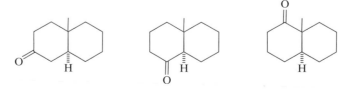

12-25 The compounds below exhibit negative (−) Cotton effects. On treatment with base, a (+) Cotton effect is produced. Explain fully.

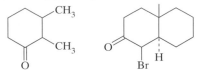

12-26 An isomer of the compound below has a weakly positive (+) Cotton effect. On treatment with base, another isomer is obtained with Δε = −0.8. Draw the configurations and conformations of the two isomers and explain the transformation by drawing the intermediate between the two. Indicate (*R*,*S*) the absolute configuration.

12-27 The partial structure and experimentally determined Cotton effect (CE) signs of cholestanone (a), lanostanone (b), β-amyrone (c), and 2,2-dimethylcholestanone (d) are given below. The partial structures given are sufficient for octant rule projection diagrams. Explain the experimental CE signs or rationalize any difference in terms of conformational structures.

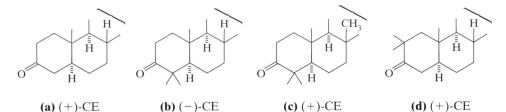

(a) (+)-CE   **(b)** (−)-CE   **(c)** (+)-CE   **(d)** (+)-CE

12-28 Explain the following data in terms of the octant rule (the absolute configurations are as given).

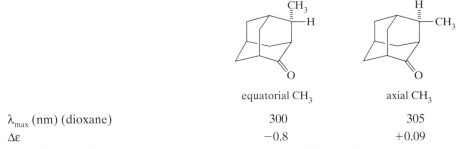

	equatorial CH$_3$	axial CH$_3$
λ$_{max}$ (nm) (dioxane)	300	305
Δε	−0.8	+0.09

12-29 Optically pure 2-methylcyclopentanone shows Δε = −1.8; draw its conformation and absolute configuration.

12-30 Optically pure 2-methylcyclopentanone shows Δε = +2.1; draw its conformation and absolute configuration.

12-31 Treatment of (+)-α-pinene with Fe(CO)$_5$ gives two isomeric ketones, **(a)** and **(b)**. **(a)** has a (−) Cotton effect (Δε −13.1, λ$_{max}$ 292 nm); **(b)** has a (+) Cotton effect (Δε +14.1, λ$_{max}$ 292 nm). Assign the absolute configurations of **(a)** and **(b)**.

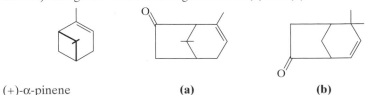

(+)-α-pinene   **(a)**   **(b)**

12-32   The figure below shows the CD curves of (+)-*trans*-6-chloro-3-methylcyclohexanone in methanol (M, ———) and isooctane (I, – – –). Rationalize in terms of conformations for the absolute configuration shown.

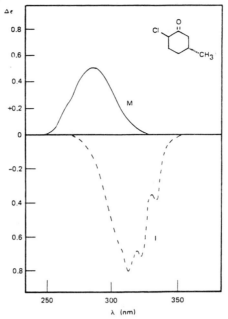

12-33   A 4-hydroxycholesterol (below) di-2-naphthoate shows a bisignate CD Cotton effect with $\Delta\varepsilon_{242}$ −214.5, $\Delta\varepsilon_{229}$ +191.9 for UV ε_{max} 124,500 at λ_{max} 232 nm. What is the configuration at C-4?

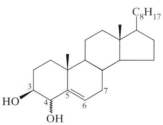

12-34   A 6-hydroxy-5α-dihydrocholesterol bis(*p*-dimethylaminobenzoate) shows a bisignate CD cotton effect with $\Delta\varepsilon_{320}$ −37.6, $\Delta\varepsilon_{295}$ +19.2 for UV $\varepsilon_{max,309}$ 53,300. What is the configuration at C-6?

12-35   The absolute configuration of succinic anhydride *trans*-2,3-diol was determined by measuring the CD Cotton effects of its dibenzoate ester: $\Delta\varepsilon_{240}$ +27.5, $\Delta\varepsilon_{223}$ −5.7 for UV $\varepsilon_{max,233}$ 29,500. Draw the structure indicating its absolute configuration.

12-36   (+)-Camphorquinone was reduced to give two different trans diols. The di(*p*-chlorobenzoate) of one gave CD Cotton effects: $\Delta\varepsilon_{248}$ −35, $\Delta\varepsilon_{230}$ +9. Draw the structure of the diol indicating absolute stereochemistry.

12-37   The absolute configuration of *trans*-7,8-dihydroxy-7,8-dihydrobenzo[*a*]pyrene, a carcinogenic metabolite of benzo[α]pyrene, was determined by the CD spectrum of its bis(*p*-dimethylaminobenzoate): $\Delta\varepsilon_{322}$ −78, $\Delta\varepsilon_{292}$ +74. Indicate the absolute configuration of the diol.

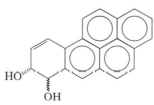

12-38   The absolute configuration of spiro[4.4]nonan-1,6-dione can be determined by CD on the bis(*p*-dimethylaminobenzoate) of the cis, trans diol (reduction product): $\Delta\varepsilon_{321}$ −37, $\Delta\varepsilon_{295}$ +16. Draw the dione, showing its absolute configuration.

12-39   The absolute configuration of *trans*-2-aminocyclohexanol was determined by the CD of its benzoate benzamide derivative: $\Delta\varepsilon_{235}$ −17, $\Delta\varepsilon_{222}$ +10 for UV $\varepsilon_{max,228}$ 4400. Draw the absolute configuration of the aminoalcohol.

12-40   1,1'-Binaphthyl exhibits a bisignate CD Cotton effect, $\Delta\varepsilon_{225}$ +250, $\Delta\varepsilon_{214}$ −179 for UV $\varepsilon_{max,220}$ 108,000. Draw a structure of binaphthyl that shows its absolute configuration.

12-41   The benzoate ester of 7-hydroxycholest-4-en-3-one shows a bisignate CD Cotton effect: $\Delta\varepsilon_{250}$ +13, $\Delta\varepsilon_{230}$ −3. What is the configuration at C-7?

12-42   In the presence of Ni(AcAc)$_2$, butane-2,3-diol shows a bisignate CD Cotton effect: (+) at 395 nm, (−) at 370 nm. Draw a structure of the diol showing its absolute configuration.

BIBLIOGRAPHY

12.1.   D.A. Lightner, in *Circular Dichroism: Interpretation and Application,* K. Nakanishi, N. Berova and R. Woody, eds., VCH Publishers, New York, 1994, pp 259-299.

12.2.   D.A. Lightner, in *Analytical Applications of Circular Dichroism,* N. Purdie and H.G. Brittain, eds., Elsevier Science Inc., Amsterdam, 1994, pp 131-174.

12.3.   C. Djerassi, *Optical Rotatory Dispersion,* McGraw-Hill Book Co., New York, 1960.

12.4.   W. Moffitt, R.B. Woodward, A. Moscowitz, W. Klyne, and C. Djerassi, *J. Am. Chem. Soc.,* **83,** 4013 (1961).

12.5.   T.D. Bouman and D.A. Lightner, *J. Am. Chem. Soc.,* **98,** 3145 (1976).

12.6.   P. Crabbé, *Optical Rotatory Dispersion and Circular Dichroism in Organic Chemistry,* Holden-Day, San Francisco, 1965.

12.7.   C. Djerassi and W. Klyne, *J. Am. Chem. Soc.,* **79,** 1506 (1957).

12.8.   J.K. Gawroński, *Tetrahedron,* **38,** 3 (1982); references therein.

12.9.   H. Harada and K. Nakanishi, *Circular Dichroism Spectroscopy—Exciton Coupling in Organic Stereochemistry,* University Science Books, Mill Valley, CA, 1983.

12.10.   Y.S. Byun and D.A. Lightner, *J. Org. Chem.* **56,** 6027-6033 (1991).

12.11.   H.H. Jaffé and M. Orchin, *Theory and Applications of Ultraviolet Spectroscopy,* John Wiley & Sons, Inc., New York, 1962.

12.12.   J.N. Murrell, *J. Chem. Soc.,* 3779 (1965).

12.13.   S.N. Vinogradov and R.H. Linnell, *Hydrogen Bonding,* Van Nostrand Reinhold, New York, 1971.

12.14.   A. Moscowitz, K. Mislow, M.A.W. Glass, and C. Djerassi, *J. Am. Chem. Soc.,* **84,** 1945 (1962).

12.15.   D.J. Sandman, K. Mislow, W.P. Giddings, J. Dirlam, and G.C. Hanson, *J. Am. Chem. Soc.,* **90,** 4877 (1968).

12.16.   See, for example: R. Foster, ed., *Molecular Complexes,* Elek Science, London, 1973; N. Mataga and T. Kubota, *Molecular Interactions and Electronic Spectra,* Marcel Dekker, New York, 1970; R.S. Mulliken and W. Person, *Molecular Complexes,* John Wiley & Sons, Inc., New York, 1969; R. Foster, *Organic Charge Transfer Complexes,* Academic Press, New York, 1969.

12.17.   D.A. Lightner, J.K. Gawroński, Aa.E. Hansen, and T.D. Bouman, *J. Am. Chem. Soc.,* **103,** 4291 (1981).

12.18.   O. Korver, *Rec. Trav. Chim. Pays-Bas,* **88,** 1070 (1969).

part IV

MASS SPECTROMETRY

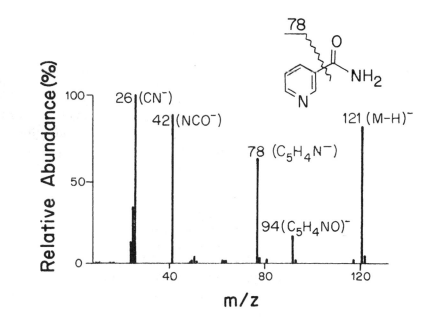

chapter 13

Ionization and Mass Analysis

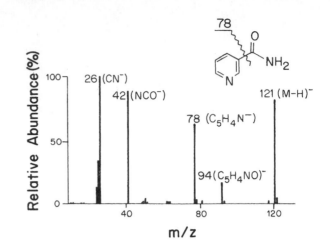

13-1 SCOPE AND APPLICATIONS OF MASS SPECTROMETRY

Mass spectrometry (MS) is a spectroscopic method for elucidating molecular structure and is one of the truly interdisciplinary methods in science. It originated in physics, has been applied throughout the biological and earth sciences and across the field of chemistry, and is of particular importance in environmental science. Mass spectrometry is both a tool for accomplishing measurements in many areas of science and the basis for a fundamental chemical science, namely, gas phase ion chemistry. A knowledge of intrinsic (gas phase) ion chemistry is of value in elucidating phenomena seen in solution. It allows direct comparisons of structure and thermochemistry with ab initio calculations, it serves to model complex biological and catalytic experiments, and it provides access to elusive species (for example, the discoveries of C_{60}, CH_5^+, and many free radical species).

Mass spectrometry is distinguished by its extremely high sensitivity and by its applicability to samples in all physical states (including aqueous solutions and solid materials) and to samples of high as well as low molecular weight. The methods available for sample ionization are paralleled by an equally wide variety of methods of mass analysis. The result, as shown in Table 13-1, is that many combinations of procedures can be used

TABLE 13-1
Sequence of Operations in a Mass Spectrometer*

Sample Introduction	Ionization	Mass Analysis	Detection
Direct insertion probe	Electron ionization	Magnetic sector	Faraday cup
Chromatograph (on-line)	Chemical ionization	Time-of-flight	Electron multiplier
Batch (vapor) inlet	Desorption ionization	Quadrupole mass filter	Image current
Membrane introduction	Spray ionization	Quadrupole ion trap	
		Ion cyclotron resonance	

* In some experiments (such as spray ionization) sample introduction and ionization occur in a single process.

to record mass spectra. Almost every combination of these operations is possible, and so there is an enormous variety of mass spectrometers, each with particular advantages.

Mass spectrometry is usually perfomed to determine the molecular weight of a compound(s). To accomplish this end, one of the several ionization methods for producing intact molecular ions must be used. These methods generate either positive or negative ions related to the original molecule by adding or subtracting an electron, or by adding or subtracting an anion or cation.

Although nominally a spectroscopic method, mass spectrometry is unlike other forms of spectroscopy because it does not involve electromagnetic radiation. Instead it involves chemical manipulations (such as ionization and fragmentation), and the relative intensity of each peak is a measure of the quantity of the corresponding ionic species. A typical 70 eV electron ionization or electron impact (EI) mass spectrum (Figure 13-1) depicts a plot of relative ion abundances vs. mass-to-charge ratios. This information can be used to deduce the molecular weight of the analyte (in this case, 120 Dalton, abbreviated Da) and to acquire structural information (here the presence of a phenyl ring and a methyl group). Note that 120 Da is the nominal molecular mass (to the nearest Da). Mass spectrometry usually gives separate signals for the various isotopic forms of an ion ($^{13}C_0$, $^{13}C_1$, $^{13}C_2$...), and the weighted average of all isotopic masses is the chemical average mass.

For samples of complex mixtures, chromatographic separation is often combined with mass spectrometry. Mass spectrometry can improve the resolution of the chromatograph by acting as a compound-specific detector. A full mass spectrum is often much more helpful than a retention time in identifying individual peaks in the chromatogram. The widespread use of gas chromatography–mass spectrometry (GC–MS) and liquid chromatography–mass spectrometry (LC–MS) in the environmental, chemical, and pharmaceutical industries have made mass spectrometers one of the most common analytical instruments.

Mass spectrometry developed in the late 19th century from studies on electrical discharges in gases and the nature of the atom. Thomson, the discoverer of the electron, accelerated ions to a high kinetic energy (10's of keV) in an evacuated tube, deflected them in a mass-selective fashion using a magnetic field, and recorded the image produced on a photographic plate. From 1919 to 1939, refined instruments capable of higher resolution were used by Aston ad others to determine the masses and abundances of the isotopes. Since 1940, MS has developed to meet the needs of successive

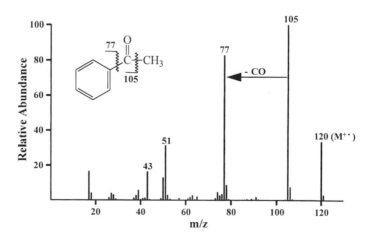

Figure 13-1 Electron ionization (EI) mass spectrum of acetophenone, which provides information on the molecular weight of the compound and some of its structural features. Note the formation of an intact radical cation, indicated as $M^{+\cdot}$, with mass-to-charge (*m/z*) 120 Da/atomic charge. (From H. Budzikiewicz, C. Djerassi, and D.H. Williams, *Mass Spectrometry of Organic Compounds*, Holden-Day, San Francisco, 1967, p. 163.)

TABLE 13-2

Uses of Mass Spectrometry in Organic and Biological Chemistry

Application	Samples	Methods	Comment
Molecular weight determination	Pure compounds, mixtures	Recognize intact molecular ion in spectrum	Several ionization methods can be used for confirmation
Molecular formula determination	Usually pure compounds	High resolution measurement on molecular ion	High resolution alone seldom gives a unique molecular formula
Molecular structure determination	Pure compounds or mixtures by LC–MS, GC–MS, and MS–MS	Spectrum-structure correlations; library comparisons	Confirmation of suspected structures is usual; de novo interpretations rare
Sequence determination	Proteins, other biopolymers	Tandem mass spectrometry (MS–MS)	Sensitive, very rapid and increasingly useful
Isotopic incorporation and fractionation	Naturally and artificially labeled compounds (^{13}C, ^2H, ^{18}O, etc.)	Ion abundance measurements	Precise isotope ratio measurements require special instruments

communities of users: first petroleum engineers, then organic chemists, environmental scientists, and the biomedical community. Some of the applications of mass spectrometry are summarized in Table 13-2.

13-2 SAMPLE INTRODUCTION

Samples introduced into the mass spectrometer in a suitable form (such as an aqueous solution), are then ionized to create gas phase ions that are analyzed for their mass-to-charge ratios *(m/z)* and then detected. Although mass spectrometers are capable of subnanogram sensitivity, solid samples are normally required to be in the microgram range. Similarly, solutions can sometimes be analyzed at the parts per trillion level, but concentrations in the parts per million level are more common. Mass spectrometers are operated under a vacuum, often 10^{-6} torr (mm Hg) or less, which is maintained by a combination of mechanical forepumps with diffusion pumps, turbomolecular pumps, or ion pumps. Many instruments can be vented to atmosphere and a working vacuum can be reachieved in a matter of minutes. Samples are introduced into the ion source by one of the methods described in Table 13-3. These methods are (1) batch gas and vapor inlets, (2) direct insertion probes, (3) membrane interfaces, and (4) gas or liquid chromatographs. Figure 13-2 illustrates schematic versions of these interfaces.

TABLE 13-3

Sample Introduction Systems*

System	Sample Type	Minimum Sample	Characteristics	Ionization Method
Batch (reservoir)	Gas, liquid, low melting solid	<1 mg	Steady sample delivery for long periods	EI, CI
Direct insertion probe	Less volatile samples	<1 µg	Sample delivery varies with probe temperature	EI, CI, DI
Membrane	Mixtures in solution	$<10^{-6}$ M	Volatiles only	EI, CI
Chromatography: GC, SFC†	Mixtures in solution	<1 µg	More volatile compounds	EI, CI
Chromatography: LC, CZE‡	Mixtures in solution	<1 µg	Less volatile compounds	SI

* See Table A-7 for acronyms.

† SFC = supercritical fluid chromatography.

‡ CZE = capillary zone electrophoresis.

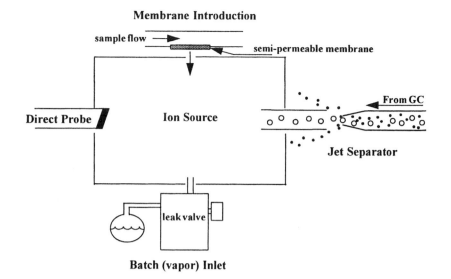

Figure 13-2 Four commonly used sample introduction systems.

13-2a Direct Introduction

Ionization when the sample is in the vapor phase can be achieved by evaporation by resistive heating from a direct insertion probe. Typically, the maximum temperatures used are <400° C. Nonvolatile or thermally unstable solids of low vapor pressure are commonly examined by a desorption ionization (DI) method. The sample is introduced into the mass spectrometer using a vacuum lock and is then irradiated by a laser or by an energetic particle beam to produce ions.

13-2b Introduction via Gas or Liquid Chromatographs

Mixtures of compounds are often introduced into mass spectrometers using a chromatograph (see Section 15-6). Efficient interfaces to GC's and LC's employ high-performance, small-diameter columns. The GC–MS interface is usually made by directly coupling the two devices. A capillary column can be fed directly into the ion source of the mass spectrometer, and its vacuum system used to remove the sample–carrier gas combination. The eluting components can be ionized by a universal method such as electron ionization (see Section 13-3a) or by a selective method such as chemical ionization (see Section 13-3b). With packed columns, it becomes desirable to use separators to remove carrier gas and enrich the sample stream in analyte. The usual method is to use a jet separator—a device that selectively removes lighter atoms and molecules such as He (Figure 13-2).

LC–MS is usually performed by electrospray ionization, discussed in more detail in Section 13-3d. This method uses an electric field to charge microdroplets of solution, and a combination of heat and pneumatic forces to cause desolvation to produce gas phase ions. Solvent removal in spray ionization (SI) methods is greatly assisted by nebulization of the solution, which is often achieved by a combination of thermal, pneumatic, and electrical means. Ionization of samples introduced at atmospheric pressure is a rapidly growing procedure that simplifies sampling procedures for GC and LC eluent introduction. The advantages of this method are high sensitivity, rapid response times, and minimal maintenance requirements. Nonaqueous solvents usually evaporate more readily than water, but the importance of aqueous solutions in biochemistry means that attention is focused on them.

Other forms of chromatography have been coupled to MS. Supercritical fluid chromatographs (SFC) can be coupled directly to mass spectrometers. Electrophoresis and other planar separation techniques are normally performed off-line and the spots extracted before examination by MS, but methods based on desorption ionization (DI) allow direct examination of electrophoresis plates or thin-layer or paper chromatograms.

13-2c   Other Sample Introduction Methods

Besides direct insertion probes and chromatographs, there are many other ways of introducing samples into a mass spectrometer. Gases or vapors can be introduced continuously and very conveniently from reservoirs through restricted flow "leak" inlet systems. Most mass spectrometers have vapor inlet systems (often simple metering valves) for introducing mass calibration compounds such as fluorocarbons. These batch inlet systems may be all-glass, variable temperature systems. This type of inlet system requires more sample than the other methods and also demands greater sample volatility and thermal stability. This once common method of introducing organic samples has the advantage that a steady sample pressure is maintained for long periods, allowing slow scanning or extensive signal averaging.

Reaction mixtures and environmental samples are often examined on-line, using vapor inlets that connect the mass spectrometer to the sample vessel via capillary tubes, which are often made of silica. Two-stage pressure reduction is often achieved by passage through a narrow orifice (micron dimensions). The capillaries are normally heated to prevent condensation, and the method requires that the sample be volatile.

Another simple introduction method, applicable to more volatile constituents of aqueous solutions and air samples, is simply to pass the gas or solution of interest across the surface of a semipermeable membrane. This method, known as membrane introduction mass spectrometry (MIMS), avoids introducing the fluid into the vacuum system. The analyte passes across the membrane, moving from the condensed phase into the vapor phase. The membrane is usually a silicone polymer, which is hydrophobic and selectively passes relatively volatile and low molecular weight organic compounds while excluding the aqueous matrix. The high resistance to fouling and the mechanical and chemical resistance of silicone membranes adds to their usefulness. MIMS is particularly suitable for low-level continuous-monitoring experiments, which are important in environmental science applications.

13-3   IONIZATION

Various procedures are used to form gas phase ions from molecules, depending on the physical state of the analyte. Choices are available as to the types of ions produced (positively and negatively charged, radical cations, protonated molecules, etc.) and the degree to which these ions are internally excited. Internally excited molecular ions dissociate to produce fragment ions, which may reveal details of molecular structure. On the other hand, an intact molecular ion (such as the protonated molecule or the radical anion) provides information on molecular weight.

The major ionization methods used for organic and biological compounds can be grouped into four categories: electron ionization (EI), chemical ionization (CI), desorption ionization (DI), and spray ionization (SI) (Table 13-4). Except for EI, each method can produce abundant positive or negative ions. With the same exception, the type of molecular ion produced can be varied by choice of the medium/matrix in

TABLE 13-4

Ionization Methods

Method	Ionizing Agent	Sample	Molecular Ion*
Electron ionization (EI)	Electrons	Vapor	$M^{+\cdot}$
Chemical ionization (CI)	Gaseous ions	Vapor	$(M+H)^+$, $(M-H)^-$, etc.
Desorption ionization (DI)	Photons, energetic particles	Solid	$(M+Na)^+$, $(M-H)^-$, etc.
Spray ionization (SI)	Electric field, heat	Aqueous solution	$(M+H)^+$, $(M-nH)^{n-}$, $(M+nH)^{n+}$, etc.

* More than one type of molecular ion is often formed, depending on the conditions chosen for each method; M = intact analyte molecule.

which ionization is performed. For example, chemical ionization can yield $(M+H)^+$, $(M+NH_4)^+$, $(M+Ag)^+$, $(M+Cl)^-$, etc., as forms of molecular ion, as well as fragment ions. Ionization necessarily involves chemical reactions, and as such is subject to the same rich variety as most other chemical processes.

The gas phase ions needed for mass spectrometry are generated in the ion source, and reactions of the initially formed ions occur there. Energetic ions fragment spontaneously by unimolecular processes, whereas ion-molecule reactions result from collisions between ions and neutral molecules. The common EI technique is concerned with unimolecular fragmentation. Ion-molecule reactions form the basis for chemical ionization, a more versatile ionization method that often provides molecular weight information missing in EI experiments. In the desorption ionization techniques, condensed phase samples are examined directly through the input of energy. These methods, typified by laser desorption, enable nonvolatile and thermally labile compounds to be ionized. So, too, do the spray ionization methods, typified by electrospray, which yields multiply-charged ions from aqueous solutions. This technique makes high-mass biological compounds accessible. Ion-molecule reactions and unimolecular fragmentations occur in CI, DI, and the SI techniques.

Given this variety of ionization methods, how does one decide which method(s) to apply? There are often several suitable methods that give complementary information on the analyte. The deciding criteria, besides availability of particular techniques, are often the following:

1. Physical state of the sample (for example, aqueous solution vs. solid sample)
2. Volatility and thermal stability of the sample
3. Type of information sought (for example, molecular structure vs. sequence analysis)

If a particular method of ionization successfully produces ions of the analyte, its performance can be evaluated in terms of (1) signal strength: the rate of ion production (on the order of nanoamperes, 1 nA $\approx 10^{10}$ ions/s; (2) background signal: some ionization methods produce large amounts of "chemical noise" (ions generated from the matrix); (3) ionization efficiency: a measure of the yield of the ionization process; and (4) the ease with which the degree of fragmentation can be varied through control over the internal energy of the ions. Desiderata for ionization methods are summarized in Table 13-5.

13-3a   Electron Ionization (EI)

EI is one of the oldest and simplest of ionization methods. Its main advantage is that it gives reproducible spectra, and its main disadvantage is that it is limited to samples that can be vaporized without dissociation. It is generally applicable only to lower molecular weight (<1000 Da), less polar organic compounds. EI is a purely physical process (eq. 13-1). The sample vapor is bombarded by electrons having sufficient energy to

TABLE 13-5
Desiderata for Ionization Methods

Characteristic Desired	Characteristics Usually Achieved*
High signals (ion current)	10^{-10} A (approximately 10^9 ions/s)
High efficiency (neutrals $\longrightarrow$ ions)	0.1%
Bipolarity (positive and negative ions)	Except for EI
High molecular weight compounds	Approximately 10^5 Da
All sample states	All physical states are compatible with MS
Control of fragmentation	By control of energy deposited in ion

* Not every method exhibits all these characteristics, even for the most favorable analytes.

$$M \text{ (vapor)} + e^- \longrightarrow M^{+\cdot} + 2e^- \tag{13-1}$$

cause ionization. Ionization energies (IE) are on the order of 10 eV for many compounds, but substantially greater electron energies, often 70 eV, are chosen to achieve higher signals and to avoid changes in the mass spectrum with small changes in electron energy. The initial product of ionization is a cation radical ($M^{+\cdot}$), which, at the electron energies used, often fragments extensively. This so-called fragmentation pattern may provide more information with which to characterize the compound, but it necessarily leads to a decrease in molecular ion abundance. Fragmentation may be so extensive that a useful molecular radical cation ($M^{+\cdot}$) is not observed.

The relatively high degree of reproducibility of EI mass spectra—its major advantage—arises largely because (1) electron ionization is a physical not a chemical process and (2) fragmentation involves exclusively gas phase unimolecular reactions. The only extensive libraries of mass spectra employ EI. (To make libraries for mass spectra based on ion-molecule reactions would be a little like making a library of Diels–Alder reaction products in solution: it can be done, but it is better to describe the factors controlling product yields.) Note, however, that all mass spectra are far less reproducible than the spectra based on interactions of electromagnetic radiation with matter (IR, NMR). Mass spectra are simply distributions of products, and the intensities of peaks depend on the original energy deposited and the time allowed for fragmentation, that is, on the exact conditions chosen to record the spectrum.

Fragmentation in EI occurs when the molecular ion, $M^{+\cdot}$, is produced with sufficient internal energy to dissociate spontaneously. The molecular radical cations, $M^{+\cdot}$, are always produced with a range of internal energies that depends on the amount of energy transferred to the molecule by the ionizing electron. Each collision event is unique, and the resulting molecular ions behave differently—some will fragment extensively, others less so, whereas still others will be observed as undissociated molecular ions.

It is possible to visualize the energy transfer occurring during electron ionization as the result a hard or a soft collision, depending on the degree of interaction of the electron with the gas phase molecule. This distinction is shown in Figure 13-3, where a soft collision event is shown to result in molecular ions that do not fragment and a harder electron-molecule collision yields unstable molecular ions. Since the nature of the collision between an electron and molecule is statistically controlled, a mixture of processes occurs. Thus electron impact mass spectra often contain both intact molecular ions as well as characteristic fragment ions. Electron impact with low-energy electrons is a softer ionization method and increases the relative abundance of intact molecular ions. However, this result is achieved only with a great loss in ionization efficiency.

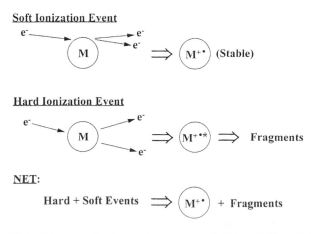

Figure 13-3 Electron ionization accompanied by different degrees of excitation of the molecular ion. Soft ionizing events transfer little excess energy to the ionized molecule, which is observed intact. Harder collisions also occur and give rise to the fragment ions frequently seen in EI mass spectra.

Continuing to use the terms "hard" and "soft" to designate the extent to which ions are internally excited, we note that EI is a much harder ionization method than most others. Relatively energetic molecular ions are created, and they therefore tend to fragment extensively. However, different compounds are differently susceptible to dissociation, even when ionized under the same conditions. This concept is illustrated in Figure 13-4, which contrasts the 70 eV EI mass spectra of two compounds of identical molecular weight and similar complexity. The aromatic compound, acenaphthalene, displays a characteristically abundant molecular ion; the only significant fragment ions are due to loss of $H \cdot$ and H_2 and the formation of the ion m/z 76, $C_6H_4^{+\cdot}$. The only other ion of more than 10% relative abundance is the ^{13}C isotope of the molecular ion, which has a m/z ratio of 153. The aliphatic, bicyclic nonane derivative, by contrast, displays a molecular ion of just 1% relative abundance, and characteristically shows abundant fragment ions, many of low mass.

The Ionization Event. EI is the result of an electronic transition in the molecule. Because the velocity of the bombarding electron (5×10^8 cm s^{-1} for 70 eV electrons) greatly exceeds the rate of intramolecular atomic motion (typical bond vibrations require $>10^{-12}$ s, whereas a 70 eV electron transits a 1 nm molecule in 2×10^{-16} s), the molecule remains frozen as the electronic excitation occurs, a condition known as the *Franck–Condon principle* (see Section 10-2c). Figure 13-5 illustrates such an ionizing event. Abundant intact ions are more likely when the internal energy deposition in the molecular ion is small and the molecular structure resists fragmentation. Obviously the shapes of the potential energy hypersurfaces as well as the electron energy chosen will control the first factor, whereas the chemical nature of the analyte will control the second.

The EI Source. The heart of any mass spectrometer is the ion source. Not only does ionization take place in this region, but most of the chemically significant events occur there, too. Experimentally, EI is performed in an ion source into which the vaporized sample is introduced and from which ions are extracted. The sample may be introduced from a gas reservoir, a heated probe, or a chromatograph, and typical sample pressures are approximately 10^{-5} torr. To avoid collisions of ions with neutral molecules (which change the character of the spectrum), the source is relatively open and much of the sample is pumped away. An electron beam of 70 eV energy, generated by thermionic emission from an incandescent tungsten or rhenium filament, is passed across the ionization region. The electron current is continuously measured on an anode and its value is held constant in order to generate a constant

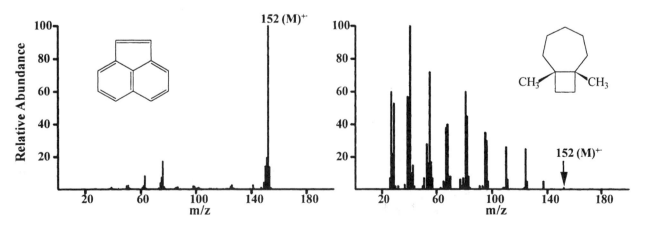

Figure 13-4 Contrasting degrees of fragmentation observed in the EI spectra of an aromatic and an alicyclic compound. The former resists dissociation and gives a molecular ion from which the molecular weight (152 Da) is measured. (From P.J. Ausloos, C. Clifton, O.V. Fateev, A.A. Levitsky, S.G. Lias, W.G. Mallard, A. Shamin, and S.E. Stein, NIST/EPA/NIH Mass Spectral Library—Version 1.5, National Institute of Standards and Technology, Gaithersburg, MD [1996].)

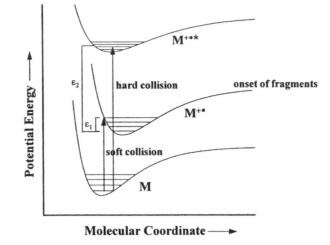

Figure 13-5 Vertical transitions associated with electron ionization ($M \longrightarrow M^{+\cdot}$) showing deposition of internal energy ε in the ion $M^{+\cdot}$. Two ionization events are shown, a soft collision leading to stable $M^{+\cdot}$ of relatively low energy (ε_1), and a hard collision leading to high-energy molecular ions (ε_2), which rapidly fragment.

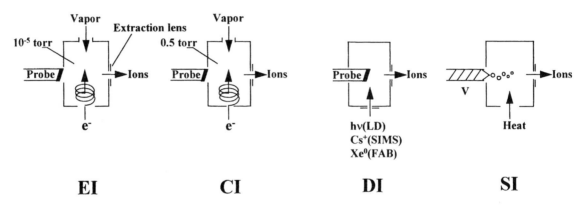

Figure 13-6 Several types of ionization sources. Sample may be introduced on the probe or by any of the methods shown in Figure 13-2.

ion current. Ions spend about 1 μs in the EI source, during which time unimolecular fragmentations occur. The resulting population of ions is extracted from the source with an electric field and passed into the mass analyzer. The ion current may be as high as 10^{-9} A (6×10^9 ions/s) after mass analysis, but frequently is much lower, sometimes as low as 10^{-17} A or less. Fortunately, electron multiplier detectors and high-gain amplifiers allow such signals to be measured. Figure 13-6 shows the EI source schematically, and compares it with other ion sources to be discussed in the following sections.

The similarities in ion sources should not mask the sharp differences in the chemical and physical phenomena associated with the several ionization techniques. The major ionization methods (EI, CI, and DI) typically yield complementary information, because they are based on different physical principles and different ion chemistry. Positive vs. negative ion spectra also provide complementary information. For these reasons it is extremely useful to obtain more than one type of spectrum when characterizing an unknown compound.

A summary of some characteristics of EI is presented in Table 13-6.

TABLE 13-6
Advantages and Disadvantages of EI

	Consequence
Advantage	
Reproducible method	Libraries of EI spectra allow compound identification
Extensive fragmentation occurs	Molecular structure information can be deduced
Ionization efficiency high	Method is sensitive: 1 in 1000 molecules is ionized
Ionization is nonselective	All vaporized molecules can be ionized
Disadvantage	
Only positive ions formed	Not ideal for some classes of compounds
Radical cations formed	Rearrangement processes complicate mass spectra
Sample must be volatile	Limited to low molecular weight compounds of approximately 600 Da or less
Ionization is nonselective	All vaporized molecules contribute to the mass spectrum
Relatively energetic (large internal energy) method	Often extensive fragmentation: limits value in molecular weight determination

13-3b   Chemical Ionization (CI)

Chemical ionization is a much more controllable method of ionization than EI. It involves reaction of the neutral analyte (M) with an ion generated by a high pressure EI process. A variety of types of molecular ion can be formed, including the protonated molecule $(M+H)^+$, the molecular radical cation $(M^{+\cdot})$, the molecular radical anion $(M^{-\cdot})$, metal ion adducts and many others. Some typical ionization reactions are charge exchange (eq. 13-2), electron capture (eq. 13-3), proton transfer (eq. 13-4) and cationization by adduct formation with various ions (eq. 13-5):

$$\text{Charge exchange:} \quad M + Ar^{+\cdot} \longrightarrow M^{+\cdot} + Ar \tag{13-2}$$

$$\text{Electron capture:} \quad M + e^- \text{ (slow)} \longrightarrow M^{-\cdot} \tag{13-3}$$

$$\text{Proton transfer:} \quad M + CH_5^+ \longrightarrow (M+H)^+ + CH_4 \tag{13-4}$$

$$\text{Adduct formation:} \quad M + TiCl_2^+ \longrightarrow (M+TiCl_2)^+ \tag{13-5}$$

In addition to the choice of the *type* of molecular ion, control over the degree of *internal excitation* of the molecular ion—and hence over the degree of fragmentation—is readily achieved. These features provide great power in establishing molecular weights and in recording spectra that contain structurally diagnostic fragment ions. These choices are available by simply selecting the appropriate reagent gases. For example, a soft ionizing reagent can generate from the intact molecule abundant ions that undergo little or no fragmentation and allow molecular weight determination. Such control is not available in EI or in the ionization methods to be discussed later. On the other hand, CI, like EI, is limited to samples that are thermally stable and have significant vapor pressure.

The primary ionization event in a CI ion source is electron ionization, but it is the reagent gas, present at a pressure that is several orders of magnitude greater than that of the sample, which is ionized. By operating at pressures in the range of 1 torr, or less commonly by working at a lower pressure but increasing the residence time of ions in the source, numerous ion-molecule collisions occur. These collisions produce reagent ions characteristic of the reagent gas chosen. For example, molecular hydrogen is ionized by electron impact to give $H_2^{+\cdot}$, and this primary product undergoes ion-molecule reactions with neutral H_2 to give H_3^+, which is the reagent ion for H_2 chemical ionization. Some proton transfer reagent gases and their corresponding reagent ions are listed in Table 13-7.

The reagent ions undergo ion-molecule reactions with the neutral analyte molecules (M) to produce analyte ions of the types listed in Table 13-7. This variety of

TABLE 13-7

Some CI Reagent Gases and Reagent Ions (Proton Transfer Reactions)

Reagent Gas	Reagent Ion	Analyte Ion	Comment
H_2	H_3^+	$(M+H)^+$	Very energetic protonating agent; produces considerable fragmentation
CH_4	CH_5^+, $C_2H_5^+$, $C_3H_5^+$	$(M+H)^+$, $(M+C_2H_5)^+$, $(M+C_3H_5)^+$	Energetic protonating agent, forms adduct ions
$i\text{-}C_4H_{10}$	$C_4H_9^+$	$(M+H)^+$, $(M+C_4H_9)^+$	Mild protonating agent; ionizes all nitrogen bases
NH_3	NH_4^+	$(M+NH_4)^+$, $(M+H)^+$	Selective ionization, little fragmentation
NH_3/CH_4	NH_4^+	$(M+H)^+$	Selective protonating agent
CH_3ONO/CH_4	CH_3O^-	$(M-H)^-$	Mild proton abstraction reagent
NF_3	F^-	$(M-H)^-$	Proton abstraction reagent

ionized molecules allows molecular weight determinations to be cross checked. Note that some reagent gases produce a mixture of reagent ions, which then react independently with the analyte molecules. For example CH_4 yields CH_5^+, $C_2H_5^+$, and $C_3H_5^+$; CH_5^+ is a proton transfer agent, whereas $C_2H_5^+$ tends to abstract H^- from neutral molecules or to form adduct ions $(M+C_2H_5)^+$.

The CI method is versatile because it depends on chemical reactions that yield particular products and deposit a controlled amount of energy into them. This process allows selective ionization of particular compounds present in a mixture. Next we will focus on just one class of CI reagents: those that react by proton transfer.

CI Based on Protonation Reactions. Only exothermic or thermoneutral ion-molecule reactions normally occur, so that CI is a selective ionization method. A weak gas phase acid like NH_4^+, for example, can protonate strong bases such as alkylamines, but weaker bases such as ethers are not ionized (eqs. 13-6 and 13-7):

$$NH_4^+ + R_3N \longrightarrow NH_3 + R_3NH^+ \qquad \Delta H_{reaction} \text{ (negative)} \qquad (13\text{-}6)$$

$$NH_4^+ + ROR \not\longrightarrow NH_3 + RO^+HR \qquad \Delta H_{reaction} \text{ (positive)} \qquad (13\text{-}7)$$

The thermochemistry of the ionization reaction also controls the amount of internal energy lodged in the protonated molecule. In this way, CI can cause extensive fragmentation, after a highly excited ion (a hard ionization event) is formed. Alternatively, a mildly exothermic reaction can be chosen so that a soft ionization process occurs and generates only the intact ionized molecule.

Two extremes among proton transfer reagent gases are H_2 and NH_3. The former is a universal and hard proton transfer reagent, the latter a selective and soft reagent. To see why, consider the reagent ions created from the two gases, H_2 and NH_3 (eqs. 13-8 and 13-9).

$$H_2 + e^- \longrightarrow H_2^{+\cdot} + 2e^-$$
$$H_2^{+\cdot} + H_2 \longrightarrow H_3^+ + H^\cdot \qquad (13\text{-}8)$$
$$H^\cdot + H_2^{+\cdot} \longrightarrow H_3^+$$

and

$$NH_3 + e^- \longrightarrow NH_3^+ + 2e^-$$
$$NH_3^{+\cdot} + NH_3 \longrightarrow NH_4^+ + NH_2^\cdot \qquad (13\text{-}9)$$

The proton affinity (PA) is an enthalpic quantity that measures the strength of binding of the proton to a neutral molecule (eq. 13-10). The PA of H_2 is low, only

$$PA \equiv -\Delta H_{reaction} \text{ for the reaction: } H^+ + M \longrightarrow (M+H)^+ \qquad (13\text{-}10)$$

422 kJ mol^{-1} (101 kcal mol^{-1}), compared to 854 kJ mol^{-1} (204 kcal mol^{-1}) for NH_3. Hence proton transfer from H_3^+ to NH_3 is exothermic by the difference between these values, or 432 kJ mol^{-1} (103 kcal mol^{-1}). Organic amines have proton affinities on the order of 900 kJ mol^{-1} (215 kcal mol^{-1}) and can be ionized using either hydrogen or ammonia as reagent gases, but with vastly different consequences (eqs. 13-11 and 13-12):

$$M + H_3^+ \rightarrow MH^+ + H_2 \qquad \Delta H_{reaction} = -(900 - 422 \text{ kJ mol}^{-1}$$
$$\text{or } 215 - 101 \text{ kcal mol}^{-1}) \quad (13\text{-}11)$$
$$= -478 \text{ kJ mol}^{-1} \text{ or } -113 \text{ kcal mol}^{-1}$$

$$M + NH_4^+ \rightarrow MH^+ + NH_3 \qquad \Delta H_{reaction} = -(900 - 854 \text{ kJ mol}^{-1}$$
$$\text{or } 215 - 204 \text{ kcal mol}^{-1}) \quad (13\text{-}12)$$
$$= -46 \text{ kJ mol}^{-1} \text{ or } -11 \text{ kcal mol}^{-1}$$

The excess energy of the protonation reaction using H_2 reagent is almost 500 kJ mol^{-1} (120 kcal mol^{-1}). It lodges principally in MH^+ and causes extensive dissociation. In the case of NH_3 reagent, the exothermicity is less than 50 kJ mol^{-1} (12 kcal mol^{-1}). This value is less than typical bond energies and is insufficient to cause dissociation. As a result, H_2 is appropriately described as a hard ionization reagent, which produces extensive fragmentation and leaves few or no intact protonated molecules to be observed. Ammonia is a soft reagent and produces principally the intact protonated molecule of the organic amine. Proton affinities of a variety of simple organic compounds are listed in Table A-3. With these data, it is possible to predict whether a particular analyte–reagent ion combination will lead to mild or extensive fragmentation.

Figure 13-7 illustrates, for a particular compound, the consequences of hard and soft ionization made possible through choice of appropriate CI reagent gases. The reagent gas isobutane yields protonated isobutene (m/z 57) as reagent ion. The proton affinity of isobutene is 802 kJ mol^{-1} (192 kcal mol^{-1}), which is lower than that of phenyl propyl ketone (estimated PA 870 kJ mol^{-1} or 208 kcal mol^{-1}), making proton transfer from $C_4H_9^+$ exothermic. Relatively little excess energy is expected to be deposited in the protonated molecule, and the experimental results show that little fragmentation occurs. On the other hand, methane has a PA of 544 kJ mol^{-1} (130 kcal mol^{-1}), making proton transfer from CH_5^+ highly exothermic. The excess energy, 326 kJ mol^{-1} (78 kcal mol^{-1}), is lodged in the protonated ketone, and extensive dissociation is expected and observed. One should note that the fragment ions are predominantly even-electron (odd-mass) species, which arise by such simple processes as cleavage of the C—C bond α to the carbonyl group to give an ion m/z 105.

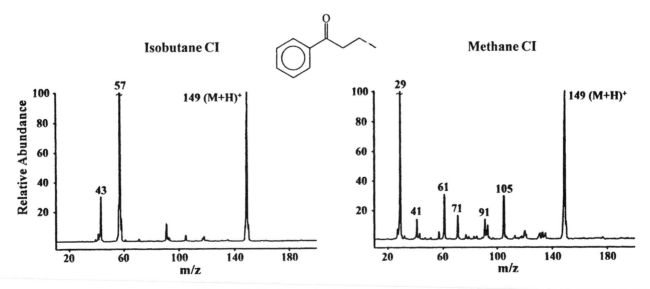

Figure 13-7 Control of degree of fragmentation of phenyl propyl ketone made possible by choice of chemical ionization reagent gas.

TABLE 13-8

Reactions Used in CI

Reaction	Reagent*	Reagent Ion*	Product	Thermochemical Property
Proton transfer	$i\text{-}C_4H_{10}$	$C_4H_9{}^+$	$(M+H)^+$	Proton affinity
Charge exchange	Ar	$Ar^{+\cdot}$	$M^{+\cdot}$	Ionization energy
Electron capture	CH_4	e^-	$M^{-\cdot}$	Electron affinity
Cl^- attachment	$CHCl_3/CH_4$	Cl^-	$(M+Cl)^-$	Cl^- affinity
Adduct formation	Biacetyl	CH_3CO^+	$(M+CH_3CO)^+$	Acetyl affinity
Cluster ion formation	Py	PyH^+	$(M + Py + H)^+$	Molecular pair affinity

* Examples given are typical cases; many other choices can be made.

Other CI Reactions. So far only proton transfer has been considered as the basis for chemical ionization, and only reactions leading to positively charged ions have been discussed. Although proton transfer reactions such as those shown in Table 13-7 are indeed the most common CI processes, a larger group is collected in Table 13-8.

The versatility of CI, especially the control over the extent of molecular ion formation, applies to all these processes, including the case of charge exchange. In charge exchange CI, the reagent ion is a radical cation, and the ion-molecule reaction produces the radical cation of the analyte, $M^{+\cdot}$. Just as in protonation, the reaction exothermicity is deposited in the products. The greater this energy, the more fragmentation it causes. Charge exchange CI can be performed using soft or hard reagents, illustrated by the cases of $Ar^{+\cdot}$ and $CS_2{}^{+\cdot}$. The ionization energy of CS_2 is only 10.1 eV, and this is the amount of energy that is available to ionize and excite a typical organic compound. If this has ionization energy 10.0 eV, then the charge exchange reaction is only weakly exothermic (eq. 13-13).

$$CS_2{}^{+\cdot} + M \longrightarrow CS_2 + M^{+\cdot} \tag{13-13}$$

$\Delta H_{\text{reaction}} = \Delta H_{\text{f}}(\text{products}) - \Delta H_{\text{f}}(\text{reactants}) = IE(M) - IE(CS_2) = -0.1$ eV (9.7 kJ/mol or 2.3 kcal mol^{-1}). By contrast, Ar has a much higher ionization energy (15.8 eV), and even after it has ionized a typical organic compound of ionization energy 10 eV, there is still approximately 6 eV of energy left in the products, which can cause extensive fragmentation (eq. 13-14).

$$Ar^{+\cdot} + M \longrightarrow Ar + M^{+\cdot *} \longrightarrow \text{fragments}$$
$$\Delta H_{\text{reaction}} = 5.8 \text{ eV } (559 \text{ kJ mol or } 134 \text{ kcal mol}^{-1}) \tag{13-14}$$

Obviously, Ar is a hard charge exchange reagent and CS_2 a soft one. The product ion generated by charge exchange CI is, of course, exactly the same type of radical cation formed by electron ionization (eq. 13-1). In EI, some ions have large internal energies whereas others do not. The wide range of internal energies deposited in EI often leads to the observation of abundant fragment ions as well as an intact molecular ion in the same spectrum. In charge exchange CI, one can choose to form either low- or high-energy molecular ions. The options are illustrated in Figure 13-8, which compares mass spectra of benzene recorded using $N_2{}^{+\cdot}$ and $CH_3OH^{+\cdot}$ reagent ions with the 70 eV EI mass spectrum. Whereas $N_2{}^{+\cdot}$ causes extensive dissociation, $CH_3OH^{+\cdot}$ causes none. The EI spectrum shows both features.

Negative Ion Formation in CI. So far, we have discussed CI processes that generate positive ions. Negative ions are formed readily in CI experiments. There are two principal modes of negative ion formation, those that depend on ion-molecule reactions (eq. 13-15) and those that involve electron-molecule reactions (eq. 13-16). The underlying physical and chemical processes are different in the two cases, as is the nature of the molecular ions. The differences are illustrated for the environmentally important compound, 2,4,6-trichlorophenol, which can form negative ions by both ion-molecule and electron-molecule reactions.

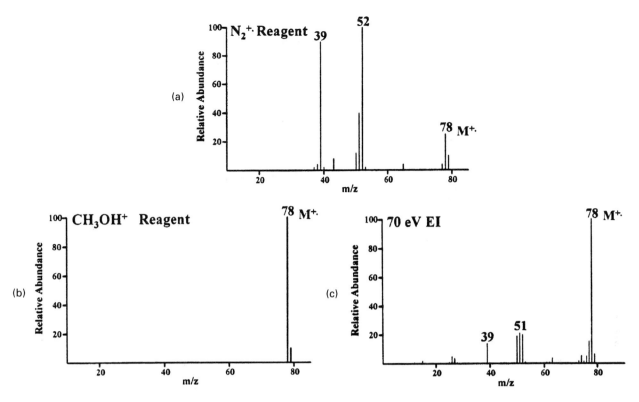

Figure 13-8 Charge exchange CI mass spectra of benzene (IE 9.2 eV) recorded using hard and soft reagents (N$_2$ and CH$_3$OH as reagent gases), compared with the 70 eV EI mass spectrum. The molecular ion (M$^{+\cdot}$) is generated with a large amount of internal energy from N$_2^{+\cdot}$ (IE 15.8 eV) compared with CH$_3$OH$^{+\cdot}$ (IE 10.9 eV).

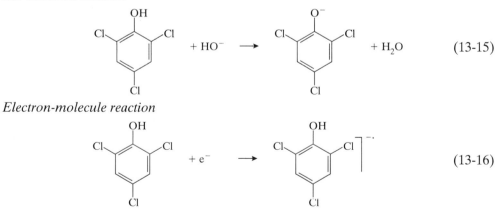

Ion-molecule reaction

$$\text{(13-15)}$$

Electron-molecule reaction

$$\text{(13-16)}$$

Proton abstraction reagents such as CH$_3$O$^-$ and F$^-$, can be generated by electron ionization of appropriate reagent gases (Table 13-7). They deprotonate the analyte in an ion-molecule reaction to yield (M−H)$^-$ (eq. 13-15). Electron capture (eq. 13-16) is favored when the molecule has a positive electron affinity (M$^{-\cdot}$ is thermodynamically stable relative to M+e$^-$), as is the case for 2,4,6-trichlorophenol. Electron capture requires low-energy electrons, so an inert gas such as methane (or ammonia, see above) is often used to thermalize the electrons in a CI source. The low energy of the electrons is also the reason why electron capture ionization cannot be performed in an EI source, where the low pressure means that electrons have high energies. The thermochemical condition for electron capture is that the analyte have a positive electron affinity, namely, that electron capture be exothermic. Exceptions are quite well known.

Electron capture occurs in the negative ion mass spectrum of 1,2,3,4-tetrachloro-naphthalene (Figure 13-9). The spectrum shows a set of ions, *m/z* 262, 264, 266, 268, whose characteristic intensity ratios and spacing point to a multiply halogenated compound. In fact, the characteristic isotopic abundance pattern for four chlorine atoms (see Section 15-1) is reproduced by this set of ions. A formula of $C_{10}H_4Cl_4$ is consistent with this molecular weight and isotopic abundance pattern, that is, the molecular ion is generated by electron attachment to the neutral molecule and has the formula, $C_{10}H_4Cl_4^{-\cdot}$.

Desorption chemical ionization (DCI) is a hybrid between CI, which requires that the sample be vaporized, and the DI techniques, discussed later, which require energetic bombardment of the condensed phase sample. DCI works well for both positive and negative ions. The advantage of DCI is that samples of low volatility can be examined using simple apparatus. The data in Figure 13-9 were recorded by DCI. Its applicability to nonvolatile samples is illustrated by the case of thymidylyl(3'-5')thymidine methyl phosphotriester, the positive and negative DCI spectra of which are shown in Figure 13-10. Note that the reagent used in this experiment, NH_3, can serve to thermalize electrons and so act as an electron capture reagent, or it can serve as a proton

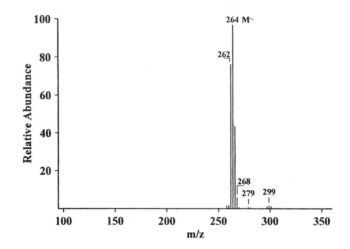

Figure 13-9 Electron attachment mass spectrum of 1,2,3,4-tetra-chloronaphthalene (methane reagent gas) showing the intact molecular ion, $M^{-\cdot}$.

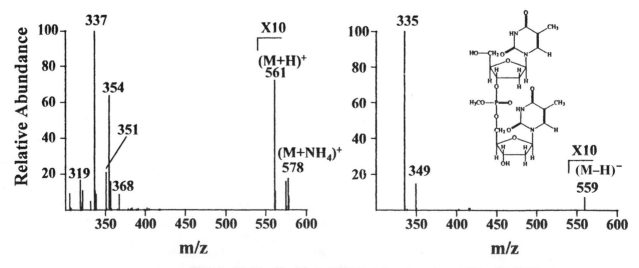

Figure 13-10 Positive and negative ion desorption CI (DCI) mass spectra of a methylated dinucleotide, ammonia reagent gas, showing various forms of the molecular ion. (From I. Isern-Flecha et al., *Biomed. Environ. Mass Spectrom.,* **14,** 17 [1987]. Reproduced with the permission of John Wiley & Sons Ltd., Chichester, UK.)

abstraction agent with suitable analytes due to formation of NH_2^-. The negative ion spectrum is much simpler than the corresponding ammonia positive ion spectrum and it yields lower detection limits. Both positive and negative ion spectra show the formation of a mononucleotide fragment ion as a major cleavage product. This is one example of the tendency of biopolymers to undergo structurally diagnostic fragmentations upon which sequence determinations can be based (compare Section 14-6).

Negative ion formation allows selective ionization. Different chemical features confer selectivity compared with those that allow selectivity in positive ion formation. For example, only compounds with acidic groups of appropriate strength give up a proton to a reagent anion. Thus ionization using the amide anion (NH_2^-) only occurs when the acidity of the analyte is greater than that of ammonia (gas phase acidity can be approximated by the enthalpic quantity, $\Delta H_{acid} \equiv$ heterolytic bond strength of AH, see Table A-4; note that higher ΔH_{acid} values correspond to weaker acids). In addition to the ability to ionize selected types of compounds present in mixtures, negative ion CI allows control over the degree of fragmentation. This result is again because more exothermic reactions lead to a greater degree of excitation and fragmentation of the product ions.

Ion-molecule and electron-molecule collisions, the two principal types of processes used to form negative ions, find analogies in the processes used to generate positively charged ions. Thus (1) radical anions M^- are the counterparts of radical cations, M^+, and (2) species such as $(M-H)^-$, obviously are the counterpart of $(M+H)^+$. The electron capture process, however, is fundamentally different from electron ionization to yield radical cations. It forms the basis for an ionization method that is extremely sensitive and selective.

Formation of radical cations is a common process, but attachment of an electron to a molecule is energetically favorable only for certain classes of compounds. Electron capture is therefore a selective method of ionization. This selectivity is valuable when measuring compounds containing any combination of P, S, Cl, and F, as well as some classes of aromatic compounds, all of which can be ionized selectively in the presence of large concentrations of other types of organic compounds. Many compounds of environmental interest (including many pesticides and herbicides) undergo electron capture. Certain groups, notably fluorinated groups such as pentafluorobenzyl, have high electron capture cross sections and hence are introduced into molecules for this purpose. Chemical derivatization in this fashion is often worthwhile if trace analysis is required. In addition to its high selectivity, electron capture also is more sensitive than most other ionization methods. Electrons have high velocities, which increase the likelihood of electron-molecule collisions leading to electron capture and increase the M^- abundance. Detection limits often are one or two orders of magnitude lower than those for other positive or other negative ion CI experiments.

Figure 13-11 summarizes some chemical ionization processes and contrasts the rich CI chemistry with the much simpler EI process. In addition, a summary of some of

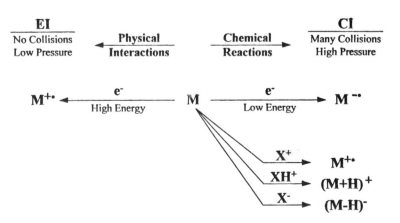

Figure 13-11   Contrasts between EI and CI, emphasizing greater versatility of the latter in terms of the type of molecular ion formed and the control achieved over its internal energy.

TABLE 13-9

Some Reagents Used in Negative Ion and Electron Capture CI

Reagent Gas	Reagent Ion/e^-	$PA(X^-)$* $kJ\ mol^{-1}$
NH_3	NH_2^-/e^-	1689
CH_4	e^-	

CH_4 and N_2O
  or CH_4, N_2O and He
  or H_2, N_2O and He
	OH^-	1635

N_2O **or** $N_2O + N_2$	$O^{-\cdot}$	1599
CH_4 (at 1 torr) + 0.1% methyl nitrite	CH_3O^-	1592
$CH_3O^- + CH_3CN$	CH_2CN^-	1560
NF_3 (0.1 torr) **or** CHF_3	F^-	1554
O_2 (1 torr, Townsend discharge) **or** mixture of (O_2 and H_2)	$O2^{-\cdot}$	1476
$F^- + CH_3CH_2CN$	CN^-	1469
$OH^- + CH_3COOCH_3$	$CH_3CO_2^-$	1459
CH_2Br_2	Br^-	1356
CH_2Cl_2 **or** $CHCl_3$ **or** CF_2Cl_2 **or** ($CH_2Cl_2 + CH_4$)	Cl^-	1395

* Proton affinity of X anion (4.18 kJ mol^{-1} = 1 kcal mol^{-1}).

TABLE 13-10

Summary of Characteristics of CI

Sample must be volatile	Limited to sample molecular weight approximately 800 Da or less
Hard or soft ionization	Molecular weight determination or fragmentation for structural information
Variety of ionization processes	Multiple checks of molecular weight, structural features
Universal or selective method	Either is available, depending on choice of reagent gas
Negative ions are readily formed	Additional tool for molecular weight and structural and thermochemical measurements

the reagent gases used in negative ion CI and electron capture experiments is given in Table 13-9.

The CI Ion Source. CI is performed in a high-pressure ion source adapted from the EI type (see Figure 13-6). Ion residence times and ion currents also are similar to EI values, despite the difference in ionization mechanism. The sample may be introduced via a chromatograph, a direct insertion probe, a membrane, or a molecular leak inlet system (compare Figure 13-2). In the DCI variant on the CI technique, the sample is deposited from solution onto a metal filament, which is introduced into the CI plasma and then heated very rapidly (100's ° C/s^{-1}). The method is not very convenient, since the sample lasts only a very short time (seconds). However, compounds of low volatility often yield spectra in this way that cannot be recorded by conventional CI methods.

A summary of some of the characteristics of CI is given in Table 13-10.

13-3c Desorption Ionization (DI) Including Matrix-Assisted Laser Desorption Ionization (MALDI)

Introduction. The ionization methods discussed so far, EI and CI, are limited to volatile samples. The DI methods remove this limitation and allow mass spectra to be recorded for samples in the condensed phase. Desorption ionization therefore provides one means to ionize nonvolatile samples with high molecular weight. An alternative is provided by the spray ionization methods discussed in Section 13-3d.

TABLE 13-11

Procedures Used in Desorption Ionization

Ionization Method	Energy Source	Flux	Matrix	Mass Analyzer	Comments
Static secondary ion MS (SIMS)	keV ions	10^{-10} A cm^{-2} (Ar$^+$, Cs$^+$, etc.)	None, solid	Any	Surface sensitive, nondestructive, low signal
Liquid SIMS	keV ions	10^{-6} A cm^{-2} (10^{13} ions cm^{-2} s^{-1}) (Cs$^+$, etc.)	Liquid	Any	Higher, longer lasting signal
Fast atom bombardment (FAB)	keV atoms	10^{13} atoms cm^{-2} s^{-1} (Xe atoms)	Liquid	Any	High, long-lasting signal, high background
Plasma desorption (PD)	MeV ions	10^3 particles cm^{-2} s^{-1}	Nitrocellulose	Time-of-flight	High ionization efficiency, low signals
Laser desorption (LD) and matrix-assisted laser desorption (MALDI)	Photons	$\geq 10^6$ watt cm^{-2}	Solid matrix absorbs radiation in MALDI	Usually time-of-flight	CW lasers can cause thermal degradation

Release of the sample into the vapor phase in the ionic form is often facilitated by its admixture into a suitable solid or liquid *matrix*, which plays a crucial role in the DI methods and profoundly influences the type and number of ions formed. Much of the art in this method is concerned with choice of matrix and understanding of its role. Energy can be supplied to the condensed phase sample by a variety of procedures, and there are actually a family of related desorption methods. Table 13-11 summarizes some of them, including the two most widely used DI techniques: (1) FAB and the closely related method of liquid SIMS and (2) matrix-assisted laser desorption. All DI techniques rely on deposition of relatively large amounts of energy into the sample in short times—described by the term *energy prompt*. The specific methods use various sources of energy: energetic primary ions (secondary ion mass spectrometry), energetic atoms (fast atom bombardment), nuclear fission fragments (plasma desorption), photons (laser desorption), and very rapid heating (desorption chemical ionization).

The following discussion is broken in two main sections. The first concentrates on FAB–liquid SIMS and uses this method to illustrate many general characteristics of all the DI methods, including matrix and internal energy effects. The second section is devoted to the MALDI method and illustrates its capabilities.

FAB–Liquid SIMS: The Prototypical DI Method.   In the FAB–liquid SIMS method, the sample is dissolved in a lower vapor pressure matrix and the mixture is bombarded by an ion beam (SIMS) or an atom beam (FAB). The diluted analyte is protected from direct impact by the energetic beam. This procedure allows high primary ion or neutral fluxes to bombard the sample and so increases ionization efficiency while maintaining relatively soft ionization conditions. Abundant protonated or deprotonated molecules characterize the spectra and provide molecular weight information. Although often considered to be a soft ionization technique, numerous fragment ions are typically generated in FAB–liquid SIMS. Matrix-derived ions occur throughout the mass spectrum, especially at lower mass-to-charge ratios, and can obscure sample ions (for example, clusters involving the common liquid matrix glycerol are abundant in FAB). For these reasons, the structural information that can be obtained by fragmentation is normally sought by using MS–MS experiments (see Section 13-4g). An ion characteristic of the intact molecule is isolated and dissociated in a controlled fashion to provide structural information. Even though FAB–liquid SIMS is only moderately sensitive, it provides relatively large ion currents that last for many minutes. Detection limits are often in the nanomole range, and compounds with molecular weights of a few thousand Da can be readily examined, yielding characteristic singly-charged ions.

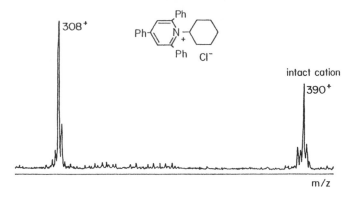

Figure 13-12 DI mass spectrum of a simple quaternary ammonium salt showing the intact cation, *m/z* 390, in an ammonium chloride matrix (1:50,000) using Ar⁺ of 4.5 keV. (From K.L. Busch et al., *Anal. Chem.,* **55,** 1157 [1983]. Copyright ©1983 American Chemical Society.)

As an introduction to the DI family of ionization techniques, it is instructive to examine the SIMS spectrum of a simple quaternary ammonium salt (Figure 13-12). The spectrum shows the intact cation, *m/z* 390. This ion is produced by *direct desorption of the precharged cation* (C^+). A fragment ion occurs at *m/z* 308 due to loss of neutral cyclohexene from this cation.

Positively and negatively charged ions are generated efficiently in DI by cation and anion attachment, proton transfer, and other ion-molecule reactions already described in the discussion of CI. Preionized compounds such as organic salts are desorbed directly into the gas phase by the sudden input of energy into the sample. The variety of methods of generating molecular ions representative of the compound of interest is illustrated by the partial mass spectra of the simple imine shown in Figure 13-13. Treating the matrix with acid forms the protonated ion and provides protons for gas phase attachment to the molecule. Correspondingly, treating the matrix with a base yields predominantly the ion $(M-H)^-$, whereas an alkali metal salt such as NaCl facilitates $(M+Na)^+$ formation. Silver salts yield adduct ions that are readily recognized by their characteristic isotopic signatures (Ag 107 and 109). In the absence of deliberately added salts, adventitious ions, especially Na^+, H^+, and K^+, are involved in cationization. The analyte can also be derivatized by standard chemical methods and converted to the pyridinium salt and, if the methyl salt is analyzed, the ion $(M+CH_3)^+$ is observed.

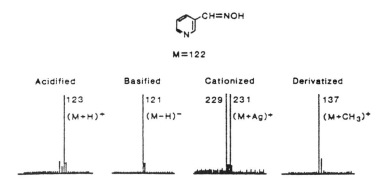

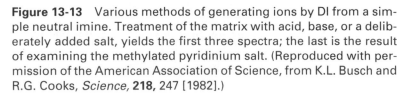

Figure 13-13 Various methods of generating ions by DI from a simple neutral imine. Treatment of the matrix with acid, base, or a deliberately added salt, yields the first three spectra; the last is the result of examining the methylated pyridinium salt. (Reproduced with permission of the American Association of Science, from K.L. Busch and R.G. Cooks, *Science,* **218,** 247 [1982].)

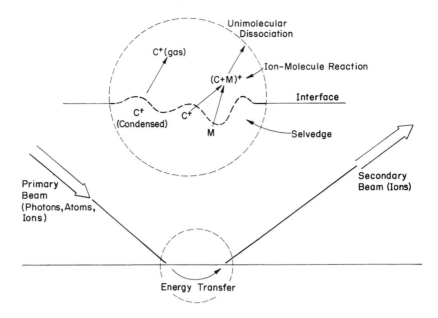

Figure 13-14 Ionization in DI, showing the disturbed interfacial region known as the *selvedge;* M is a neutral molecule of analyte, C⁺ is a preformed cation, such as adventitious Na⁺, which attaches to M to generate the ion-molecule reaction product (M+C)⁺. Alternatively, the analyte may be present as a salt, such as a protonated peptide in acidic solution, and C⁺ then represents this form of analyte, that can be directly released into the vacuum as C⁺.

Mechanism of DI. The various DI techniques all rely on deposition of relatively large amounts of energy into the sample in short times. They differ in the form in which the energy is delivered. Although the spectra recorded certainly depend on the method chosen, on the amount of energy supplied, and on the sample preparation, the similarities among DI spectra are such as to suggest that the key processes are those that follow energy input. Figure 13-14 provides a schematic description of the desorption event.

Energy is absorbed by the matrix in a localized region of the surface. As a result of this local, very rapid heating, the analyte acquires translational energy without being internally excited to a corresponding extent. The translational excitation occurs by a series of momentum transfer collisions in FAB–liquid SIMS and by the expansion of the vaporized matrix in MALDI and accounts for the release of the analyte molecules from the surface of the condensed phase sample into vacuum. To explain ionization, two distinct mechanisms can be distinguished in DI. In the case of salts, the ions are preformed, and their transfer from the condensed phase into the gas phase yields gaseous ions. For example, the DI mass spectrum of quaternary ammonium salts display the intact cation in the positive ion spectrum and the counterion in the negative ion mass spectrum. In both cases these observations are due to the direct ion desorption into the gas phase by the sudden energy input. On the other hand, neutral analytes must be complexed with an ion, frequently a metal ion or a proton, to be observable as gas phase ions. The ion with which they complex may be deliberately added to the matrix to serve in this role (Figure 13-13). Note the connection between ionization of a neutral analyte in DI and in conventional chemical ionization; in both situations ion-molecule reactions are involved. Like CI too, the DI methods are ambiphilic, both positively and negatively charged ions are generated almost equally efficiently. The DI ionization processes are summarized in Table 13-12.

Internal Energy Deposition and Fragmentation in DI. The delivery of intact molecules from condensed phase samples into the gas phase as ions is the objective and indeed the great success of DI. The fact that this process occurs without the need to vaporize the sample is what distinguishes this method from EI and CI. Depending on

TABLE 13-12

Ionization Mechanisms in Desorption Ionization

Direct emission	C^+(condensed) $\longrightarrow$ C^+ (gas)
	A^-(condensed) $\longrightarrow$ A^- (gas)
Ion-molecule reactions	$C^+ + M$ (selvedge) $\longrightarrow$ $(C+M)^+$ (gas)
	$A^- + M$ (selvedge) $\longrightarrow$ $(A+M)^-$ (gas)
	$A^- + MH$ (selvedge) $\longrightarrow$ $(A+MH)^-$ (gas)

the experimental conditions chosen for DI, the desorbed ions contain different amounts of internal energy. Fragmentation can be controlled to some extent by controlling the energy input, such as the flux or kinetic energy of bombarding particles. The population of ions formed in a typical DI experiment has a broad range of internal energies, so that some ions appear as intact molecular ions and others as fragment ions. DI is therefore simultaneously a soft and a hard ionization method.

The formation of both intact molecular ions and characteristic fragment ions in DI can be illustrated by the simple case of the *N*-cyclohexylpyridinium salt shown in Figure 13-12. This compound gives a weak spectrum with considerable fragmentation in the absence of a matrix. However, when examined in a matrix, the energy provided by the primary beam is transferred to the matrix, which is present in overwhelmingly greater amounts. The absorption of energy by the matrix liberates intact molecules of analyte. Not only does the matrix increase ion yields, but it decreases fragmentation and avoids intermolecular reactions between analyte molecules that might otherwise complicate the spectrum. The simplicity of matrix-assisted DI, and the fact that it is simultaneously a soft ionization method while also providing energy to cause fragmentation of some molecular ions, is illustrated in this spectrum. The high-mass region is dominated by the intact cation, m/z 390, while the fragment ion due to elimination of neutral cyclohexene and formation of the protonated pyridine, m/z 308, is the base peak.

Matrices in DI. The choice of matrix is key to successful DI experiments. Ideally, the matrix should interfere only minimally in the mass spectrum, analyte molecules should be isolated from each other (hence the common choice of liquid matrices in FAB–liquid SIMS), and the matrix should not react with the sample except to provide (or accept) protons or other ionic species to generate the molecular ion. The ideal matrix for DI should be a material that strongly absorbs the energy provided, contributes few ions to the mass spectrum, interacts with the analyte to produce ions from ionic or neutral compounds, and effectively transfers energy to the ionized analyte to cause its release into vacuum. In addition, the analyte should be mobile in the matrix so that damaged material, that has been chemically altered, should not be resampled but returned to the bulk. There are also physical properties, including vacuum compatibility, that must be satisfied. It should not be surprising that matrices sometimes work particularly well for one class of compounds but not for another. A number of different materials have been used as matrices in the various DI methods; the most important examples for FAB and MALDI experiments are listed in Table 13-13.

Efficiency of DI. The sensitivity of DI for analysis of ionic compounds is much greater than for neutral compounds, because ionization in the former case simply requires a phase change (eq. 13-17), whereas an ion-molecule association reaction is

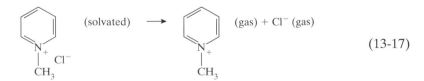

$$(13\text{-}17)$$

required in the latter case. For this reason, neutral compounds of interest are often derivatized before analysis to improve ionization efficiency. For example, simple acidi-

TABLE 13-13

Some Matrices Used in Desorption Ionization

FAB–liquid SIMS

Glycerol	General purpose
1:1 Mixture of dithiothreitol and dithioerythritol	Most useful
1:1 Mixture of glycerol and thioglycerol	For peptides
Diethanolamine	When a basic matrix is needed
m-Nitrobenzyl alcohol	For aromatics and less polar compounds

Matrix-assisted laser desorption ionization (MALDI)

Sinapinic acid	For proteins with a molecular weight >10,000 Da
α-Cyano-4-hydroxycinnamic acid	For peptides and proteins with a molecular weight <10,000 Da
2,5-Dihydroxybenzoic acid	For carbohydrates, aromatics, and small organic compounds
Mixtures of 3-hydroxypicolinic acid, picolinic acid, and ammonium citrate	For oligonucleotides

fication can increase $(M+H)^+$ production, and examination of materials from more basic solutions can promote $(M-H)^-$ formation. When such procedures are not available, it is often useful to add a characteristic ion to the matrix to generate easily recognized adduct ions. Derivatization procedures used for desorption ionization have been referred to as "reverse derivatizations" because they make the analyte more ionic, compared with conventional derivatization, which makes the analyte more volatile for EI or CI. The conversion of pyridine to the methylpyridinium chloride derivative, before mass spectrometry, exemplifies this process. Girard's reagent is used to derivatize steroids, primary amines can be converted to pyrylium salts, and many other acid–base and charge transfer reagents can be used to convert the analyte into a charged derivative.

Applications to Biological Compounds. The success of FAB–liquid SIMS in the characterization of peptides in the early 1980s was an important impetus that launched many contemporary applications of mass spectrometry in the biomedical sciences. Problem 13-3 illustrates an application of this type.

Matrix-Assisted Laser Desorption Ionization (MALDI). MALDI is the newest of the DI methods. Although it shares many characteristics with other DI techniques, it deserves special attention because of its remarkable efficiency in producing intact molecular ions (often $[M+H]^+$, $[M+Na]^+$) of large biological compounds. An even more remarkable characteristic of MALDI is its extraordinary sensitivity. Total amounts of sample loaded onto the target surface are often in the picomole to femtomole range and much of this sample is not used. The experiment yields singly charged ions from compounds having molecular weights in excess of 100,000 Da. This high range has led to the choice of time-of-flight (TOF) mass analyzers (Section 13-4b) for use with this ionization technique because of their compatibility with high-mass ions and pulsed ion production. The speed and simplicity of operating laser–TOF instruments is also noteworthy.

MALDI Matrices. The MALDI matrix compounds are usually organic acids that have strong electronic absorption features in the region of the laser wavelength used. Samples are usually present in very high dilution in the matrix, often 10^4 or more, to prevent analyte-analyte interactions. Proteins and peptides are often successfully examined using cinnamic acid derivatives, whereas nucleic acid derivatives are usually more successfully studied using picolinic acid matrices. Table 13-13 summarizes a number of matrices used in MALDI. The pK_a of the matrix correlates with the degree of fragmentation observed—highly acidic matrices tend to cause more fragmentation. An unwelcome characteristic is that the analyte-matrix mixture must yield good crystals, presumably those in which intimate mixing of the analyte and matrix occurs. Some experience in preparing samples may be required.

Ionization Mechanisms in MALDI. The dilute analyte is present in a solid matrix chosen to absorb radiation at the laser wavelength, often 308 nm, but occasionally in the IR range. The analyte, which is present in high dilution in the matrix, is not energized directly (the high dilutions and indirect ionization are a feature of the gas phase CI as well as the liquid phase DI experiments). As the matrix molecules vaporize, the intermolecular forces that bind the analyte break and the analyte molecules desorb into the gaseous state. The analyte ions leave the surface with significant kinetic energies, entrained as they are in a microsupersonic molecular beam of expanding matrix vapor. The analyte may be precharged, such as a salt, and the intact cation, C^+, may simply be transferred as an ion from the solid to the vapor state on laser irradiation of the matrix (eq. 13-18). Alternatively, a neutral analyte may be ionized through ion-molecule

$$C^+ (s) + energy \longrightarrow C^+(v) \qquad (13\text{-}18)$$

reactions occurring in the energized selvedge. In broad outline, the MALDI experiment and ionization processes are very similar to those encountered in FAB–liquid SIMS, even though the method of delivering energy to the sample is very different.

Applications of MALDI to Biomolecules. Figure 13-15 illustrates the determination of molecular weights of large biomolecules by MALDI. The sample, a heme protein having a molecular weight (chemical average) of 9853.2 Da, was analyzed using α-cyano-4-hydroxycinnamic acid and irradiated with a nitrogen laser (337 nm). Data were acquired in just a few seconds. The spectrum is dominated by the protonated molecule $(M+H)^+$, m/z 9849 Da/charge (expected 9854 Da/charge), giving an error of 0.05%. The other major ion is due to the doubly protonated compound, $(M+2H)^{+2}$, m/z 4924.6 (expected 4927.6 Da/charge).

Post-Source Decay. The structural information obtained in MALDI can be increased by causing fragmentation. The experiment involves examining the spontaneous (metastable ion) or collision-induced dissociations that occur after the ions have left the ion source. This MS–MS technique, known as post-source decay (PSD), is easy to implement and gives very high sensitivity.

Delayed Extraction. By delaying the extraction of ions from a MALDI source, the neutral matrix is pumped away and the ions can be observed at much improved resolution since collisions are minimized.

   DI Ion Sources and Sample Introduction. All the DI experiments employ relatively simple ion sources in which the sample is mounted on a probe, or multiple samples are mounted on a sample stage, and the desorbed ions are extracted into the mass

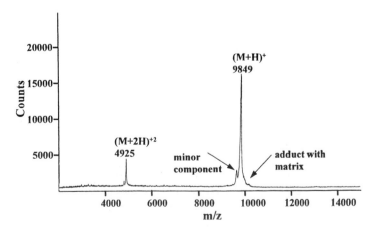

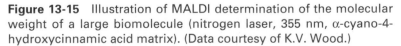

Figure 13-15 Illustration of MALDI determination of the molecular weight of a large biomolecule (nitrogen laser, 355 nm, α-cyano-4-hydroxycinnamic acid matrix). (Data courtesy of K.V. Wood.)

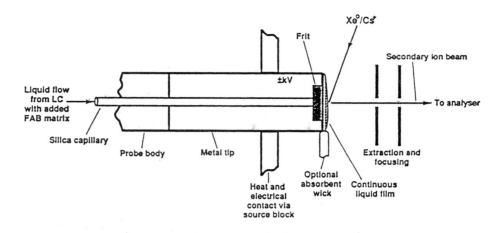

Figure 13-16 The flow FAB interface that allows on-line monitoring by DI. (From J.R. Chapman, *Practical Organic Mass Spectrometry*, 2nd ed., John Wiley & Sons Ltd., Chichester, UK, [1993] p. 58. Reproduced with permission.)

analyzer. For liquid matrices, loss by evaporation restricts measurements to a period of some minutes. Multiple samples are prepared and loaded together into the instrument in the MALDI technique, and the laser beam is then used to examine each sample sequentially. The SIMS and FAB ionization methods couple well with almost any type of mass analyzer, although the evaporating matrix makes this a relatively dirty technique in terms of its effects on the vacuum system. Because of its compatibility with pulsed sources, time-of-flight mass analyzers are used with MALDI sources. Either reflector or linear instruments can be used (Section 13-4b), with the former typically giving higher resolution. Trapping instruments are also well suited to MALDI experiments.

Introducing samples individually into the mass spectrometer is tedious and does not permit continuous monitoring of an analyte stream, such as an LC effluent. One solution to this problem is to add glycerol (or other matrix material) to the eluant from the LC and to pass this mixture into the mass spectrometer through a porous frit (Figure 13-16). Good coupling between the two instruments is achieved. This variant of the FAB experiment is known as flow FAB, and its great advantage is that it allows on-line monitoring. Flow injection methods of sample control can be used to monitor a sample stream that is changing with time, for instance, due to an enzymatic reaction or some other chemical or biological process. Continuous sample introduction methods for MALDI are becoming available.

Summary of DI Characteristics. FAB and liquid SIMS provide long lasting signals (some minutes) of relatively large ion currents, which also contain relatively abundant matrix ions. Detection limits are often in the nanomole range, and compounds with molecular weights of a few thousand Da can be readily examined, yielding characteristic singly charged ions of both polarities. Depending on the presence of salts in the sample, the molecular ion may take the form $(M+H)^+$, $(M+Na)^+$, $(M-H)^-$, etc. This ionization method couples well with almost any type of mass analyzer.

MALDI is extremely sensitive and applicable to high molecular weight compounds, even those present in relatively complex mixtures. The method is very simple, although matrix preparation requires experience. It too gives predominantly singly charged ions. The most important features of the DI methods are collected in Table 13-14.

13-3d Spray Ionization (SI) Including Electrospray Ionization (ESI)

Introduction. As just discussed, condensed phase samples, including solids, can be ionized by mixing them with a suitable matrix and energizing this matrix by particle or photon bombardment. It is often desirable to examine organic and particularly bio-

TABLE 13-14

Summary of Characteristics of DI methods

Sample type	Nonvolatile, thermally labile
Basic technique	Prompt delivery of energy to sample by energetic beam
Matrix	Organic liquid or solid; enhances ionization, minimizes fragmentation
Ionization mechanisms	Direct ion emission
	Cationization and ion-molecule reactions
Energy deposition	Broad; both a soft and a hard method
Ionization efficiency	10^{-1}% in FAB to 10^{-4}% in MALDI (sample can be recovered)
Mass range	10^4 Da (FAB–liquid SIMS) to 10^6 Da (MALDI)

logical compounds directly in aqueous solution. This approach minimizes manipulation of the sample and so facilitates coupling LC's with mass spectrometers. The family of ionization techniques known as spray ionization (SI) achieves the direct conversion of solvated molecules into gas phase ions. Table 13-15 summarizes the three main spray ionization methods.

Each of the SI procedures produce a mist of fine droplets, that are then dried to yield, improbably, isolated gas phase ions of intact molecules! The process of droplet formation is achieved by spraying the aqueous solution from a fine capillary. The techniques differ in the way in which energy is supplied to the droplets to cause solvent evaporation and electrical charging. Pneumatic heating, thermal energy, and an electrical potential are used in some combination in each of the various methods.

Thermospray Ionization (TS). Thermospray was the first successful SI method. Although TS is no longer widely used, a knowledge of the method helps us understand the SI phenomenon. In the thermospray method, radiative heating of a fine spray of droplets causes solvent evaporation. Appropriate salts present in the sprayed solution allow the analyte molecules to be observed in their cationized forms, such as $(M+K)^+$. Multiply charged cations are also formed, for example, $(M+4K)^{+4}$. Singly and, less commonly, multiply charged anions are also generated, typically by deprotonation; for instance, nucleotides give ions $(M-2H)^{-2}$. TS is most successful in ionizing precharged compounds, but is also useful for uncharged analytes, especially if a supplementary electron ionization filament or corona (needle) is used to produce a plasma via a direct current (DC) discharge. In such cases the radical cation $(M^{+\cdot})$ or radical anion $(M^{-\cdot})$ can also be observed, and the ionic processes occurring are closely related to those already discussed in CI. When employed, TS, like the other SI techniques, is normally used in conjunction with LC sample introduction to the mass spectrometer.

TABLE 13-15

Procedures Used in Spray Ionization Techniques*

Ionization Method	Energy Source	Operating Conditions	Ion Formation	Comments
Thermospray (TS)	Thermal energy	1-10 torr, elevated temperature	Gas phase ion-molecule reactions	Temperature dependent spectra; interface for LC–MS
Electrospray (ES)	Electric field	Reduced pressure, ambient temperature	Ion-molecule reactions in solution, droplets	Forms multiply charged ions for high molecular weight determination; interface for LC–MS and capillary zone electrophoresis (CZE)–MS
Atmospheric pressure chemical ionization (APCI)	Corona discharge	Atmospheric pressure, elevated temperature	Gas and solution phase ion-molecule reactions	Multiply charged ions; interface for LC–MS and GC–MS

* There is considerable overlap in techniques and phenomena between these methods.

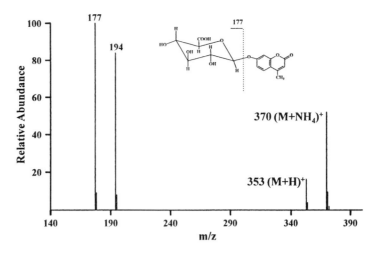

Figure 13-17 Thermospray mass spectrum of 4-methylumbelliferyl glucuronide examined from an aqueous solution containing ammonium acetate and introduced at a flow rate of 1 ml/min. Note the ions due to the intact molecule and the fragment ions due to glycoside cleavage. (Reproduced with permission from D.J. Liberato et al., *Anal. Chem.*, **55**, 1741 [1983]. Copyright 1983 American Chemical Society.)

A typical thermospray spectrum is shown in Figure 13-17. In this spectrum, the protonated molecule is evident, but the more abundant form of molecular ion is the molecule cationized by an ammonium ion, derived from the buffer. Note the presence of abundant fragment ions, that are characteristic of the structure of the compound. Thermospray often gives intact molecular ions, but it is not a particularly soft method of ionization. The role of the heating process in ion production is shown by the fact that ionization is often only successful over a narrow range of temperatures.

Electrospray Ionization. Commercial spray painting uses electric fields to spray uniform-sized droplets. When sufficient energy is provided, the charged droplets formed from aqueous solution can be caused to desolvate and subdivide to yield, ultimately, single ionized molecules. In the process known as electrospray ionization (ES), less commonly as ion spray, energy is provided by the field and by pneumatic heating using a warm countercurrent gas stream. Evaporation terminates when single molecules remain. Typically, the molecules are associated with several charges, such as multiple deprotonation in a biopolymer, which yields $(M-nH)^{-n}$ or, in acidic solution, $(M+nH)^{+n}$. The amount of heating is easily adjusted so that the analyte is completely free of water or other solvent in order to record accurate molecular weights. The extremely high ionization efficiency in ES leads to very low detection limits—in exceptional cases down to the low attomole range.

High molecular weight compounds give ions that contain many charges. For example, on the order of 100 protons can be attached to a bovine albumin dimer with a molecular weight of about 133,000 Da to give a molecular ion with m/z 1330 Da/charge. These ions are easily studied using conventional mass spectrometers so that we can make accurate determinations of the molecular weights of large biomolecules. Because of the statistical nature of the charging process, the number of charges is not fixed but varies smoothly over a range of values. For example, the 229 amino acid protein, cytochrome *c*, with a chemical average molecular weight 12,360 Da, gives the spectrum shown in Figure 13-18, when examined from an acidic solution. Note that the most probable charge state is +14 . The fact that this ion can acquire a variable number of charges accounts for the appearance of the mass spectrum in Figure 13-18(a).

Generally one does not know the number of charges associated with a particular ion. However, the molecular weight can be determined from these data, provided one knows the interval between charge states (invariably one charge unit). It is then

13-3e Ionization Summary

The newer methods of ionization (CI, DI, and SI) address the three main disadvantages of EI. Chemical ionization still requires that the sample be vaporized, but it allows a wide choice as to the type of ion formed, such as $(M+H)^+$ and $(M-H)^-$. Even more significantly, the energy deposition accompanying ionization can be controlled through choice of reagent. By making an appropriate choice, one can achieve efficient ionization with small degrees of internal excitation (soft ionization). One can therefore select conditions that give abundant ions from which molecular weights can be deduced. Such control is not available to the same extent in spray ionization or in desorption ionization. These are both relatively soft methods, at least when compared with 70 eV electron impact, and so intact molecular ions are produced. An equally significant advantage is that either solids or solutions can be examined. Thus DI and SI do not share the limitation of EI mass spectrometry to thermally stable, volatile samples.

Interestingly, both CI and DI have ionization efficiencies similar to that of EI, and all three methods are suitable for use in high resolution mass spectrometers. CI, DI, and SI spectra do not have the high reproducibility of EI spectra, but they do have the additional advantage of generating negative ions in comparable quantities to positive ions. A second advantage provided by CI, SI, and DI is that the ion formed directly from the molecule is not normally a radical ion, so that simpler fragmentation results than in EI. However, all these techniques are normally used to measure molecular weights. When structural information is desired with DI or SI, tandem mass spectrometry is used.

13-4 MASS ANALYSIS

Ions can be separated on the basis of their mass-to-charge ratios using electric or magnetic fields arranged so as to spread them in time or space. Among the many distinct types of mass analyzers we consider five: (1) sector magnetic fields, (2) time-of-flight analyzers, (3) quadrupole mass filters, (4) quadrupole ion traps, and (5) ion cyclotron resonance. Figure 13-20 illustrates the operation of each type. In some forms of analyzer, physical separation of ions in space is achieved, in others the mass-dependent frequency of ion motion is examined, and in still others ion velocity is measured using timing circuitry. The principal characteristics of the various mass analyzers are summarized in Table 13-17.

13-4a Magnetic Sectors

Magnetic sectors deflect accelerated beams of ions, with the degree of deflection depending on the mass, the charge, and the velocity of the ion beam. For ions of the same kinetic energy, the deflection depends on the mass-to-charge ratio. Sectors instruments are relatively large, but have the advantage of giving precise peak shapes and reproducible ion abundances. For these reasons, as well as their historical importance, they have been very widely used as service mass spectrometers. They can be scanned quite rapidly (1 second for a factor of 10 change in mass-to-charge) and their resolution is mass independent.

The sector mass spectrometer employs ions in the keV range of translational energies. Ions accelerated from rest to this energy provide a beam of approximately constant energy. Such a beam, passing at right angles through a magnetic field B, experiences a deflection that just balances the centripetal force for ions of a fixed m/z, which therefore move along a path of constant radius r. In this example, the balance of forces is given by eqs. 13-19 and 13-20.

$$Bzv = \frac{mv^2}{r} \qquad (13\text{-}19)$$

$$\frac{mv}{z} = Br \qquad (13\text{-}20)$$

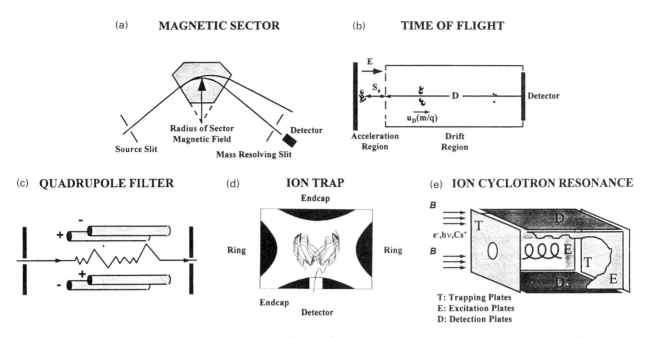

(a)   **MAGNETIC SECTOR**   (b)   **TIME OF FLIGHT**

(c)   **QUADRUPOLE FILTER**   (d)   **ION TRAP**   (e)   **ION CYCLOTRON RESONANCE**

Figure 13-20 Schematic diagram showing operation of the major types of mass analyzers: (a) ion deflection in a magnetic sector, (b) measurement of flight times in a time-of-flight, (c) selection of ions with stable trajectories in a quadrupole mass filter, (d) ejection of ions with characteristic frequencies from a quadrupole ion trap and (e) trapping of ions with characteristic frequencies in an ion cyclotron resonance spectrometer.

TABLE 13-17

Characteristics of Mass Analyzers

Method	Quantity Measured	Mass Analysis Equation	Mass-to-Charge Range*	Reso-lution*†	Mass Measure-ment Accuracy*	Dynamic Range‡	Oper-ating Pressure (torr)
Sector magnet	Momentum/charge	14–24	10^4	10^5	<5 ppm	10^7	10^{-6}
Time of flight	Flight time	14–27	10^6	10^3–10^4	0.1–0.01%	10^4	10^{-6}
Quadrupole ion trap	Frequency	14–32	10^4–10^5	10^3–10^4	0.1%	10^4	10^{-3}
Quadrupole	Filters for m/z	14–29	10^3–10^4	10^3	0.1%	10^5	10^{-5}
Cyclotron resonance	Frequency	14–35	10^5	10^6	<10 ppm	10^4	10^{-9}

* At 1000 Da/charge.
† Mass/peak width.
‡ Number of orders of magnitude of concentration over which response varies linearly.

As is evident from eq. 13-20, the device acts as a momentum/charge analyzer. In the special case when all the ions are accelerated through a constant potential difference, *V*, and have the same kinetic energy, we eliminate the velocity, *v*, from eqs. 13-19 and 13-20 using eqs. 13-21 and 13-22 to yield eq. 13-23, or, in common units, eq. 13-24

$$\frac{1}{2}mv^2 = zV \qquad (13\text{-}21)$$

$$v = \left(\frac{2zV}{m}\right)^{1/2} \qquad (13\text{-}22)$$

chromatograph, the result is a series of mass spectra recorded as the components elute.

We obtain a fuller understanding of the operation of quadrupole mass analyzers by recognizing that ions are subjected to forces associated with the quadrupole field that depend on ion position and the field direction. As the rf field changes sign, ions at a particular position alternately are pushed away from and then drawn toward the center of the device, the magnitude of the force being proportional to ion displacement from the center. A complex motion occurs that depends on the operating conditions and the mass-to-charge ratio of the ion. Quadrupole mass filters are normally operated so that only ions with a narrow range of masses (given by eq. 13-28) undergo stable motion. All

$$m/z = \frac{2V}{0.706\omega_{rf}^2 r_0^2} \qquad (13\text{-}28)$$

other ions have unstable trajectories, that take them further and further from the center line of the device. The factor 0.706 represents the value of a dimensionless parameter representing the condition for stability, ω_{rf} is the angular frequency of the rf field, V is the rf amplitude, and r_0 is the internal radius of the device. The expression is given in common units by eq. 13-29, in which f_{rf} is the rf frequency in megahertz.

$$m/z \ (\text{Da/charge}) = 0.069 \frac{V(\text{volts})}{f_{rf}^2 r_0^2 (cm^2)} \qquad (13\text{-}29)$$

From eq. 13-28 and 13-29 it is clear that ions of different masses can be brought into stability by adjusting the magnitude of the applied rf field, V. Note that the applied DC voltage is kept proportional to V so it does not appear in the mass analysis equation.

The stability diagram of Figure 13-22 summarizes conditions required for stable ion motion in terms of dimensionless parameters a_z and q_z, which depend on the m/z ratio and the operating conditions, specifically the applied DC (a_z) and rf (q_z) voltages, respectively. By operating at a constant DC and rf voltage ratio, the inclined scan line shown in the figure is utilized; ions of different mass-to-charge ratio fall along the line with lighter ions falling further from the origin. Only the ions indicated as m_2 are stable under the conditions illustrated. However, as the rf and DC voltages are increased, maintaining the fixed ratio, more massive ions such as m_3 will move into the stable region and be transmitted. A mass spectrum can then be recorded.

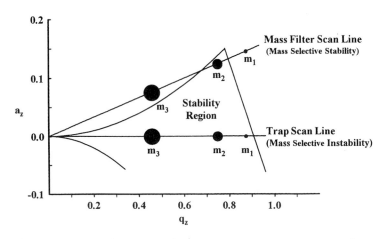

Figure 13-22 Stability conditions for ion traps and quadrupole mass filters expressed in terms of applied DC and rf voltages, which are expressed in terms of the dimensionless parameters a_z and q_z, respectively. The sizes of the symbols represent the m/z value of ions that occur at the indicated points along the scan lines for the two devices under fixed experimental conditions. This figure is schematic, as there are actually differences in these stability diagrams.

13-4d Quadrupole Ion Traps

Ion traps are three-dimensional versions of the quadrupole analyzer. Because forces operate in three directions, the electric fields can be used to store ions in an "electric bottle." These inexpensive devices are built to low mechanical tolerances, operate at relatively high pressures (10^{-3} torr of He), and can trap mass-selected ions for many seconds. Their low cost and high sensitivity has led to their wide use as detectors for liquid and gas chromatography, and their small size makes them suitable for field analysis.

Trapping is achieved by applying an rf voltage to the ring electrode, with the end cap electrodes grounded (Figure 13-23). Mass selection is achieved by scanning the applied rf voltage, V, so that ions of successively increasing mass-to-charge ratio go into unstable motion, leave the trap, and impinge on an external multiplier. Figure 13-22 illustrates, using symbols of different size, the relationship between m/z and the stability parameter q_z, defined in eq. 13-30, where V is the zero-to-peak rf voltage applied between the ring and

$$q_z = \frac{4zV}{mr_0^2\omega_{rf}^2} \qquad (13\text{-}30)$$

the endcap electrodes and ω_{rf} is the angular frequency. Instability occurs at $q_z = 0.908$ so the ion trap mass analysis equation, which is similar to that for the quadrupole mass filter (eq. 13-28), is given by eq. 13-31, or in more convenient units by eq. 13-32, in which

$$m/z = \frac{4V}{r_0^2\omega_{rf}^2\,0.908} \qquad (13\text{-}31)$$

$$m/z \,(\text{Da/charge}) = 0.1075\,\frac{V(\text{volts})}{f_{rf}^2 r_0^2(cm^2)} \qquad (13\text{-}32)$$

the frequency (f) is in MHz. For a typical device of 1 cm internal radius (r_0), operated at a frequency of 1 MHz, the maximum mass-to-charge value that is accessible when the rf field is scanned to a maximum reasonable value of 7.5 kV, is 806 Da/charge. The mass range can be increased by decreasing the frequency or the size of the device, or by resonantly ejecting ions at q_z values lower than 0.908.

The operation of the quadrupole ion trap can be illustrated with the same stability diagram (Figure 13-22) used to show quadrupole mass filter operation (note that this figure is schematic; there are differences between the two stability diagrams). The ion trap is operated using an rf voltage only, that is, by operating along the $a_z = 0$ axis. As shown in Figure 13-22, when operating at fixed rf voltage, ions of different mass-to-charge ratios have different q_z values and are spread out across this axis in mass. Ions of a wide range of masses are stable and the boundary between stability and instability

ION TRAP

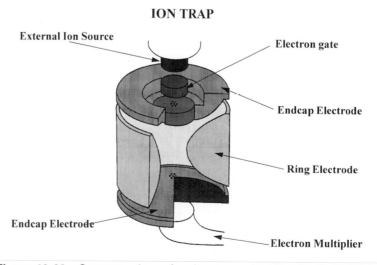

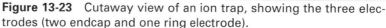

Figure 13-23 Cutaway view of an ion trap, showing the three electrodes (two endcap and one ring electrode).

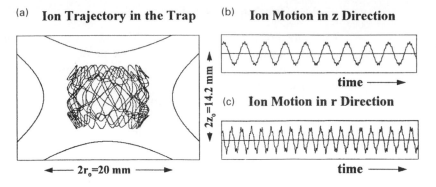

Figure 13-24 Simulation of motion of a single trapped ion in an ion trap: (a) shows the overall motion within the trap in cross section, (b) shows the time-dependent component of motion in the axial direction, and (c) shows the radial-direction component.

is at $q_z = 0.908$. Ions of different mass can be brought to this point by raising the rf voltage, as indicated in the figure and by eq. 13-32.

Some appreciation for the motion of trapped ions can be had by simulating their behavior as in Figure 13-24, which shows the trap in cross section and the path traced out by a single ion. To facilitate examination of the path, motion in the orthogonal axial and radial directions is also shown as a function of time in the boxes. It is apparent that the motion is complex and that there are characteristic frequencies involved. These frequencies can be used to manipulate the ion population in a mass-selective fashion.

Recently rapid progress has been made in the development of ion traps and methods of operation based on resonance between an external signal and the characteristic frequency of motion of stable trapped ions. These methods have allowed the mass range and mass resolution to be greatly extended. Ion traps have many features in common with ion cyclotron resonance instruments, including the ability to study sequences of reactions by MS–MS methods (Section 13-4g) and nondestructive broadband detection and ion remeasurement.

13-4e   Ion Cyclotron Resonance (ICR)

Ion cyclotron resonance is the mass spectrometric experiment that most closely resembles NMR. The technique is capable of high mass resolution and access to high molecular weight ions, and has the unusual feature of allowing nondestructive detection and ion remeasurement (Section 13-4h). It is extremely sensitive and readily compatible with tandem mass spectrometry.

The principles of ion cyclotron resonance can be derived from consideration of the forces experienced by ions in magnetic and electric fields. Ions move in circular orbits in a magnetic field of fixed strength and do so with fixed cyclotron frequencies that depend on their mass-to-charge ratios but are independent of kinetic energy. An ion of mass m and velocity v, in the plane normal to a magnetic field of strength B, experiences a force Bzv normal to both the field direction and to the direction of motion (eq. 13-19). Hence the ion describes a circle of radius r in the plane normal to the fields, so that the cyclotron frequency is given by eq. 13-33. Thus every ion has a characteristic cyclotron angular frequency (ω_c), which is inversely proportional to its mass-to-charge

$$\omega_c = Bz/m \tag{13-33}$$

ratio (m/z) and directly proportional to the applied magnetic field strength (B). Although ω_c is independent of the velocity of the ion, the radius of the orbit is directly

Time Domain   **Frequency Domain**

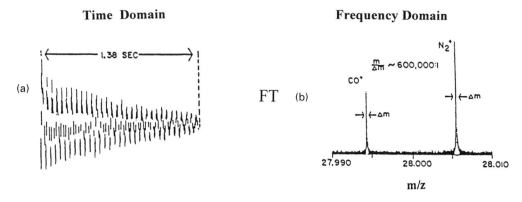

Figure 13-25 (a) Time domain signal (transient) recorded for a mixture of $CO^{+\cdot}$ and $N_2^{+\cdot}$ ions of nominal m/z 28 and (b) the corresponding frequency (mass) domain signal. (Courtesy of A.G. Marshall.)

proportional to ion velocity. The mass analysis equation is obtained by substituting the cyclotron frequency f_c ($2\pi f_c = \omega_c$) for angular frequency and writing eq. 13-33 as eq. 13-34, or, using convenient units of Da/charge, Tesla, and MHz, as eq. 13-35.

$$m/z = B/2\pi f_c \tag{13-34}$$

$$m/z \text{ (Da/charge)} = 15.36\, B \text{ (Tesla)}/f_c \text{ (MHz)} \tag{13-35}$$

A spectrum is recorded by exciting the trapped ions of interest by applying an oscillating electric field (frequency f) normal to the direction of B. At certain frequencies this signal will be in resonance with the frequencies of the ions. At resonance, energy is absorbed by the ions, increasing their velocities and orbital radii and causing ions of the same mass to move coherently, as a packet. Then the imaging current method described in Section 13-4h can be used to detect the ions without destroying them. This method is sensitive to as few as 10^2 to 10^3 charges. The coherently moving ions give rise to a transient, time-varying signal (Figure 13-25) that decays with time, just as in NMR. Typical transient times, which are strongly pressure dependent, are some seconds at ultrahigh vacuum ($<10^{-9}$ torr).

Because frequencies can be selected and measured with great accuracy, the mass resolution obtainable using ICR is very large indeed, especially at low mass, for which ions have high frequencies. This capability is illustrated in Figure 13-25, which shows nominal mass m/z 28 resolved into the $CO^{+\cdot}$ and $N_2^{+\cdot}$ contributions. The maximum mass-to-charge ratios that can be measured by ICR increase linearly with the magnetic field strength. Superconducting magnets of field strength up to 7 Tesla or more are therefore used for high performance work. Such magnets, also used in NMR, are characterized by high field stability.

The heart of the ICR instrument, the ICR cell, can have a variety of geometries, among which the cubic cell, shown in Figure 13-20, is typical. Ions are confined in the direction parallel to B by small DC trapping potentials, that establish a small electric field. They are confined in the other directions by the magnetic field itself. The rf signal used to resonantly excite ions is applied through one set of plates, and detection is accomplished through the remaining set as shown in Figure 13-20. Note that ions can either be generated in the ICR cell or introduced into it from some external ion source. The former instance requires that the sample be introduced into the high vacuum region of the instrument. The external ionization scheme allows conventional ion sources to be used under their usual operating conditions. Differential pumping is used to couple high pressure sources (atmosphere in the case of ESI, 1 torr in the case of CI) to the low pressure ($<10^{-9}$ torr) analyzer.

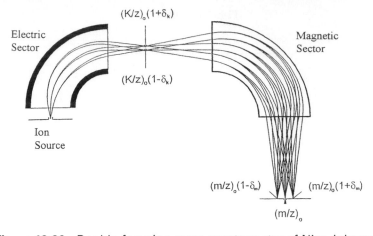

Figure 13-26 Double-focusing mass spectrometer of Nier-Johnson geometry. The magnetic field is normal to the plane and the electric field lies in the plane. Ions of three different mass-to-charge ratios (δ_m) and a range of angles emerge from an ion source with a spread in kinetic energies and are focused for angle and kinetic energy but dispersed for m/z ratio, at the point detector. The electric and magnetic sectors both disperse ions according to their momentum-to-charge *(K/z)* ratios. (From R.G. Cooks et al., *Encyclopedia for Applied Physics,* VCH Publishers, New York, vol. 19, p. 289 [1997].)

13-4f High Resolution Mass Spectrometers

High-resolution mass spectrometry is important since it allows precise mass measurements that help to identify the elemental composition of compounds (Section 15-3). Magnetic sectors, combined with appropriate sector electric fields, form the basis for one type of high resolution mass spectrometer. The combination of elements focuses ions of the same m/z to the same point, even when they have slightly different kinetic energies (Figure 13-26). The double-focusing mass spectrometer produces a resolution well in excess of 10^5 (defined as the mass *m*, divided by the full width at half maximum of the peak, Δm). It also provides access to ions with mass-to-charge ratios of up to 10^4 Da/charge. Geometries other than that illustrated here are capable of simultaneously focusing ions of a wide range of masses along a single plane. In such cases, sensitivity is greatly increased by using an imaging detector (Section 13-4h) capable of simultaneously recording all the ions.

High resolution is also achieved using FT–ICR instruments. One simply zooms in and scans over a relatively narrow range of frequencies, observing the transient (time-domain signal) for as long as possible. In the case of ICR, resolution is mass dependent, falling with increasing mass. However, a resolution of 10^6 (FWHM definition) is available at a m/z value as high as 10^4 Da/charge.

13-4g Tandem Mass Spectrometers

Tandem mass spectrometry is used to characterize individual compounds in a complex mixture or to identify a compound's structure. These goals are achieved by separating the ionization step from the fragmentation process and so controlling the degree of fragmentation. Mass analysis must be performed twice in a tandem instrument—to identify both the parent ion and the product ion. This objective can be achieved in two distinct ways (1) by separating the mass analysis operations in space or (2) by separating them in time (Figure 13-27).

Separation in space can be achieved by coupling two mass analyzers. For example, a sector magnet can be coupled to a quadrupole mass filter. Parent ions are selected by the sector magnet, and so separated from all other ions generated from the sample.

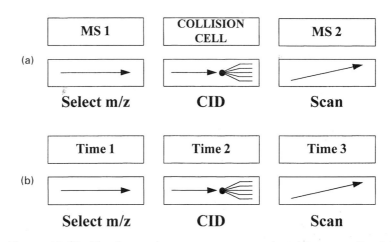

MS 1	COLLISION CELL	MS 2

(a)

Select m/z	CID	Scan

Time 1	Time 2	Time 3

(b)

Select m/z	CID	Scan

Figure 13-27 Tandem mass spectrometry using (a) separation in space and (b) separation in time. Mass-selected ions are dissociated using collisions and their products are mass-analyzed again. (Adapted from E. de Hoffmann, *J. Mass Spectrom.,* **31,** 129 [1996]. Reproduced with the permission of John Wiley & Sons Ltd., Chichester, UK.)

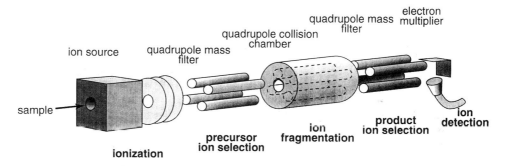
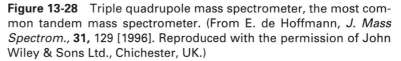

Figure 13-28 Triple quadrupole mass spectrometer, the most common tandem mass spectrometer. (From E. de Hoffmann, *J. Mass Spectrom.,* **31,** 129 [1996]. Reproduced with the permission of John Wiley & Sons Ltd., Chichester, UK.)

These selected ions are activated by a collision process, and the resulting set of product ions is subjected to mass analysis with the quadrupole mass filter. Fragmentation is achieved by raising the internal energy of the ions by collision-induced dissociation (CID). This process involves passing the energetic beam of parent ions through a cell containing a collision gas, such as Ne, He, N_2, or Ar.

The most important separation-in-space tandem mass spectrometer is the triple quadrupole (Figure 13-28). In this tandem analyzer instrument, an intermediate quadrupole is used to confine ions to the axis in the presence of a collision gas. This quadrupole is not operated in a mass analysis mode but is set to transmit all ions. The products of CID are passed into the third quadrupole (the second mass analyzer) for mass analysis. Triple quadrupoles give unit resolution for parent and product ions with good dissociation efficiencies for ions of less than 1000 to 2000 Da. They operate up to *m/z* 4000 and are highly sensitive.

The second major type of tandem mass spectrometer uses a single mass analyzer to perform two steps of mass analysis, that are separated from each other in time. Both the quadrupole ion trap and the ICR instrument can be used in this way. The ion of interest is first isolated in the trap, then activated by collision, and finally its dissociation products are mass analyzed.

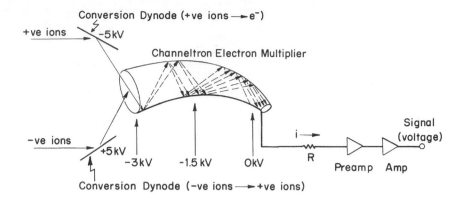

Figure 13-29   Channel electron multiplier with conversion dynodes for measurement of ions of either polarity.

13-4h   Ion Detection and Data Handling

Electron Multiplier Detectors.   When energetic ions collide with a suitable solid surface, one of low work function, electrons are liberated. The conversion of ions to electrons and subsequent electron amplification can be achieved using discrete collisions on separate electrodes (dynodes), the potentials of which are arranged so that electrons are accelerated from one stage to the next. Alternatively, one can apply a potential across a semiconductor material so that electrons produced in one collision event are accelerated into the device to undergo further electron-surface collisions. Such a channel electron multiplier or channeltron is shown in Figure 13-29.

The velocity of the incoming ions determines the probability and number of electrons released. As a result, high mass ions, which are relatively slow-moving even at the energies of a few keV normally used in electron multipliers, often need to be converted into smaller ions. This initial ion-to-ion conversion point is achieved by impact of the primary ions on a suitable *conversion dynode* surface, at which point they dissociate or sputter ions from the surface (Figure 13-29). Primary electrons are generated by a subsequent ion-surface collision, and in turn they are accelerated and collide with another surface in order to generate secondary electrons to multiply the number of electrons produced. Typically 10 to 20 stages of electron amplification are used. A single electron gives 10^6 electrons in just 12 stages of gain, when the average gain factor on each collision is 3.3, a typical number. Negative ions can be detected with an electron multiplier operating at the same potentials as for positive ion detection, but using a conversion dynode with the opposite bias (Figure 13-29).

Ion Counting.   The ability to convert a single ion into 10^5 to 10^8 electrons in a time on the order of 10^{-7} s is one factor that is responsible for the sensitivity of mass spectrometers. In most organic and biological work, ion currents are sufficiently large that the analog, averaged output of such detectors can be recorded. When necessary, however, *ion counting* experiments can be performed in which the signal from each individual ion is converted into a measurable signal. Ion currents of the order of 10^{-19} A (approximately 1 ion/s) can be counted in this way; note that this is ten orders of magnitude smaller than the typical current produced by an ion source (see Table 13-5).

Point and Array Detectors.   *Point detectors,* such as electron multipliers, can be used to observe ions of a single mass-to-charge ratio, and a scan of the mass analyzer can be used to record a mass spectrum. Channeltrons have been miniaturized to dimensions of less than 100 microns. Arrays of these miniature electron multipliers, known as microchannel plates, produce amplified images of ion beams. These devices and the still more recent charge couple devices (CCDs) convert ions into electric signals that measure both the position and the intensity of an ion beam. Such *array detectors* are particularly useful for analyzers such as magnetic sectors, that disperse ion beams in space.

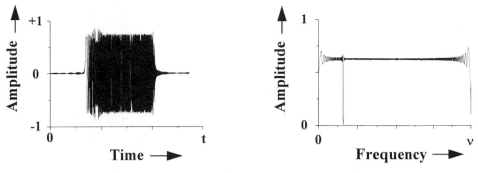

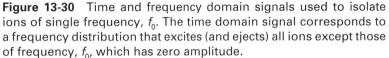

Figure 13-30 Time and frequency domain signals used to isolate ions of single frequency, f_0. The time domain signal corresponds to a frequency distribution that excites (and ejects) all ions except those of frequency, f_0, which has zero amplitude.

When the detector is located in the focal plane of the instrument, simultaneous recording of the abundances of ions of a wide range of mass-to-charge ratios can be made with greatly increased sensitivity.

Image Charge Detection. In ICR, ions of a given mass move together as a packet of charges. As they approach and then move away from the metal walls of the ICR instrument, they polarize the metal and induce an image potential of opposite charge. The image charge alternates in magnitude at the frequency of ion motion, and this time-domain signal can readily be amplified. Note that the method of detection does not destroy the ion packet; not only can its motion be observed until it loses coherence (decays) due to collisions with residual gas in the analyzer, but the resulting ion cloud can be re-excited and *remeasured*. This important capability increases the sensitivity of the ICR experiment. Typical transient times, which are strongly pressure dependent, are seconds at very low pressure (approximately 10^{-9} torr).

Broadband Detection. In the ICR mass spectrometer, ions of a single mass-to-charge ratio can be excited and detected nondestructively. However, it is more convenient to use an electrical signal that contains a whole range of frequencies and to excite ions of a range of masses simultaneously. This method involves using a chirp signal in which different frequencies are applied in rapid succession. Better results are obtained by deciding on the desired set of frequencies and then using an inverse Fourier transform to determine the time domain signal needed to produce this set of frequencies. This procedure, termed stored waveform inverse fourier transform (SWIFT), is extremely flexible and in addition to allowing ion detection, allows particular ions or groups of ions to be excited while adjacent ions are not. By using appropriate excitation amplitudes, selected ions can be collisionally dissociated or ejected from the ICR cell while others are retained. This capability is of enormous value in studies on particular compounds in mixtures. An example of the time and frequency domain signals used to isolate ions of a single mass-to-charge ratio is shown in Figure 13-30.

Excitation of ions of different mass-to-charge ratios is followed by simultaneous detection. This process involves recording the signal in the time domain and transforming it into the frequency domain. The time-domain signal consists of the sum of all the individual signals with their fixed frequencies, and these frequencies are recovered on performing the FT operation. Note the close similarity between FT–NMR and these FT–mass spectrometry experiments.

Chemical Noise. Detection limits in mass spectrometers often are controlled not by detector response but by the interference produced by other compounds present in the sample and in the instrument. This signal, termed *chemical noise,* occurs, for example, in the FAB method of ionization in the form of matrix-derived ions. It is removed efficiently by multiple operations on the ion beam as is done in tandem mass spectrometry.

Full Scans and Ion Chromatograms. Mass spectrometric data can be acquired by recording full scan mass spectra or by observing selected ions as a function of time. In the case of scanning mass spectrometers, which normally use point detectors, the full scan experiment is wasteful of sample since most of the time is spent *between* the peaks. When sample amounts allow, the ion abundances in such scans can be summed to record the total abundance vs. time. Such a plot can be used to represent the output of a chromatograph and identify the times of elution of chromatographic peaks. However, individual ion abundances can also be measured as a function of time, and these single ion chromatograms are more specific than are total ion chromatograms. For example, only ions characteristic of certain compounds might be measured. In such a case, chromatographic resolution is actually improved, since an overlapping chromatographic peak lacking the characteristic ion does not give a signal. More information on chromatography–mass spectrometry is found in Section 15-4a.

Duty Cycle, Scan Speed and Data Burden. The performance of an instrument is affected by the duty cycle of the analyzer and source. These cycles should be matched: if the source produces ions continuously, then the analyzer should continuously monitor all these ions simultaneously by means of an array or imaging detector. If ions are produced in short bursts, the analyzer should operate in the same way. The pulsed laser/time-of-flight combination falls into this latter category.

The performance of the mass spectrometer in chromatographic and other experiments requiring the examination of transient signals is determined by the scanning speed of the instrument. In chromatographic work one should take 10 measurements to define peak shapes and hence a scan speed of 0.1 s per spectrum is needed. Most analyzers can achieve such speeds when covering a normal mass range for an organic compound. Time-of-flight instruments can acquire as many as 10^3 full mass spectra per second and so can be used to monitor rapidly reacting systems.

Finally, it should be noted that enormous amounts of data are acquired and processed in mass spectrometry. In a chromatography run of 10 minutes, a complete mass spectrum often is taken each second. These data might represent 10^3 units of *m/z* information and the abundance data is stored to a similar precision. The uncompressed data file therefore contains almost 10^9 bits of information.

13-5   WORKED PROBLEMS

These five problems are worked. The problems in the following section are left for the reader to work. Answers are provided in the accompanying Solutions Manual.

Problem 13-1

The CI (isobutane) mass spectrum of 4-methylbenzophenone (Figure 13-31a) shows only the protonated molecule and its ^{13}C isotopic peak. When methane is employed as a

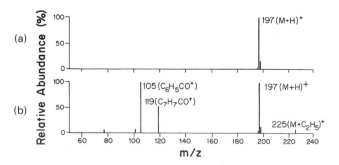

Figure 13-31   CI mass spectra of 4-methylbenzophenone using (a) isobutane and (b) methane as reagent gases. (From J. Michnowitz and B. Munson, *Org. Mass Spectrom.*, **4**, 481 [1970]. Reproduced with the permission of John Wiley & Sons Ltd., Chichester, UK.)

reagent, ions occur at both lower and high mass (Figure 13-31b). Explain their origin and usefulness. The proton affinities of methane, isobutene, and 4-methylbenzophenone are 554, 801, and 886 kJ mol^{-1} (130, 192, and 212 kcal mol^{-1}). Compare with Table A-3.

Answer. Protonation by CH_5^+ is more exothermic by $801 - 544 = 257$ kJ mol^{-1} (62 kcal mol^{-1}) than is protonation by $C_4H_9^+$. The ions below $(M+H)^+$ are due to unimolecular fragmentation of $(M+H)^+$ because of its large internal energy, $886 - 544 = 342$ kJ mol^{-1} (82 kcal mol^{-1}). The fragments are indicators of structure.

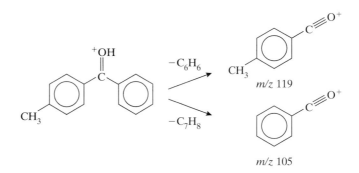

The ion of greater mass than $(M+H)^+$ is the result of the ionization reaction $M + C_2H_5^+ \longrightarrow (M+C_2H_5)^+$ (methane gives CH_5^+, $C_2H_5^+$, and $C_3H_5^+$ as reagent ions, Table 13-7). The presence of two molecular ions $(M+H)^+$ and $(M+C_2H_5)^+$, separated by 28 mass units, helps in checking the molecular weight assignment.

Problem 13-2

1,2,3,4-Tetrafluorobenzene is ionized by charge exchange with argon. If the ionization energy of the organic molecule is 9.6 eV and that of argon is 15.8 eV, what is the internal energy of the molecular ion?

Answer. Charge exchange proceeds as follows:

$$Ar^{+\cdot} + C_6F_4H_2 \longrightarrow Ar + C_6F_4H_2^{+\cdot}$$

$$\Delta H_{reaction} = \Delta H_f(Ar) - \Delta H_f(Ar^{+\cdot}) + \Delta H_f(C_6F_4H_2^{+\cdot}) - \Delta H_f(C_6F_4H_2)$$

but by definition $IE(M) = \Delta H_f(M^{+\cdot}) - \Delta H_f(M)$

$$\therefore \Delta H_{reaction} = IE(C_6F_4H_2) - IE(Ar)$$

$$= 9.6 - 15.8 \text{ eV}$$

$$= -6.2 \text{ eV}$$

This 6.2 eV appears as internal energy of $C_6F_4H_2^{+\cdot}$ and leads to a spectrum that shows considerable fragmentation. The base peak is due to CF_2 loss, common in fluorinated compounds, whereas losses of CHF and CHF_2 also give intense peaks. The molecular ion abundance is 22% of the base peak.

Problem 13-3

Predict the results of attempted negative chemical ionization of nitrobenzene and biacetyl (electron affinities of 0.4 and 1.1 eV, respectively) (1) if the experiment is done under conditions that promote electron attachment (e$^-$ reagent) or (2) if the reagent ion is HO$^-$ (see Table 13-9).

Answer. (1) Electron attachment is often successful when the compound has a positive electron affinity. Both compounds should yield molecular anions, M$^-$. (2) Biacetyl is relatively acidic and should yield the deprotonated molecule, (M−H)$^-$, with hydroxyl reagent. Nitrobenzene will not be ionized by this reagent unless it forms an adduct.

Problem 13-4

The CI (methane) mass spectra of alkanes show prominent $(M-H)^+$ ions in the molecular ion region. Explain this observation.

Answer. The major methane reagent ion is CH_5^+ (Table 13-7). The reaction $RH + CH_5^+ \longrightarrow RH_2^+ + CH_4$ is only slightly exothermic for alkanes (it is thermoneutral for $R = CH_3$). Hence $(R-H)^+$ does *not* arise by fragmentation of the usual molecular ion, $(M+H)^+$. Rather, other ion-molecular reactions can compete with protonation. In particular, hydride abstraction occurs readily for alkanes.

$$C_2H_5^+ + RH \longrightarrow C_2H_6 + R^+$$

Problem 13-5

Figure 13-32 shows two mass spectra recorded by one of the SI methods. What are the reactions that produce the observed ions?

Answer. In the spray ionization methods, droplets of the solution evaporate and subdivide, ultimately yielding single charged molecules. The overall reaction is therefore $(M + H)^+_{solution} \longrightarrow (M+H)^+_{gas}$. This process gives the ion m/z 363 in cortisol. The other ions in this spectrum result from subsequent gas phase dissociation: loss of water (m/z 345), loss of H_2CO from the hydroxylmethyl group (m/z 333), and loss of the C_{17} side-chain with hydrogen transfer, $HCOCH_2OH$ (m/z 303).

The benzophenone derivative has several groups that can capture electrons: the aromatic ring, the nitro group, and the chlorine atom. However, it has no Brønsted acid groups. Hence its reactions are $M_{gas} + e^- \longrightarrow M^-$. In this experiment, SI is used to produce the neutral molecule in the vapor phase.

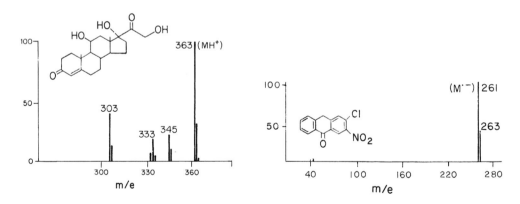

Figure 13-32   LC–MS spectrum obtained by thermospray of (a) cortisol and (b) a substituted benzophenone examined using an auxiliary filament to facilitate electron capture. (Data courtesy of Vestec Corp.)

PROBLEMS

13-1   A sector mass spectrometer of radius 30 cm and maximum field strength 0.8 Tesla is normally operated at an accelerating voltage of 8000 V. What is its mass range for singly charged ions? What accelerating potential should be used to reach an upper mass of 6000 Da?

13-2   The DCI method works well for the antineoplastic drug paclitaxel (Taxol). Discuss the spectra of Taxol,* recorded using ammonia as the reagent gas, and the observation of (a) positive and (b) negative ions, as shown below.

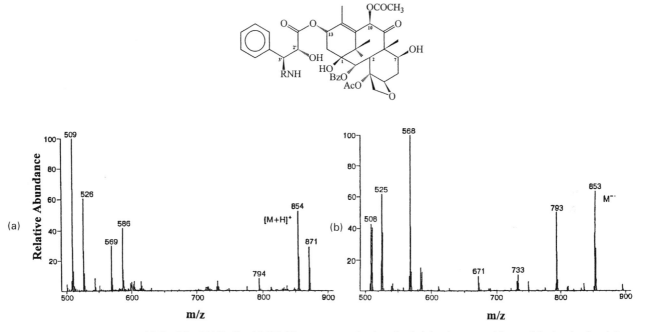

(a)   (b)

13-3   The FAB–liquid SIMS spectrum of teicoplanin[†] (a glycopeptide antibiotic obtained from cultures of *Actinoplanes teichomyceticus*) shown below depicts the intact protonated molecule $(M + H)^+$, the isotopic cluster of which is centered at *m/z* 1880.6 and other adduct and fragment ions which appear at higher and lower *m/z*, respectively. What information does this spectrum provide?

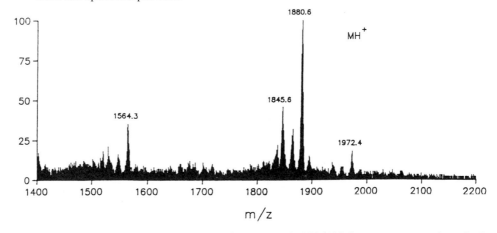

*From P. Heinstein et al., *J. Chem. Soc. Perkin Trans.*, **1**, 845 (1996). Royal Society (London).

†From J.E. Coutant, T-M. Chen, and B.L. Ackermann, *J. Chrom.*, **529,** 265 Elsevier Science Publishers, Amsterdam (1990).

13-4 The spectra below are (a) the APCI mass spectrum of the protein cytochrome *c* and (b) the deconvoluted spectrum that displays ion mass, not mass-to-charge ratio. What can be learned from these spectra?

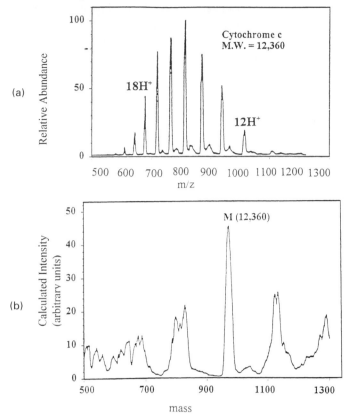

BIBLIOGRAPHY

13.1.   Historical: J.J. Thomson, *Rays of Positive Electricity,* Longmans, Green, and Co., London, 1913; F.W. Aston, *Mass Spectra and Istopes,* 2nd ed., Edward Arnold and Co., London, 1942; R.W. Kiser, *Introduction to Mass Spectrometry and Its Applications,* Prentice-Hall, Englewood Cliffs, NJ, 1965; J.H. Beynon and R.P. Morgan, *Int. J. Mass Spectrom. Ion Processes.,* **27,** 1 (1978); R.G. Cooks, *J. Mass Spectrom.,* **30,** 1215 (1995).

13.2.   General: E. de Hoffmann, J. Charette, and V. Stroobant, *Mass Spectrometry, Principles and Applications,* John Wiley & Sons Ltd., Chichester, UK, 1996; J.T. Watson, *Introduction to Mass Spectrometry,* Raven Press, New York, 1985. J.R. Chapman, *Practical Organic Mass Spectrometry,* John Wiley & Sons Ltd., Chichester, UK, 1993.

13.3.   Industrial and engineering aspects of mass spectrometry: F.A. White and G.M. Wood, *Mass Spectrometry: Applications in Science and Engineering,* John Wiley & Sons, Inc., New York, 1986; P. Nicholas, *Spectroscopy,* **6,** 36 (1991); K. Rollins, J.H. Scrivens, R.C.K. Jennings, W.E. Modern, J.K. Welby, and R.H. Bateman, *Rapid Commun. Mass Spectrom.,* **4,** 454 (1990).

13.4.   Mass analyzers: C. Brunnée, *Int. J. Mass Spectrom. Ion Processes.,* **76,** 125 (1987).

13.5.   Tandem mass spectrometry (MS–MS): K.L. Busch, G.L. Glish, and S.A. McLuckey, *Mass Spectrometry/Mass Spectrometry: Techniques and Applications of Tandem Mass Spectrometry,* VCH Publishers, New York, 1988; S.J. Gaskell, *Biol. Mass Spectrom.,* **21,** 413 (1992); E. de Hoffmann, *J. Mass Spectrom.,* **31,** 129 (1996).

13.5.   Chemical ionization: M. Vairamani, U.A. Mirza, and R. Srinivas, *Mass Spectrom. Rev.,* **9,** 235 (1990); A.G. Harrison, *Chemical Ionization Mass Spectrometry,* ed. 2, CRC Press, Boca Raton, FL, 1992; E. Uggard, *Mass Spectrom. Rev.,* **11,** 389 (1992); E.P. Burrows, *Mass Spectrom. Rev.,* **14,** 107 (1995).

13.6. Negative ions: C.E. Melton, *Principles of Mass Spectrometry and Negative Ions,* Marcel Dekker, New York, 1970; J. Jalonen, *Biol. Mass Spectrom.,* **19,** 253 (1990); J.H. Bowie, *Org. Mass Sprectrom.,* **28,** 1407 (1993).

13.7. Desorption ionization: R.G. Cooks and K.L. Busch, *J. Chem. Educ.,* **59,** 926 (1982); P.A. Lyon (ed.), *Desorption Mass Spectrometry,* ACS Symposium Series, No. 291, American Chemical Society, Washington, DC, 1985; A.S. Woods, J.C. Buchsbaum, T.A. Worrall, J.M. Berg, R.J. Cotter, *Anal. Chem.,* **67,** 4462 (1995); K.L. Busch, *J. Mass Spectrom.,* **30,** 233 (1995).

13.8. Matrix-assisted laser desorption ionization (MALDI): M. Karas and F. Hillenkamp, *Anal. Chem.,* **60,** 229 (1989); M. Kraras, U. Bahr, and U. Giessmann, *Mass Spectrom., Rev.,* **10,** 335 (1991); A.S. Woods, J.C. Buchsbaum, T.A. Worrall, J.M. Berg, R.J. Cotter, *Anal. Chem.,* **67,** 4462 (1995); A.M. Belu, J.M. DeSimone, R.W. Linton, G.W. Lange, and R.M. Friedman, *J. Am. Soc. Mass Spectrom.,* **7,** 11 (1996).

13.9. Electrospray ionization (ESI): S. Gaskell, *J. Mass Spectrom.,* **32,** 677 (1997).

13.10. Quadrupoles and ion traps: P. Dawson, *Quadrupole Mass Spectrometry and Its Applications,* Elsevier Science Inc., Amsterdam, 1976; R.E. March and R.J. Hughes, *Quadrupole Storage Mass Spectrometry,* John Wiley & Sons, Inc., New York (1989); J.F.J. Todd, *Mass Spectrom. Rev.,* **10,** 3 (1991); R.E. March and J.F.J. Todd, (eds.), *Practical Aspects of Ion Trap Mass Spectrometry,* vols. 1-3, CRC Press, New York, 1995.

13.11. Ion cyclotron resonance (ICR) spectrometry: T.A. Lehman and M.M. Bursey, *Ion Cyclotron Resonance Spectrometry,* John Wiley & Sons, Inc., New York, 1976; A.G. Marshall and P.B. Grosshans, *Anal. Chem.,* **63,** 215A (1991); V.H. Vartaniain, J.S. Anderson, and D.A. Laude, *Mass Spectrom. Rev.,* **14,** 1 (1995).

13.12. Time-of-flight (TOF) spectrometry: J.F. Holland, C.G. Enke, J. Allison, J.T. Stults, J.D. Pinkston, B. Newcome, and J.T. Watson, *Anal. Chem.,* **55,** 997A (1983); R.J. Cotter, (ed.) *Time-of-Flight Mass Spectrometry,* American Chemical Society, Washington, DC, 1994; H. Wollnik, *Mass Spectrom. Rev.,* **12,** 89 (1993); M. Guilhaus, *J. Mass Spectrom.,* **30,** 1519 (1995).

13.13. Interfaces and data processing: F.W. McLafferty, St. Y. Loh, and D.B. Stauffer, *Comput. Enhanced Anal. Spectrosc.,* **2,** 163 (1990); K.J. Hart, and C.C. Enke, *Comput. Enhanced Anal. Spectrosc.,* **3,** 149 (1992); F.H. Strobel, L.M. Preston, K.S. Washburn, and D.H. Russell, *Anal. Chem.,* **64,** 754 (1992).

13.14. Ion-molecule reactions: T.H. Morton, *Tetrahedron,* **38,** 3195 (1982); N.M.M. Nibbering, *Acc. Chem. Res.,* **23,** 279 (1990); Z. Herman and D. Smith, *Chem. Rev.,* **92,** 1471 (1992).

13.15. Biological applications: M.E. Rose and R.A.W. Johnstone, *Mass Spectrometry for Chemists and Biochemists,* Cambridge University Press, Cambridge, UK, 1982; V.N. Reinhold and S.A. Carr, *Mass Spectrom. Rev.,* **2,** 153 (1983); A.L. Burlingame, K.M. Straub, and T.A. Baillie, *Mass Spectrom. Rev.,* **2,** 331 (1983); R.M. Caprioli, *Mass Spectrom. Rev.,* **6,** 237 (1987); K. Biemann and S.A. Martin, *Mass Spectrom. Rev.,* **6,** 1 (1987); J.A. McCloskey (ed.), *Mass Spectrometry, Methods in Enzymology,* vol. 193, Academic Press, San Diego, 1990; R.D. Smith, J.A. Loo, R.R. Ogorzalek Loo, M. Busman, and H.R. Udseth, *Mass Spectrom. Rev.,* **10,** 359 (1991); R.B. Van Breemen, *J. Med. Chem.,* **35,** 4919 (1992).

13.16. Web resources: Murray's Mass Spectrometry Page at http://userwww.service.emory.edu/~kmurray/mslist.html This site contains a wealth of well-organized information on mass spectrometry, including software for isotope ratio, molecular formula, peptide sequence, ion trajectories, and other calculations.

chapter 14

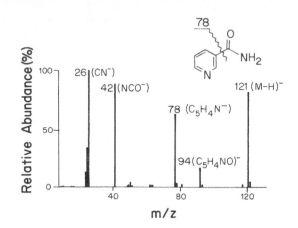

Fragmentation and Ion Chemistry

14-1 *GENERAL PRINCIPLES OF FRAGMENTATION*

Qualitative analysis by mass spectrometry is a matter of identifying the molecular ion in order to determine the molecular weight of a compound. In addition, the fragmentation observed in the mass spectrum can provide the chemical structure by comparison with a library of EI mass spectra or by interpretation of the spectrum using the molecular structure–fragmentation relationships that are dealt with here. First, the factors that control mass spectral fragmentation are considered and then examples are given of the behavior of compounds with various functional groups.

Strict rules are not possible for mass spectral fragmentation, just as they are not for describing product yields in solution phase reactions. What follows, therefore, are guidelines to the fragmentation behavior of functional groups. Data refer to EI mass spectra unless otherwise indicated, but examples are also given from other ionization techniques. The principal features common to all fragmentations are found in EI spectra, which serve as an ideal foundation for the more general subject.

Even in experiments where a single molecular ion with a particular internal energy is generated, a complex mixture of ions normally results within a microsecond of ionization. These ions are formed by a network of competing and consecutive individual reactions. The mass spectrum approximates an instantaneous picture of this ion collection. Fortunately, only a few reactions are usually dominant, and these processes can often be rationalized from a knowledge of solution chemistry.

The internal energy distribution $P(\varepsilon)$ of a population of ions is the most important factor controlling the appearance of the mass spectrum. This distribution can be changed by altering the ionization conditions. Thus lower energy electrons in EI, different reagent gases in CI, or higher collision energies in DI have important effects on $P(\varepsilon)$.

14-1a Energetics of Dissociation

The energy available in a fragmenting ion is normally quite small, not much greater than a typical bond energy. Consequently, weak bonds in the ion, usually the same bonds that are weak in the corresponding neutral molecule, are more likely to cleave than are strong bonds. Reactions that lead to electron unpairing are relatively unlikely, but favored reactions are those that result in conjugation of charge or radical or that otherwise lead to stabilized products. Since the ionic rather than the neutral product is the high enthalpy species, one can generalize that mass spectral fragmentation is *governed by product ion stability.*

The relative stabilities of different ions are based on chemical principles that are also applicable in solution. Maintenance (whenever possible) of an octet of electrons, localization of charge on the most favorable site available, resonance delocalization, and absence of electron unpairing are fundamental. These principles can be illustrated by the following specific examples.

1. Dialkyl ethers tend to fragment to give the resonance stabilized α-cleavage* product $R—O—CH_2^+ \longleftrightarrow R—O^+=CH_2$, not to give RO^+ or $R—O—CH_2—CH_2^+$ ions.
2. Dialkyl thioethers, however, do give RS^+ ions, which are stabilized by the large polarizability of sulfur.
3. Aryl ethers yield the resonance stabilized C—O cleavage products ArO^+.
4. Alkyl arenes give $ArCH_2^+$ in strong preference to Ar^+ or $Ar(CH_2)_n^+$.
5. The bifunctional compound $R^1—O—CH_2—NR^2R^3$ yields $CH_2=N^+R^2R^3$ rather than $R^1—O^+=CH_2$, since the charge is preferentially localized on the more electropositive element.

Some of the above points, including reasons for the preferred stability order, are summarized in equations 14-1 to 14-4.

$$\underset{R}{\overset{R}{\diagdown}}\underset{R}{\overset{R}{N^+}}\underset{R}{\diagup} \quad > \quad \underset{R}{\overset{R}{\diagdown}}\overset{R}{C^+}\diagup \qquad \text{Octet rule} \qquad (14\text{-}1)$$

$$\diagup\!\!\!\diagdown\cdot_+ \quad > \quad \diagup\!\!=\!\!\diagdown^+ \qquad \text{Resonance delocalization} \qquad (14\text{-}2)$$

$$\underset{R}{\overset{R}{\diagdown}}\overset{R}{C^+}\diagup \quad > \quad \underset{H}{\overset{R}{\diagdown}}\overset{R}{C^+}\diagup \quad > \quad \underset{H}{\overset{R}{\diagdown}}\overset{H}{C^+}\diagup \qquad \text{Polarizability and hyperconjugation} \qquad (14\text{-}3)$$

$$\underset{R}{\overset{R}{\diagdown}}\underset{R}{\overset{R}{N^+}}\diagup \quad > \quad \underset{R}{\overset{R}{\diagdown}}\overset{R}{O^+}\diagup \quad > \quad \underset{R}{:\overset{R}{F}:^+} \qquad \text{Electronegativity} \qquad (14\text{-}4)$$

The importance of charge delocalization in driving the fragmentation of dioxolanes as shown in eq. 14-5.

$$(14\text{-}5)$$

14-1b Odd– and Even–Electron Ions

Odd-electron ions necessarily contain unpaired electrons. They include the molecular radical cations formed by EI. Those odd-electron ions that contain sufficient energy to fragment may eliminate either a radical or an even-electron neutral fragment (eqs. 14-6 and 14-7), but an even-electron ion normally fragments only to yield another even-

---

*C—C bonds attached to a functionality are described in order as α-bonds, β-bonds, etc.

$$\text{odd}^{+\cdot} \longrightarrow \text{even}^+ + \text{R}^\cdot \qquad (14\text{-}6)$$

$$\text{odd}^{+\cdot} \longrightarrow \text{odd}^{+\cdot} + \text{N} \qquad (14\text{-}7)$$

electron ion (eq. 14-8); it will not usually eliminate a radical to form an odd-electron ion (eq. 14-9). Exceptions to this last rule occur when the bond cleaved in the even-electron

$$\text{even}^+ \longrightarrow \text{even}^+ + \text{N} \qquad (14\text{-}8)$$

$$\text{even}^+ \nrightarrow \text{odd}^{+\cdot} + \text{R}^\cdot \qquad (14\text{-}9)$$

ion is particularly weak (polybrominated compounds, for example, can undergo multiple successive Br· losses) or when a particularly stable odd-electron ion results as in eq. 14-10. Consistent with the odd–even electron generalization, the important even-electron

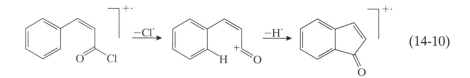

$$(14\text{-}10)$$

fragment ions $\text{R}-\text{C}\equiv\text{O}^+$, $\text{C}_6\text{H}_5\text{CH}_2{}^+$, and $\text{R}-\text{O}^+=\text{CH}_2$ lose small neutral molecules in the further fragmentations shown in eqs. 14-11 to 14-13. Note in each case that the

$$\text{RC}\equiv\overset{+}{\text{O}} \longrightarrow \text{R}^+ + \text{CO} \qquad (14\text{-}11)$$

$$\text{ArCH}_2^+ \longrightarrow \text{C}_5\text{H}_5^+ + \text{C}_2\text{H}_2 \ \ (\text{Ar}{=}\text{C}_6\text{H}_5) \qquad (14\text{-}12)$$

$$\text{R}-\overset{+}{\text{O}}{=}\text{CH}_2 \longrightarrow \text{R}^+ + \text{H}_2\text{CO} \qquad (14\text{-}13)$$

neutrals lost are stable, closed-shell molecules.

14-1c Stevenson's Rule

$$\text{A}-\text{B}^{+\cdot} \begin{array}{c} \nearrow \text{A}^\cdot + \text{B}^+ \\ \searrow \text{A}^+ + \text{B}^\cdot \end{array}$$

$$(14\text{-}14)$$

The control of product ion enthalpy over the extent of dissociation by two competitive pathways (eq. 14-14) is summarized in Stevenson's rule. This rule states that for a simple bond cleavage, the fragment with the lowest ionization energy will preferentially take the charge. Since the same bond is broken and only competition for the charge is involved, entropic factors cancel each other out. From the definition of ionization energy (heat of the reaction $\text{A}^\cdot \longrightarrow \text{A}^+$), the difference in activation energies equals the difference $\text{IE}(\text{A}) - \text{IE}(\text{B})$. Hence the rule suggests that fragment ion abundances are governed by the energetics of fragmentation. Elimination reactions also follow the rule. For example, ionized cyclohexene undergoes a retro Diels-Alder reaction (eq. 14-15) and yields $\text{C}_4\text{H}_6^{+\cdot} + \text{C}_2\text{H}_4$ in preference to $\text{C}_4\text{H}_6 + \text{C}_2\text{H}_4^{+\cdot}$. (The ionization energy of 1,3-butadiene is 9 eV, that of ethylene is 10 eV.)

$$(14\text{-}15)$$

14-1d Rearrangement vs. Simple Cleavage

Ions of very low internal energy, generated by electron impact at low ionizing energy, tend to fragment by low activation energy processes that generate stable products. Some entropically favored processes, such as simple bond cleavages, are no longer allowed energetically, and lower energy but intrinsically slower processes, often involving molecular rearrangement, increase in importance. The choice of low energy conditions, therefore, has marked effects on the appearance of the mass spectrum (see Figure 14-7).

Rearrangements can compete with simple bond cleavages only if they have low activation energies and, even then, they are favored only if the ions examined have low internal energies. EI yields ions of relatively high internal energy, for which only simple cleavages or the simpler rearrangements need be considered. The latter often involve hydrogen migration and/or the elimination of a stable neutral molecule (H_2O, N_2, CO_2, CO, an alkene, or an alcohol). For example, diaryl carbonate molecular ions eliminate CO_2 to give ions that show all the properties of the diaryl ether molecular ions (eq. 14-16). Similarly, 1-hexanol eliminates water via the six-centered intermediate shown in eq. 14-17. Note single- and double-headed arrows are used to show the movements of single electrons and electron pairs, respectively.

$$Ar-O-Ar + CO_2 \qquad (14\text{-}16)$$

$$C_6H_{12}{}^{+\cdot} + H_2O \qquad (14\text{-}17)$$

Under low-energy conditions more than one new bond can be formed, as in the double hydrogen rearrangement shown in eq. 14-18. The protonated enol ion and the accompanying allylic C_4H_7 radical are both very stable products, even though their formation involves considerable rearrangement.

$$CH_3\overset{O^{+\cdot}}{C}-C_5H_{11} \longrightarrow CH_3-\overset{H_2O^+}{C}=CH_2 + C_4H_7{}^{\cdot} \qquad (14\text{-}18)$$

14-1e Ions of High Internal Energy

A small fraction of the ions generated by electron impact have very high internal energies. These ions do not dissociate by familiar chemical processes; for example, they fragment to form $C_n{}^{+\cdot}$, C_nH^+, and other high enthalpy products. Minor signals are observed at almost every mass in EI mass spectra, partly because of the presence of these ions that undergo indiscriminate, high-energy bond cleavages.

14-1f Proximate vs. Remote Fragmentation

Many fragmentations of gas phase ions are triggered by the charge site. Bonds in the vicinity of the atoms carrying low electron density are weakened and dissociation is promoted. In some ions, however, the charge is a spectator to dissociation, which occurs by a process that resembles thermolysis. An example of such a charge remote fragmentation is C—C cleavage in the *n*-alkyl carboxylate anion (eq. 14-19). The reaction is

$$RCH_2CH_2(CH_2)_nCO_2{}^- \longrightarrow CH_2=CH(CH_2)_nCO_2{}^- + H_2 + (R-H) \qquad (14\text{-}19)$$

known from neutralization–reionization experiments to yield H_2 and an alkene as neutral fragments. Deuterium isotopic labeling shows it to occur via a six-centered transition state. The reaction occurs with approximately equal facility at every bond along the chain and terminates with cleavage of the γ-bond to yield the product $CH_2=CH-(CH_2)_nCO_2{}^-$. Similar processes occur in quaternary ammonium cations when functional groups such as ethers, epoxides, or olefins are embedded in the chain, the interruption of the homologous series of product ions associated with remote fragmentation indicates the position of the functionality. These types of reactions have proved valuable in natural product structural elucidation. They are normally performed by MS–MS, which isolates the precursor ion and provides it with the necessary internal energy.

14-1g   Distonic Ions

Distonic ions are species in which the radical site is formally separated from the charge site. They may be more or less stable than their conventional radical cation isomers, but they are a distinct chemical species whose chemistry is of much interest since it is often dominated by the radical rather than the charged site. For example, the molecular ion of methanol, $CH_3OH^{+\cdot}$, is much less stable than its distonic isomer, $\cdot CH_2OH_2^+$. The latter is conveniently generated by electron impact on 1,2-ethanediol. Calculations and experiments show that both species exist in deep potential wells and are noninterconverting. By contrast, the neutral ylid $^-CH_2OH_2^+$ is unstable. The chemistry of distonic ions is investigated by studying both their ion-molecule reactions and their fragmentations. For example, the ion $\cdot H_2CCH_2C\equiv O^+$ is characterized by reaction with dimethyl disulfide to give the thiomethyl abstraction product and by fragmentation to give $C_2H_4^{+\cdot}$. Fragmentation is often driven by the radical rather than the charge site.

14-1h   Charge Localization

Much of what has been discussed so far can be summarized as follows: dissociation of an ion can be rationalized as proceeding from its most stable, charge-localized structure. The total energy of an isolated ion is fixed on ionization, so the structure that is thermodynamically most stable also has the maximum internal energy. Hence it should also dissociate most readily. The corresponding functional group also is particularly stable in the product. The charge localization concept, therefore, implies control of fragmentation by product ion stability.

Consider the α,ω-amino alcohol shown in eq. 14-20 and eq. 14-21. The two charge-

$$\overset{+\cdot}{N}H_2-(CH_2)_n-OH \longrightarrow \overset{+}{N}H_2{=}CH_2 + \cdot(CH_2)_{n-1}OH \qquad (14\text{-}20)$$

$$NH_2-(CH_2)_n-\overset{+\cdot}{O}H \longrightarrow NH_2(CH_2)_{n-1}^{\cdot} + CH_2{=}\overset{+}{O}H \qquad (14\text{-}21)$$

localized forms of the molecular ion differ in internal energy by approximately 1.2 eV, which is the difference between typical alkylamine and alkanol ionization energies. Hence the structure with charge localized on the more electropositive atom (N) should have the greater internal energy and be more likely to fragment. The observed product is $H_2N^+{=}CH_2$, a particularly stable product in which the charge is retained on nitrogen.

Reactions 14-22 and 14-23, each of which are major processes in the normal EI mass spectra, illustrate further the charge localization hypothesis in rationalizing fragmentation.

$$C_2H_5-\overset{\frown}{C}H-\overset{+\cdot}{O}-CH_3 \longrightarrow CH_3CH{=}\overset{+}{O}-CH_3 + C_2H_5^{\cdot}$$
$$\overset{|}{C}H_3 \qquad (14\text{-}22)$$

$$C_6H_5-\overset{+\cdot}{S}-CH_3 \longrightarrow C_6H_5S^+ + CH_3^{\cdot} \qquad (14\text{-}23)$$

14-1i   Characteristic Fragment Ions and Neutral Fragments

Mass spectra often display ions that are characteristic of particular functional groups. Some of these are collected for convenience in Table 14-1. Note the overlap of ions of several different structures and formulas at the same mass-to-charge ratios. Note also that these characteristic ions occur as homologous series. Even more valuable in structure elucidation is a knowledge of the neutral fragments that are lost from the molecular ion. This information is deduced from the appearance of the spectrum. Obviously,

TABLE 14-1

Some Characteristic Fragment Ions*

Mass	Ion	Possible Functionality	Mass	Ion	Possible Functionality
15	CH_3^+	Methyl, alkane	50	$C_4H_2^{+\cdot}$	Aryl
29	$C_2H_5^+$, HCO^+	Alkane, aldehyde	51	$C_4H_3^+$	Aryl
30	$CH_2{=}NH_2^+$	Amine	77	$C_6H_5^+$	Phenyl
31	$CH_2{=}OH^+$	Ether or alcohol	83	$C_6H_{11}^+$	Cyclohexyl
39	$C_3H_3^+$	Aryl	91	$C_7H_7^+$	Benzyl
43	$C_3H_7^+$, CH_3CO^+	Alkane, ketone	105	$C_6H_5C_2H_4^+$	Substituted benzene
45	CO_2H^+, CHS^+	Carboxylic acid, thiophene		$CH_3C_6H_4CH_2^+$	Disubstituted benzene
47	CH_3S^+	Thioether		$C_6H_5CO^+$	Benzoyl

*Lowest number of homologous series is listed.

TABLE 14-2

Some Characteristic Neutral Losses

Mass	Composition	Possible Functionality	Mass	Composition	Possible Functionality
14	Impurity, homologue		29	C_2H_5	Alkyl
			30	CH_2O	Methoxy
15	CH_3	Methyl		NO	Aromatic nitro
16	CH_4	Methyl		C_2H_6	Alkyl (CI)
	O (rarely)	Amine oxide	31	CH_3O	Methoxy
	NH_2	Amide	32	CH_3OH	Methyl ester
17	NH_3	Amine (CI)	33	$H_2O^+ CH_3$	Alcohol
	OH	Acid, tertiary alcohol		HS	Mercaptan
18	H_2O	Alcohol, aldehyde, acid (CI)	35	Cl	Chloro compound
			36	HCl	Chloro compound
19	F	Fluoride	42	CH_2CO	Acetate
20	HF	Fluoride	43	C_3H_7	Propyl
26	C_2H_2	Aromatic	44	CO_2	Anhydride
27	HCN	Nitrile, hetero-aromatic	46	NO_2	Aromatic nitro
			50	CF_2	Fluoride
28	CO	Phenol			
	C_2H_4	Ether			
	N_2	Azo			

the first fragmentation step, from the molecular ion, gives more structural information than subsequent steps that may follow rearrangements. A collection of characteristic neutral losses is provided in Table 14-2.

14-2 AMINE AND ETHER TYPE FRAGMENTATION

The fundamental principles of fragmentation, described previously, are evident in the mass spectra of simple aliphatic and aromatic compounds containing the amine and ether functional groups. These functional groups are representative of all those in which a heteroatom is singly bonded to the carbon skeleton. They display fragmentation in EI that is driven by the radical cation, the charge being localized on the heteroatom ($-X^{+\cdot}-$). Amines invariably yield fragment ions that contain nitrogen because of its electropositive nature.

14-2a Amines

Aliphatic amines yield molecular radical cations and those with excess energy fragment predominantly by α-cleavage to generate a free radical and a closed-shell iminium ion (eq. 14-24).

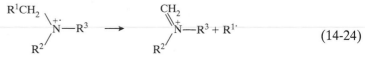

$$(14\text{-}24)$$

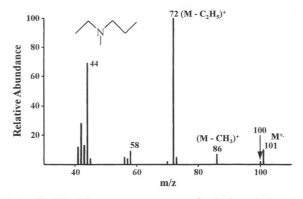

Figure 14-1 EI (70 eV) mass spectrum of ethylmethylpropylamine showing the intact molecular ion, m/z 101, α-cleavage fragments m/z 100, 86, and 72, and characteristic iminium ions, m/z 44 and 58. (From P.J. Ausloos, C. Clifton, O.V. Fateev, A.A. Levitsky, S.G. Lias, W.G. Mallard, A. Shamin, and S.E. Stein, NIST/EPA/NIH Mass Spectral Library—Version 1.5, National Institute of Standards and Technology, Gaithersburg, MD [1996].)

If R^2 and R^3 contain C—C bonds, α-cleavage might involve any of the three C—C bonds. Competition among the various primary processes depends on the stability of the resulting ions: the higher the stability the more favored the process. This general point was made previously in the discussion of the energetics of dissociation and is illustrated for the particular case of ethylmethylpropylamine in Figure 14-1. The spectrum shows three α-cleavage products associated with loss of H·, methyl, and ethyl radicals, the last of which (m/z 72) is the base peak (the most intense ion) in the spectrum.

After α-cleavage, the product ion may be stable and appear as such in the mass spectrum, or it may retain sufficient internal energy to undergo further fragmentation. Secondary fragmentation follows the odd–even electron law, that is, a closed-shell neutral is released and another closed-shell ion of lower m/z is generated. The liberation of a neutral molecule from an iminium ion necessarily involves rearrangement—the formation of new bonds—but this process can simply involve hydrogen migration. In fact, alkene loss is the dominant further fragmentation step, and it generates a product ion of the same structural type, another iminium ion (eq. 14-25).

$$\underset{R^2}{\overset{CH_2}{\underset{\diagdown}{\overset{\diagup}{N}}}}\!\!-\!\!R^3 \longrightarrow \underset{R^2}{\overset{CH_2}{\underset{\diagdown}{\overset{\diagup}{N}}}}\!\!-\!\!H + (R^3 - H) \tag{14-25}$$

Note that this reaction requires at least a C_2-alkyl substituent and occurs with hydrogen rearrangement to the charged site, nitrogen. As with the primary alkyl loss fragmentation, different alkene losses can occur competitively. In the mass spectrum of ethylmethylpropylamine the ions m/z 44 (eq. 14-26) and m/z 58 result from such further fragmentation.

$$R^2\!\!-\!\!\overset{+}{N}\!\!\overset{\diagup CH_2}{\underset{\diagdown CH_3}{}} \longrightarrow H\!\!-\!\!\overset{+}{N}\!\!\overset{\diagup CH_2}{\underset{\diagdown H}{}} + (R^2 - H) \tag{14-26}$$
$$\underset{m/z\ 30}{}$$

In general, competitive fragmentation leads to a family of characteristic iminium ions from aliphatic amines. These ions form the series m/z 30, 44, 58, 72, 86, . . . and are well represented in Figure 14-1. As indicated in Table 14-1 they provide evidence that the compound is an alkylamine.

Protonated aliphatic amines, as generated in CI and some DI experiments, do not readily undergo α-cleavage. This observation is also consistent with the odd–even electron rule. Like their analogues, the iminium ions just discussed, they undergo alkene loss

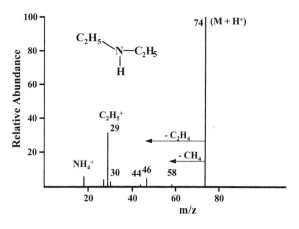

Figure 14-2 Fragmentation of protonated diethylamine generated by CI and dissociated in an MS–MS experiment. Note the iminium ion series at m/z 30, 44, and 58, as well as the abundant $(M+H)^+$ ion.

(one of the most widespread neutral losses) to generate a lower homologue (eq. 14-27). In addition, an alkane molecule can be lost to generate an iminium ion (eq. 14-28). These

$$\underset{\substack{\text{H} \quad \text{R}^3}}{\overset{\substack{\text{R}^1\text{CH}_2 \quad \text{R}^2}}{\text{N}^+}} \longrightarrow \underset{\substack{\text{H} \quad \text{H}}}{\overset{\substack{\text{R}^1\text{CH}_2 \quad \text{R}^2}}{\text{N}^+}} + (R^3 - H) \qquad (14\text{-}27)$$

$$\underset{\substack{\text{H} \quad \text{R}^3}}{\overset{\substack{\text{R}^1\text{CH}_2 \quad \text{R}^2}}{\text{N}^+}} \longrightarrow R^1CH{=}\underset{\substack{\text{H}}}{\overset{\substack{\text{R}^3}}{\text{N}^+}} + R^2H \qquad (14\text{-}28)$$

features are illustrated in Figure 14-2, which displays the fragmentation behavior of protonated diethylamine, generated by CI and dissociated by collision in an MS–MS experiment. The spectrum shows prominent ions due to loss of alkane and alkene molecules as just predicted. Depending on the other functional groups present, protonated primary amines can lose NH_3 by a simple heterolytic bond cleavage process (eq. 14-29), although this reaction is favored only if R^+ is a stabilized ion.

$$R\overset{\curvearrowleft}{-}\overset{+}{N}H_3 \longrightarrow R^+ + NH_3 \qquad (14\text{-}29)$$

If attention is turned from aliphatic to aromatic amines, a relatively small number of fragments of high relative abundances are observed in EI. If we limit attention to aromatics with just a few simple substituents to discern the behavior of amines, we record data such as that shown in Figure 14-3. All three compounds show either the molecular ion or a high mass fragment ion as the base peak. The odd mass of the molecular ion indicates a single (or odd number) of nitrogen atoms in each molecule. (This conclusion is general and arises simply from the valences and masses of C, H, N, and O.) A somewhat unexpected feature of the EI mass spectrum of aniline is the loss of HNC, which occurs to give what is known to be a linear isomer of $C_5H_6^{+\cdot}$, an ion that undergoes further H· loss to generate the more stable even-electron ion, $C_5H_5^+$. The 4-isopropyl-aniline spectrum shows α-cleavage directed by the phenyl group rather than the nitrogen atom. In N-ethyl N-methylaniline, β-cleavage yields m/z 120, which dominates the spectrum. Note that there are also rearrangement processes evident in this spectrum, including that leading to the ion of m/z 91.

Elimination of HCN occurs from heteroaromatic radical cations such as pyridines. The dissociation of the radical cation of 3-ethylpyridine and the protonated molecule is compared in Figure 14-4. Fragmentation occurs readily from the radical cation to form a stable ion $C_5H_5^+$ at m/z 65. Protonated 3-ethylpyridine fragments little; methyl loss forms m/z 93, a distonic ion, in which the radical and charge sites are separated.

Other information on amine fragmentation appears elsewhere in this text. For example, other reactions typical of protonated and quaternary amines are discussed in Section 14-6. The worked problems provide more information on the EI spectra of aromatic nitrogen compounds. All the examples are worth consideration, since many other functional groups also show characteristic α-cleavage and alkene loss fragmentations. Therefore, amines serve as a prototypical case.

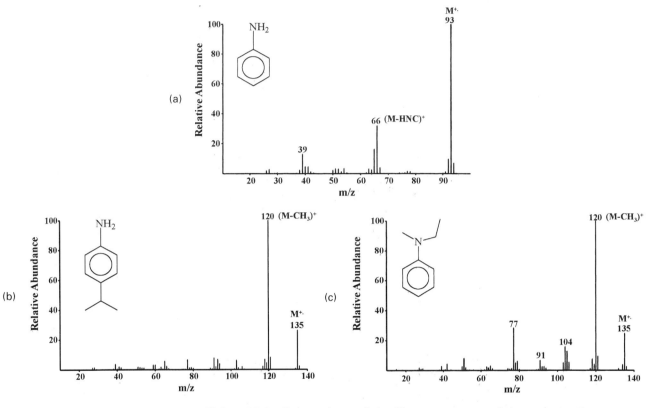

Figure 14-3 Comparison of the EI mass spectra of three aromatic amines, all of which show abundant, odd-mass, molecular ions. The ions m/z 120 are due to $CH_3^.$ loss by cleavage of the β-bond to the phenyl ring (b) and the α-bond to the amine nitrogen (c), respectively. (From P.J. Ausloos, C. Clifton, O.V. Fateev, A.A. Levitsky, S.G. Lias, W.G. Mallard, A. Shamin, and S.E. Stein, NIST/EPA/NIH Mass Spectral Library—Version 1.5, National Institute of Standards and Technology, Gaithersburg, MD [1996].)

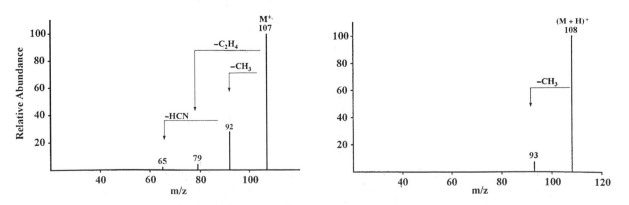

Figure 14-4 Comparison of the fragmentation undergone by the molecular cation (M+·) and the protonated molecule (M+H)+ of 3-ethylpyridine. Data are from collision-induced dissociation experiments.

14-2b   Ethers

Primary Fragmentations.   Alkyl ethers, like amines, typify the behavior of compounds containing functional groups with singly-bonded heteroatoms. The less electropositive heteroatom (oxygen vs. nitrogen) means that additional characteristic processes are observed compared with those seen for amines. Consider first the EI mass spectra of aliphatic ethers. *Primary fragmentations* are (1) α-cleavage, (2) C—O cleavage, and (3) fragmentation with simple H rearrangement.

The spectrum of ethyl 1-phenylethyl ether (Figure 14-5) displays ions due to each of these primary reactions of the molecular ion. These processes are typical of ethers, although the charge-stabilizing character of the aryl ring affects ion abundances. The three primary reactions are the following:

1. α-Cleavage, that is, cleavage of the first C—C bond counting from the functionality, which gives rise to *m/z* 135, the base peak in the 70 eV spectrum (eq. 14-30).

$$\text{α-Cleavage:} \quad H_5C_2 \overset{+\cdot}{-\!O} \!-\! \underset{\underset{CH_3}{|}}{CH} \!-\! C_6H_5 \quad \longrightarrow \quad H_5C_2 \overset{+}{-\!O} \!=\! CH \!-\! C_6H_5 \qquad (14\text{-}30)$$
$$m/z\ 135$$

2. C—O cleavage with charge retention by the hydrocarbon and formation of an oxy radical produces *m/z* 105 (eq. 14-31).

$$\text{C—O Cleavage:} \quad H_5C_2 \overset{+\cdot}{-\!O} \underset{\underset{CH_3}{|}}{CH} \!-\! C_6H_5 \quad \longrightarrow \quad CH_3 \!-\! \overset{+}{CH} \!-\! C_6H_5 \qquad (14\text{-}31)$$
$$m/z\ 105$$

3. The rearrangement process in which ethane is lost and the low abundance ion, *m/z* 120, is produced (eq. 14-32).

$$\text{Elimination:} \quad H_5C_2 \overset{+\cdot}{-\!O} \!=\! \underset{\underset{CH_3}{|}}{CH} \!-\! C_6H_5 \quad \longrightarrow \quad \overset{+\cdot}{O}\!=\!C \!\!\begin{array}{l} C_6H_5 \\ \\ CH_3 \end{array} \qquad (14\text{-}32)$$
$$m/z\ 120$$

If α-cleavage can occur at several alternative positions, the process leading to the most stable carbenium ion will occur preferentially. If different radicals can be lost by α-cleavage, then the larger radical is preferentially lost, provided the rule concerning product ion stabilities is not contradicted. Thus ethyl isobutyl ether loses the $C_3H_7\!\cdot$ radical to give an ion of 8% relative abundance and the $CH_3\!\cdot$ radical to give an ion of less than 1% relative abundance (eq. 14-33). If formation of a stable ion and formation of a stable radical are competitive processes, then ionic stabilization is more important. Thus isobutyl iso-

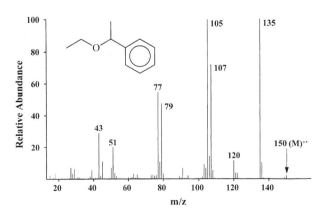

Figure 14-5   EI (70 eV) mass spectrum of ethyl 1-phenylethyl ether (1-ethoxyethyl benzene). (From P.J. Ausloos, C. Clifton, O.V. Fateev, A.A. Levitsky, S.G. Lias, W.G. Mallard, A. Shamin, and S.E. Stein, NIST/EPA/NIH Mass Spectral Library—Version 1.5, National Institute of Standards and Technology, Gaithersburg, MD [1996].)

propyl ether loses a secondary propyl radical to give $CH_2{=}O^+{-}iPr$ ion to a lesser extent than it loses a primary methyl radical to give $iBu{-}O^+{=}CH{-}CH_3$. These generalizations apply to other functional groups as well as ethers.

$$
\begin{array}{c}
\\
\text{CH}_3\text{CH}_2\overset{+\cdot}{-}\text{O}\text{---}\text{CH}_2\text{---}\text{CH}\begin{smallmatrix}\diagup\text{CH}_3\\[2pt]\diagdown\text{CH}_3\end{smallmatrix}\\
m/z\ 102
\end{array}
\qquad
\begin{array}{c}
\xrightarrow{-\text{C}_3\text{H}_7\cdot}\quad
\begin{array}{c}\text{CH}_3\text{CH}_2\overset{+}{-}\text{O}{=}\text{CH}_2\\ m/z\ 59\end{array}
\xrightarrow{-\text{C}_2\text{H}_4}\\[18pt]
\begin{array}{c}\overset{+}{\text{HO}}{=}\text{CH}_2\\ m/z\ 31\end{array}\\[18pt]
\xrightarrow{-\text{CH}_3\cdot}\qquad
\begin{array}{c}\text{CH}_2{=}\overset{+}{\text{O}}\text{---}\text{CH}_2\text{---}\text{CH}\begin{smallmatrix}\diagup\text{CH}_3\\[2pt]\diagdown\text{CH}_3\end{smallmatrix}\\ m/z\ 87\end{array}
\xrightarrow{-\text{C}_4\text{H}_8}
\end{array}
\tag{14-33}
$$

Secondary Fragmentations.   Secondary fragmentations occur because some primary fragment ions retain sufficient internal energy to dissociate further. For the α-cleavage products, formation of an even-electron product ion with elimination of an even-electron neutral is favored on energetic grounds (see eq. 14-8). The tendency to form the most stable product ion favors retention of the charge on the heteroatom. Frequently when these two considerations hold, a neutral hydrocarbon molecule is lost. Thus, the α-cleavage product, m/z 135, in ethyl 1-phenylethyl ether (Figure 14-5) fragments further by loss of ethylene to yield the abundant ion of mass m/z 107. Similarly, ethyl isobutyl ether forms an ion with m/z 31 of 76% relative abundance, by the sequence of reactions shown in eq. 14-33. As an alternative mode of secondary fragmentation, neutral formaldehyde may be eliminated as shown in eq. 14-34, which is analogous to eq. 14-29. Naturally, this C—O cleavage pathway is most important when R^+ is a stable ion, such as a tertiary, benzylic, allylic, or other stabilized carbocation.

$$
\text{R}\overset{+}{-}\text{O}{=}\text{CH}_2 \longrightarrow \text{R}^+ + \text{CH}_2\text{O} \tag{14-34}
$$

Secondary fragment ions that retain enough energy will again expel a neutral molecule. Thus $CH_2{=}O^+ - H$ (m/z 31) loses H_2 to give $H{-}C{\equiv}O^+$ (m/z 29). These higher order fragmentations yield low mass ions, which provide little specific information on molecular structure. Nevertheless, low mass ions, especially homologous series of ions, can indicate the types of *functional groups* present (Table 14-1). Most mass spectra, including those of ethers, are more easily interpreted by considering the *neutral fragments* that are lost from the molecular ion. In other words, the information-rich region of a mass spectrum is the high and not the low mass region. Table 14-2 lists some neutral losses and their structural implications.

Further fragmentation of the cation (R^+) follows a pattern typical of the spectra of alkanes and other compounds which yield alkyl cations: H_2 loss is ubiquitous, and CH_4 and larger alkane losses also can be observed. Highly unsaturated alkyl cations such as $C_7H_7^+$ tend to lose carbon-containing fragments (eq. 14-35) rather than H_2. The

$$
\begin{array}{c}
\text{C}_7\text{H}_7^+\\ m/z\ 91
\end{array}
\longrightarrow
\begin{array}{c}
\begin{array}{c}\text{C}_5\text{H}_5^+ + \text{C}_2\text{H}_2\\ m/z\ 65\end{array}\\
\searrow\\
\begin{array}{c}\text{C}_3\text{H}_3^+ + \text{C}_2\text{H}_2\\ m/z\ 39\end{array}
\end{array}
\tag{14-35}
$$

spectrum of ethyl 1-phenylethyl ether (Figure 14-5) shows many features from hydrocarbon ions. For example, the 1-methylbenzyl cation (m/z 105) decomposes through loss of C_2H_4, and then of C_2H_2, to give m/z 77 and 51, respectively.

The relative importance of ions with and without heteroatoms in ethers is intermediate between that for compounds containing more electropositive heteroatoms such as nitrogen, for which hydrocarbon ions are of low total abundance, and less electropositive atoms such as the halogens, for which there is a strong tendency to form hydrocarbon ions.

Rearrangements.   Processes in which bonds are formed as well as broken are often favored by their lower energy requirements. In some cases it is obvious from the nature of the products that rearrangements have occurred; in other cases, bond formation is only revealed after detailed study of the structure of the product ion. The fragmentation of amines produces a number of examples of molecular rearrangement, including the ion m/z 91 in the spectrum of N-ethyl N-methylaniline (Figure 14-3c). Fortunately, for the ease of interpretation of mass spectra, it is hydrogen atoms often that migrate during rearrangements. These processes make important contributions to the mass specta of ethers.

Some bond-forming reactions, particularly simple eliminations occurring in the molecular ion, are immediately evident from their even-mass (odd-electron) nature. They are of value in characterizing the analyte. These processes often give rather low abundance ions in 70 eV mass spectra, but become more important at lower electron energies. In ethers, loss of an alkene or loss of an alcohol can occur by hydrogen rearrangement. The reactions are complementary; only the location of the charge is changed (eq. 14-36). The alcohol has the lower ionization energy and retains the charge,

$$H-\overset{|}{\underset{|}{C}}-\overset{+\cdot}{\underset{|}{C}}-\overset{+\cdot}{O}-R \quad \begin{matrix} \nearrow \\ \searrow\!\!\!\!\! \times \end{matrix} \quad \begin{matrix} \underset{/}{\overset{\backslash}{C}}\!\!=\!\!\underset{\backslash}{\overset{/}{C}} \;+\; \overset{+\cdot}{ROH} \\[2ex] \underset{/}{\overset{\backslash}{C}}\!\!\overset{+\cdot}{=}\!\!\underset{\backslash}{\overset{/}{C}} \;+\; ROH \end{matrix} \qquad (14\text{-}36)$$

according to Stevenson's rule (see Section 14-1c). The butyl ethyl ether spectrum (Figure 14-6) shows only the ion at m/z 56, composition $C_4H_8^{+\cdot}$, instead of $C_2H_5OH^{+\cdot}$ at m/z 46 (ionization energies are 9.2 and 10.5 eV, respectively).

The species ROH_2^+ is another important fragment ion produced by rearrangement. It is not abundant in 70 eV spectra of ethers, but it is the dominant fragment ion at very low electron energies. Obviously, several bonds must be cleaved to form this very stable ion. Similar stable ions formed by multiple hydrogen transfers occur for other functional groups, including $RCO_2H_2^+$ for alkyl esters (compare also eq. 14-18). The spectrum of dihexyl ether taken at 70 eV (Figure 14-7) is dominated by formation and further fragmentation of the alkyl cation (m/z 85). Oxygen-containing ions are almost absent. At 12 eV further fragmentation of the ionized alkene (m/z 84) and of m/z 85 is virtually eliminated, and the relative abundances of the molecular ion and the ROH_2^+ rearrangement ion (m/z 103) increase dramatically.

Aryl Ethers.   Alkyl aryl ethers typically show molecular ions of much greater abundance than dialkyl ethers. The stabilizing role of the aryl group is a general phenomenon. Butyl phenyl ether can be compared with dibutyl ether for purposes of illustration. The molecular ion abundance of the aromatic compound is 20%, that of the dialkyl ether <1%. Aryl methyl ethers differ from dialkyl ethers in two of their most important primary fragmentations. One of these reactions is loss of the methyl radical

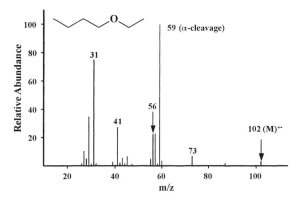

Figure 14-6   EI (70 eV) mass spectrum of butyl ethyl ether showing primary α-cleavage fragment due to propyl radical loss and the alkene rearrangement ion at even mass, m/z 56.

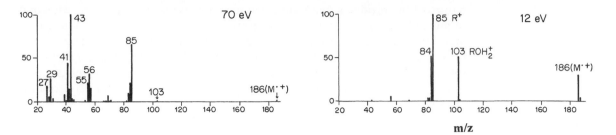

Figure 14-7 High (70 eV) and low energy (12 eV) EI mass spectra of dihexyl ether. Note the stable rearrangement ions generated by low energy fragmentation. (From G. Spiteller and M. Spiteller-Freidmann, in R. Bonnett and J.G. Davis, eds., *Some Newer Physical Methods in Structural Chemistry,* United Trade Press, London, 1967.)

to give the aryloxy ion (ArO^+), a far more stable species than are alkoxy cations. Further fragmentation (eq. 14-37) of this even-electron cation occurs by elimination of CO. The second primary fragmentation—loss of the heteroatom as formaldehyde to give the molecular ion of the parent aromatic compound (eq. 14-38)—is an illustration

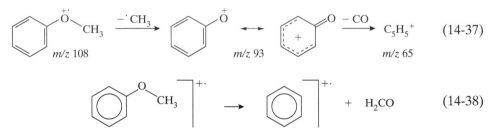

of the fact that *simple* rerrangement reactions can lead to relatively abundant ions in 70 eV spectra.

The arene radical cation formed by fragmentation undergoes the same further fragmentation reactions as do $ArH^{+\cdot}$ molecular ions formed by direct electron ionization. The subsequent reactions, ignoring those due to substituents that may be present, are $H^\cdot$ loss, $C_3H_3^\cdot$ radical loss, and loss of C_2H_2 as a closed-shell neutral fragment. The spectrum of 2-methoxypyridine (Figure 14-8) shows the fragmentation sequence $M^{+\cdot} \longrightarrow (M-H_2CO)^{+\cdot} \longrightarrow (M-H_2CO-HCN)^{+\cdot}$, in which the $(M-H_2CO)^{+\cdot}$ ion apparently has the same structure as the pyridine molecular ion. The benzene molecular ion itself shows losses of two and three hydrogen atoms to give *m/z* 76 and 75, fur-

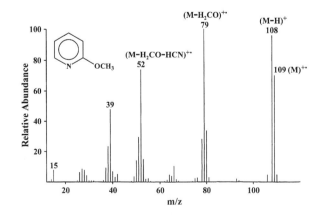

Figure 14-8 EI (70 eV) mass spectrum of 2-methoxypyridine showing formaldehyde elimination and hydrogen atom loss.(From P.J. Ausloos, C. Clifton, O.V. Fateev, A.A. Levitsky, S.G. Lias, W.G. Mallard, A. Shamin, and S.E. Stein, NIST/EPA/NIH Mass Spectral Library—Version 1.5, National Institute of Standards and Technology, Gaithersburg, MD [1996].)

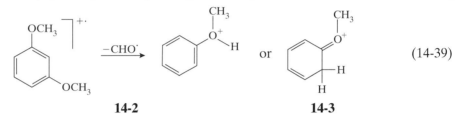

14-1

ther fragmentation of *m/z* 77 to give *m/z* 51, and of *m/z* 76 to give *m/z* 50, as well as other minor reactions.

This spectrum is also important because it provides an example of an *ortho effect.* The molecular ion loses H· to a far greater extent than do most methyl ethers; here $(M-1)^+ > M^{+\cdot}$, whereas in anisole $(M-1)^+$ is 3% of $M^{+\cdot}$. Direct interaction of the amino and methoxy groups occurs, forming a unique stable $(M-1)^+$ ion, perhaps the quinonoid ion **14-1.**

Another primary reaction of substituted anisoles is loss of the alkoxyl radical. This process is important because the ion Ar^+ (*m/z* 77 in phenyl ethers) provides a means of recognizing the aromatic group.

In certain substituted anisoles the loss of the formyl radical occurs to yield the species ArH_2^+. Why should the ion undergo a double hydrogen transfer to yield this product? The answer is product stability. The process is restricted to anisoles that bear oxygen- or nitrogen-containing substituents to which hydrogen can be transferred. It is also restricted to anisoles in which the competitive methyl elimination process is not facilitated. These points are well illustrated by comparing the mass spectra of the *m*- and *p*-dimethoxybenzenes (Figure 14-9). The $(M-CH_3)^+$ ion in the para isomer is resonance stabilized, so that it gives the base peak in the spectrum (37% of the total ion current); formyl radical loss is not observed. In the meta isomer, methyl radical loss barely occurs, and formyl radical loss is the dominant fragmentation with 11% of the total ion current. The formyl loss product probably has either of the stable structures shown in **14-2** or **14-3** (eq. 14-39). Either pathway has a requirement of a second heteroatom,

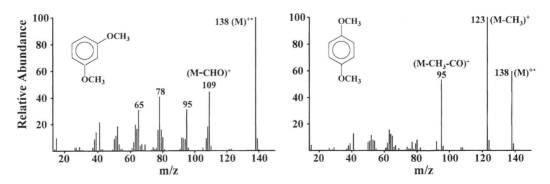

14-2   **14-3**   (14-39)

which explains why this pathway is absent in anisole itself. This process is comparable to ROH_2^+ formation from long chain alkyl ethers (compare also eq. 14-18).

The third isomer, *o*-dimethoxybenzene, exhibits a mass spectrum that differs from the spectra of the meta and para isomers. In particular, interaction of the ortho substituents occurs with formation of $C_6H_5^+$. Therefore, in this particular case, the mass spectra distinguish the isomeric disubstituted benzenes. Some other isomeric substituted anisoles give identical mass spectra. Again, this general situation applies far more widely than just to this class of compounds.

Turning from anisoles to other alkyl aryl ethers, one finds that alkene elimination to give ionized phenol is the dominant reaction. For example, ethyl *p*-tolyl ether (Fig-

Figure 14-9 Easily distinguished EI (70 eV) mass spectra of isomeric dimethoxybenzenes. (From P.J. Ausloos, C. Clifton, O.V. Fateev, A.A. Levitsky, S.G. Lias, W.G. Mallard, A. Shamin, and S.E. Stein, NIST/EPA/NIH Mass Spectral Library—Version 1.5, National Institute of Standards and Technology, Gaithersburg, MD [1996].)

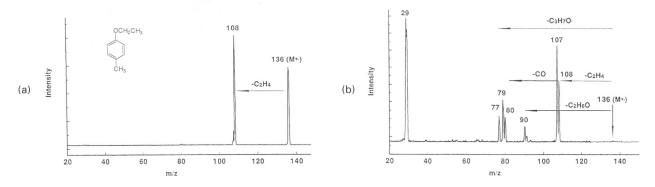

Figure 14-10 Fragmentation of the molecular cation radical of *p*-tolyl ethyl ether at (a) low and (b) high internal energy (MS–MS data). Alkene elimination with hydrogen rearrangement occurs at low energy.

ure 14-10) displays the base peak due to ethene elimination at low energy. As the internal energy deposition increases, further fragmentation occurs. Diaryl ethers are notable for their stable molecular ions and for the occurrence of a complex skeletal rearrangement resulting in CO elimination (see Worked Problem 14-3). Although this particular rearrangement could hardly have been predicted, guidelines have been discussed that rationalize most skeletal rearrangements. These complex rearrangements are minor processes in analytical 70 eV spectra. The major fragments observed in the diphenyl ether mass spectrum are the unexceptional ions $C_6H_5^+$ (55%) and its fragment (acetylene loss) $C_4H_3^+$ (58%).

Protonated Ethers. Ethers are easily protonated under CI and DI conditions, and the fragmentation of the protonated molecule determines the nature of the mass spectrum. Based on the behavior of protonated amines already discussed and on energetics, one expects that elimination reactions will be significant. This is the case: the loss of a neutral alcohol molecule gives the carbenium ion (eq. 14-40) and alkene elimination gives the protonated alcohol (eq. 14-41). Both reactions follow the even-

$$R^1 \overset{+}{\underset{H}{O}} R^2 \longrightarrow R^1OH + (R^2)^+ \tag{14-40}$$

$$R^1 \overset{+}{\underset{H}{O}} R^2 \longrightarrow R^1OH_2^+ + (R^2 - H) \tag{14-41}$$

electron → even-electron rule (eq. 14-8), as does alkane loss (eq. 14-42), a formal 1,2-elimination process that is important in some protonated ethers.

$$R^1CH_2 \overset{+}{\underset{H}{O}} R^2 \longrightarrow R^1CH = \overset{+}{O}H + R^2H \tag{14-42}$$

In addition, alkyl radical loss, an electron unpairing reaction, is observed, especially under high energy conditions (Table 14-3). The fragmentation of other cationized ethers, although less well characterized, follows similar principles.

14-2c Alcohols and Phenols

The principles of alcohol fragmentation parallel those of ethers. Molecular ions of aliphatic alcohols are only of moderate abundance. Aliphatic alcohols undergo α-bond cleavage as a major primary fragmentation to form oxonium ions (*m/z* 31, 45, 59, . . .), and this homologous series helps to identify the alcohol functional group, provided that it is remembered that alkyl ethers give the same set of ions.

TABLE 14-3

Reactions of Protonated *i*-Butyl Ethyl Ether

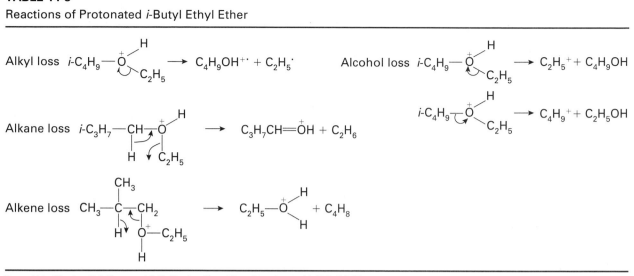

From M.L. Sigsby, R.J. Day, and R.G. Cooks, *Org. Mass Spectrom.,* **14,** 278 (1979).

The importance of product ion stability explains the occurrence of α-cleavage if one considers the charge localized form **(14-4)** of the molecular ion (eq. 14-43).

$$R^2 \overset{\displaystyle R^1}{\underset{\displaystyle R^3}{-\!\!\!\overset{|}{\underset{|}{C}}\!\!\!-}} \overset{+\cdot}{O}\!\!-\!\!H \quad \longrightarrow \quad \overset{\displaystyle R^2}{\underset{\displaystyle R^3}{>}}C\!=\!\overset{+}{O}\!\!-\!\!H + R^{1\cdot}$$

(14-43)

14-4 R^1, R^2, R^3 = alkyl, substituted alkyl, or H

The formation of stable fragments by valence expansion at oxygen is a unifying feature of most of the mass spectrometry of hydroxyl compounds. A systematic study of a variety of primary, secondary, and tertiary aliphatic alcohols suggests the following generalizations: (1) alkyl radicals are more readily lost than hydrogen, (2) α-cleavage is most important in tertiary and least important in primary alcohols, and (3) larger primary radicals are more readily lost than smaller ones.

The spectra of the typical primary, secondary, and tertiary alcohols, shown in Figure 14-11, illustrate many features of alcohol spectra. The molecular ions are of extremely low abundance. Major fragment ions arise as a result of α-cleavages, which produce primary fragment ions at *m/z* 31, 45, and 59, respectively. In the secondary and tertiary alcohols, these ions form the base peaks in the spectra. The alternative α-cleavage product, *m/z* 87 is also observed in the secondary and tertiary compounds.

In the spectra of cyclic alcohols, α-cleavage results in ring opening. Unexpected but easily rationalized processes often follow, such as ethyl radical loss from cyclopentanol, which is clearly driven by the formation of a highly stable fragment ion (eq. 14-44).

(14-44)

The products of α-cleavage are even-electron protonated ketones or aldehydes, which, when they fragment further, eliminate stable neutral molecules, including alkenes.

Returning to the spectra shown in Figure 14-11, one notes that the other major primary fragmentations of aliphatic alcohols, besides α-cleavage, are loss of ·OH and elimination of water. The loss of ·OH is of importance in cyclic and acyclic tertiary alcohols. In the latter compounds it often gives a more abundant ion than does loss of H_2O. In *tert*-butanol, for example, the increased stability of the *tert*-butyl cation and the limitation of dehydration to 1,2-elimination processes combine to make ·OH elimination (eq. 14-45) important relative to loss of water. α-Cleavage nevertheless remains the

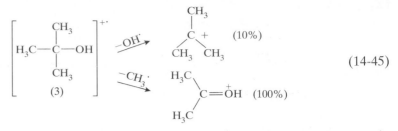

$$(14\text{-}45)$$

dominant process. Hydroxyl radical elimination is not observed in phenols. Even in benzyl alcohol, where the extremely stable $C_7H_7^+$ ion results, it is not a major process.

Further fragmentation of the ·OH loss product is characteristic of carbenium ions generated by other routes, for example, halogen loss from alkyl halides. Loss of H_2 and of alkenes dominates, and these processes yield lower mass, even-electron ions.

The 1-hexanol spectrum (Figure 14-11a) is dominated by dehydration and is very similar to that of its dehydration product, 1-hexene. Isotopic labeling demonstrates that water loss often occurs by a 1,4-elimination mechanism. Water elimination in fused ring systems can provide information on the relative stereochemistry of the hydroxyl group and the active hydrogen. Especially with sterols, this method has been valuable in assigning epimeric configurations. Typically, significant H_2O loss occurs only when the two groups are, or can approach to, within 0.17 nm of each other. Further fragmentation of the dehydration product is often very similar to the primary fragmentations of the corresponding alkenes.

Hydroxyl compounds also fragment by a variety of other characteristic routes, some of which will be mentioned here. Several types of hydrogen rearrangement have been identified in the spectra of alcohols, notably those resulting in formation of $CH_3OH_2^+$ (*m/z*

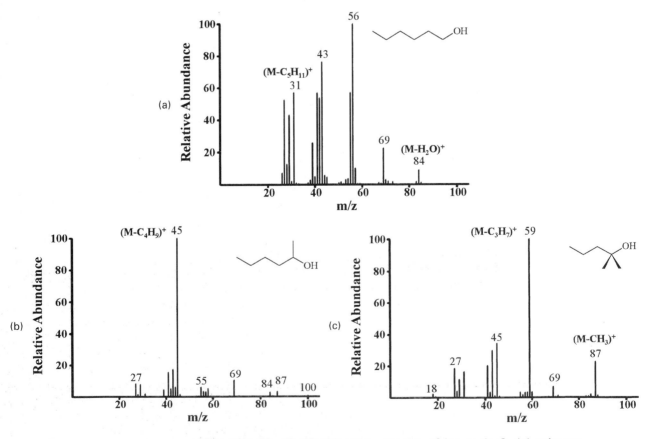

Figure 14-11 EI (70 eV) mass spectra of isomeric C_6 (a) primary, (b) secondary, and (c) tertiary alcohols, M = 102 Da. The primary α-cleavage products are marked. (From P.J. Ausloos, C. Clifton, O.V. Fateev, A.A. Levitsky, S.G. Lias, W.G. Mallard, A. Shamin, and S.E. Stein, NIST/EPA/NIH Mass Spectral Library—Version 1.5, National Institute of Standards and Technology, Gaithersburg, MD [1996].)

33) and H_3O^+ (m/z 19). These rather complex processes involve, once again, the formation of trivalent oxonium ions, although the stability of the radical that is lost may provide much of the driving force for reaction (compare $ROH^{+\cdot}$ formation from alkyl ethers, eq. 14-36).

As with most functional groups, deep-seated skeletal rearrangements also are evident in the mass spectra of particular alcohols. An example that involves OH migration is the molecular ion isomerization of 4-hydroxycyclohexanone, a complex process that begins with α-cleavage and leads to significant product ions that are shown by exact mass measurement to contain both oxygen atoms (eq. 14-46).

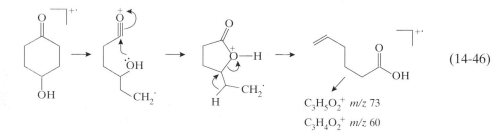

$$C_3H_5O_2^+ \ m/z \ 73$$
$$C_3H_4O_2^+ \ m/z \ 60$$

(14-46)

Phenols show EI mass spectra that contrast strongly with those of aliphatic alcohols. The molecular ions are far more abundant, and, as is typical of functionalized aromatics, phenols do not undergo α-cleavage or dehydration. Their most significant fragmentation is loss of CO, following keto-enol tautomerism, to give a mixture of cyclic and acyclic $C_5H_6^{+\cdot}$ ions and their analogues. Loss of CHO· to give $C_5H_5^+$ is also a major process. In these respects, their behavior parallels that of aromatic amines. Substituted phenols show spectra that can differ considerably from that of phenol itself since ring expansion, proximity effects, and fragmentation through the substituent can occur. Some of these features are evident in the mass spectra of the isomeric methoxyphenols shown in Figure 14-12.

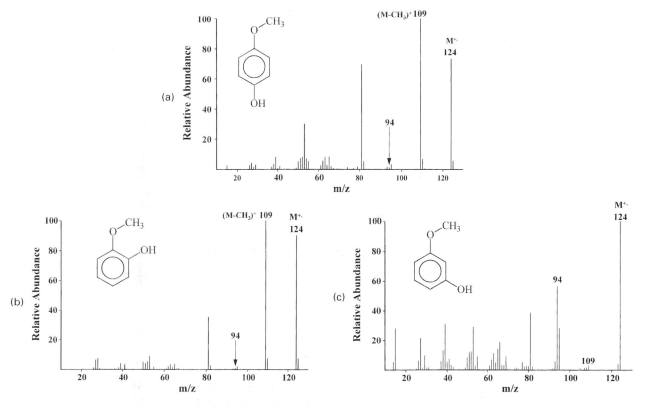

Figure 14-12 EI (70 eV) mass spectra of (a) 2-methoxyphenol, (b) 3-methoxyphenol, and (c) 4-methoxyphenol, showing loss of $CH_3\cdot$ (m/z 109) and H_2CO (m/z 94). (From P.J. Ausloos, C. Clifton, O.V. Fateev, A.A. Levitsky, S.G. Lias, W.G. Mallard, A. Shamin, and S.E. Stein, NIST/EPA/NIH Mass Spectral Library—Version 1.5, National Institute of Standards and Technology, Gaithersburg, MD [1996].)

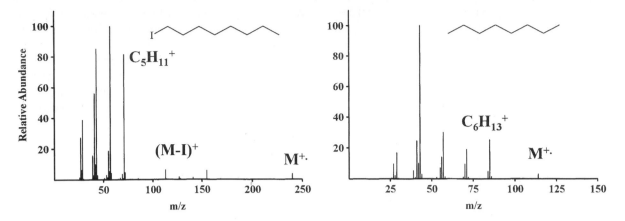

Figure 14-13 Comparison of EI (70 eV) mass spectra of 1-iodooctane (molecular weight 240) and octane (molecular weight 114), both dominated by hydrocarbon ions. (From P.J. Ausloos, C. Clifton, O.V. Fateev, A.A. Levitsky, S.G. Lias, W.G. Mallard, A. Shamin, and S.E. Stein, NIST/EPA/NIH Mass Spectral Library—Version 1.5, National Institute of Standards and Technology, Gaithersburg, MD [1996].)

These isomers all display abundant molecular ions. The ortho and para isomers show similar spectra dominated by the loss of a methyl radical to give the aroxy cation. Clearly, this process is favored by resonance stabilization of the product, m/z 109, since the meta isomer fragments to a smaller extent and does not display a peak at m/z 109. The major primary fragmentation in the meta isomer is loss of formaldehyde, to give m/z 94, a rearrangement reaction (compare with eq. 14-38) that is significant in this compound since the available simple cleavage processes do not lead to stabilized products.

14-2d   Halides

Many halogenated compounds are readily recognized by their characteristic isotopic signatures (see Section 15-4) or by the masses of characteristic neutrals lost (127 Da for I, 50 for CF_2, 20 for HI, 36/38 for HCl, etc.). The different aliphatic halides show a remarkable range of behavior, consistent with the range of C—X bond energies (from approximately 200 kJ mol^{-1} [45 kcal mol^{-1}] for R—I bonds to more than 400 kJ mol^{-1} [96 kcal mol^{-1}] for R—F bonds). Not surprisingly, loss of the halogen is facile for the bromides and iodides, and these radical cations yield carbenium ions as their primary products. Weak molecular ions and hydrocarbon-like mass spectra are therefore typical of bromides and iodides. The spectrum of 1-iodooctane (Figure 14-13) is typical. The fact that C—X bond cleavage is the dominant fragmentation pathway is evident from a comparison of its EI mass spectrum with that of the normal alkane, n-octane.

Chlorides lose HCl to generate ionized alkenes. Carbon–carbon bond cleavage, even α-cleavage, is rare except when γ- or δ-cleavage can lead to cyclic halonium ions, which are then favored. Aryl chlorides and bromides show facile loss of the halogen atom; iodides fragment less readily and fluorides very little by this route. Fluorinated compounds undergo unique reactions, including CF_2 elimination with accompanying F atom rearrangement. By contrast, CH_2 loss is energetically unfavorable and rarely occurs.

14-3   KETONE AND ALDEHYDE TYPE FRAGMENTATION

Ketones represent a second major functional type, that possessing a multiple bond to a heteroatom, just as amines and ethers represent functional groups in which the heteroatom is singly bonded. Once again, EI mass spectra are emphasized, but the fragmentation of protonated ketones is also considered and the information provided covers fragmentations associated with all the ionization methods.

A major primary fragmentation is acylium ion formation, which occurs via cleavage of the α-C—C bond. Further fragmentation of the acylium ions by CO loss is facile, and the ions R_L^+, R_S^+, $R_L CO^+$, and $R_S CO^+$ are easily recognized in almost all ketone spectra (R_S and R_L refer to smaller and larger alkyl substituents). The second major fragmentation of ketones is six-membered cyclic hydrogen transfer and associated β-cleavage leading to loss of an alkene molecule. The reaction is termed the *McLafferty rearrangement* (eq. 14-47). Although some hydrogen exchange occurs in ketone molec-

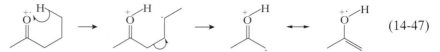

(14-47)

ular ions even at 70 eV, the rearrangement is highly specific. It is the most studied rearrangement in mass spectrometry, and the evidence favors a stepwise reaction initiated by the radical site on oxygen. Note that the intermediate is a distonic ion (Section 14-7b). The product ions are unusually abundant for a rearrangement fragmentation. The McLafferty rearrangement product is the enolic species shown. Evidence for the structure comes from isotopic labeling, metastable ion studies, and molecular orbital calculations. Further fragmentation of the ion product occurs by α-cleavage, after ketonization, or directly by further alkene elimination.

There are several other notable features of the McLafferty rearrangement, including the following:

1. Related reactions occur in other C=X systems, for example, oximes.
2. Secondary hydrogen atoms in the γ position are more readily abstracted than are primary.
3. Methyl groups do not rearrange when substituted for hydrogen; the reaction does not occur in compounds that lack γ-hydrogens.
4. An analogous process occurs in even-electron ions.
5. If both alkyl groups can undergo the McLafferty rearrangement, the more abundant product is due to reaction from the larger group. In addition, the product of two successive alkene eliminations (double McLafferty rearrangement) usually appears as an abundant ion. Many of the major features of ketone mass spectra are illustrated in the spectra of butyrophenone (Figure 14-14) and 4-nonanone (Figure 14-15).

The butyrophenone spectrum is very simple. The base peak, m/z 105 ($C_6H_5CO^+$), is formed by methyl radical loss from the McLafferty rearrangement product, m/z 120, as well as by simple α-cleavage in the molecular ion. It is interesting to note that methyl loss involves the CH_2 group of the enol ion and an ortho ring hydrogen, *not* the hydroxyl hydrogen. In other words, a hydrogen atom is abstracted via a five-membered cyclic transition state (eq. 14-48). The acetophenone spectrum is even simpler, being dominated by the benzoyl ion (m/z 105) and its fragmentation products m/z 77 and 51.

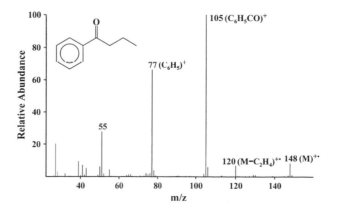

Figure 14-14 EI (70 eV) mass spectrum of butyrophenone (phenyl propyl ketone), which includes the even-mass McLafferty rearrangement radical ion at m/z 120.

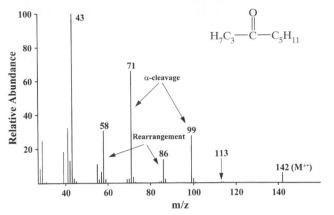

Figure 14-15 EI (70 eV) mass spectrum of 4-nonanone. (From G. Eadon, C. Djerassi, J.H. Beynon, and R.M. Caprioli, *Org. Mass Spectrom.*, **5**, 917 [1971]. Reproduced with the permission of John Wiley & Sons Ltd., Chichester, UK.)

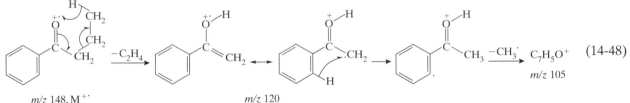

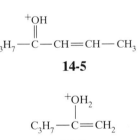

4-Nonanone fragments by α–cleavage to give the acylium ions at *m/z* 71 and 99. These ions fragment further by CO loss to give the alkyl cations at *m/z* 43 and 71, the latter of which is composed of both $C_5H_{11}^+$ and $C_3H_7CO^+$. Of the three possible odd-electron ions due to McLafferty rearrangement, that associated with elimination of the smaller alkene is not observed (loss of C_2H_4 would give *m/z* 114), but loss of C_4H_8 gives the fragment ion at *m/z* 86 and the double McLafferty rearrangement gives rise to the peak at *m/z* 58. Both of these ions fragment further by α-cleavage and then CO elimination, and so contribute to the acylium and alkyl ions already mentioned. The only other ion that appears in the spectrum besides hydrocarbon fragments at *m/z* 55, 41, etc., is a low abundance ion, *m/z* 113, due to loss of C_2H_5˙ from the molecular ion. This ion comes from a rearrangement reaction and probably has the stable structure **14-5,** although a cyclized structure cannot be excluded. Although not evident in the 70 eV mass spectrum of 4-nonanone, several other low energy rearrangement processes do appear in ketone spectra. One of these is analogous to the McLafferty rearrangement, except that two hydrogen atoms are transferred to the carbonyl group, probably to give the stable ion **14-6**. It will be recognized that this process (eq. 14-18) is analogous to ROH_2^+ formation from ethers.

Thus these rather complex rearrangements do not interfere with the analytical applications of mass spectrometry, nor do they occur haphazardly. Rather, they are controlled by product ion stability, and they increase the structural information available in a mass spectrum, particularly when low energy ions are deliberately examined.

Cyclic ketones show behavior that is typical of other substituted cycloalkanes. They undergo α-cleavage but, because of the cyclic nature of the ion, this process leads to an isomeric form of the molecular ion rather than directly to a fragment ion. Hydrogen rearrangement then occurs and is often followed by loss of an alkyl radical. Cyclopentanone shows the base peak at *m/z* 55 due to ethyl radical loss (Figure 14-16 and eq. 14-49); compare ethyl radical loss from cyclopentanol (eq. 14-44).

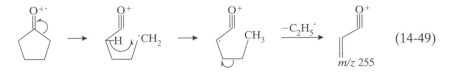

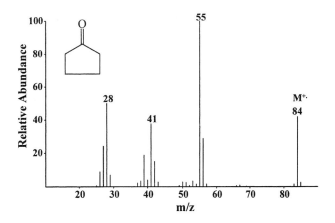

Figure 14-16 EI (70 eV) Mass spectrum of cyclopentanone showing the base peak at *m/z* 55 due to α-cleavage fragmentation leading to ethyl radical loss. (From P.J. Ausloos, C. Clifton, O.V. Fateev, A.A. Levitsky, S.G. Lias, W.G. Mallard, A. Shamin, and S.E. Stein, NIST/EPA/NIH Mass Spectral Library—Version 1.5, National Institute of Standards and Technology, Gaithersburg, MD [1996].)

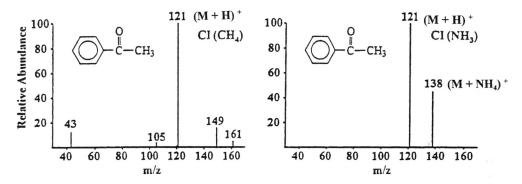

Figure 14-17 Thermochemical factors control CI mass spectrum of acetophenone (*m/z* 149 and 161 are the adducts $[M+C_2H_5]^+$ and $[M+C_3H_5]^+$ common to methane CI). Note the lack of fragmentation of the protonated molecule generated by the soft CI reagent, NH_4^+, compared with the hard reagent, CH_5^+.

The chemical ionization mass spectra of ketones, recorded using protonating reagents, show alkane and alkene eliminations analogous to those observed for protonated ethers. For example, protonated 4-heptanone fragments via loss of propane when using methane as reagent gas. The thermochemistry that underlies chemical ionization controls fragmentation. This phenomenon was illustrated for the particular case of ionization by proton transfer in Section 13-3b. Remember that the proton affinity (PA) of a molecule M is defined as the exothermicity of protonation (eq. 14-50).

$$PA(M) \equiv \Delta H_f(M) + \Delta H_f(H^+) - \Delta H_f(MH^+) \qquad (14\text{-}50)$$

When a proton is transferred between two species, A and B, then the enthalpy of reaction is simply the difference in proton affinities ΔPA. The excess energy must appear as internal energy of the products. If this energy (ΔPA) is large, extensive fragmentation is seen. Large energy deposition often occurs in electron ionization too, although the different form of the molecular ion, $(M+H)^+$ vs. $M^{+\cdot}$, results in formation of different fragment ions in the two cases. On the other hand, if the internal energy deposited is small, the protonated product molecule is abundant. Although ΔPA describes the sum of the internal energies of both the ionic and the neutral products, the larger ionic product typically takes almost all the excess energy. The result is that the ion internal energy distribution $P(\varepsilon)$ for chemical ionization is rather narrow, compared with that for EI, and can be set to high or low values by choosing the appropriate reagent gas.

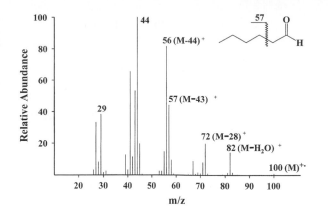

Figure 14-18 EI (70 eV) mass spectrum of hexanal showing dehydration and ethylene elimination. (From H. Budzikiewicz, C. Djerassi, and D.H. Williams, *Mass Spectrometry of Organic Compounds,* Holden-Day, San Francisco [1967].)

This capability is illustrated in Figure 14-17, which shows CI mass spectra of acetophenone recorded with hard and soft CI reagents, methane and ammonia, respectively. Note the analogy to the 70 eV and 15 eV EI mass spectra of acetophenone shown later in Figure 14-31, even though the protonated molecule, not the radical cation, is involved.

A typical aliphatic aldehyde, hexanal, shows characteristically different behavior to ketones on EI, fragmenting chiefly by dehydration (Figure 14-18). Interestingly, dehydration is also a significant fragmentation process for the molecular radical cations of aldehydes. This observation is evident from the 70 eV EI mass spectrum of hexanal, which shows (1) the α-cleavage process, which is also characteristic of ketones and here yields m/z 71 with the charge retained by the alkyl group; (2) the McLafferty rearrangement leading to the enolic ion, m/z 44; (3) the complementary process, loss of the neutral enol, leading to the ionized alkene, m/z 56; (4) elimination of ethene by hydrogen rearrangement to give m/z 72; and finally (5) dehydration to yield m/z 82. Each of these primary products can undergo further fragmentation under the energetic conditions of 70 eV EI spectra. In many cases aldehydes display moderate H· loss to yield stable acylium ions.

14-4 FRAGMENTATION OF OTHER FUNCTIONAL GROUPS

14-4a Carboxylic Acids

EI of carboxylic acids generates positive molecular ions of moderate abundance. Fragmentation by ·OH loss is not facile except in aromatic acids, nor is H_2O loss a general process. Ions corresponding to $COOH^+$ (m/z 45) or the loss the radical COOH· are observed for short-chain aliphatic acids. The McLafferty rearrangement gives abundant fragments, as does C—C cleavage some distance from the functionality, such as δ-cleavage. The most characteristic rearrangement ions occur at m/z 60, 74, 88, etc., and these ions, formally $RCO_2H^{+·}$, can be used to characterize the position and extent of branching along the alkyl chain. Aromatic acids yield benzoyl ions by HO· loss, whereas ortho effects can result in dehydration. Carboxylic acids are best examined through their negative ion spectra, where $(M-H)^-$ is dominant (such as by CI using HO^- as reagent ion).

14-4b Esters

Esters fragment by routes that are characteristic of both ethers and ketones, as well as showing distinctive processes. This behavior is illustrated in the mass spectrum of methyl stearate (Figure 14-19). The molecular ions of lower molecular weight esters are moderately abundant in EI mass spectra. Fragmentation at the ether oxygen gives acylium ion (RCO^+) formation, resulting in the loss of methoxyl radical in the case of methyl esters. The product ion subsequently undergoes CO loss to form a carbenium

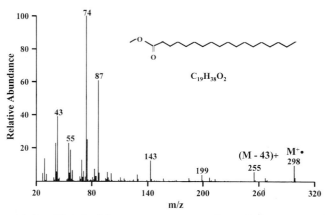

Figure 14-19 EI (70 eV) mass spectrum of methyl stearate. (From P.J. Ausloos, C. Clifton, O.V. Fateev, A.A. Levitsky, S.G. Lias, W.G. Mallard, A. Shamin, and S.E. Stein, NIST/EPA/NIH Mass Spectral Library—Version 1.5, National Institute of Standards and Technology, Gaithersburg, MD [1996].)

ion (R^+), especially when R^+ is secondary or tertiary. Acylium ion formation is a dominant process in esters of aromatic acids. In the cases of ethyl and higher esters, there is a second major fragmentation mode proceeding through the alkoxyl group: McLafferty rearrangement yields the enolic form of the ionized carboxylic acid. Fragmentation via the alkyl substituent on the carbonyl group also occurs by McLafferty rearrangement (β-cleavage, with H transfer, to give m/z 74 in methyl esters). δ-Cleavage yields m/z 87 in methyl esters, which is accompanied by ions at m/z 143, 199, 255. . . . This set of ions results from hydrogen atom abstraction by the carbonyl group followed by radical migration and cleavage in the alkyl chain (eq. 14-51). The radical site can migrate

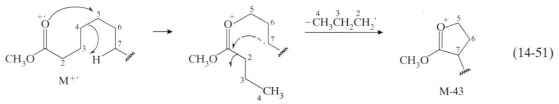

$$(14\text{-}51)$$

down a long alkyl chain and does so preferentially through six-centered intermediates, yielding the higher members of this series.

The 70 eV mass spectrum shown in Figure 14-20 is that of an aromatic acetate. Even if the structure were unknown, the spectrum could be interpreted by applying the simple fragmentation guidelines already discussed. The arguments are as follows:

1. The abundant ion at highest mass (m/z 181, ignoring the isotopic peak at 182) represents the molecular ion, $M^{+\cdot}$.
2. There is no evidence of the characteristic isotopic signatures for elements other than C, H, N, and O.
3. Because $M^{+\cdot}$ has an odd mass, the molecule must have an odd number of nitrogens.
4. Therefore, reasonable possibilities for the molecular formula are $C_{13}H_{11}N$, $C_{12}H_{23}N$, $C_{12}H_7NO$, $C_{11}H_{19}NO$, $C_{10}H_{15}NO_2$, and $C_9H_{11}NO_3$. Some are highly unlikely; for example, the spectrum is clearly that of an aromatic compound, which rules out the second structure as not possessing a sufficient number of unsaturation units.
5. Elements of the structure evident in the spectrum include an acetyl group (loss of 42 from $M^{+\cdot}$ and formation of m/z 43); nitro group (loss of 46 and 30 from $[M-CH_2CO]^{+\cdot}$, m/z 139); and the phenoxyl cation, m/z 93.
6. These possibilities restrict the formula to the last candidate, and assembly of the structural units yields nitrophenyl acetate as the only reasonable possibility. Note that the ring isomers are not expected to be readily distinguished in this case.

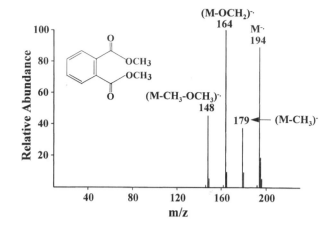

Figure 14-22   Negative ion CI mass spectrum of dimethylphthalate (methane reagent gas) showing the intact radical anion, M⁻·and the rearrangement fragments, *m/z* 148 and 164 (From E.A. Stemmler and R.J. Hites, *Biomed. Envir. Mass Spectrom.,* **17,** 311 [1988]. Reproduced with permission of John Wiley & Sons Ltd., Chichester, UK.)

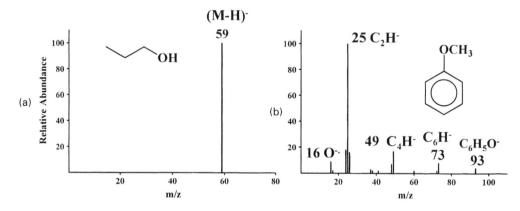

Figure 14-23   (a) Negative ion CI spectrum (HO⁻) of propanol contrasted with (b) negative ion EI spectrum (70 eV electrons) of anisole. (Spectrum b from R.T. Aplin, H. Budzikiewicz, and C. Djerassi, *J. Amer. Chem. Soc.,* **87,** 3180 [1965]. Copyright © 1965 American Chemical Society.)

Another example occurs in the negative ion CI mass spectrum of dimethyl phthalate (Figure 14-22), in which the proximity of the ester groups promotes rearrangement. The ion at *m/z* 148 in this spectrum corresponds to the molecular ion of phthalic anhydride.

Conjugate anions of acidic compounds, $(M-H)^-$, are commonly generated by CI and DI. For example, propanol can be ionized by CI using HO⁻ as reagent ion, and it gives the deprotonated molecule as the only ion in the spectrum (Figure 14-23a). This behavior is in contrast with attempts to generate negative ions using EI, as in the case of anisole (Figure 14-23b); such experiments give mostly low mass fragments of little value in structure elucidation.

Even-electron anions tend to fragment by simple elimination processes, as illustrated by the negative ion DI (SIMS) spectrum of nicotinamide (Figure 14-24), which shows loss of HCN and HNCO from $(M-H)^-$ to yield *m/z* 94 and 78, respectively. Other examples of negative ion fragmentation already encountered include charge remote fragmentations in carboxylates (eq. 14-19). Negative ion spectra give abundant $(M-H)^-$ ions for alcohols and phenols.

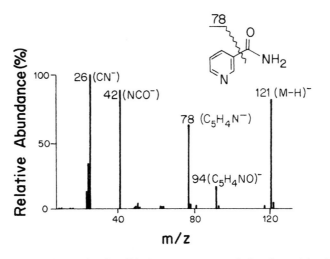

Figure 14-24 Negative ion DI mass spectrum of nicotinamide showing intact (M−H)⁻ ion and structurally diagnostic fragment ions (Ar⁺ bombardment, solid sample). (Reproduced with the permission of Pergamon Press, from L.K. Liu, S.E. Unger, and R.G. Cooks, *Tetrahedron*, **37**, 1067 [1981].)

14-6 FRAGMENTATION IN CI, DI, AND SI

Examples of fragmentation of protonated and deprotonated molecules have been discussed. Fragmentation in DI and SI is largely associated with closed-shell ions, both positively and negatively charged. The fragmentation processes resemble those that occur in other ionization methods. Typical cases are provided by the nicotinamide spectrum (Figure 14-24) and by the behavior of the quaternary alkaloid candicine (Figure 14-25). Figure 14-25 shows the intact cation (*m/z* 180) as an intense peak as well as fragmentations from this even-electron ion to give even-electron product ions with elimination of stable neutral molecules.

The DI spectra of simple quaternary ammonium compounds illustrate many of the rules for fragmentations encountered previously. For example, *N*-ethyl

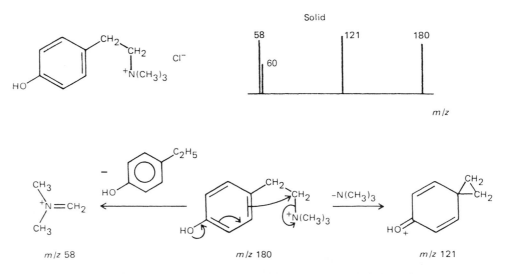

Figure 14-25 DI mass spectrum and fragmentation of the salt candicine chloride showing the intact cation at *m/z* 180 and characteristic fragment ions (Ar⁺ ion bombardment). (Reproduced with the permission of the American Association for the Advancement of Science, from K.L. Busch and R.G. Cooks, *Science*, **218**, 247 [1982].)

N-methylpiperidinium iodide is readily ionized by any of the DI techniques. If the intact cation is selected, excited by collision, and allowed to fragment, it yields the products shown in eq. 14-54. Note that radical loss occurs under the conditions of this particular experiment, in contravention of the even $\longrightarrow$ even electron rule (eqs. 14-8 and 14-9).

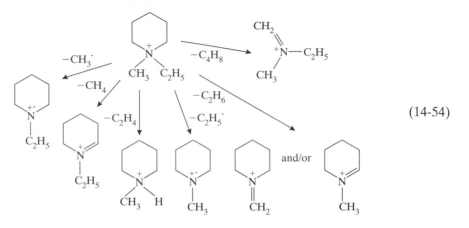

(14-54)

In DI the molecule can be ionized in a variety of ways, and each molecular ion displays its own unique set of fragment ions. Another difference between DI and EI spectra is that intermolecular reactions can sometimes occur. These are often simple trans-methylations, giving peaks 14 mass units greater than expected.

Fragmentation of biological compounds is just beginning to be studied systematically. Considerable information is available on protonated peptides generated by DI and SI. The site of protonation is usually the most basic amino acid, invariably arginine when present, although mobility of the proton is evident. The main fragmentation channels are directed by the protonated site and two particularly important fragmentations are known as b- and y-type fragmentations (eq. 14-55). Both of these processes involve

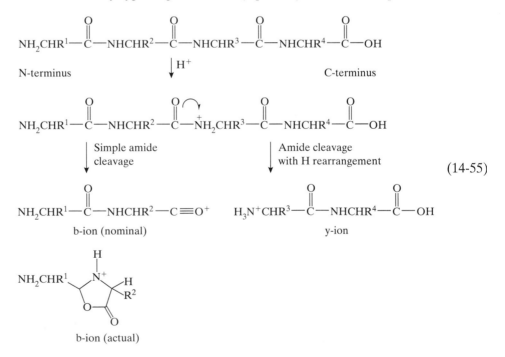

(14-55)

amide bond cleavage, although one is nominally a simple cleavage leading to an acylium ion, which characterizes the N-terminus of the peptide (b-type fragmentation). The other involves hydrogen transfer and leads to a protonated amine, which characterizes the C-terminus. The actual mechanisms of these processes are more complex, and the b-ions may be the oxazolone ions also shown in eq. 14-55. See Worked Problem 14-7 for an example of peptide sequencing by mass spectrometry.

14-7 ION CHEMISTRY

14-7a Unimolecular Fragmentation

Kinetics. The use of mass spectrometry as a structural tool (as opposed to its use simply for molecular weight measurements) primarily depends on observation of the unimolecular fragmentations that excited ions undergo. In this section, the principles that underlie ionic reactions in the *isolated* phase environment are discussed. The controlling factors are the distribution of internal energies the ions possess and the thermochemistry of the dissociation processes available to them.

Time Scale of Fragmentation. The mass spectrometer time scale is illustrated in Figure 14-26. Note that longer time scales apply to quadrupole instruments than magnetic sectors, and that quadrupole ion traps and ICR instruments access extremely long times. The mass spectrum is a time-integrated picture of all reactions occurring after ionization and before ions leave the source. Although this length of time is as short as $\sim 10^{-6}$ s, it is long compared with bond vibrational periods of 10^{-10} to 10^{-12} s. In almost all excited ions, fragmentation does not occur within a vibrational period; rather, internal energy is shuttled around the excited, isolated ion until it collects in the appropriate modes for fragmentation to occur. The rate at which fragmentation occurs increases rapidly with excess internal energy above the activation energy. This characteristic is the main reason why thermochemistry controls mass spectrometric fragmentation behavior.

Potential Energy Surfaces. Unimolecular fragmentations are normally endothermic reactions and, even when exothermic, they often have substantial activation energies. Potential energy surfaces are shown schematically in Figure 14-27 for the two main types of unimolecular reactions. These are *simple bond cleavages*, which have small or *no reverse activation energies*, and those fragmentations that involve rearrangement of bonds and in which reverse activation energies can be substantial. Note that rearrangements must have low activation energies to compete with simple bond cleavages since they have substantial entropic requirements, which limit reaction rates. Therefore, rearrangements tend to be favored for ions of low internal energy, for example ions generated by soft ionization methods or by low energy EI, or for ions examined at long intervals after ionization when only low internal energy ions survive. Relatively little energy in excess of the activation energy for fragmentation is required to observe products of simple cleavage on the time scale of the mass spectrometer. The additional energy required for rearrangement fragmentation to proceed at an observable rate can be substantial. The different forms of the rate curves associated with these reaction types are illustrated in the lower half of Figure 14-27. Note how the rate constants depend on internal energy for (1) simple cleavage and (2) a rearrangement process. However, the important, practical conclusion is that only when rearrangements have substantially lower activation energies can they compete with simple cleavages, because of the greater excess energy required to cause observable fragmentation on the time

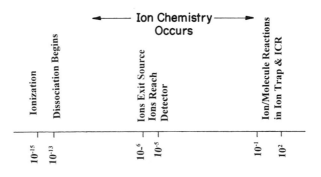

Figure 14-26 Time scale for EI and CI (seconds).

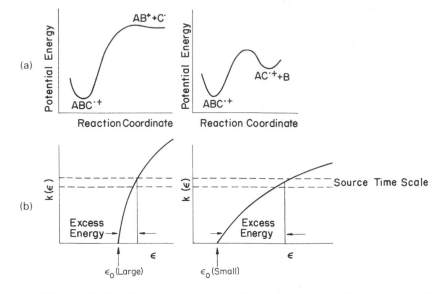

Figure 14-27 Potential energy surfaces and energy dependence of unimolecular dissociation, showing typical simple bond cleavage (a) and rearrangement (b). ε is internal energy and $k(\varepsilon)$ is reaction rate constant.

scale of the mass spectrometer. The lower part of Figure 14-27 illustrates competition between rearrangement and simple cleavage reactions and shows that the former only compete if they have low activation energies *and* if the ions examined are of low internal energy. Figure 14-31 illustrates the point just made by showing how rearrangements can be favored by simply lowering the energy deposition in EI.

Internal Energy Distributions P(ε). Ions are generated with a distribution of internal energies on electron ionization. Transfer of the full kinetic energy of the electron to the molecule is unlikely; the amount of energy transferred depends on the nature of each individual electron-molecule collision. When a population of molecular ions is considered, there is an associated distribution of ion internal energies, termed $P(\varepsilon)$ (Figure 14-28). Since electron ionization is usually achieved under high vacuum, i.e., collision-free conditions, the distribution of internal energies of molecular ions $M^{+\cdot}$ is *not* altered by collision and it certainly is not a Boltzmann distribution. The shape of the distribution associated with electron impact depends on the probability of excitation to various excited states of the ion. It can be controlled to some extent by varying the ionizing electron energy, since the maximum energy the ion $M^{+\cdot}$ can have is the difference between the electron energy and the minimum energy required for ionization, namely, the ionization energy.

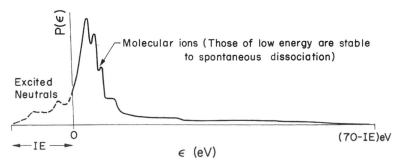
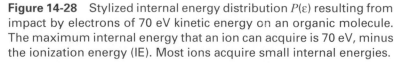

Figure 14-28 Stylized internal energy distribution $P(\varepsilon)$ resulting from impact by electrons of 70 eV kinetic energy on an organic molecule. The maximum internal energy that an ion can acquire is 70 eV, minus the ionization energy (IE). Most ions acquire small internal energies.

The internal energy distribution $P(\varepsilon)$ is important because it controls the subsequent unimolecular dissociation behavior of the ion population, just as the temperature of a system with a Boltzmann distribution of internal energies controls its chemistry.

Thermochemical Control. Fragmentation behavior of isolated ions is controlled by the thermochemical factors (activation energies) already discussed as well as by the kinetic factors just discussed. These can be considered together to produce a *breakdown curve.* Such a curve shows the dependence of the fragmentation behavior on the internal energy. For example, ionized 1-propanol molecular ion undergoes H· loss to give *m/z* 59 as the lowest energy process, only to be replaced in turn by higher energy consecutive fragmentations, yielding *m/z* 31 and 29. The breakdown curve demonstrates the behavior of molecular ions with particular internal energies. For example, ions that have exactly 4.4 eV of internal energy yield 96% of 31^+ and 4% of 59^+. Such ions can be generated by charge exchange with $N_2^{+\cdot}$ reagent (IE = 15.6 eV, IE [propanol] = 10.2 eV, difference 4.4 eV).

A mass spectrum is simply a breakdown curve with appropriate weighting of the internal energy axis. In other words, depending on the method and conditions of ionization, ions have a particular internal energy distribution $P(\varepsilon)$. By convoluting this distribution with the breakdown curve, the mass spectrum is obtained (Figure 14-29). This figure, which should be studied carefully, illustrates how different methods of generating a set of molecular ions result in different unimolecular product distributions, that is, different mass spectra.

Fragmentation occurs unimolecularly for ions whose internal energy ε equals or exceeds a minimum value, ε_0, the activation energy for fragmentation. Ions with less

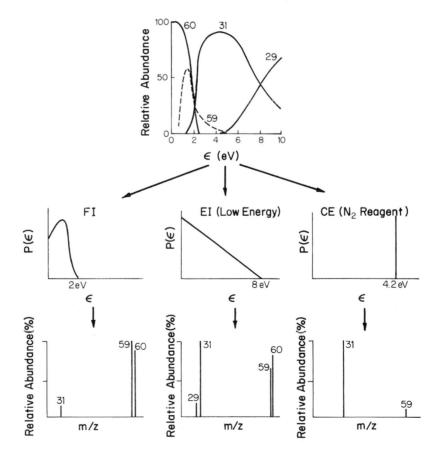

Figure 14-29 Convolution of the breakdown curve (top, an intrinsic property) with the internal energy distribution ($P[\varepsilon]$) resulting from different ionization techniques under particular experimental conditions to produce mass spectra of 1-propanol. FI indicates field ionization, a soft ionization method used for vapor phase samples.

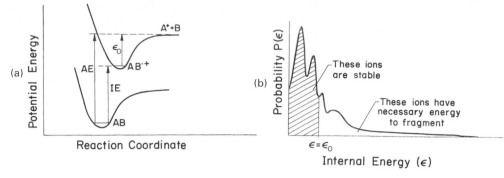

Figure 14-30 (a) Thermochemical quantities that control unimolecular dissociation and (b) their effects on the internal energy distribution $P(\varepsilon)$.

than this internal energy are stable indefinitely, whereas those with greater energies fragment at a rate that rapidly increases with energy in excess of the activation energy. As discussed previously, vertical electronic transitions can convert neutral molecules into molecular ions with a range of internal energy states (Section 13-3). In some collisions the minimum energy required to cause ionization, the ionization energy IE, is transferred to the molecule. This situation leads to formation of the ion $AB^{+\cdot}$ in its ground state, with zero internal energy. In other collision events the energy transferred might exceed the minimum energy required to cause ionic dissociation, namely, the appearance energy AE. Figure 14-30 illustrates these quantities in addition to their difference, the *activation energy*, ε_0 $(=AE-IE)$. Both the potential energy surfaces and the internal energy distribution of the ionic population are shown. In this figure, it is possible to divide the internal energy distribution into two regions, based on whether the initially formed ions have or do not have sufficient energy to fragment unimolecularly. This division provides an indication of the molecular ion abundance relative to fragment ion abundances recorded in the mass spectrum.

Consider the behavior of acetophenone on electron ionization. Energetic electrons interact with the vapor phase molecules producing the molecular ion as a result of electronic transitions (Figure 14-30a). In a few cases, the minimum energy required for ionization, or the ionization energy, may be delivered to the molecules giving molecular ions in their ground electronic and vibrational states. In many more cases, each molecular ion has a particular excess internal energy ε, and the population has an energy distribution $P(\varepsilon)$ (Figure 14-30b). The more highly excited ions rapidly fragment to give ionic fragments that are excited and can fragment further. Ions of lower energy may fragment but by different routes: the favored fragmentation mode depends on the degree of internal excitation of each ion. A network of competing and consecutive unimolecular reactions exists and leads to a set of products that evolves with time. The mass spectrum is simply a sampling of this product distribution at a particular time after ionization, often a time of about 1 μs.

The main electron impact chemistry of acetophenone is shown in eq. 14-56 with

EI chemistry:

$$M \xrightarrow{e^-} M^{+\cdot} \longrightarrow (M-CH_3)^+ \longrightarrow C_6H_5^+ \longrightarrow C_4H_3^+$$

$$\qquad\quad m/z\ 120 \qquad\quad m/z\ 105 \qquad m/z\ 77 \qquad m/z\ 51$$

Chemical species:

$$\underset{C_6H_5CCH_3}{\overset{\displaystyle\overset{O}{\|}}{}} \longrightarrow \underset{(C_6H_5CCH_3)^{+\cdot}}{\overset{\displaystyle\overset{O}{\|}}{}} \longrightarrow C_6H_5C\equiv O^+ \longrightarrow C_6H_5^+ \longrightarrow C_4H_3^+ \qquad (14\text{-}56)$$

Energy requirements:

$$0\ eV \qquad\qquad 9\ eV \qquad\qquad 10\ eV \qquad 13\ eV \qquad 16\ eV$$

$$\qquad\qquad\qquad\qquad\qquad\qquad\qquad \varepsilon_0 = 1\ eV \qquad \varepsilon_0 = 4\ eV \qquad \varepsilon_0 = 7\ eV$$

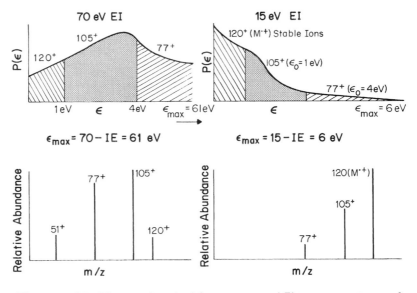

Figure 14-31 Thermochemical factors control EI mass spectrum of acetophenone.

the energy requirements for each product referenced to the neutral molecule. A linear sequence of reactions is involved in this particular molecule. If one assumes that a fragment ion is generated from all ions that have sufficient energy to make this reaction accessible, then the mass spectrum can be inferred from a knowledge of the molecular ion internal energy distribution $P(\varepsilon)$ and the thermochemistry of fragmentation. To a first approximation, the mass spectrum is determined by such thermochemical factors. Figure 14-31 illustrates this principle for acetophenone. For example, all molecular ions with internal energies between 1 eV (the onset of 105^+ formation) and 4 eV (the onset of 77^+ formation) can be assigned as fragmenting to yield stable 105^+ ions. The ion abundance in the mass spectrum is taken simply as this area in the $P(\varepsilon)$ distribution. Figure 14-31 also shows the increase in molecular ion abundance as the ionizing electron energy is reduced. The practical advantage of this relative increase in molecular ion abundance is offset by a decrease in ionization efficiency. Those molecular ions in the distribution $P(\varepsilon)$ that are generated with insufficient internal energy to surmount the barrier for even the lowest energy fragmentation process *must* appear in the mass spectrum as molecular ions. Those with greater internal energies fragment.

Acetophenone behaves straightforwardly because competitive processes are not important in the breakdown pattern. All molecular ions of energy 1 to 4 eV behave the same way and lose $CH_3{}^{\cdot}$ to give benzoyl fragment ions of mass 105. When competitive reactions are possible, it is still often true that thermochemical factors dominate and only the lowest energy processes available to a given ion are observed.

The control of thermochemistry over the type and degree of fragmentation can be illustrated by considering a set of monosubstituted benzenes, each of which undergoes a simple cleavage reaction as its main fragmentation pathway. Table 14-4 is a collection of this thermochemical data and illustrates how the abundance of the fragment ion increases with decreasing activation energy (ε_0). If the molecular ions of all these compounds have roughly similar internal energy distributions, this inverse dependence of ion abundance on ε_0 is exactly the behavior expected from a consideration of Figure 14-30. Summarizing this approach, if one knows (1) the energy requirements for fragmentation and (2) the internal energy distribution $P(\varepsilon)$, one can approximate the mass spectrum without considering the kinetics of individual reactions. *This approximate thermochemical approach to mass spectra works best when complex rearrangements are not important.*

Similar considerations apply to CI and other ionization methods.

TABLE 14-4

Thermochemical Control of Fragmentation of Substituted Benzenes*

Substituent	Neutral Fragment	IE (eV)	AE (eV)	ε_0 (eV)	Fragment Ions/M+†
$COCH_3$	CH_3	9.3	10.0	0.7	2.2
$C(CH_3)_3$	CH_3	8.7	10.3	1.6	2.1
$(CO)OCH_3$	OCH_3	9.3	10.8	1.5	1.5
C_2H_5	CH_3	8.8	11.3	2.5	1.0
NO_2	NO_2	9.9	12.2	2.3	0.7
CH_3	H	8.8	11.8	3.0	0.5
I	I	8.7	11.5	2.8	0.4
Br	Br	9.0	12.0	3.0	0.4
Cl	Cl	9.1	13.2	4.1	0.2

* Table adapted from I. Howe and D.H. Williams, *Principles of Organic Mass Spectrometry*, McGraw-Hill Book Co., New York, 1972.

† Ratio of fragment to molecular ion abundance, 70 eV EI.

14-7b   Ion-Molecule Reactions

Ion-molecule reactions occur in the course of ionization by all methods, except EI. They can be used as an alternative to fragmentation, to elucidate molecular structures. The subject of gas phase ion chemistry, which includes the study of ion-molecule reactions, is important both for the information it yields on solution chemistry through characterization of analogous processes occurring in the absence of solvent, as well as for its intrinsic interest. Gas phase acidities and basicities, as well as other types of reactivities, often show more systematic trends than do the corresponding solution properties. An early and well-known example is the order of basicity of primary, secondary, and tertiary amines, which increases in this simple sequence in the gas phase but not in solution where solvent effects lead to the order tertiary > primary > secondary. By studying ion-molecule reactions in the gas phase, one is studying systems in the same state as examined in ab initio calculations.

Potential Energy Surfaces.   A key to gas phase ion chemistry is that gaseous ions have high enthalpies and are stabilized by interaction with *any* molecule. Polarization stabilizes the charge, and ion-molecule association complexes are normally more stable than the free reagents. However, if the collision complex does not have an available reaction channel, it will revert to reactants if excess energy is not removed, usually by collision with a third body. The very different forms of the potential energy surfaces for the same ion-molecule reaction in solution and in the gas phase are illustrated in Figure 14-32. The differences are entirely due to solvent effects. In solution the ionic reagent (AB+) is already stabilized by solvation, but its gas phase counterpart is not. Hence the interaction with C is strongly stabilizing in the

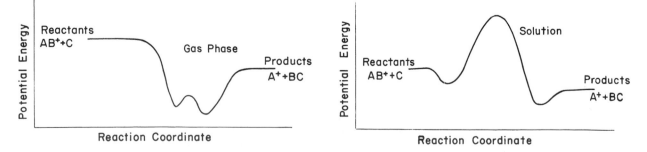

Figure 14-32   Potential energy surfaces for ion-molecule reactions in the gas phase and in solution.

latter but not the former case. In both systems the structural changes necessary to achieve the activated complex configuration are endothermic with respect to the surrounding regions of the potential energy surface. Large energy changes occur across the gas phase reaction surface because energy differences between unsolvated and monosolvated ions are large.

Rates of Ion-Molecule Reactions.   Although important exceptions exist, many exothermic ion-molecule reactions occur on every collision. In other words, they proceed without an activation barrier, as already shown in the potential energy surfaces illustrated in Figure 14-32. Ion-dipole attractive forces increase the rate of approach of the ion and neutral molecule and this makes ion-molecule reaction rates even greater than typical collision rates among neutral molecules. The rates of the reactions are therefore very high, 10^{-9} cm^3 molecule^{-1} s^{-1} (6×10^{11} liter mol^{-1} s^{-1}).

Structural Elucidation Using Ion-Molecule Reactions.   Instead of employing unimolecular fragmentation, it is possible to obtain structural information through a study of reactions with structurally diagnostic ions. Despite its obviousness—given the analogy with solution reactions—this approach has seen relatively little application. Ion-molecule reactions are used to generate characteristic ions from the analytes of interest in CI, DI, and SI, as described in Section 13-30. Nevertheless, even in these cases, structural information still arises predominantly from the *unimolecular* dissociation of the ion-molecule reaction product. This situation occurs partly because the ion population is not at thermal equilibrium in most mass spectrometers, and hence the effects of internal energies on ion-molecule reactions may exceed the structural effects of interest. Nevertheless, it is possible to perform gentle ion-molecule reactions that are chemically finely controlled in the mass spectrometer. Not only are the reactions analogous in type to those seen in solution (acid-base reactions, nucleophilic substitution, cycloaddition) but also high degrees of regio-, stereo-, and even enantioselectivity can be achieved.

Although ion-molecule reactions can be studied in the ion source of a single stage mass spectrometer, MS–MS techniques are normally used since this approach allows definition of the nature of both the ionic and the neutral reactant. An example of bimolecular ionic chemistry applied to structural identification is found by considering gas phase Diels-Alder reactions (eq. 14-57).

$$CH_3C\equiv O^+ \;+\; \underset{\substack{\\ m/z\ 43}}{} \quad \overset{[4+2]^+}{\underset{CID}{\rightleftharpoons}} \quad \underset{\substack{\\ m/z\ 97}}{}\!\!\!-CH_3 \tag{14-57}$$

$$H_3C-C\equiv O^+$$
14-7

$$CH_2=C=\overset{|}{O}H$$
14-8

14-9

The above reaction has been used to distinguish three isomeric $C_2H_3O^+$ ions **(14-7 to 14-9)**. When reacting with isoprene, the acetyl cation (CH_3CO^+) is unique in that it displays a [4+2$^+$] Diels-Alder cycloaddition product (eq. 14-57), whereas the other isomers undergo proton-transfer reactions. Further distinction of $C_2H_3O^+$ isomers is achieved in reactions with methyl anisoles. The acetyl ion preferentially forms the intact adduct ion, whereas the other isomers fail to do so, their spectra being dominated by characteristic charge-exchange, proton-transfer, and adduct-fragmentation products.

Enantioselectivity is shown in gas phase ion-molecule reactions and can be used to distinguish optical isomers. Chiral reagents show different reactivity towards enantiomers. An example is the measurement of equilibrium constants for reactions involving transfer of the *R*- and *S*-enantiomers of α-(1-naphthyl)ethylammonium (NapEt$^+$) cation between (*S,S*)-dimethyldiketopyridino-18-crown- 6 (S,S-14-10) and the achiral host 18-crown-6 (eq. 14-58).

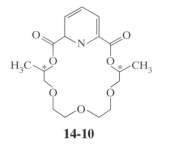

14-10

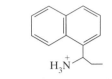

α-(1-naphthyl)ethylammonium cation

$$(S,S\text{-}14\text{-}10)\cdot(R \text{ or } S)-\text{NapEt}^+ + 18\text{-Crown-6} \longleftrightarrow (S,S\text{-}14\text{-}10) + 18-\text{Crown}-6\cdot(R \text{ or } S)-\text{NapEt}^+ \qquad (14\text{-}58)$$

The equilibrium constants are 567 ± 68 and 130 ± 15 for the reactions involving $S-\text{NapEt}^+$ and $R-\text{NapEt}^+$, respectively. It is now feasible, in many cases, to distinguish enantiomers using mass spectrometric methods.

Ion-Molecule Reactions Accompanying Ionization (CI, DI, and SI). One of the principal advantages of EI in chemical analysis is the absence of ion-molecule reactions, a feature that accounts for the highly reproducible nature of EI mass spectra. The fact that ion-molecule reactions may occur in all of the other major types of ionization methods deserves emphasis. Consider, for example, the electron attachment CI mass spectrum of tetrachloronaphthalene, shown in Figure 13-9. As noted previously, the molecular anion $(M^{-\cdot})$ is readily recognized in this mass spectrum, however, under some conditions there is an abundant set of ions that lie 19 Da below the molecular anion (that is, the all ^{35}Cl-version of this isotopic cluster is at m/z 243 compared with m/z 262 in the molecular ion). The isotopic cluster, m/z 243, 245, 247 clearly suggests a trichloro ion, yet the mass difference of just 19 Da is inconsistent with this conclusion. In fact, an ion-molecule reaction must occur, and it has been shown that hydroxylation is possible when CI mass spectra are recorded. The chemistry involves chloride displacement by hydroxide, which is shown in eq. 14-59. It is interesting to note that an

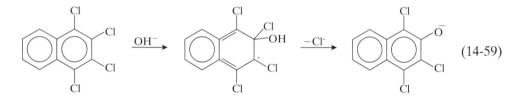

$$(14\text{-}59)$$

even closer examination of this same CI mass spectrum reveals the occurrence of other ion-molecule reaction products. Low-abundance ions are centered at m/z 279 and m/z 299, which result from hydroxylation and chlorination of tetrachloronaphthalene, as illustrated by the reactions in eq. 14-60. Such features do not normally make a major

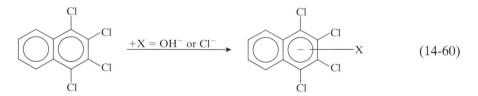

$$(14\text{-}60)$$

contribution to CI mass spectra run under conditions where analytical information is sought. The chlorine substitution process is a prominent exception.

Intact cations can be generated by any of a variety of DI and SI methods and delivered into the gas phase with high efficiency as preformed ions. These ions possess different amounts of internal energy and dissociate to different extents, depending on the amount of energy deposited during ionization. For example, when choline chloride is examined from a glycerol matrix by energetic Xe atom bombardment (FAB), the spectrum is dominated by the intact cation, m/z 104. However, if the solid material is examined directly without using a diluting matrix, either by laser desorption or by energetic ion impact (SIMS), then the intact cation C^+ is accompanied by a higher homologue. This product is due to ion-molecule reactions in the energized selvedge. Mechanistic studies using isotopic labeling have established that the process involves intermolecular transalkylation.

14-7c Advanced Ion Chemistry

Gas phase ion chemistry is extremely rich in terms of the variety of structures encountered and the reactivity displayed, and it is relatively easily accessible experimentally. In this treatment of mass spectrometry, we have rationalized the processes observed rather than enquiring deeply into the underlying physical and chemical principles. This short section briefly highlights ion chemistry in the hope that some students will be intrigued. Liturature references are given for further information.

Fragmentation Accompanied by Rearrangement. Fragmentation tends to occur at particular bonds that correspond to the formation of energetically favored products. Thus the 2-alkyl substituted pyridines (but not the 3- or 4-substituted isomers) undergo facile δ-C—C cleavage in EI to give the cyclic product ion shown in eq. 14-61.

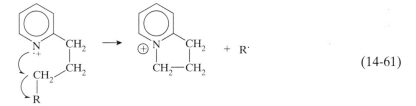

$$ (14\text{-}61) $$

A similar case is the fragmentation of the alkaloid candicine, the DI spectrum and fragmentation mechanism of which is shown in Figure 14-25. Note that the neutral molecule lost is very stable, and the ion is resonance stabilized.

Regioselectivity. Reactions of gas phase ions possess the same possibilities for stereo- and regiocontrol that exist in solution. Much work is being done in this area and it can be briefly illustrated by considering the regioselectivity of deprotonation of sulfones, which has been investigated by negative ion chemical ionization. In principle, the deprotonation of asymmetric sulfones can result in a mixture of different α-sulfonyl carbanions, as exemplified in eq. 14-62 for methyl-d_3 isopropyl sulfone. With this analyte a moderate preference for the formation of the primary $[M-D]^-$ carbanion is

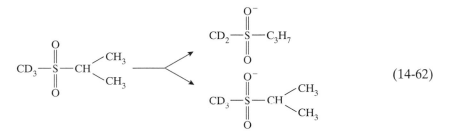

$$ (14\text{-}62) $$

observed, thus indicating a slightly higher stability of the anion.* On the other hand, the deprotonation of benzyl methyl-d_3 sulfone proceeds by proton abstraction from the benzylic position.

In cases where an ion-molecule reaction proceeds to generate a chiral center, the least hindered face is normally the site of attack, as expected. For example, axial orientation of the new C—H bond in 4-*tert*-butylcyclohexanone is strongly favored on reduction using a pentacoordinate silicon hydride reagent. By contrast, 3,3,5-trimethyl-

*T. Surig and H.F. Grutzmacher, *Org. Mass Spectrom.*, **24**, 851 (1989).

cyclohexanone shows almost exclusive equatorial hydride attack, the axial face being blocked by the axial methyl at C-3 (eq. 14-63)*

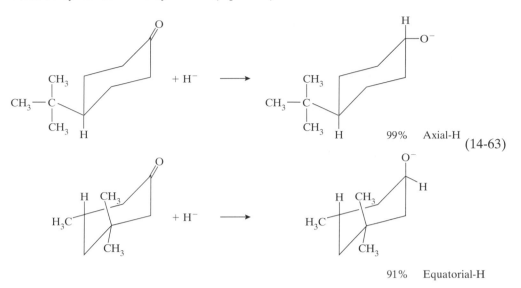

$$99\% \quad \text{Axial-H} \tag{14-63}$$

$$91\% \quad \text{Equatorial-H}$$

Otherwise Inaccessible Neutral Molecules.   Mass spectrometric methods provide access to many hypervalent, energy-rich, or previously unknown neutral compounds. Some of these compounds are remarkably simple, including HNC, H_2CClH, $CH_3C(OH){=}CH_2$, H_2Cl, CH_3OH_2, and RNH_3, all prepared from the corresponding ions by neutralization of the fast beam. These compounds were shown to be stable by using the neutralization–reionization technique, an experimental tool in which ions are neutralized by charge exchange and then reionized. An intact reionized signal indicates that the corresponding neutral species is a stable compound. In the case of ethylenedione, a beam of ions of structure $[O{=}C{=}C{=}O]^{+\cdot}$, was generated either by CI of carbon monoxide or by dissociative electron impact on squaric acid. This species was neutralized by charge exchange with xenon. Several microseconds later, the neutral beam was reionized by collision with oxygen. The fact that an intact ion was observed at m/z 56 demonstrates that the neutral $O{=}C{=}C{=}O$ is stable.[†]

In related experiments, the neutral fragments of dissociation are separated from the ionic constituents of the beam and then ionized by charge exchange and characterized by their fragmentation behavior. An instructive case is the neutral fragment generated when the molecular ion of methyl acetate dissociates by α-cleavage in one of its principal pathways, to yield the acetyl cation, m/z 43. The expected neutral fragment, the methoxyl radical ($CH_3O^\cdot$) is generated but is accompanied by an isomer, the hydroxymethyl radical ($HOCH_2^\cdot$), which is more stable by about 42 kJ mol^{-1} (10 kcal/mol^{-1}).[‡]

Ion-Neutral Complexes.   Some fragmentation processes proceed through ion-neutral complexes in the exit channel of the potential energy surface. These loosely bound complexes survive for long enough to allow rearrangement and exchange of groups between the ionic and neutral components. This is one way in which rearrangements occur in the course of ionic fragmentations. Note that the ion-neutral complexes are similar to the ion-molecule complexes produced in the course of ion-molecule reactions (Figure 14-32). Ion-neutral intermediates in fragmentation, therefore, provide an important bridge between the two main processes in ion chemistry, unimolecular dissociation and ion-molecule reactions. Loss of the stable neutral molecule CH_2O by oxonium ions to yield carbenium ions proceeds through an ion-neutral complex (eq. 14-64). Isomerization of the carbenium ion R^+ can occur in the course of this apparently simple bond cleavage, for example, n-RCO^+ yields sec-R^+. Other examples include H/D

*Y. Ho and R.R. Squires, *J. Am. Chem. Soc.,* **114,** 10961 (1992).
†D. Sülzle, T. Wieske, and H. Schwarz, *Int. J. Mass Spectrom. Ion Proc.,* **125,** 75 (1993).
‡J.K. Terlouw and H. Schwarz, *Angew. Chem. Int. Ed. Engl.,* **26,** 811 (1987).

$$CH_2\!=\!O^+\!-\!C_4H_9{}^n \longrightarrow \left[CH_2\!=\!O\cdots{}^+CH_2C_3H_7\right]$$

(14-64)

exchange in ion-molecule reactions. The extent of exchange in such reactions depends on the depth of the potential well.*

Dissociation of Multiply Charged Ions. With the advent of electrospray mass spectrometry, the dissociation of multiply charged ions has become of practical interest. The major difference between singly and multiply charged ions is the release of coulombic energy on dissociation. The expectation that multiply charged ion dissociation is facilitated has been confirmed in studies of peptides. The charges are seldom closer than 0.5 nm apart, but the coulombic energy associated with two charges separated by this amount is 278 kJ mol^{-1} (66.5 kcal mol^{-1})! This large amount of energy assists in dissociation, but short-range chemical bonding forces are stronger than the coulombic forces so that the ions are metastable to dissociation and can live indefinitely. (The coulombic energy [kJ mol^{-1}] = 138.8/r[nm].) When fragmentation does occur, the coulombic energy appears, at least in part, as kinetic energy of separation of the two fragments. From this energy the distances between the two charges in the fragmenting ion can be estimated. For example, the ion $C_6H_6{}^{2+}$ isomerizes to the charge-separated form $H_3C\!-\!{}^+C\!=\!C\!=\!C\!=\!C^+\!-\!CH_3$, which fragments to give $CH_3{}^+$ and $C_5H_3{}^+$ accompanied by a kinetic energy release of 2.6 eV, which is consistent with an intercharge distance of 0.55 nm. This value is consistent with the expected bond lengths in the linear ion.[†]

Analogies With Solution Chemistry. Many mass spectrometric reactions display analogies with solution reactions, whereas in other cases there are sharp differences due to the effects of solvent in the solution case. For example, the Fischer indole synthesis can be performed in the chemical ionization source of a mass spectrometer and it proceeds by deamination of the protonated hydrazone, exactly as in solution. Eq. 14-65 shows the Fischer indole synthesis in a CI source, when a mixture of phenylhy-

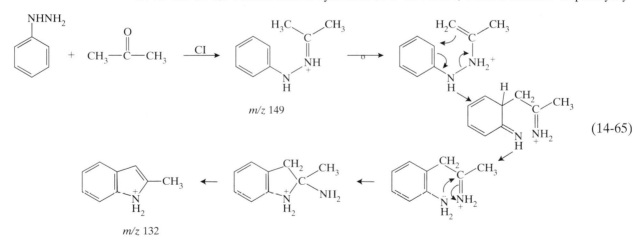

(14-65)

drazine and acetone are introduced. Product analysis is by isolation of the protonated hydrazone (m/z 149) and its characterization by collision with nitrogen (MS–MS). The major process is loss of 17 Da (NH_3) from the protonated hydrazone. These observations confirm the mechanism and its similarity to the solution process.[‡]

*T.H. Morton, *Tetrahedron*, **38**, 3195 (1982).
[†]R.G. Cooks, J.H. Beynon, R.M. Caprioli, and G.R. Lester, *Metastable Ions*, Elsevier, Amsterdam (1973).
[‡]G.L. Glish, R.G. Cooks, *J. Amer. Chem. Soc.*, **100**, 6720 (1978).

Problem 14-5

Account for the major fragment ions at m/z 304 and 207 in the negative ion FAB spectrum of the antibiotic ampicillin **14-15**.

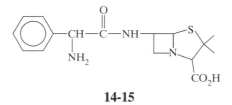

14-15

Answer. This acidic compound yields an abundant $(M-H)^-$ ion. This ion is readily desorbed from solution into the gas phase, where it fragments in a straightforward fashion. (Gas phase fragmentation is a characteristic of DI spectra [see Section 13-3c].) Simple decarboxylation yields m/z 304. The ion at m/z 207 is the result of cleavage of the β-lactam ring.

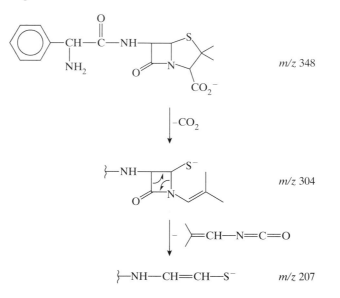

Problem 14-6

The EI (70 eV) mass spectrum of leucine is shown below.
(a) Account for the major ions.
(b) Suggest three specific methods for obtaining a spectrum that shows an ion characteristic of the molecular weight compare (compare with Section 15-1).

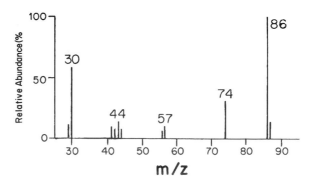

Figure 14-34 EI (70 eV) mass spectrum of leucine.

Answer

(a)

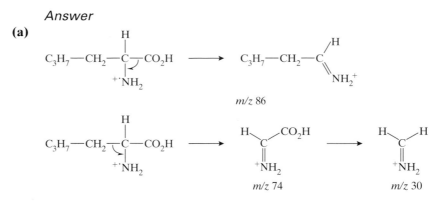

m/z 86

m/z 74   m/z 30

(b)   (1) Chemical ionization, with ammonia or isobutane reagent. (The amino acid is strongly basic, so almost any gas phase acid will protonate it.) (2) Electron impact after derivatization to form the methyl ester. (3) Desorption ionization, recording either positive or negative ions, for example, negative ion FAB.

Problem 14-7

Protonated peptides undergo fragmentations at the C—C and C—N bonds, with and without hydrogen rearrangement. These cleavages yield ions that are classified in the Roepstorff notation as a, b, c, x, y, and z types.

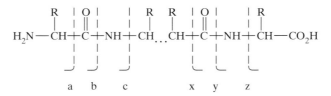

a   b   c   x   y   z

The most significant of these, the b- and y-types, are illustrated in eq. 14-55. Figure 14-35 shows the product ion MS–MS spectrum (Section 15-6) of protonated Tyr-Gly-Gly-Phe-Leu, a small peptide analyzed by the DI method. Comment on the origin of some of the fragment ions, given that the internal masses of the peptides, i.e., of the group HN—CHR—CO, are as follows:

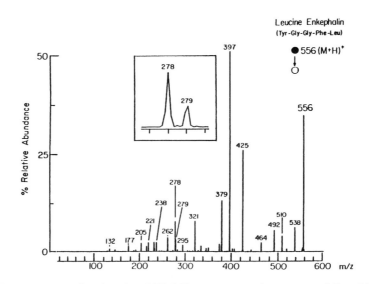

Figure 14-35 Product ion MS–MS spectrum of protonated Tyr-Gly-Gly-Phe-Leu, (M+H)$^+$ = 556, which was ionized by SIMS from a liquid matrix, mass-selected, and dissociated in an ion trap mass spectrometer. (From R.E. Kaiser, Jr., Ph.D. thesis, Purdue University, 1990.)

Peptide	Internal Residue Mass	Peptide	Internal Residue Mass
alanine	71	leucine	113
arginine	156	lysine	128
asparagine	114	methionine	131
aspartic acid	115	phenylalanine	147
cystine	222	proline	97
glutamic acid	129	serine	87
glycine	57	threonine	101
histidine	137	tryptophan	186
hydroxyproline	113	tyrosine	163
isoleucine	113	valine	99

Answer. The DI (SIMS) mass spectrum gives the molecular weight but relatively little information on structure because of the high background from the glycerol/thioglycerol matrix used. The MS–MS spectrum shows better signal-to-noise ratios and structurally diagnostic fragment ions. The selected parent ion consists of two main isotopes, the all ^{12}C and the $^{13}C_1$ versions, which have masses after protonation of 556 and 557, respectively. Both are mass-selected and contribute to the fragmentations shown. The b-ions are present from b_2 (m/z 221) through b_3 (m/z 278) to b_4 (m/z 425) and b_5 (m/z 538). This series alone goes a long way toward establishing the sequence of amino acids in the peptide. a-Ions are generated from b-ions by loss of CO. The prominent ions at m/z 397 and 510 are assigned as a_4 and a_5, respectively. The y-series is not so well developed, but y_1 (m/z 132) and y_2 (m/z 279) are represented. Note the use of slow scans in the inset to the figure, to resolve the ions at m/z 278 and 279. The z-series is represented among the remaining ions, as are dehydration products of some of the ions listed.

PROBLEMS

14-1 The EI mass spectrum of butylbenzene recorded at 14 eV ionizing energy shows three ions of significant relative abundance, the base peak at m/z 134, an ion of 70% relative abundance at m/z 92, and one of 40% relative abundance at m/z 91. Considering the effects of internal energy distributions on mass spectra (Section 14-7a), predict how the spectrum changes as the ionizing energy is raised.

14-2 Sulfur and phosphorus compounds are particularly prone to molecular rearrangements in EI mass spectrometry. Account for the observation of ions at m/z 109 and 111 (100/5 ratio) in the spectrum of **14-16**.

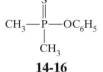

14-16

14-3 The CI mass spectrum of methyl cinnamate, recorded with a protonating reagent ion shows fragments at m/z 121 and 131. Account for these with charge-localized structures.

14-4 Predict the extent to which compounds **14-17** and **14-18** lose (a) H· and (b) CH_3· upon 70 eV electron impact. Give reasons.

14-17

14-18

14-5 Methyl *m*- and *o*-methylbenzoates give identical EI 70 eV mass spectra with the exception of an ion at m/z 118. Of what general circumstances is this behavior a particular case?

14-6 An EI (70 eV) spectrum of an organic compound is shown at the top of p. 291.* **(a)** Is the compound pure? **(b)** What is the compound's molecular weight? **(c)** Are heteroatoms present? **(d)** Is the compound aromatic or aliphatic? **(e)** What is the probable molecular formula? **(f)** Why is there a peak at m/z 77.5?

*From R.E. Kaiser, Jr., Ph.D. thesis, Purdue University, 1990.

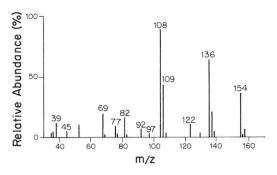

14-7 A compound of molecular weight 113 Da gives ions in its EI (70 eV) mass spectrum at m/z 84, 70, 56, 41, 20, 44, and 98 (in decreasing abundance). Give a structure that is consistent with the data.

14-8 Hernandulcin, an intensely sweet compound discovered by reviewing conquistador literature (*Science,* **277,** 417 [1985]), gives a high resolution mass spectrum that includes a molecular ion at 236.18005 Da and prominent fragment ions at m/z 110 ($C_7H_{10}O^+$), 95 ($C_6H_7O^+$) and 82 ($C_5H_6O^+$). Discuss the claim that these data are consistent with structure **14-19** ($^{12}C = 12.0000$, $^1H = 1.0078$, $^{16}O = 15.9949$).

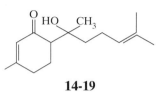

14-19

BIBLIOGRAPHY

14.1. Fundamentals and theory: H. Budzikiewicz, C. Djerassi, and D.H. Williams, *Mass Spectrometry of Organic Compounds,* Holden-Day, San Francisco, 1967; K. Levsen, *Fundamental Aspects of Organic Mass Spectrometry,* Verlag Chemie, Weinheim, Germany, 1978; R.G. Cooks, ed., *Collision Spectroscopy,* Plenum Press, New York, 1978; I. Howe, D.H. Williams, and R.D. Bowen, *Mass Spectrometry, Principles and Applications,* 2nd ed., McGraw-Hill Book Co., London, 1981; K. Vekey, *J. Mass Spectrom.,* **31,** 445 (1996).

14.2. Metastable ions: H.E. Audier and T.B. McMahon, *J. Am. Chem. Soc.,* **116,** 8294 (1994); F. Cacace, F. Grandinetti, and F. Pepi, *Angew. Chem.,* **33,** 123 (1994); R.G. Cooks, J.H. Beynon, R.M. Caprioli, and G.R. Lester, *Metastable Ions,* Elsevier Science Inc., Amsterdam, 1973.

14.3. Spectral interpretation: R.W. Kiser, *Introduction to Mass Spectrometry and Its Applications,* Prentice-Hall, Englewood Cliffs, NJ, 1965; F.W. McLafferty and F. Tureček, *Interpretation of Mass Spectra,* 4th ed., University Science Books, Mill Valley, CA, 1993; I.A. Papayannopoulos, *Mass Spectrom. Rev.,* **14,** 49 (1995).

14.4. Remote and proximate fragmentation: N. Jensen, K. Tomer, and M.L. Gross, *J. Am. Chem. Soc.,* **107,** 1863 (1985); J. Adams, *Mass Spectrom. Rev.,* **9,** 141 (1990).

14.5. Stereochemistry by mass spectrometry; M. Vairamani and M. Saraswathi, *Mass Spectrom. Rev.,* **10,** 491 (1991); J.S. Splitter and F. Tureček, eds., *Applications of Mass Spectrometry to Organic Stereochemistry,* VCH, Weinheim, Germany, 1994; G. Smith and J.A. Leary, *J. Am. Chem. Soc.,* **118,** 3293 (1996).

14.6. Data collections: E. Stenhagen, S. Abrahamsson, and F.W. McLafferty, *Registry of Mass Spectral Data,* John Wiley & Sons, Inc., New York, 1974; S.R. Heller and G.W.A. Milne, *EPA-NIH Mass Spectral Data Base,* vols. 1-3, NSRDS-NBS 63, National Bureau of Standards, Washington, DC, 1978; R.H. Hites, *Handbook of Mass Spectra of Environmental Contaminants,* 2nd ed., CRC Press, Boca Raton, FL, 1992; K. Pfleger, H.H. Maurer, and A. Weber, *Mass Spectral and GC Data of Drugs, Pollutants, Pesticides and Metabolites,* 2nd ed., VCH, Weinheim, Germany, 1992; F.W. McLafferty and D.B. Stauffer, *Wiley/NBS Registry of Mass Spectral Data,* John Wiley & Sons, Inc., New York, 1994; NIST Mass Spectral Search Program for the NIST/EPA/NIH Mass Spectral Library, Windows 1.5, distributed by the Standard Reference Data Program of NIST, 1996.

chapter 15

Chemical Analysis

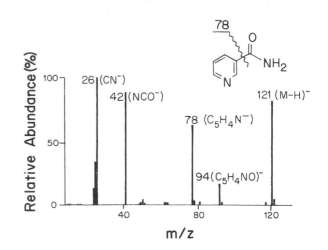

Mass spectrometry can be used to obtain information on organic compounds of all types. The information includes molecular weight, molecular formula, molecular structure, isotopic enrichment, and concentrations of components in mixtures. Each of these topics is dealt with in this chapter.

15-1   MOLECULAR WEIGHT DETERMINATION

The single most valuable item of information that the mass spectrometer can provide is the molecular weight of a compound. Mass spectrometry is the preferred procedure for this determination because of its speed and accuracy. A convenient starting point in making this measurement for organic compounds is to record a conventional EI (70 eV) spectrum. Figure 15-1 illustrates typical results of this approach. When the sample is pure and the compound is sufficiently volatile, a molecular ion is often observed and the molecular weight can be determined. This is the case for the derivatized pentasaccharide, molecular weight 1614 Da, shown in Figure 15-1a. The spectrum in Figure 15-1b shows an ion 4 mass units higher than the probable molecular ion (m/z 123), indicating an impurity or an incorrect molecular ion assignment. (The ion I^+ from an iodine-containing impurity gives rise to this peak at m/z 127.) In some cases, as illustrated by Figure 15-1c, with an expected molecular ion at 243, the thermal lability of the sample or ease of fragmentation results in the absence of molecular ions. This case is difficult to interpret using EI.

The molecular ion is usually the highest mass ion in the EI spectrum, and it may occur in several isotopic forms. However, some compounds fail to give molecular ions. The mass spectrum needs to be considered carefully to evaluate whether an ion is the molecular ion, and additional experiments may be necessary to secure the molecular weight. Figure 15-2 suggests steps that can be taken. Each ionization method offers choices with regard to the chemical state of the analyte, as well as the ionization conditions. Among the many possibilities, a selection can be made based on the nature of the sample. For example, acidic substances often give good negative ion spectra. High

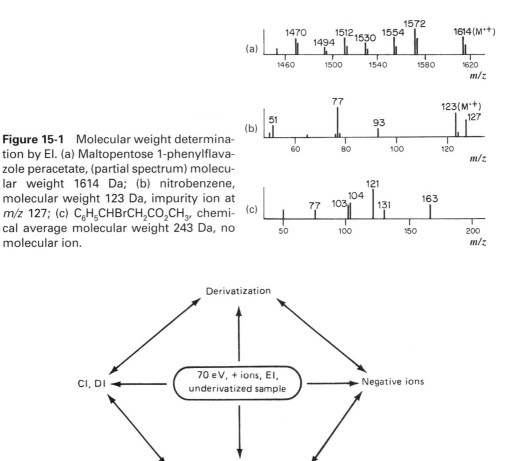

Figure 15-1 Molecular weight determination by EI. (a) Maltopentose 1-phenylflavazole peracetate, (partial spectrum) molecular weight 1614 Da; (b) nitrobenzene, molecular weight 123 Da, impurity ion at m/z 127; (c) $C_6H_5CHBrCH_2CO_2CH_3$, chemical average molecular weight 243 Da, no molecular ion.

Figure 15-2 Options for confirming molecular weight.

molecular weight compounds should be examined by desorption or spray ionization. Easily fragmented molecules can be examined by reducing the ionizing energy (electron energy in EI, matrix selection in DI, ionizing reagent acidity or basicity in CI) to increase the yield of molecular ions relative to fragment ions.

The variety of experiments used to determine molecular weight is illustrated by the case of the amino acid arginine. Figure 15-3 shows arginine mass spectra of several types, any one of which could be used to characterize the molecular weight as 174 Da. Three of the experiments employ DI (Section 13-3c), specifically laser desorption, fast atom bombardment, and field desorption, the last being a procedure in which the solid sample is heated in a high electric field. Conventional EI and methane or isobutane CI do not provide molecular ions.

It is because *chemical reactions* underlie mass spectrometry that the arginine mass spectra display such variety. Ionized arginine occurs as $M^{+\cdot}$, M^-, $(M+H)^+$, $(M-H)^+$, and $(M+C)^+$. Electron and proton transfer, as well as formation of adducts with metal and other ions (C^+), generate these species. The effectiveness of molecular weight determination often depends on the choice of ionizing agent. This is illustrated in the case of D-glucopyranose in Figure 15-4; the sample yields a molecular ion only when the *mild* protonating reagent NH_4^+ is employed; otherwise, dissociation is extensive.

15-2 MOLECULAR STRUCTURE DETERMINATION

Determination of molecular structure by mass spectrometry is based on interpretation of fragmentation patterns. This material is covered in detail in Chapter 14. In some cases the mass spectrum does not provide this information, either because the fragments are

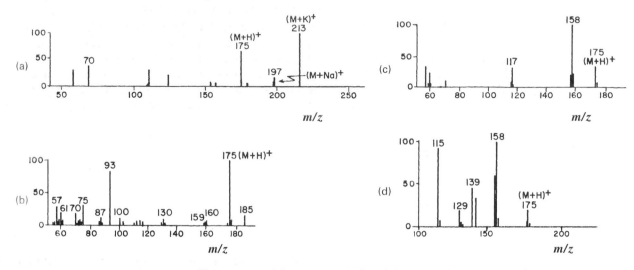

Figure 15-3 Mass spectra of arginine recorded to determine molecular weight. (a) Laser desorption mass spectrum (CO_2 laser, 0.1 J pulse^{-1}, 0.15 µs pulse width). (b) Fast atom bombardment (argon, 5 keV, sample in glycerol). (c) Field desorption (emitter temperature 220° C). (d) Chemical ionization (NH_4^+ reagent, mass-selected). ([a] Reprinted with permission from M.A. Posthumus et al., *Anal. Chem.*, **50**, 985 [1978], copyright 1978 American Chemical Society; [b] from J.J. Zwinselman et al., *Org. Mass. Spectrom.*, **18,** 525 [1983], copyrighted 1983 John Wiley & Sons Ltd., reprinted by permission; [c] from H.U. Winkler and H.D. Beckey, *Org. Mass Spectrom.*, **6,** 655 [1972], copyright 1972 John Wiley & Sons Ltd., reprinted by permission; [d] reprinted with permission from R.J. Beuhler et al., *J. Am. Chem. Soc.*, **96,** 3990 [1974], copyright 1974 American Chemical Society.)

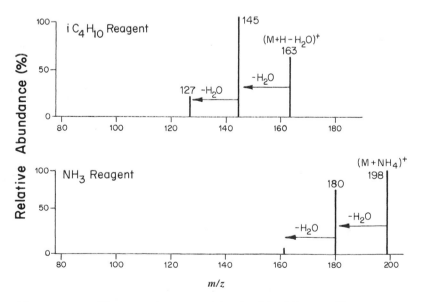

Figure 15-4 Glucopyranose examined with different CI reagents. The soft CI reagent NH_3 gives the intact $(M+NH_4)^+$ ion, whereas the more energetic reagent isobutane yields only fragment ions. (Reprinted with permission from A.G. Harrison, *Chemical Ionization Mass Spectrometry*, CRC Press, Inc., Boca Raton, FL, 1983, p. 119.

obscured by ions due to the matrix or other constituents of the sample, or because the ionization method used does not provide adequate fragmentation. In such circumstances, tandem mass spectrometry can be used, as discussed in Section 15-4b. Structure assignments can be made by de novo interpretation or by comparison of the mass spectrum with a library of spectra.

15-3 MOLECULAR FORMULA DETERMINATION (EXACT MASS MEASUREMENT)

A molecular formula is normally derived from an exact mass measurement. Exact mass measurements frequently can be made to an accuracy of better than 10^{-3} Da using an internal standard of known exact mass. Even with this accuracy, however, a unique fit is seldom obtained when all possible elemental compositions at any nominal mass are considered. The number of possibilities increases with the number of elements that may be present, with the number of atoms of each element possible, and with the molecular weight. For organic compounds that have only the common heteroatoms (O, N, F, Cl, Br, I, S, and Si) and molecular weights less than 500 Da, only a few possibilities need be considered (see Appendix A-5 for masses and abundances of the isotopes). Thus, while exact mass measurement seldom gives a single formula, it often gives a single *reasonable* formula, especially when other information on the sample is considered. Some of this ancillary information may come from the mass spectrum itself including isotopic abundance distributions, which are considered in the next section.

Exact mass measurements are performed with mass spectrometers capable of high mass resolution, namely double-focusing sector (Section 13-4f) or Fourier transform instruments (Section 13-4e). High-resolution measurements can be made after ionization by any of the common methods. The resolution required to separate ions that differ by the units CH_4/O (both nominally 16 Da) (mass difference 32×10^{-3} Da) is approximately at 10^5 at 3300 Da. This value illustrates the need for supplementary data to assign molecular formulas confidently. Frequently, only the composition of the molecular ion and several of the more abundant fragment ions are measured in solving structural problems.

To illustrate the application of high-resolution measurements to the mass spectrum of a natural product available in small amounts, a portion of the high-resolution spectrum of the ergot alkaloid dihydroelmyoclavine ($C_{16}H_{20}N_2O$) is reproduced in Table 15-1. The mass scale is calibrated using ions from perfluorokerosine, an internal standard. The assignments shown in the table are not the only ones possible within the uncertainties of the experiment. They are, however, the only reasonable ones (for example, the mass of the molecular ion, 256.1579 Da, also fits the composition $C_{14}H_{18}N_5$, which, from simple valence considerations, cannot be a stable organic compound).

TABLE 15-1

Part of the High Resolution Mass Spectrum of an Alkaloid

Measured	Calculated	Error ($\times 10^3$ Da)	Composition
235.9866	235.98722	−0.61	C_7F_8
237.1388	237.13917	−0.35	$C_{16}H_{17}N_2$
238.1464	238.14700	−0.55	$C_{16}H_{18}N_2$
241.1347	241.13409	0.59	$C_{15}H_{17}N_2O$
242.1430	242.14191	1.13	$C_{15}H_{18}N_2O$
242.9856	242.98562	0.02	C_6F_9
243.1445	243.14527	−0.80	$C_{14}(^{13}C)H_{18}N_2O$
254.9856	254.98562	0.02	C_7F_9
256.1579	256.15756	0.37	$C_{16}H_{20}N_2O$

15-4   ISOTOPIC ANALYSIS

One of the earliest applications of mass spectrometry was the determination of the isotopic ratios of the elements as they occur in nature. Specialized instruments are now used to measure precisely (often to much better than 0.01%) isotope abundance ratios for use in solving problems in geology, physics, biology, and other sciences. Nuclear materials custody, geochronology, biochemical pathways, and many other applications depend on this measurement. Isotope ratios can only be measured to a precision of about 1% relative standard deviation using conventional mass spectrometers, but this measurement also provides valuable information, as is now shown.

15-4a   Recognition of Elements: Isotopic Signatures

The natural distribution of the isotopes of individual elements means that molecules have characteristic isotopic ratios. Even a qualitative inspection of isotopic ratios is often valuable in indicating the presence of particular elements in an ion. Some elements give characteristic isotopic signatures, which characterize the nature and number of the isotopes. These cases include chlorine, bromine, boron, sulfur, silicon, and many of the metals. Appendix A-5 contains the masses and abundances of the isotopes commonly encountered in organic chemistry. Remember that any mass spectrum shows characteristic isotopic patterns for fragment ions as well as the molecular ion. Figure 15-5 illustrates the characteristic peaks that can occur in fragment ions.

Among the common elements with characteristic isotopic signatures are Cl, Br, S, Si, and B. Table A-6 reproduces these signatures for the cases of a small number of atoms of each element. Many metals also have easily recognized signatures. The common elements, C, H, N, and O, give more subtle effects. The presence of a single nitrogen atom can usually be recognized from the odd mass molecular weight, and the presence of two or more nitrogen atoms is usually indicated by other evidence. A common problem is to decide between C and O.

15-4b   Number of Carbon Atoms in a Molecule

Information on the number of carbon atoms in a molecule can be obtained from the ratios of the isotopic peaks of the molecular ion. The ratio of the $(M+1)^+$ to $M^{+\cdot}$ abundances is given by eq. 15-1, since each C atom has a 1.1% chance of being ^{13}C instead of ^{12}C. Since protonation may contribute to the $(M+1)^+$ ion, i.e., it might include $(^{12}M+H)^+$ as well as $^{13}M^{+\cdot}$, the upper limit of the number of carbon atoms is obtained from eq. 15-1.

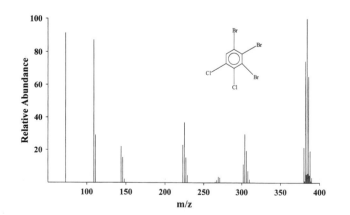

Figure 15-5 EI (70 eV) mass spectrum of 1,2,3-tribromo-4,5,-dichlorobenzene showing characteristic isotopic abundance ratios (isotopic signatures). (From P.J. Ausloos, C. Clifton, O.V. Fateev, A.A. Levitsky, S.G. Lias, W.G. Mallard, A. Shamin, and S.E. Stein, NIST/EPA/NIH Mass Spectral Library—Version 1.5, National Institute of Standards and Technology, Gaithersburg, MD [1996].)

$$\frac{[(M+1)^+]}{[M^+]} = 0.0111 \text{ (number of C atoms)} \tag{15-1}$$

For example, a compound of molecular weight 162 Da showed an $[M^+]/[(M+2)^+]$ ratio of 100/13.0/1.2. The maximum possible number of carbon atoms is 11, since this value requires $(M+1)^+ \geq 12.1\%$ of M^+. Formulas with 10 carbon atoms, $(M+1)^+ \geq 11.0\%$, should also be considered. Hence the probable molecular formulas are $C_{11}H_{14}O$, $C_{10}H_{14}N_2$, and $C_{10}H_{10}O_2$. It is extremely valuable to take the units of unsaturation (rings and double bonds) into account when considering possible molecular formulas, by examining the corresponding hydrocarbons, which in this case are $C_{11}H_{14}$, $C_{10}H_{12}$, and $C_{10}H_{10}$, respectively. These formulas have 5, 5, and 6 units of unsaturation, respectively. Exact mass or other measurements are needed to decide among these possibile formulas.

Consider an unknown compound, molecular weight 322 Da, which shows an $[(M+1)^+]/[M^+]$ ion abundance ratio of only 17% (Figure 15-6). Since the natural abundance of ^{13}C is 1.1% of ^{12}C, the number of carbon atoms must be less than 15 and, as a maximum, the CH content can only account for 212 ($C_{15}H_{32}$) of the 322 Da mass. A 6% $(M+2)^+$ to M^+ ion suggests the presence of one sulfur atom, but ≥ 78 Da are still left to be accounted for by other heteroatoms. Such a situation suggests the possible presence of fluorine, iodine, or phosphorus atoms. The actual molecular formula is $C_{14}H_8F_6S$ (measured mass-to-charge ratio 322.032 Da/charge; expected 322.033). Expected isotopic abundance ratio for $M^+/(M+1)^+/(M+2)^+$ is 100/16/6.

15-4c Calculation of Ion Abundance Ratios From the Molecular Formula

It is a straightforward combinatorial problem to calculate the mass–abundance pattern for a collection of atoms if one knows the masses and abundances of the isotopes of the elements. For example, natural carbon has isotopes ^{12}C and ^{13}C, which occur in the ratio 100/1.11, whereas naturally occuring chlorine has isotopes ^{35}Cl and ^{37}Cl, which occur in the ratio 100/32. In a collection of six carbons (C_6) there is a statistical distribution of isotopes made by random combinations of the atoms. Each carbon atom chosen has a 1.1% probability of being ^{13}C instead of ^{12}C, so the choice of six carbon atoms means that there is a 6.6% probability of a single heavy carbon in the molecule, that is, the M+1 peak corresponding to $^{13}C_1{}^{12}C_5$, is 6.6% of $M^+,{}^{12}C_6$. There are also smaller probabilities of molecules with two or three or more heavy carbon atoms.

Calculated isotopic distributions for typical organic compounds containing various numbers of heteroatoms are shown in Figure 15-7. Note the negligible effect of N and O, the small but recognizable effect of S and Si, the large and characteristic effect of Cl and Br, and the effect of the multi-isotopic metals Sm and Mo. The isotopic distributions shown in Figure 15-7 are obtained by calculation based on binomial expansions. Programs are commonly available that can be used to calculate isotopic patterns for given molecular formulas.

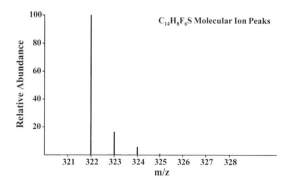

Figure 15-6 Isotope pattern for the molecular ion of a compound. The $^{13}C_1/^{13}C_0$ versions of the molecular ion at m/z 323 and 322, respectively, have an abundance ratio of 17%, which allows no more than 15 carbons.

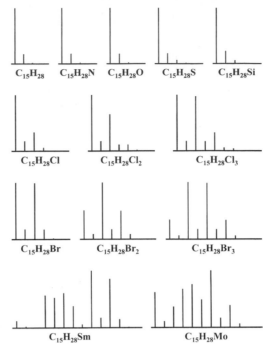

Figure 15-7 Characteristic isotope abundances for ions having the molecular compositions indicated. The abundance distributions were calculated from the binomial expansion using the program ISOMASS. (Program from J.S. Grossert, Dalhousie University, Halifax, NS, Canada.)

15-4d Typical Molecular Ion Isotopic Profiles for Organic and Biological Compounds

As the number of polyisotopic elements in an ion increases, the isotopic distribution becomes more complex. Consider the collection of atoms represented by the formula C_6Cl_6 (hexachlorobenzene). There can be anywhere from zero to six ^{37}Cl atoms in the molecule, and these seven possibilities have a mass range of 12 Da. Thus the molecular ion peak is at least 12 Da wide (from 282 to 294 for ^{12}C only). A separation of 2 Da between intense peaks is characteristic of chlorine-containing compounds (among others). If one only considers the carbon isotopes, seven different isotopomers of the molecular ion, covering mass-to-charge ratios that extend from m/z 282 to 288, are recognized for ^{35}Cl only. Combining carbon and chlorine isotopes means the full range is m/z 282 to 300.

The ion abundance distribution due to the isotopes (often termed the *isotopic distribution*) for C_{60} is shown in Figure 15-8. Note that even in this relatively low molecular weight compound, four different ions have significant abundances (>1%), and the $^{13}C_1$−molecule at m/z 721 is almost as abundant as the all-^{12}C isotopomer at 720. Hydrogens have only minor effects on the overall isotopic distribution, and the isotopic distribution for $C_{66}H_4$, the benzoderivative of C_{60}, is very similar to that for C_{60}.

The isotopic distributions in high molecular weight biological compounds are complex. Calculated data for singly protonated bovine insulin are shown in Figure 15-9. The most abundant ion, nominal mass-to-charge ratio (rounding down as usual) 5733 but actually 5733.6 Da/charge, is composed of five major isobaric ions, the formulas of which are shown in Table 15-2. In biological compounds of this size, the number of carbon atoms effectively determines the width of the isotopic distribution, and heteroatoms, such as S and Cl, make minor contributions to the overall pattern. Note too that the isotope that we normally think of as dominating the mass spectra of organic compounds—that made up of the most abundant isotopes ^{12}C, ^1H, ^{16}O, ^{14}N—occurs at m/z 5728.6 and is of vanishingly small abundance! The center of the distribution corre-

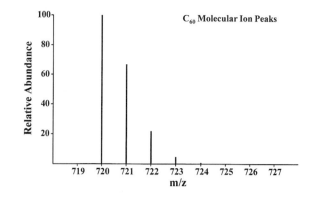

Figure 15-8 Calculated isotopic abundance distribution for the molecular ion of [60]-fullerene.

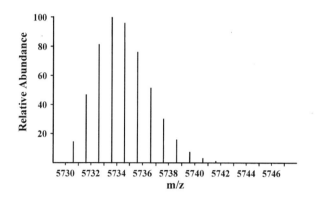

Figure 15-9 Calculated isotopic distribution for protonated bovine insulin. The most important isotopomers have multiple ^{13}C atoms in the molecule. Note that actual not nominal *m/z* values are plotted.

TABLE 15-2

Compositions of Major Isobars in the Base Peak
(*m/z* 5733.6) of Protonated Insulin

$$^{12}C_{251}\ ^{13}C_3\ ^1H_{378}\ ^{14}N_{65}\ ^{16}O_{75}\ ^{32}S_6$$
$$^{12}C_{254}\ ^1H_{378}\ ^{14}N_{64}\ ^{15}N_1\ ^{16}O_{74}\ ^{18}O_1\ ^{32}S_6$$
$$^{12}C_{253}\ ^{13}C_1\ ^1H_{378}\ ^{14}N_{65}\ ^{16}O_{75}\ ^{32}S_5\ ^{34}S_1$$
$$^{12}C_{254}\ ^1H_{378}\ ^{14}N_{64}\ ^{15}N_1\ ^{16}O_{75}\ ^{32}S_4\ ^{33}S_2$$
$$^{12}C_{253}\ ^{13}C_1\ ^1H_{376}\ ^2H_2\ ^{14}N_{65}\ ^{16}O_{75}\ ^{32}S_6$$

sponds to ions with multiple ^{13}C atoms. It also corresponds quite closely to the chemical molecular weight of protonated insulin.

These facts raise interesting questions about the definition of molecular weight and the molecular ion. At least three different definitions can be used: (1) the chemical molecular weight based on elemental atomic weights, (2) the molecular weight based on the most abundant isotope of each element, and (3) the molecular weight based on the most abundant isotopic form of the molecule. The chemical molecular weight of protonated bovine insulin is 5734.6 Da, the molecular weight based on the most abundant element of each isotope is 5728.6 Da, and that based on the most abundant isotopic form of the molecule is 5733.6 Da.

At very high mass, or when low resolution measurements are made, isotopic peaks are not resolved and chemical averaging is inevitable. An early FAB spectrum of human proinsulin (Figure 15-10) illustrates this point. In cases like this, the chemical average molecular weight is the only accessible quantity and it is often quite accurately measurable, despite the relatively low resolution. Note the readily desorbed inorganic cluster ions, which are used as mass standards.

Individual isotopic peaks are resolved to unit mass in the bovine insulin spectrum shown in Figure 15-11. At sufficiently high mass, even unit mass/charge separation constitutes a high resolution measurement. Resolution of individual isobars is not only experimentally difficult (for example, a resolution of 2×10^6 is needed to separate the five ions constituting the base peak of protonated bovine insulin), but an extremely complicated pattern would result in which each nominal mass is compromised of various isotopic forms. Remember that all these isotopic forms are associated with a single chemical species!

Given the smooth distribution of ion abundances in biological compounds resolved to unit mass, it is often relatively easy to recognize the presence of an impurity. For example, the presence of adduct ions or the products of ionic reactions, such as the reduction of disulfide bonds in peptides, can often be observed.

15-4e Determination of Molecular Formula From Isotopic Abundances

The isotopic abundance distribution is characteristic of the molecular formula of a compound. If there is no overlap between the peaks due to ions with different chemical formulas, the pattern of intensities yields information on the molecular formula of the ion.

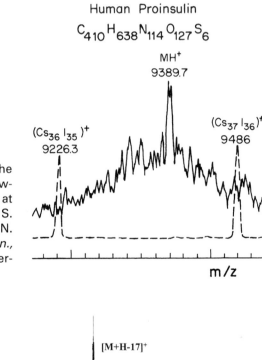

Figure 15-10 Molecular ion region of the FAB spectrum of human proinsulin showing unresolved isotopic cluster centered at 9389.7 Da. (Adapted from M. Barber, S. Bordoli, G.J. Elliott, N.J. Horoch and B.N. Green, *Biochem. Biophys. Res. Commun.*, **110**, 753 [1983]. Reproduced with the permission of Academic Press, New York.)

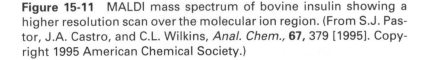

Figure 15-11 MALDI mass spectrum of bovine insulin showing a higher resolution scan over the molecular ion region. (From S.J. Pastor, J.A. Castro, and C.L. Wilkins, *Anal. Chem.*, **67**, 379 [1995]. Copyright 1995 American Chemical Society.)

TABLE 15-3

Calculated and Measured Isotope
Abundance Pattern for Ferrocene

| Mass, (m/z) | Relative Abundance (%) | |
	Calculated	Observed
562	1.1	1.1
563	4.7	4.4
564	18.8	19.0
565	29.6	28.5
566	100	100
567	39.2	39.2
568	8.7	7.6
569	1.3	1.1

In principle, therefore, isotope ratios can to assist in assigning molecular formulas. It is often possible to arrive at a probable molecular formula before undertaking a detailed interpretation of the low resolution mass spectrum. This method can be supplemented by high resolution mass spectrometry (Section 15-3).

For organometallics the *isotope cluster method* can even become competitive with *exact mass measurements* for determining elemental compositions. For example, the molecular ion of triferrocenylborane ($C_{30}H_{27}BFe_3$) gives the calculated and observed isotope pattern shown in Table 15-3. However, the procedure is applicable only when there are no interfering ions, especially neighboring ions such as $(M-H)^+$ and $(M+H)^+$.

15-4f Isotopic Purity Determination

As just discussed, compounds have characteristic isotopic distributions associated with natural isotopic abundance ratios. Unnatural isotopic compositions appear as changes in the isotopic profile. As a result, the isotopic composition of a sample can often be measured by mass spectrometry. Such information is often needed in mechanistic studies in which stable isotopic labeling of particular positions in a reagent is used to elucidate the reaction pathway. Ionic reaction mechanisms, as well as processes occurring in solution, can be examined using this type of information. In the following discussion, we emphasize determination of the *extent* of incorporation, but it should be noted that the *site* of incorporation also can be determined from a knowledge of the fragmentation pattern of the compound.

The most common stable isotopic labels are ^2H, ^{15}N, ^{13}C, and ^{18}O. The mass spectrometric method of determining isotopic incorporation assumes the absence of interference in the ion chosen for measurement. This is frequently the molecular ion in EI or a fragment known to be formed without loss of isotopic label. Ions that can interfere with $M^{+\cdot}$ include $(M+H)^+$ and $(M-H)^+$; since isotopic forms of these species have the same nominal mass as isotopic forms of $M^{+\cdot}$.

The analytical method involves obtaining spectra of the labeled compound and the unlabeled analogue under conditions as identical as possible. The spectrum of the unlabeled compound includes contributions from natural labeled isotopic species, ion-molecule reaction products, and $(M-H)^+$ and $(M-H_2)^{+\cdot}$ ions. If these last two can be completely removed, the determination becomes trivial. The determination of deuterium incorporation may have appreciable uncertainty if processes leading to $(M-H)^+$ cannot be removed, since (1) accurate correction for primary isotope effects cannot be made and (2) the site of origin of the protium atom lost from the unlabeled ion is not always known, so its distribution between protium and deuterium loss in the labeled compound is uncertain.

Consider the molecular ion region of a pure compound, 1,3-dimethoxybenzene, molecular weight 138 Da, ionized by EI. Ions at m/z 138, 139, and 140 occur in relative abundances of 100, 9.6, and 0.2%. Even in this simple case, each nominal mass does not consist of a single isobaric ion (for instance, m/z 139 includes the ^{13}C and ^2H forms of the molecular ion as well as a contribution from the small amount of the protonated

TABLE 15-4

Calculation of the Mass Spectrum of a Labeled Compound (1,3-Dimethoxybenzene)

m/z	138	139	140	141	142	143
d_0	100	9.6	0.2			
d_3				100	9.6	0.2
d_2			100	9.6	0.2	
d_1		100	9.6	0.2		
80% d_3				80.0	7.7	0.2
15% d_2			15.0	1.4	0	
5% d_1		5.0	0.5			
Sum,normalized to m/z 141		6.1	19.0	100.0	9.5	0.2
(80 % d_3, 15% d_2, 5% d_1)						

TABLE 15-5

Calculation of Isotopic Incorporation From Mass Spectra

m/z	138	139	140	141	142	143
d_0	100	9.6	0.2			
Labeled		0.5	2.6	100	9.6	0.2
Assigning 139 to d_1		0.5	0.05			
Subtracting d_1 (0.05 units)		0	2.55	100	9.6	0.2
Assigning 140 to d_2			2.55	0.24		
Subtracting d_2 (2.55 units)			0.0	99.76	9.6	0.2
Assigning 141 to d_3				99.76	9.58	0.2
Subtracting d_3 (99.76 units)				0.0	0.02	0.0

Conclusion: Labeled compound is 99.76 units d_3, 2.55 units d_2, and 0.5 units d_1 that is 97.0% d_3, 2.5% d_2, and 0.5% d_1.

molecule generated in EI). The simplest possible assumption is that the pure mono-deuterated compound contains exactly the same relative abundance ratios, each shifted by 1 Da to higher mass. On this basis the mass spectrum of a sample that is 80% d_3, 15% d_2, and 5% d_1 is calculated as shown in Table 15-4.

To determine isotopic compositions from a mass spectrum, one must reverse the procedure just described. For example, consider an experiment in which catechol is exchanged in NaOD—D_2O at 80° C and subsequently methylated to give 1,3-dimethoxybenzene-2,4,6-d_3. The problem is to determine the isotopic purity of this product. The molecular ion is the base peak in the mass spectrum, and the unlabeled compound shows no $(M-1)^+$ ions at low ionizing energy. The molecular ion regions of the unlabeled (d_0) and labeled compounds are measured as:

m/z	138	139	140	141	142	143
Unlabeled	100	9.6	0.2			
Labeled		0.5	2.6	100	9.6	0.2

Compositions are assigned to the ions in the d_3 spectrum, heavy isotope and protonated species are subtracted in the ratio found for the d_0 compound, and the isotopic composition is calculated as shown in Table 15-5.

From the data in Table 15-5, the isotopically enriched sample has 97% d_3, 2.5% d_2-, and 0.5% d_1 compound. Cases in which the contributions of an $(M-H)^+$ ion cannot be removed contain additional uncertainty, because the assignment of H relative to D loss in the labeled compound is not usually known.

15-5 *QUANTITATIVE ANALYSIS*

It is often important to measure the amount of a particular compound present in a sample. This is a common situation in biological samples. One can employ an internal standard and measure the ratio of the abundance of a characteristic ion of the standard to an ion characteristic of the analyte. Since the ions due to the analyte and internal standard occur in the same mass spectrum, many of the errors associated with ion abundance measurements are canceled. The more similar the standard and analyte, the more accurate the determination; therefore, the best reference compounds are *isotopically labeled* versions of the analyte. In this way, the problem of quantitation becomes one of isotopic abundance measurements, namely, the determination of isotopic purity. An alternative to the use of an internal standard is the standard addition method. In this experiment, the spectrum is recorded before and after the addition.

Quantitation can be achieved by recording simple mass spectra, but for complex samples it is more common that chromatography–mass spectrometry or mass spectrometry–mass spectrometry be employed. In such cases, selected ion monitoring (SIM) experiments are often used (see Figure 15-12). These procedures increase sensitivity by examining only the ions of interest. In a typical case, such as that of the quantitation of the barbiturate phenobarb, present as a component of a complex mixture, the sample is spiked with a known amount of phenobarb-d_3 and introduced from a gas chromatograph (GC). The mass spectrometer records only characteristic ions of phenobarb and its d_3 analogue; in a CI experiment, these are the protonated molecules at m/z 233 and 236. These two characteristic ions occur at the same time in the GC analysis, and during the elution of the chromatographic peak, SIM will be done repeatedly so that the abundances of m/z 233 and m/z 236 are each measured multiple times. Note that although there are many other ions present, the increased specificity achieved using the selected ion monitoring SIM experiment translates into improved detection limits. The ratio of the abundances of the ions due to the d_0 and d_3 compounds is a direct measure of the amount of the analyte (A), since the amount of the added d_3 internal standard (I) is known. Thus the quantity Q is related to the signal S by eq. 15-2.

$$Q_A/Q_I = S_A/S_I \qquad (15\text{-}2)$$

The advantages to this procedure are (1) ratios of isotopic ions minimize analytical errors, (2) once the labeled compound has been added, any losses by whatever process, chemical or physical, as long as not subject to significant isotope effects, does not affect the result, and (3) the selected ion monitoring experiment increases the time spent monitoring ions of interest and so increases signal-to-noise ratios by 10^2 to 10^3 times.

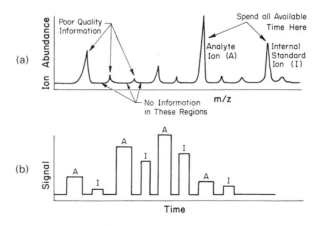

Figure 15-12 Quantitation using selected ion monitoring (SIM) with an internal standard. (a) Mass spectrum, including peaks due to the analyte (A) and the internal standard (I), a labeled form of the analyte. (b) SIM output during elution of the compound of interest, the signals due to A and I being interrogated alternately.

TABLE 15-6

Summary of LC–MS Interfaces

Interface	Ionization Mode	Maximum Flow Rate (μL/nm)	Characteristics
Moving belt or wire	EI/CI/FAB	2000	Limited to volatile organic mobile phases, high efficiency; memory effects
Direct liquid introduction	EI/CI	50	Mechanically simple, no memory effect; inefficient
MAGIC (monodisperse aerosol generation interface for chromatography)	EI/CI	500	Limited to relatively volatile compounds; not very sensitive; good for EI
Thermospray	TS	2000	Used for volatile and some nonvolatile compounds; also an independent ionization method
Electrospray	ES	10 max.; min. <0.02	Interface for LC–MS and CZE–MS; multiply charged ions; high molecular weight determination; also an independent ionization method

Adapted from R.G. Cooks et al., *Encyclopedia of Applied Physics,* vol, 19, VCH, New York, 1997, p. 289.

15-6 MIXTURE ANALYSIS

15-6a Chromatography–Mass Spectrometry

Mass spectrometry is applicable to analysis of mixtures through the various chromatograph–mass spectrometer combinations. The combined techniques retain the separating power of chromatography combined with the molecular specificity of mass spectrometry. The chromatography step also serves the further important function of preconcentrating samples and delivering them to the mass spectrometer in a continuous automated fashion. Mixture analysis using two instruments—a mass spectrometer as chemical separator and a second mass spectrometer as the spectroscopic identification tool—forms the basis for the conceptually related technique of tandem mass spectrometry (Section 15-6b).

The main chromatography–mass spectrometry instruments, the GC–MS and the LC–MS, can each employ a variety of interfaces (Section 13-2b), although as column dimensions have decreased, direct coupling has become standard. Table 15-6 summarizes some characteristics of LC–MS interfaces.

The mass spectrometer can serve either a universal or a selective detector for chromatography. In the latter role it increases the intrinsic resolution of the chromatograph, since many overlapping compounds do not show up in the selective ion detection experiment. Moreover, this experiment of *selected ion monitoring* (SIM), shows greatly improved sensitivity, as noted above, because all the time available during the analysis is spent examining ions relevant to the problem. The increased sensitivity is often a factor of 100 over full-scan mass spectra, making this an extremely important experiment in environmental monitoring and trace biological (such as metabolic) analysis. The plot of the abundance of a single ion, the selected ion chromatogram, records all the chromatographic peaks in which components of interest might fall.

An example is to be found in the analysis of barbiturates by GC–MS. Figure 15-13 illustrates both the total ion chromatogram and a selected ion chromatogram (for the fragment ion, m/z 195) in the case of a sample consisting of a mixture of barbiturates. The total ion chromatogram shows signals due to each component in the mixture, whereas the selected ion chromatogram responds only when allyl-substituted barbiturates are encountered.

As an example of LC–MS analysis, Figure 15-14 shows a chromatogram recorded by electrospray MS for a reduced and then alkylated tryptic digest of a glycoprotein. The chromatogram was recorded using acetonitrile–water as solvent. The information available on the nature of the components that represent individual peaks is illustrated by the representative mass spectrum of a single chromatography peak, which is also shown in the Figure 15-14.

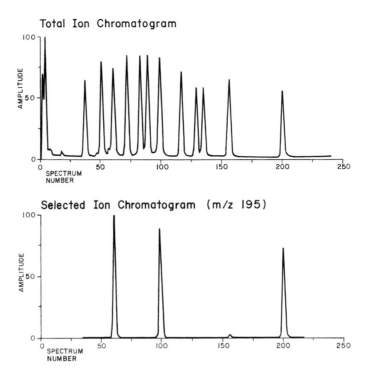

Figure 15-13 Universal and selective detection in GC–MS of a barbiturate mixture, 0.2 μg per component: (a) total ion chromatogram, and (b) selected ion chromatogram recorded by selected ion monitoring. (Courtesy of Finnigan Corp.)

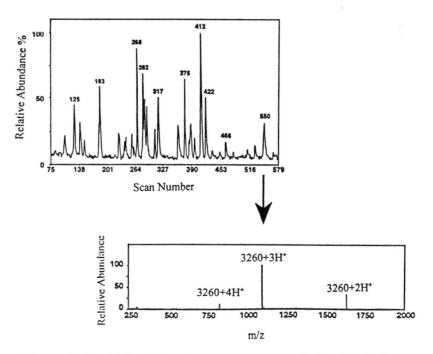

Figure 15-14 (a) Total ion chromatogram recorded by ESMS for a derivatized protein digest and (b) mass spectrum of the component at scan 550 in the chromatogram, showing a component of molecular weight of 3260 Da. (From S.A. Carr, M.E. Hemling, M.F. Bean, and G.D. Roberts, *Anal. Chem.,* **63,** 2802 [1991]. Copyright 1991 American Chemical Society.)

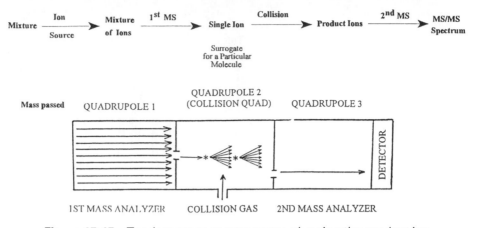

Figure 15-15 Tandem mass spectrometry, showing the product ion scan used to characterize particular compounds in mixtures.

These data clearly help in identifying the individual components and hence in reassembling the structure of the original protein. The selected peptide is of molecular weight 3260 ± 1 Da. Characterization of its amino acid sequence is possible by further experiments using tandem mass spectrometry.

15-6b Tandem Mass Spectrometry (MS–MS)

Concept.   Mass spectrometry–mass spectrometry is a procedure for examining individual ions in a mixture of ions and was briefly introduced in Section 13-4g. It can be compared with the mass spectrometry–chromatography methods. Figure 15-15 summarizes one version of the experiment, the product ion scan, emphasizing its role in mixture analysis. MS–MS employs two stages of mass spectrometry, the first to isolate the ion of interest and the second to characterize it via the spectrum of fragments generated in a collision process. An ion that retains the structure of the neutral molecule being analyzed is selected from other components of the mixture by mass analysis. The selected ion is then excited and dissociated by CID to yield characteristic fragments. A second mass analyzer is employed to record these fragments. In a further parallel with GC–MS and LC–MS, quantitation in MS–MS uses a procedure (selected reaction monitoring) that is the analogue of selected ion monitoring.

Tandem mass spectrometry comprises a number of two-dimensional experiments in principle analogous to two-dimensional NMR. There are various ways of implementing these experiments, depending on the types of mass analyzers used (Section 13-4g). The main advantage of MS–MS experiments is enhanced specificity and lower detection limits. An example is shown in Figure 15-16, which compares the poor signal-to-noise ratio in a conventional mass spectrum of a modified nucleoside sample with the good signal-to-noise ratios observed in the MS–MS experiment. The ion of interest, the protonated free base at *m/z* 166, is obscured by other components of the biological sample in the mass spectrum but gives characteristic product ions in the MS–MS spectrum.

Collisional Activation.   The key requirement to perform MS–MS is that two separate stages of mass analysis be performed, sometimes using two independent mass analyzers through which the ion beam moves in succession (tandem-in-space), or using the same mass analyzer but manipulating the ion population between successive mass analysis operations (tandem-in-time). The former case is typified by the widely used triple quadrupole mass spectrometer, illustrated in Figure 13-28 and the latter by the ion trap mass spectrometer (see Figure 13-23).

A key aspect of any MS–MS experiment is that the mass-selected ion population (the *parent* or *precursor ion*) be subjected to a reaction, usually collisional activation, so

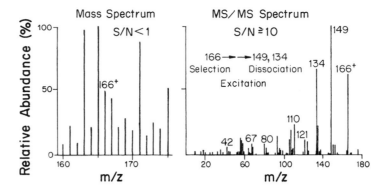

Figure 15-16 Tandem mass spectrometry for a sample containing O^6-methyldeoxyguanosine $(M+H)^+ = 166$ Da. (From D.J. Ashworth et al., *Biomed. Mass Spectrom.,* **12**, 309 [1985]. Reproduced with the permission of John Wiley & Sons Ltd., Chichester, UK.)

that a set of characteristic *product ions* is produced. In conventional mass spectrometry, conversion from precursor to product ions normally occurs in the ion source in conjunction with ionization. The separation of these two steps is a key feature of the MS–MS experiment. When the activation step is achieved by collision with a target gas, the overall process is known as collision-induced dissociation (CID). This collision occurs in a suitable region between the two analyzers in the case of the tandem-in-space experiment. For example, in a triple quadrupole mass spectrometer, there are only two mass analyzers, and the center quadrupole is used as a gas cell in which collisions can occur and their products can be collected and transmitted to the subsequent analyzer quadrupole. In the tandem-in-space experiment, such as is performed in ICR, the collisions are performed in the analyzer itself.

Product Scans. There are a number of different MS–MS experiments, but most applications employ a single type, that in which ions of a given *m/z* value are selected from a mixture. This experiment is known as the *product ion scan,* or simply product scan. Two applications are common: (1) the characterization of particular compounds or groups of related compounds present in a complex mixture and (2) structural analysis of a particular compound through collisional activation. In either case, MS–MS may be used with or without chromatographic separation. In the first type of experiment, the molecular ions of the compound(s) of interest are isolated from other ions formed in the source, so that they can be dissociated and their fragmentation pattern(s) recorded to allow structural, isotopic, or other analysis. The selected ion serves as a surrogate for the neutral molecule from which it is formed, and structural information deduced from the fragmentation of the ion is applied in drawing conclusions about the structure of the neutral precursor. The second experiment is often performed on pure compounds, rather than on mixtures. In such cases, MS–MS is used when the compound either does not fragment sufficiently in the ion source to allow the desired information to be obtained or when its fragmentation is obscured by matrix or background ions associated with the ionization method used. Some of the DI methods, such as fast atom bombardment (FAB) and matrix-assisted laser desorption ionization (MALDI), produce copious amounts of matrix ions over the entire mass spectrum, and the separation of analyte ions from such matrix ions is achieved by tandem mass spectrometry.

An example of the application of a product ion scan to the analysis of a complex mixture of biological origin is shown in Figure 15-17. The mixture was ionized by FAB, and protonated octanoylcarnitine was mass selected in the first quadrupole of a triple quadrupole, and dissociated using collisions in the second quadrupole. The product ion spectrum then was recorded using the third quadrupole. This spectrum is characteristic of the structure of the compound: note particularly the loss of trimethylamine (59 Da)

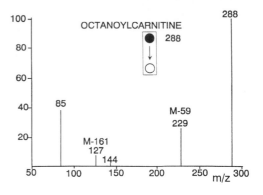

Figure 15-17 Product ion mass spectrum of a component (*m/z* 288, protonated octanoylcarnitine) present in a mixture. The symbol • ⟶ ∘ indicates the product spectrum. Data taken using a triple quadrupole mass spectrometer and 2 eV collisions. (From E. deHoffmann, *J. Mass Spectrom.*, **31**, 129 [1996]. Reproduced with the permission of John Wiley & Sons Ltd., Chichester, UK.)

and the formation of the product ion of *m/z* 85. The product ion mass spectrum of octanoylcarnitine suggests the fragmentation scheme shown in eq. 15-4.

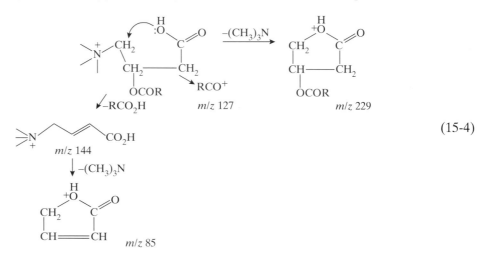

$$(15\text{-}4)$$

One can use MS–MS to lower detection limits in mass spectrometry and perform quantitation at low levels. Qualitative analysis is an equally important application of the method. With advances in ionization methods, many biological compounds can be converted into intact gas phase ions and their molecular weights determined. The delivery of intact molecules into the gas phase as ions is the objective of DI and is the great success of the method. Depending on the experimental conditions chosen for DI, the desorbed ions will contain different amounts of internal energy. This situation leads to fragmentation, which can be more or less extensive. DI is often employed with tandem mass spectrometry so that the ion characteristic of the intact molecule can be isolated and dissociated in a controlled fashion to provide structural information.

Other Scan Modes. It is possible to operate a triple quadrupole mass spectrometer so as to produce other types of MS–MS scans. If a particular product ion is selected by setting the second mass analyzer (the third quadrupole) to pass ions of one particular mass-to-charge ratio and the first mass analyzer is then scanned, one records a *parent ion scan*. In this experiment, all the ions m_1^+ that fragment to yield a selected product ion m_2^+ are monitored.

Returning to the example of the carnitine derivatives, one notes that in a biological matrix, the octanoyl ester is likely to be accompanied by congeners. Hence one may wish to recognize and measure the entire set of carnitine derivatives in a single experiment. This experiment involves using the parent scan and selecting for observation all ions in a mixture that produce the fragment ion of *m/z* 85, characteristic of carnitines. The result of applying such a scan to a biological extract of carnitines is shown in Figure 15-18. Clearly, this experiment shows an improved signal-to-noise ratio when compared with results of the conventional FAB spectrum taken on the whole mixture,

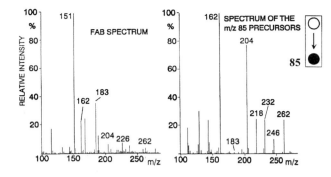

Figure 15-18 Parent ion scan used to recognize the presence and establish the molecular weights of all carnitine derivatives in a complex biological mixture. The fragmentation to yield *m/z* 85 is used as a criterion. The symbol ○—→ • represents the parent scan. (From E. deHoffmann, *J. Mass Spectrom.*, **31,** 129 [1996]. Reproduced with the permission of John Wiley & Sons Ltd., Chichester, UK.)

which is also shown in Figure 15-18. This increase in sensitivity and specificity is one of the most valuable characteristics of tandem mass spectrometry.

There are two additional types of MS—MS scan modes: the constant neutral loss scan and the reaction monitoring experiment. The reaction monitoring experiment is different from the other three, since the output is not a mass spectrum but rather the abundance of a particular product ion arising from a particular precursor ion. Such a specific reaction maximizes the sensitivity of the mass spectrometer, since all the available time is spent monitoring the reaction of interest. One of the main applications of the reaction monitoring scan experiment is the quantitation of drug metabolites in biological fluids.

15-7 THERMOCHEMICAL DETERMINATIONS: HEATS OF FORMATION AND ION AFFINITIES

The importance of thermochemical properties in controlling ionic reactions has already been noted, for example, in the reactions that underlie chemical ionization. A great deal of fundamental thermochemical data is obtainable using mass spectrometry. Early efforts in this direction used electron impact. The ionization efficiency curve, a plot of the abundance of an ion against the ionizing electron energy, provides the fundamental properties of electron affinity (EA), ionization energy (IE), and appearance energy (AE) (see Table A-6 for definitions of terms). When these values are combined with known thermochemical data on neutral species, ionic heats of formation and bond dissociation energies can be determined. A valuable feature of the mass spectrometric approach is the fact that unstable species, such as free radicals, can be studied and bonds that cannot otherwise be cleaved can be investigated.

Ionic heats of formation and bond dissociation energies of gas phase ions are essential information for an understanding of ion chemistry, including the characterization of product ions. For example, an authentic benzoyl cation can be generated by ionizing and dissociating neutral benzoyl bromide (eq. 15-5). If neither of the products is

$$C_6H_5COBr \xrightarrow{-e^-} C_6H_5CO^+ + Br\cdot \tag{15-5}$$

generated with excess energy at threshold, then, from the experimental AE (ΔH_{rxn}) and the known neutral thermochemistry, the value for the ionic heat of formation $\Delta H_f(C_6H_5CO^+)$ of the ion $C_7H_5O^+$ (*m/z* 105) is found from eqs. 15-6 and 15-7.

$$\Delta H_{\mathrm{rxn}} = AE(C_6H_5CO^+) = \Delta H_f(C_6H_5CO^+) + \Delta H_f(Br\cdot) - \Delta H_f(C_6H_5COBr) \tag{15-6}$$

$$\Delta H_f(C_6H_5CO^+) = AE(C_6H_5CO^+) + \Delta H_f(C_6H_5COBr) - \Delta H_f(Br\cdot) \tag{15-7}$$

This value, 706 ±5 kJ mol^{-1} (169 ±1 kcal mol^{-1}), can be used to characterize the benzoyl ion and to establish whether ions of molecular formula $C_7H_5O^+$ generated from other sources have the same or a different structure based on their heat of formation. One simply measures ΔH_f of the ion of interest and compares it to the value of the authentic ion.

One of the most important energetic properties is the affinity of a molecule for a particular ion. For example, the proton affinity of a compound, B, is a measure of the binding energy of a proton (eq. 15-8, compare with eq. 13-10). Similarly, one has analogous

$$PA = -\Delta H_{reaction} \ (B + H^+ \longrightarrow BH^+) \qquad (15\text{-}8)$$

definitions for electron affinity, metal ion affinity, etc. Note that these are all enthalpic properties. Relative values of affinities for particular compounds can be estimated by establishing an equilibrium between the compound of unknown affinity and a reference compound of known affinity. Alternatively, measurement of the onset for a reaction ($BM^+ \longrightarrow B + M^+$) provides the bond energy or affinity values. The fragment ion (M^+) intensity is measured as a function of the energy supplied.

15-8   OTHER APPLICATIONS OF MASS SPECTROMETRY

Mass spectrometry finds an enormous range of applications throughout the sciences. This short section addresses some applications lying outside the immediate areas of organic and biomolecular structural elucidation and quantitation. Topics include (1) industrial applications and process monitoring, (2) isotope ratio measurements, the molecular counterparts of which are also discussed in Section 15-4, (3) elemental analysis, and (4) surface analysis.

15-8a   Industrial Applications and Process Control

Much of the original impetus for the development of mass spectrometry arose from the needs of a single area of industrial production—namely, quality control in petroleum refining. This application provided a powerful stimulus for structural analysis and led to the introduction of commercial instruments in the 1940s. Mass spectrometers, fitted with a wide range of ion sources, can now be used to create ions representative of aqueous solutions, refractory solids, and gases, among other types of samples. The process control application is developing rapidly. Mass spectrometry is also widely used in the chemical and pharmaceutical industries. Even in nonchemical industries such as semiconductors, it is used for trace element determination, microscopic imaging of surfaces, and doping of semiconductor devices.

15-8b   Isotope Ratio Mass Spectrometry

Isotope ratio mass spectrometry (IRMS) is one of the oldest forms of mass spectrometry. It finds use in accurate determinations of isotopic compositions of the elements and of low molecular weight, permanent gases such as H_2, CO_2, N_2 and SO_2, as well as more complex compounds. These values provide a rich source of information on the prior physical and chemical history of the sample. The isotope ratios of hydrogen, carbon, nitrogen, oxygen, and sulfur in organic compounds are determined after combustion. This experiment can reveal a great deal about the previous history of the compounds; for example, the biological sources of marker compounds such as steranes used in petroleum exploration can be revealed.

Magnetic sectors are typically used for their stability and precision, and EI is used to avoid chemical effects arising from ion-molecule reactions. The isotope ratio of the sample is measured by direct comparison with a standard gas of known isotopic composition. The individual ion beams are focused into separate Faraday collectors so that multiple isotopes of a single element may be collected simultaneously (for example, $^{12}CO_2$, $^{13}CO_2$, $^{14}CO_2$).

IRMS has a high sensitivity to differences in isotopic abundances. A relative standard deviation of 0.001% can be achieved under optimal conditions. The widespread use of stable isotopes has led to increasing development of mass spectrometric assays of isotopic abundances in fluids, products of combustion, individual organic compounds, and mineral elements. For example, stable isotope-labeled analogues have been used as tracers to follow the metabolism and excretion of biological compounds and drugs. Different $^{13}C/^{12}C$ ratios of natural carbonaceous materials can be measured by IRMS at the level of differences in parts per thousand from standard samples. This measurement allows the differentation of early European biological samples from native North American ones based on human diet (wheat vs. maize based).

Accelerator mass spectrometry (AMS) is another technique that involves measurement of ratios of abundances of stable or radioisotopes. An accelerator operated at several million electron volts (MeV) is used to produce an ion beam, which is subjected to successive mass-to-charge ratio, momentum-to-charge ratio, or energy-to-charge ratio analyses. It is purified of even the smallest traces of molecular ion contaminants before the isotope ratio of interest is measured. The entire instrument provides extremely high abundance sensitivity, namely, the ability to make measurements on specific isotopes even in the presence of other isotopes of the same element whose abundances differ by factors as large as $1:10^{15}$. The best-known application of AMS is the measurement of radiocarbon (^{14}C) to determine the age of archaeological and geological samples. The procedure requires only a small amount of sample (~1 mg), and with $^{14}C/^{12}C$ ratios of up to 10^{15}, age determinations up to 100,000 years can be made. Note that the method measures undecayed ^{14}C, unlike conventional dating methods, which observe the number of decays of ^{14}C per unit time.

15-8c   Elemental Analysis

Glow discharge (GD) is an ionization method that is most often employed in trace elemental analysis of samples in the solid state. A glow discharge results from the passage of electric current (typically in the microampere range) through a gaseous medium (0.1 to 10 torr), traditionally using a DC voltage of several hundred volts. Inherent in the operation of the GD source is the cathodic sputtering process that produces a plasma. Sampling of the plasma with a mass spectrometer then produces a mass spectrum from which elemental identities and abundances are determined. This aspect of the experiment has much in common with atmospheric pressure ionization (Section 13-3d).

Quantification is performed with certified standards. The detection limit for most elements is in the parts-per-billion range. The method has been widely applied to the analysis of metals, alloys, high-purity materials, and thin films and coatings, and also for surface studies and in material science applications. Thin film analysis can be done as a function of depth, since the material is sputtered away in the plasma at a rate of the order of a few nanometers per minute.

The inductively coupled plasma (ICP) ion source is used for elemental analysis of aqueous solutions. The main body of the ICP torch (5000 to 10,000 K), consists of a quartz tube surrounded by an induction coil that is connected to a radiofrequency (rf) generator. The sample solution is aspirated into the tube via an argon gas stream. The high temperature produces a high ionization efficiency for most elements. ICP–MS is often used in geological, environmental, clinical, and food analysis. The method has very low detection limits (low parts-per-trillion range), excellent quantification, and a wide linear dynamic range (about six orders of magnitude). The attainable precision in quantitation in ICP–MS is about 0.1% to 0.5% relative standard deviation, by using internal standards or isotope dilution techniques. Isobaric interferences and sample matrix interferences cause difficulties in the analysis of a few elements.

15-8d   Surface Elemental Analysis

The desorption ionization methods have seen wide application in the elemental analysis of surfaces. The sputtering process used in DI (Section 13-3c) releases neutral and ionized species from a surface, which can be used for elemental as well as molecular analysis. In addition, depth profiles are obtained by monitoring the abundance of particular ions as a function of the time of sputtering. Commonly, static secondary ion mass spectometry (SIMS) is used to examine the outermost monolayers of a surface. This allows nondestructive qualitative and quantitative surface elemental analysis. A complementary technique to SIMS is Rutherford backscattering spectroscopy (RBS), which employs scattering of high-energy ions, usually $^4\text{He}^+$ beams in the MeV energy range, to study the near-surface regions of solids, including implantation profiles, impurity distributions, and diffusion phenomena.

Two surface imaging techniques based on DI, the ion microprobe and ion microscope, have long been used for surface analysis. In the ion microprobe the desired spatial resolution is achieved by rastering (systematically moving) a focused primary beam (≤ 100 nm) across a surface and analyzing the desorbed ions of interest using a mass spectrometer, often a time-of-flight instrument. The lateral resolution in the ion microprobe is essentially given by the beam size. The ion microscope utilizes primary ion beams of relatively large dimension ($\sim$ 5 to 500 μm). A one-to-one object to image transfer is achieved by an ion optical system that transports the desorbed ions from the sample to a detector while maintaining their original spatial relationships.

15-9 WORKED PROBLEMS

These fourteen problems are worked. The problems in the following section are left for the reader to work. Answers are provided in the accompanying Solutions Manual.

Problem 15-1

How many isotopic forms of the molecular ion of $C_{11}Cl_{11}^+$ exist? How many nominal masses are represented? Carbon has natural isotopes ^{12}C and ^{13}C, chlorine has ^{35}Cl and ^{37}Cl (see Table A-5).

Answer. C_{11} exists in 12 forms, ranging from all ^{12}C through $^{12}C-^{13}C$ mixtures to all ^{13}C. Similarly, Cl_{11} must exist in 12 isotopic forms. All combinations of the C_{11} and Cl_{11} species are allowed, that is $12!/2!10! = (12 \cdot 11)/2 = 66$. This is the total number of isotopic forms. Some of these have the same nominal masses. All integral masses between the two extremes $^{12}C_{11}{}^{35}Cl_{11}^+$ and $^{13}C_{11}{}^{37}Cl_{11}^+$ must be represented, since there is a 1-Da interval between the masses of the carbon isotopes and a 2-Da interval between the chlorine isotope masses. Hence the integral masses range from m/z 517 to m/z 550 and include 34 nominal masses.

Problem 15-2

The region of a Cl mass spectrum containing the protonated molecule shows relative abundances as follows:

m/z 321	2.03%
m/z 322	100.00%
m/z 323	18.38%
m/z 324	34.06%
m/z 325	3.83%

Deduce as much as possible about the molecular formula (see Section 15-4).

Answer. The presence of one Cl atom is evident from the 3/1 (100/34) ratio of m/z 322 and 324. The presence of an odd number of N atoms is inferred from the even mass after protonation. The number of carbons can be determined from the ratio of 322/323 after appropriate isotopic corrections have been made. The ion at 323 must be corrected for $H^{\cdot}$ loss from 324 [^{37}Cl form of $(M+H)^+$]. That at 322 must be corrected for ^{13}C contributions from 321^+. The ^{13}C contribution of 321 is approximately the same as that of 322, which has a heavy isotope peak of 18.38%. Hence, taking 18.4% of 2.03% (0.4%), as the approximate value of $^{13}C\text{-}M^+$ at m/z 322, one is left with 99.6% of ^{12}C, $^{35}Cl\text{-}(M-H)^{+\cdot}$. The $^{37}Cl\text{-}M^{+\cdot}$ contribution to m/z 323 is in the same proportion to 324 as the ^{35}Cl-$M^{+\cdot}$ ion at 321 is to 322 (2.03%, ignoring isotope effects and second order corrections). Hence 2.03% of 34.06% (0.69%) is the correction, leaving 18.38% − 0.67% (17.67%) as the magnitude of $^{35}Cl,^{13}C$-$(M+H)^{+\cdot}$. This ion therefore occurs in its ^{12}C and ^{13}C forms in the ratio 99.61 to 17.7 units (100/17.9). The ^{13}C to ^{12}C ratio is 1.1% per atom, therefore the maximum number of carbon atoms in the molecule is 16 (requires 17.6%, found 17.9%). The partial formula is therefore C = 16, Cl = 1, N = 1. Even with the maximum complement of hydrogen allowed (34) a considerable fraction of the mass is missing, suggesting the presence of oxygen or monovalent elements. The actual formula is $C_{16}H_{17}NO_2PCl$.

Problem 15-3

The neurotransmitter acetylcholine (**15-1**) is a quaternary ammonium ester that can be characterized by its DI spectrum. The SIMS spectrum shows the intact cation at m/z 146 and thus provides molecular weight information. Structural information is available from fragment ions, the most abundant of which corresponds to loss of 59 Da (m/z 87). There are two groups in the molecule that have this mass. Which is the group that is lost?

$$(CH_3)_3NCH_2CH_2OCOCH_3Cl^-$$

15-1

Answer. $N(CH_3)_3$ not $O(CO)CH_3$; simple cleavage of either group would yield relatively high energy products.

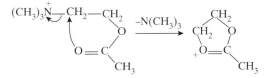

However, rearrangement with loss of $N(CH_3)_3$, a stable neutral molecule, can yield a stable product:

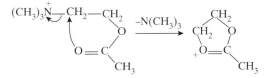

Problem 15-4

What is the isotopic incorporation in butyl ethyl ether synthesized from butanol-1,1-d_2? The 70 eV mass spectrum of the unlabeled ether is complicated since it shows an $(M-H)^+$ ion that has 13% the abundance of the molecular ion. Even at 16 eV, $(M-H)^+$ is still abundant, but the following experimental data are obtained for the labeled (d_2) and unlabeled (d_0) compounds:

m/z	100	101	102	103	104	105	106
unlabeled	0.2	4.6	100	6.5	0.3		
labeled			1.9	3.1	100	6.3	0.3

Answer. From the d_0 spectrum, m/z 100 (0.2%) can only be due to loss of H_2 from the molecular ion. Its ^{13}C isotope contributes 6.6% (<0.01 unit) to m/z 101 and can be ignored. The signal at m/z 101 (4.6 units) must be entirely due to H' loss from the molecular ion, m/z 102, since H_2 loss from m/z 103 is negligible, namely, 0.2% of 6.5 units or 0.01 unit. The ^{13}C isotope of $C_6H_{13}O^+$ (m/z 101) contributes 6.6% of 4.6 units, or 0.3 unit, to m/z 102; hence m/z 102 represents 99.7 units of $C_6H_{14}O$ and 0.3 unit of $^{13}CC_5H_{13}O$. The ions at masses m/z 103 and m/z 104 due to higher isotopes of $C_6H_{14}O$ have abundances of 6.6 units and 0.4 unit, respectively (considering ^{13}C and ^{18}O). Therefore within experimental error (0.1 unit), m/z 103 and 104 are due entirely to isotopic contributions of $C_6H_{14}O$. Thus 99.7 units of $C_6H_{14}O^{+\cdot}$ gives 4.6 units of $C_6H_{14}O^+$ and 0.2 unit of $C_6H_{14}O^{+\cdot}$, or, normalizing, 100 units of $C_6H_{14}O^{+\cdot}$ gives 4.6 units of $C_6H_{13}O^+$ and 0.2 unit of $C_6H_{12}O^{+\cdot}$.

In analyzing the d_2 ether, assumptions regarding the relative losses of H$\cdot$ and D$\cdot$ must be made. To a good approximation, all H$\cdot$ loss from alkyl ethers occurs from the α-position (section 14-2b). Hence, if we assume a k_H/k_D isotope effect of unity for the fast ion source reactions under study, the labeled compound should suffer equal H$\cdot$ and D$\cdot$ loss. Now if the labeled compound were 100% butyl-1,1-d_2 ethyl ether, the molecular ion region would show the following distribution:

m/z	102	103	104	105	106
	0.2+2.3	2.3	100	6.3	0.3

In this computation all molecular hydrogen loss is assigned to H_2, since loss of D_2 or HD is statisically much less likely, and $(M+1)^+$ and $(M+2)^+$ ion abundances are found experimentally for the d_0 compound rather than being calculated (in this case, the difference between the two is negligible). Agreement with experiment is optimized if the calculation includes some d_1 ether. The calculation for 99.0% $d_2 - 1.0\%$ d_1 gives the following result:

m/z	102	103	104	105	106
	2.4	3.1	100	6.1	0.3

These values are in good agreement with the experimental data.

Problem 15-5

Explain the data in Figure 15-19 and describe the key steps that would be needed for quantitative assay for the pesticide DDD at subnanogram levels (see Section. 15-4).

Answer. The data show a gas chromatogram recorded using mass spectrometric detection (full mass spectra). The data are also replotted to show the SIM chromatographic profile for a selected ion (m/z 237). The ion chosen for SIM is characteristic of DDD and DDT. Additional sensitivity is available if the latter experiment is done while spending all the available time measuring a few selected ions instead of extracting the SIM data from the full mass spectrum. Such a selected ion monitoring experiment allows one to achieve subnanogram detection limits. For quantitative accuracy a labeled analog of DDD should be synthesized. The ring d_4 or d_8 compound would be suitable. If the former were chosen, SIM data would be taken for m/z 237 and 241. From the known amount of the labeled compound, the analyte would be quantitated by using eq. 15-2.

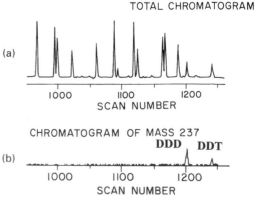

(a)

TOTAL CHROMATOGRAM

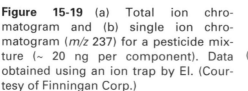

(b)

Figure 15-19 (a) Total ion chromatogram and (b) single ion chromatogram (m/z 237) for a pesticide mixture (~ 20 ng per component). Data obtained using an ion trap by EI. (Courtesy of Finningan Corp.)

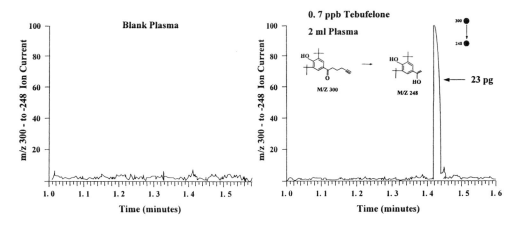

Figure 15-20 Single-reaction monitoring of tebufelone in plasma. (a) Control plasma sample, and (b) a sample spiked with 23 pg of the drug at a level of <1 ppb. The experiment was performed using a triple quadrupole; and the reaction monitored was m/z 300 $\longrightarrow$ m/z 248. (From R.L.M. Dobson, et al., *Anal. Chem.,* **62,** 1819 [1990]. Copyright 1990 American Chemical Society.)

Problem 15-6

Tebufelone (TE) is an antiinflammatory/antirheumatic drug that can be characterized by the fragmentation m/z 300 $\longrightarrow$ m/z 248. Note that the ion m/z 300 is itself a fragment ion. How would its quantitation in plasma be accomplished?

Answer. The fact that low levels of the drug must be examined in a complex matrix means that a very sensitive and selective method is needed. The reaction m/z 300 $\longrightarrow$ m/z 248 serves as a selective process and by performing a single reaction monitoring MS–MS experiment, sensitivity is maximized. This expectation is met by the observations shown in Figure 15-20. A blank plasma sample shows no response when this reaction is monitored. On the other hand, a single peak is observed in an LC experiment for a sample of 23 pg of TE in plasma (concentration 0.7 μg μl^{-1}). Quantitation is best done with an internal standard that is as similar as possible to the analyte. An isotopic form of the drug would be ideal: the d_3 analogue should undergo the characteristic reaction m/z 303 $\longrightarrow$ m/z 251, and by monitoring both reactions, the relative peak height yields the ratio of concentrations of internal standard and unknown.

Problem 15-7

Comment on the selectivity of ionization by charge exchange with the reagent NO.

Answer. From the ionization energy (Table A-2) we note that NO has an EI of 9.3 eV. Hence NO$^+$ undergoes exothermic charge exchange with any compound with a lower ionization energy. Given a mixture, only these compounds are ionized. From the same table, we note that the ionization energy of benzene is also 9.3 eV; substituted benzenes have lower ionization energies. On the other hand, aliphatic compounds, including many oxygenated compounds, have ionization energies greater than that of NO$^+$. Hence NO$^+$ is a selective reagent for the ionization of aromatic compounds in the presence of aliphatics, such as is the case in hydrocarbon mixtures derived from petroleum.

Problem 15-8

Figure 15-21 depicts a mass spectrum of the metal carbonyl, $W(CO)_6$. The compound was vaporized into the ion source. Explain the main features of this spectrum.

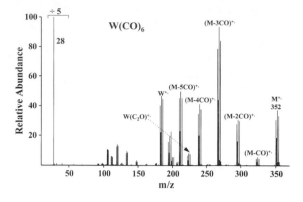

Figure 15-21   EI (70 eV) mass spectrum of the metal carbonyl, $W(CO)_6$.

Answer.   The main ionization process is:

$$W(CO)_6 + e^- \longrightarrow W(CO)_6^{+\cdot} + 2\,e^-$$

but a set of doubly charged ions is also evident at m/z 90–150, and they must arise directly by electron impact on the molecule:

$$W(CO)_6 + e^- \longrightarrow W(CO)_6^{2+} + 3\,e^-$$

Tungsten is a polyisotopic element with abundant isotopes of mass 182, 183, 184, and 186 Da. The isotopic abundances are 26%, 14%, 31%, and 29%, respectively, and the intensity ratios seen in the individual clusters of ions in the spectrum closely match these patterns. This characteristic isotopic signature makes it easy to recognize W-containing ions and to assign the composition of the major ions. As shown in the figure, abundant fragment ions are formed simply by loss of intact CO ligands. The fragmentation pattern of energetic $W(CO)_6^{+\cdot}$ ions involves successive loss of single ligands. The greater the internal energy deposited upon ionization, the more CO ligands are boiled off. (Note that there are minor ions in which ligand dissociation has occurred; the ions at *m/z* 252 fall into this group. These are high energy processes.)

Formation of the fragment $W(CO)_3^{+\cdot}$ is favored, but ions that require more energy to form, like $W(CO)_2^{+}$, as well as those that require less, like $W(CO)_4^{+}$ are also generated in considerable abundance. One concludes that although the average energy deposited in the electron impact experiments matches that required by the $W(CO)_3^{+\cdot}$ fragment ion, there is a distribution of energies that allows the formation of other fragments.

Problem 15-9

How many isotopic forms exist for the molecular ion of dichloroethene?

Answer.   If we consider only the Cl isotopes, the molecule, $C_2H_2Cl_2$, has two chlorine atoms, which can occur in three forms, $^{35}Cl_2$, $^{35}Cl^{37}Cl$, and $^{37}Cl_2$. In general, the number of isotopic forms of a particular element is given by the number of combinations ^n_rC, where *r* is the number of isotopic forms of the element and *n* is the number in the compound of interest. In the case of compounds containing more than one multiple isotopic element, the number of forms for each element is multiplied by that for every other element. Hence, if one considers the isotopes of carbon as well as chlorine, there are also three carbon isotopic forms and thus nine isotopic forms of the molecule. (Obviously, if one considers the hydrogen isotopes, which are of very low abundance, the number is increased further.) The three major isotopic forms of dichloroethene are separated by two mass units each, the separation between the chlorine isotopes. The forms that contain different carbon isotopes are separated by one mass unit. The nine isotopic forms containing various Cl and C isotopes cover a total of seven nominal masses, stretching from the isotope of lowest mass (which also happens in this case to be the most abundant isotopic form), $^{12}C_2H_2^{35}Cl_2$ at $m/z = 96$, to the highest mass isotope of the molecule, $^{13}C_2H_2^{37}Cl_2$ at $m/z = 102$. Since there are nine isotopic forms that fall at seven integral mass values, there are obviously mixtures of isobaric ions (ions of nominally the same mass but different exact masses) at one or more of the masses. For example, the ions $^{13}C_2H_2{}^{35}Cl_2$ and $^{12}C_2H_2{}^{33}Cl^{37}Cl$ both have nominal mass 98 and a mass difference of 0.015 Da. High resolution mass spectrometry (Section 15-3) is the usual method used to distinguish isobaric ions. The resolution needed to separate the two isobars at m/z 98 is $m/\delta m = 65,000$.

Problem 15-10

Fragmentation mechanisms in mass spectrometry are often complicated, and the detailed behavior of even simple systems can only be elucidated by studies using isotopic labeling, tandem mass spectrometry, and ab initio calibrations. This situation is evident by considering the case of dimethyl sulfoxide, the EI mass spectrum of which is shown below (Figure 15-22, 70 eV EI). Start by suggesting a fragmentation pattern to explain these observations.

Answer.　Note how difficult it is to explain the ion at m/z 45, CH_3S^+. In fact, the molecular radical cation of dimethyl sulfoxide can rearrange to two lower energy isomeric structures, the thioenol isomer generated by 1,3-hydrogen rearrangment and the sulfinic acid derivative, CH_3—S—O—$CH_3{}^{+\cdot}$, accessible by 1,2-methyl group migration. Ab initio calculations (Figure 15-23) show that both rearrangments are energetically more favorable than the simple cleavage process leading to methyl radical loss. In low-energy ions, methyl loss proceeds by the rearrangement route, whereas simple cleavage is predicted in ions of higher internal energy.

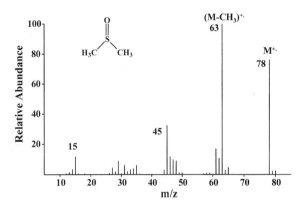

Figure 15-22　EI (70 eV) mass spectrum of dimethyl sulfoxide. (From P.J. Ausloos, C. Clifton, O.V. Fateev, A.A. Levitsky, S.G. Lias, W.G. Mallard, A. Shamin, and S.E. Stein, NIST/EPA/NIH Mass Spectral Library—Version 1.5, National Institute of Standards and Technology, Gaithersburg, MD [1996].)

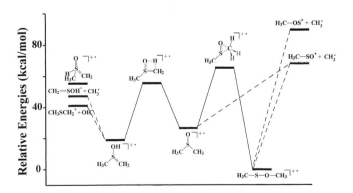

Figure 15-23　Ab initio calculations on the system CH_3—SO—$CH_3{}^{+\cdot}$ (From F.C. Gozzo and M.N. Eberlin, *J. Mass Spectrom.,* **30,** 1553 [1995]. Reproduced with permission of John Wiley & Sons Ltd., Chichester, UK.)

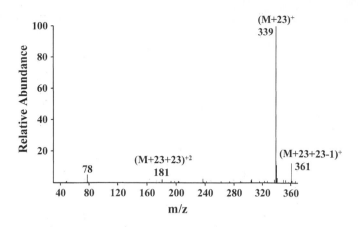

Figure 15-24 Desorption ionization (field desorption) mass spectrum of a glucoside.

Problem 15-11

The DI methods are good choices for the identification of trace amounts of biological compounds. One such method, field desorption, involves the application of a strong electric field to the material held on a heated probe. This procedure assists in vaporization of the intact compound as the cationized molecule. Figure 15-24 shows a case in which sodium is present and attaches to the molecule to generate the $(M+Na)^+$ cationized species. Deduce as much as possible about the biological compound.

Answer. The abundant ion at high mass, m/z 339, provides information on the molecular weight of the compound. The attached cation must be sodium because the mass interval between m/z 361 and 339, 22 Da, represents the replacement of hydrogen by sodium. In general, the first cation simply attaches to the molecule while additional cations can only be attached if the charge increases; otherwise, hydrogens must be replaced. These three processes, with the masses that would be formed from a compound of molecular weight 316 Da are given by the following equations:

$$M + Na^+ \longrightarrow (M + Na)^+ \ (m/z \ 339)$$

$$(M + Na)^+ + Na^+ \longrightarrow (M + 2Na)^{2+} \ (m/z \ 181)$$

$$(M + Na)^+ + Na^+ \longrightarrow (M + 2Na - H)^+ \ (m/z \ 361)$$

Clearly the compound of interest has molecular weight 316.

Field desorption is a very gentle method, so little fragmentation is observed. Nevertheless, the low abundance ion of m/z 181 might also be due to C_6-sugar (protonated glucose). The MS–MS data confirm that the compound is the glycoside shown below.

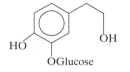

Problem 15-12

Electron ionization mass spectra are more reproducible than other types because they are based on a physical process (electron impact on a molecule) and subsequent *unimolecular* fragmentations. As a result, extensive libraries of EI mass spectra have been compiled. These large data bases ($>10^5$ spectra) can be searched in seconds to seek a match between a recorded spectrum and the library spectra. Figure 15-25 shows an EI spectrum of an unknown compound and the three closest matches as determined by an automatic search. Discuss the quality of the evidence for the identification of the unknown.

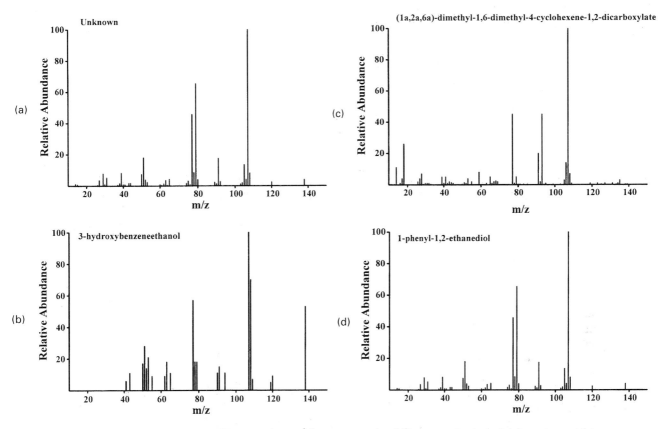

Figure 15-25 Library search of EI mass spectra: (a) spectrum of an unknown, and (b) to (d) are the closest matches found in a search of $>10^5$ spectra in the NIST/EPA/NIH Library of Mass Spectra. (From P.J. Ausloos, C. Clifton, O.V. Fateev, A.A. Levitsky, S.G. Lias, W.G. Mallard, A. Shamin, and S.E. Stein, NIST/EPA/NIH Mass Spectral Library—Version 1.5, National Institute of Standards and Technology, Gaithersburg, MD [1996].)

Answer

1. Electron ionization spectra are often rich in information, with ions at many m/z values and this pattern of information can be highly characteristic of individual compounds.

2. However, mass spectra are representations of the abundances of reaction products (Section 13-3a) and for this reason, ion abundances are much less significant than are the m/z values that are represented. Automated search routines recognize this distinction.

3. In addition, it must be noted that isomerization of gas phase ions can weaken the neutral molecule–mass spectrum relationship so that different compounds, such as positional isomers on an aromatic ring, can give identical spectra.

4. Finally, it is worth emphasizing that the region of the mass spectrum that is most rich in information is the high mass region; the molecular ion and the fragments generated directly from it typically provide the best structural information.

Problem 15-13

Deduce the molecular structure of an unknown amine from its CI (ammonia) mass spectrum (Figure 15-26).

Answer. Amines can be expected to have proton affinities that are greater than those of ammonia; hence exothermic protonation of an amine is expected with ammo-

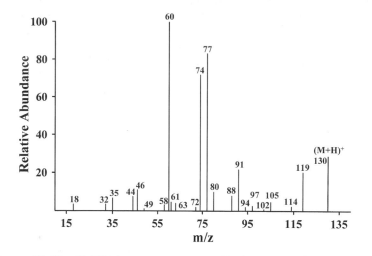

Figure 15-26 CI (NH$_3$) mass spectrum of unknown amine.

nia as the CI reagent gas. One can confidently deduce that the ion m/z 130 is (M+H)$^+$ and that the molecular weight of the compound is 129 Da. This conclusion is consistent with the presence of a single nitrogen atom and with a formula C$_8$H$_{19}$N, which corresponds to a saturated compound (the hydrocarbon analogue is C$_8$H$_{18}$). This identification is tentative, and the fragment ions must be examined for confirmation. Looking at the fragmentation pattern, we immediately encounter a problem—the presence of an ion at m/z 119, corresponding to loss of 11 Da from the putative protonated molecule. This loss is not possible, given the masses of the elements, and must be due to either an impurity or to more complex processes (remember that CI spectra are the result of a combination of fragmentations and ion/molecule reactions). Seeking evidence for the amine structure, one notes the presence of ions belonging to the characteristic amine series m/z 44, 58, 72, observed in EI spectra. One also notes a second and more abundant series at m/z 32, 46, 60, 74, 88. The former ions are singly unsaturated and the latter are saturated ions. The first members of each series are: CH$_2$=N$^+$H(CH$_3$) (m/z 44), CH$_2$=N$^+$(CH$_3$)$_2$ (m/z 58) etc., CH$_3$N$^+$H$_3$ (m/z 32), H$_2$N$^+$(CH$_3$)$_2$ (m/z 46), etc. These are the expected products of the fragmentation of protonated aliphatic amines by loss of alkane and alkene molecules, respectively. One might be tempted to conclude that the data show the compound to be a C$_8$ saturated amine except for the unexplained ion at m/z 119. This ion is accompanied by members of another complete homologous series: m/z 35, 49, 63, 77, 91, 105, 119. Almost the entire mass spectrum is encompassed in the three sets of homologous series. The clue to the identity of the third homologous series is found by recognizing that m/z 35 is the ammonia-solvated form of m/z 18, the ammonium reagent ion, so that m/z 35 corresponds to the ion (NH$_4$$^+$)NH$_3$. Each of the other ions in the series is the ammonia adduct of members of the alkene loss fragment ions. Hence the C$_8$ saturated amine interpretation still stands. Note that CI spectra are not

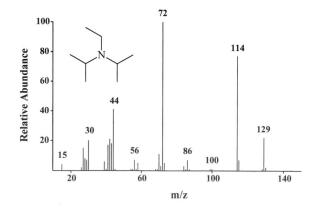

Figure 15-27 EI (70 eV) mass spectrum of diisopropylethylamine.

normally complicated by cluster ions, which can be avoided by diluting the ammonia with methane or another inert gas. (The reagent ion NH_4^+ is still the favored species, but its collisions with ammonia are far less frequent, thus simplifying the mass spectrum.) The exact nature of the C_8 amine is more difficult to deduce from the CI spectrum, although it is more easily deduced from the 70 eV EI mass spectrum (Figure 15-27) to be the diisopropylethylamine.

Problem 15-14

Predict the major features of the EI mass spectrum of the drug dapsone.

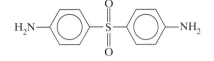

Answer. The compound is expected to give an abundant molecular ion, m/z 248, since anilines are relatively stable to fragmentation. The major cleavage is expected to occur at the sulfur-phenyl bond, since this yields the following stable product ion:

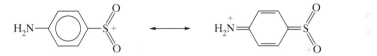

Loss of SO_2, SO, or S occurs from this ion. The loss of SO is unexpected and requires isomerization of $M^{+\cdot}$ to the sulfinic acid:

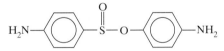

A characteristic fragmentation is the loss of SO_2 from the molecular ion. This rearrangement process is observed quite generally in diaryl compounds.

PROBLEMS

15-1 Below is a laser desorption MS–MS spectrum of the Na^+ adduction of adenosine (m/z 290) examined from the solid state using a YAG laser at 355 nm, 10 ns pulses, and 20 Hz repetition rate. Describe the information provided by this spectrum.

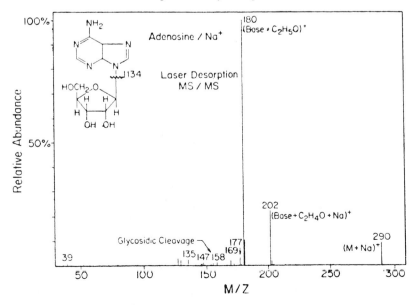

15-2 SIM (EI) scans of m/z 324 ($M^{+\cdot}$ of TCDD) and m/z 332 ($M^{+\cdot}$ of $^{13}C_8$-TCDD)* (at top of p. 322) show results of an experiment to quantitate 2,3,7,8-tetrachlorodibenzodioxin (2,3,7,8-TCDD) in an industrial fly ash sample. The internal standard is the $^{13}C_8$-labeled analogue, which is 85% pure (isotopic and chemical purity). If a 2 µl sample injected into a GC–MS contains 500 pg/µl of the labeled compound, what is the concentration of 2,3,7,8-TCDD in the fly ash, if the 2 µl sample is a 10% aliquot of the extract of a 5 g fly ash sample?

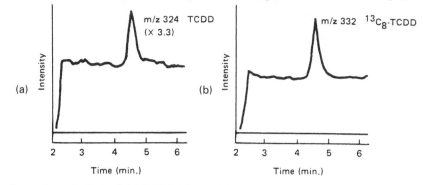

15-3 Secondary alcohols yield $C_2H_5O^+$ ions with structures CH_3—CH=OH^+ characterized by heats of formation of ~590 kJ mol^{-1} (141 kcal mol^{-1}). Does $C_2H_5O^+$ generated from dimethyl ether have the same structure if AE ($C_2H_5O^+$) = 10.7 eV = 1031 kJ mol^{-1}, (246.7 kcal mol^{-1}), ΔH_f ($H^\cdot$) = 217 kJ mol^{-1} (51.9 kcal mol^{-1}) and ΔH_f (CH_3OCH_3) = − 184 kJ mol^{-1} (−44.0 kcal mol^{-1})?

15-4 The proton affinity of water is 691 kJ mol^{-1}, (165 kcal mol^{-1}), CH_3OH is 754 kJ mol^{-1} (180 kcal mol^{-1}), and dimethyl ether is 792.0 kJ mol^{-1} (189.3 kcal mol^{-1}).
   (a) Write an equation defining proton affinity.
   (b) Comment on trends in the above three values.
   (c) Which of the three reagents would you use if you wished to ionize a compound of PA 836 kJ mol^{-1} (200 kcal mol^{-1}) while producing maximum fragmentation?

15-5 An EI mass spectrum of a polynuclear aromatic hydrocarbon is shown below.† Identify the major ions in the spectrum and comment on the processes by which they are generated.

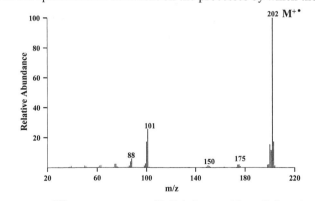

15-6 An electron capture CI mass spectrum of 3,3',4,4'-tetrachlorodiphenyl ether (top of p. 469) was recorded using methane to moderate electron energies.‡ Electron capture (eq. 13-16) is often used to study trace amounts of biological compounds in complex matrices because this ionization method is both highly sensitive and specific to compounds with particular functional groups. In other words, ionization is selective for a small group of compounds, those with high electron affinities (electron affinity is the negative of the enthalpy change for electron attachment). The method is also extremely sensitive (that is, it gives very low detection limits), due to the high mobility of electrons, which facilitates their capture by suitable compounds. Comment on (a) the data and (b) the means used to acquire it.

---

*From L.C. Lamparski and T.J. Nestrick, *Anal. Chem.,* **52,** 2045 (1980).

†From P.J. Ausloos, C. Clifton, O.V. Fateev, A.A. Levitsky, S.G. Lias, W.G. Mallard, A. Shamin, and S.E. Stein, NIST/EPA/NIH Mass Spectral Library—Version 1.5, National Institute of Standards and Technology, Gaithersburg, MD [1996].

‡From E.A. Stemmler and R.A. Hites, *Biomed. Environ. Mass Spectrom.,* **17,** 311 (1988). Reproduced with the permission of John Wiley & Sons.

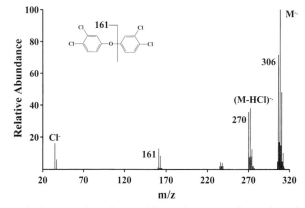

15-7 Comment on the information obtained from the comparison of product ion MS–MS spectra (see below), recorded using a triple quadrupole mass spectrometer on m/z 143 generated by CI from (a) a complex diesel particulate sample and (b) 2-methylnaphthalene.

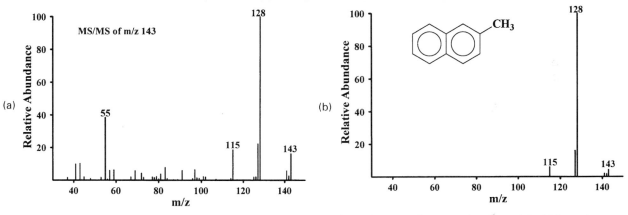

(a)

(b)

15-8 The atmospheric pollutant PAN (acetyl peroxynitrate, molecular weight 121) can be monitored by CI using a membrane-introduction system (Section 13-1). The figure shows the major ions observed in the MS–MS spectrum of the radical anion, m/z 121, formed by electron attachment in the negative ion mode. In this case, too, MS–MS is necessary to characterize the structure. Suggest routes to formation of the major product ions.

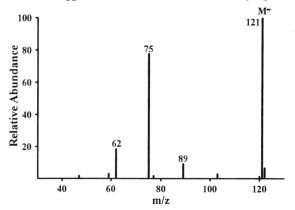

15-9 The neurotransmitter, acetylcholine, is a quaternary ammonium ester with the structure shown in Worked Problem 15-3 (p. 459). The SIMS and FAB spectra display several fragment ions, given the fact that these DI methods are not particularly *soft* methods of ionization. Suggest a route to the fragment ion at m/z 58 and it structure.

15-10 Interpretation of fragmentation pathways is facilitated by the observation of metastable ion peaks, which serve to establish parent-product ion relationships (eq. 13-25). What does one learn about the mechanisms by which the benzene radical cation fragments from the fact that metastable peaks are observed at m/z 32.9. 33.8, 34.7, 74.0, and 76.0 Da/charge?

15-11   The ten problems presented here are intended for practice in deducing molecular structures from mass spectra. The examples used, while simple, all represent real structural problems, and the mass spectra were determined to provide structural information. No source of information besides mass spectrometry was employed in solving these problems. Of course, this is seldom the case in practice, so that much more complex structures than these examples are actually amenable to analysis.

    Molecular ions are assigned to facilitate problem solving. This assignment was trivial in almost all cases. Nevertheless, it was checked by the methods discussed in Section 15-1. Finally, the spectra include only ions with >1% relative abundance.

(a)

(b)

(c)

(d)

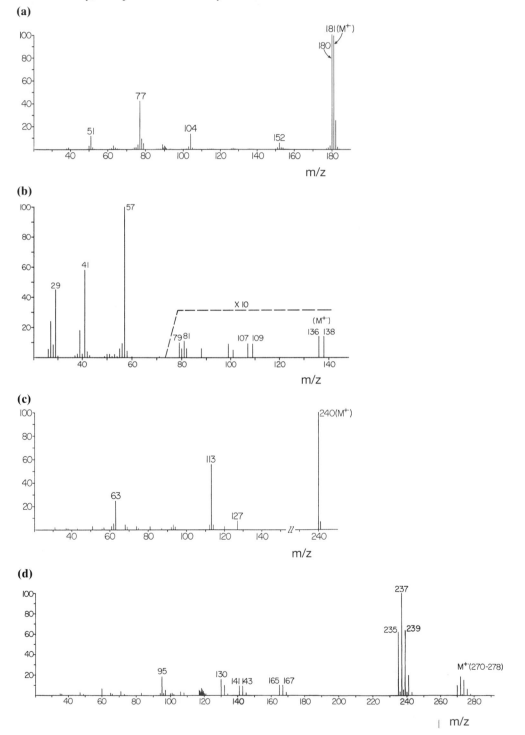

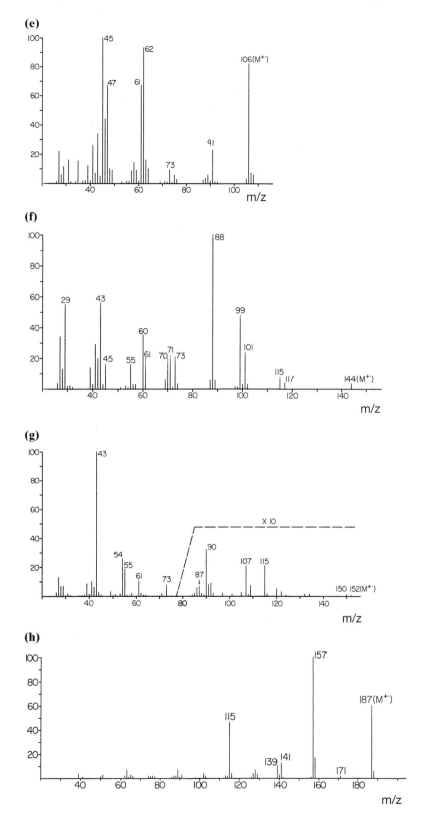

(i)

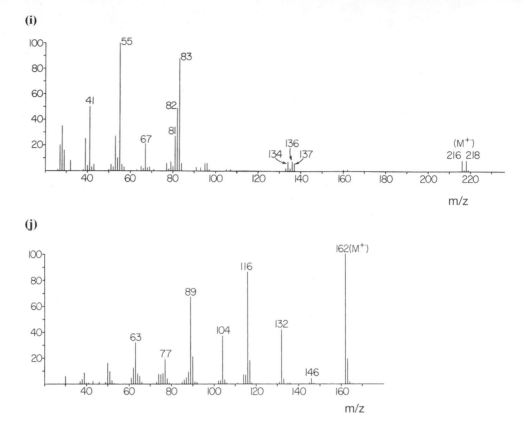

(j)

BIBLIOGRAPHY

15.1.   Gas chromatography–mass spectrometry: G.M. Message, *Practical Aspects of Gas Chromatography/Mass Spectrometry,* John Wiley & Sons, Inc., New York, 1984; M.A. Grayson, *J. Chromatographic Sci.,* **24,** 529 (1986); F.W. Karasek and R.E. Clement, *Basic Gas Chromatograph–Mass Spectrometry: Principles and Techniques,* Elsevier Science Inc., Amsterdam, 1988; A.I. Mikaya, and V.G. Zaikin, *Mass Spectrom. Rev.,* **9,** 115 (1990).

15.2.   Liquid chromatography–mass spectrometry: W.M.A. Niessen and J. Van der Greef, *Liquid Chromatography–Mass Spectrometry: Principles and Applications,* Marcel Dekker, New York, 1992; R.E. Ardrey, *Liquid Chromatography–Mass Spectrometry,* VCH, Wienheim, Germany, 1993; K.B. Tomer, M.A. Moseley, L.J. Deterding, and C.E. Parker, *Mass Spectrom. Rev.,* **13,** 431 (1994); E. Gelpi, *J. Chromatogr. A.,* **703,** 59 (1995).

15.3.   Quantitative analysis: B.I. Millard, *Quantitative Mass Spectrometry,* Heyden & Son Ltd., London, 1978; J.S. Wishnok, *Meth. Enzymol.,* **231,** 632 (1994).

15.4.   Isotopic analysis: J.S. Grossert, ISOMABS program, Dalhousie University, Halifax, N.S., Canada, B3H 4J3. Procomp ver. 1.2, contact P.C. Andrews, University of Michigan Medical School, Ann Arbor, MI 48109.

15.5.   Biological applications: K.R. Jennings, *J. Am. Soc. Mass Spectrom.,* **3,** 867 (1992); M.L. Gross, *Acc. Chem. Res.,* **27,** 361 (1994); K. Biemann and I.A. Papayannopoulos, *Acc. Chem. Res.,* **27,** 370 (1994); F.W. McLafferty, *Acc. Chem. Res.,* **27,** 379 (1994); K. Eckart, *Mass Spectrom, Rev.,* **13,** 23 (1994); D.N. Nguyen, G.W. Becker, and R.M. Riggin, *J. Chromatogr. A.,* **705,** 21 (1995); R.C. Murphy, *J. Mass Spectrom.,* **35,** 5 (1995); G.J. Feistner, K.F. Faull, D.F. Barofsky, and P. Roepstorff, *J. Mass Spectrom.,* **30,** 519 (1995); W. Kleinekofort, J. Avdiev, and B. Brutschy, *Int. J. Mass Spectrom. Ion Processes,* **152,** 135 (1996); A. Margalit, K.L. Duffin, and P.C. Isakson, *Anal. Biochem.,* **235,** 73 (1996); A.R. Donge, A. Somogyi, and V.H. Wysocki, *J. Mass Spectrom.,* **31,** 339 (1996).

15.6.   Web resources: Murray's Mass Spectrometry Page at http://userwww.service.emory.edu/ ~kmurray/mslist.html This site contains a wealth of well-organized information on mass spectrometry, including software for isotope ratio, molecular formula, peptide sequence, ion trajectories, and other calculations.

Appendix to Part IV

TABLE A-1

Conversion Factors in Mass Spectrometry

Charge	1 electronic charge $= 1.6 \times 10^{-19}$ coulomb
Current	1 ion s^{-1} $= 1.60 \times 10^{-19}$ A
Energy	1 eV $= 23.06$ kcal mol^{-1}; 1 kcal mol^{-1} $= 4.18$ kJ mol^{-1};
	1 eV $= 96.4$ kJ mol^{-1}
Mass	1 Da $= 1.66 \times 10^{-24}$ g
Pressure	1 torr $= 133$ Pascal $= 1$ mm Hg $= 1.33$ mbar $= 3.2 \times$
	10^{19} molecules cm^{-3}
Rate constant	
Unimolecular	s^{-1}
Bimolecular	1 liter mol^{-1} s^{-1} $= 2 \times 10^{-21}$ cm^{-3} molecule^{-1} s^{-1}
Velocity	Particle of 1 Da accelerated to 1 eV $= 1.1 \times 10^{-6}$ cm s^{-1}

TABLE A-2

Some Ionization Energies

Element	IE (eV)	Compound	IE (eV)	Compound	IE (eV)
He	24.6	HF	16.0	NH_3	10.2
Ne	21.6	H_2	15.4	$(CH_3)_2C{=}O$	9.7
Ar	15.8	CO_2	13.8	$(C_2H_5)_2O$	9.5
Kr	14.0	CH_4	12.5	NO	9.26
Xe	12.1	H_2O	12.6	Benzene	9.25
H	13.6	O_2	12.1	Pyridine	9.3
C	11.2	C_2H_2	11.4	n-Propylamine	8.8
N	14.5	CH_3OH	10.9	Nitrobenzene	8.7
O	13.6	CH_3CO_2H	10.7	Phenol	8.5
F	17.4	n-Butane	10.5	Aniline	7.7
		H_2S	10.4		

$M \rightarrow M^{+\cdot} + e^-$: $IE(M) = \Delta H_f(M^{+\cdot}) + \Delta H_f(e^-) - \Delta H_f(M) = \Delta H_f(M^{+\cdot}) - \Delta H_f(M)$

*D.R. Lide (ed.), CRC Handbook of Chemistry and Physics, 73rd ed., CRC Press, Boca Raton, FL, 1992.

TABLE A-3

Some Values of Proton Affinities*

Base	Proton Affinity		Base	Proton Affinity	
	kJ mol^{-1}	kcal mol^{-1}		kJ mol^{-1}	kcal mol^{-1}
He	177.7	42.5	Toluene	784.0	187.4
Ar	369.2	88.2	$i\text{-}C_3H_7OH$	795.4	190.1
O_2	421.0	100.6	$i\text{-}C_4H_8$	802.1	191.7
H_2	422.3	100.9	$tert\text{-}C_4H_9OH$	807.3	192.9
HF	484.0	115.7	$(CH_3)_2CO$	812.0	194.1
CH_4	543.5	129.9	Biphenyl	813.6	194.5
NF_3	567.9	135.7	$sec\text{-}C_4H_9OH$	815.0	194.8
C_2H_6	596.3	142.5	Phenol	816.0	195.0
$i\text{-}C_4H_{10}$	675.9	161.5	Styrene	838.2	200.3
CS_2	683.0	163.2	NH_3	854.0	204.1
Cyclohexane	687.5	164.3	Anthracene	870.1	208.0
H_2O	691.0	165.2	$C_6H_5NH_2$	882.5	210.9
H_2S	705.0	168.5	CH_3NH_2	899.0	214.9
HCN	712.9	170.4	$(CH_3)_2NH$	929.5	222.2
Benzene	750.4	179.3	Pyridine	930.0	222.3
CH_3NO_2	752.8	179.9	$(CH_3)_3N$	948.9	226.8
CH_3OH	754.3	180.3	$(CH_3)_2N(CH_2)_2N(CH_3)_2$	1013.6	242.3

*From E.P. Hunter and S.G. Lias, Evaluated Gas Phase Basicities and Proton Affinities of Molecules: An Update, *J. Phys. Chem. Ref. Data* (at press), 1997. From NIST Chemistry Webbook, NIST Standard Reference Database No. 69, August, 1997 release [http://webbook.nist.gov/chemistry].

TABLE A-4

Some Gas Phase Acid Strengths (Enthalpies)*†

Compound	ΔH_{acid} (kcal mol^{-1})	ΔH_{acid} (kJ mol^{-1})	Compound	ΔH_{acid} (kcal mol^{-1})	ΔH_{acid} (kJ mol^{-1})
CH_4	417	1743	PH_3	368	1538
NH_3	404	1689	CH_3SH	357	1492
H_2	400	1672	H_2S	351	1467
H_2O	391	1634	CH_3CO_2H	349	1459
$CH_2{=}CHCH_3$	391	1634	HCl	333	1392
CH_3OH	381	1593	HBr	324	1354
C_2H_2	379	1584	CF_3CO_2H	323	1350
HF	372	1555	HI	314	1313
$(CH_3)_2C{=}O$	370	1547			

*Data from S.G. Lias, J.E. Bartmess, J.F. Liebman, J.L. Holmes, D.R. Levin, and W.G. Mallard, Gas Phase Ion and Neutral Thermochemistry, *J. Phys. Chem. Ref. Data,* **17,** (S)1 (1988).

†Enthalpies defined for $HA \rightarrow A^- + H^+$ as $\Delta H_{acid}(HA) = \Delta H_f(A^-) + \Delta H_f(H^+) - \Delta H_f(HA)$

TABLE A-5

Natural Isotopic Masses and Abundances*

Element	Isotope	Mass	Natural Abundance	Element	Isotope	Mass	Natural Abundance
Hydrogen	^1H	1.0078	99.985	Silicon	^{28}Si	27.9769	92.23
	^2H	2.0140	0.015		^{29}Si	28.9765	4.67
Boron	^{10}B	10.0129	19.9		^{30}Si	29.9738	3.10
	^{11}B	11.0093	80.1	Phosphorus	^{31}P	30.9738	100
Carbon	^{12}C	12.0000	98.90	Sulfur	^{32}S	31.9721	95.02
	^{13}C	13.0034	1.10		^{33}S	32.9715	0.75
Nitrogen	^{14}N	14.0031	99.63		^{34}S	33.9679	4.22
	^{15}N	15.0001	0.37	Chlorine	^{35}Cl	34.9689	75.78
Oxygen	^{16}O	15.9949	99.78		^{37}Cl	36.9659	24.24
	^{17}O	16.9991	0.04	Bromine	^{79}Br	78.9183	50.70
	^{18}O	17.9992	0.20		^{81}Br	80.9163	49.32
Fluorine	^{19}F	18.9984	100	Iodine	^{127}I	126.9045	100

*From D.R. Lide (ed.), *CRC Handbook of Chemistry and Physics,* 73rd ed., CRC Press, Boca Raton, FL, 1992.

TABLE A-6
Calculated Isotopic Abundances

Ion	Ion Masses	Abundances (%)	Ion	Ion Masses	Abundances %	Ion	Ion Masses	Abundances (%)
Cl	35	100.0	ClBr	114	77.0	Br$_2$S	190	50.1
	37	32.3		116	100		191	0.4
Cl$_2$	70	100.0		118	24.3		192	100.0
	72	64.7	Cl$_2$Br	149	61.6		193	0.8
	74	10.5		151	100.0		194	52.0
Cl$_3$	105	100.0		153	45.3		195	0.4
	107	97.2		155	6.2		196	2.1
	109	31.5	Cl$_3$Br	184	51.3	ClSi	63	100.0
	111	3.3		186	100.0		64	5.0
Br	79	100.0		188	64.7		65	35.7
	81	97.5		190	17.5		66	1.6
Br$_2$	158	51.2		192	1.6		67	1.1
	160	100.0	ClBr$_2$	193	44.0	BrSi	107	99.0
	162	48.7		195	100.0		108	5.0
Br$_3$	237	34.2		197	69.5		109	100.0
	239	100.0		199	13.5		110	4.9
	241	97.5	ClBr$_3$	272	26.3		111	3.2
	243	31.6		274	85.5	ClB	45	24.6
S	32	100.0		276	100.0		46	100.0
	33	0.8		278	48.7		47	8.0
	34	4.4		280	7.9		48	32.3
S$_2$	64	100.0	ClS	67	100.0	BrB	89	24.6
	65	1.6		68	0.8		90	100.0
	66	8.9		69	36.7		91	24.1
	68	0.2		70	0.2		92	97.5
Si	28	100.0		71	1.5	SSi	60	100.0
	29	5.0	Cl$_2$S	102	100.0		61	5.9
	30	3.3		103	0.8		62	7.7
Si$_2$	56	100.0		104	69.2		63	0.2
	57	10.0		105	0.5		64	0.2
	58	7.0		106	13.4	SB	42	24.6
	59	0.2		108	0.5		43	100.0
	60	0.1	BrS	111	98.0		44	1.8
B	10	24.6		112	0.8		45	4.4
	11	100.0		113	100.0	SiB	38	24.3
B$_2$	20	6.0		114	0.8		39	100.0
	21	49.3		115	4.2		40	5.7
	22	100.0					41	3.2

TABLE A-7

Glossary

Appearance energy: endothermicity of process: $AB + e^- \longrightarrow A^+ + B + 2e^-$.

Atmospheric pressure chemical ionization: a variant of chemical ionization performed at atmospheric pressure.

Base peak: the most intense peak in the mass spectrum, hence 100% relative abundance.

Breakdown curve: plot of ion abundance vs. ion internal energy (normalized at each energy); shows mass spectrum as a function of internal energy.

Charge exchange: process whereby one particle transfers an electron to another (such as $M + A^{+\cdot} \longrightarrow M^{+\cdot} + A$); used in chemical ionization.

Chemical ionization: method in which neutral molecules are ionized by ion-molecule reactions to generate an ionized form of the molecule at a pressure of about 1 torr.

Collision-induced dissociation: process whereby a mass-selected ion is excited and caused to fragment by collision with a target gas, especially in MS–MS.

Cyclotron motion: cyclic rotation of an ion in a fixed magnetic field.

Desorption ionization: method for ionizing nonvolatile solid samples, in which molecules are subjected to the impact of energetic particles or photon beams.

Distonic ion: radical ion in which the charge and radical sites are formally located on different atoms in the molecule.

Distribution of internal energy (P[ε]):   analogue of Boltzmann distribution for molecules not in thermal equilibrium.

Double focusing: a combination of direction and velocity focusing in sector instruments, used to achieve high resolution.

Electron affinity: enthalpy change for the process: $M^- \longrightarrow M + e^-$.

Electron capture: ionization process in which a molecule or atom captures a thermal energy electron, typically in a CI source, and generates the molecular radical anion, $M^{-\cdot}$.

Electron impact: ionization method in which molecules are ionized directly by energetic electrons (usually 70 eV) at low pressure ($\sim 10^{-5}$ torr).

Electrospray ionization: method used to ionize samples from solution by combination of electric field, heat, and pneumatic force.

Even-electron ion: ion with even number of electrons, commonly with a closed-shell electronic configuration.

Fragmentation pattern: set of reactions leading from the molecular ion to fragment ions.

Fragment ion: ion generated by fragmentation not directly by ionization of a neutral molecule.

Gas chromatography–mass spectrometry (GC–MS):   combined technique for mixture analysis, in which the separated GC components are passed continuously into the MS.

Glow discharge: method used to ionize solid samples for elemental analysis by applying an electric field to create an energetic plasma.

Inductively coupled plasma: method used to ionize solution samples for elemental analysis by a plasma.

Inelastic collision: collision in which internal energy is not conserved.

Ion internal energy: total electronic, vibrational, and internal rotational energy, referenced to ground state of the ion.

Ion source: device used to generate sample ions by electron impact, chemical ionization, etc.

Ionization energy (IE):   minimum energy required to remove an electron from a molecule; endothermicity of the process: $M \longrightarrow M^{+\cdot}$.

Isobaric peak (ion):   peaks or ions of the same nominal (integral) mass, but different exact mass and composition.

Isotopic peak (ion):   peaks or ions due to other isotopes of the same chemical, but different isotopic composition.

Liquid chromatography–mass spectrometry (LC–MS):   combined technique for mixture analysis.

Magnetron motion: slow circular drift of an ion along a path of constant electrostatic potential; magnetron motion occurs in ICR as a result of the crossed radial electric field and axial magnetic field.

Mass resolving power: $m/\Delta m$, in which m is ionic mass and Δm is the width of the mass peak (typically taken as the full width at half-maximum mass spectral peak height).

Mass spectrum: Plot of ion abundance vs. mass-to-charge ratio normalized to most abundant ion.

Mass-to-charge ratio (m/z):   Daltons/electronic charge.

Mathieu stability diagram: diagram showing the solutions to the Mathieu equation that correspond to stable ion trajectories, displayed as a function of parameters related to operating voltages, mass and charge of the trapped ions.

Matrix-assisted laser desorption ionization: method of ionization that uses laser irradiation of solid analyte present in high dilution in a matrix.

Metastable ion: ion that fragments slowly after emergence from the ion source but before it reaches the detector; in sector instruments, metastable ions give rise to signals that appear at unique m/z values related to the parent and product ion masses.

Molecular ion: ion derived from the neutral molecule by loss or gain of an electron or other simple unit, such as $(M+H)^+$, $(M+Cl)^-$, $(M-H)^-$.

Multiply charged ion: ion bearing more than a single charge and having a correspondingly reduced mass-to-charge ratio.

Neutral loss scan: an MS–MS experiment that records all parent ions that lose a particular neutral fragment.

Odd-electron ion: *see* Radical ion.

Parent ion (m$_1^+$):   any ion (including negatively and doubly charged ions) that fragments to product ions.

Parent ion scan: an MS–MS experiment that records all parent ions that produce a particular product ion.

Photodissociation: process in which an ion fragments by absorption of one or more photons.

Product ion (m$_2^+$):   ion generated by fragmentation of any parent ion.

Product ion scan: an MS–MS experiment that records all product ions derived from a single parent ion.

Proton affinity: enthalpy change for the process: $MH^+ \longrightarrow M + H^+$.

Radical ion (odd-electron ion):   charged, open-shell molecule with at least one unpaired electron.

Relative abundance (RA):   abundance normalized relative to the base peak.

Secondary ion mass spectrometry: mass spectrometry based on analysis of particles emitted when a surface, usually a solid although sometimes a liquid, is bombarded by energetic ($\sim$ keV) primary particles (such as Ar^+ and Cs^+).

Selected ion monitoring (SIM):   experiment in which a mass analyzer is used to detect one or a few ions as a function of time.

Spray ionization: methods used to ionize liquid samples directly by electrical, thermal, and pneumatic energy, by means of a spray of fine droplets.

Surface-induced dissociation: process whereby a mass-selected ion is excited and caused to fragment by collision with a target surface.

Tandem mass spectrometry (MS–MS):   two-stage mass analysis experiment; used to study the chemistry of selected ions or individual components in mixtures.

Thermal ionization: method whereby solid samples are ionized on the hot surface of a metal filament.

Unimolecular rate constant (k$|\varepsilon|$):   the rate constant k (in s^{-1}) is dependent on the internal energy (ε, always shown explicitly) at which it is measured.

TABLE A-8

Acronyms

AMS	Accelerator Mass Spectrometry
APCI	Atmospheric Pressure Chemical Ionization
CI	Chemical Ionization
CID	Collision-Induced Dissociation
CZE–MS	Capillary Zone Electrophoresis–Mass Spectrometry
DI	Desorption Ionization
EI	Electron Impact (Ionization)
ES	Electrospray Ionization
FAB	Fast Atom Bombardment
FD	Field Desorption
FT–ICR	Fourier Transform–Ion Cyclotron Resonance
GC–MS	Gas Chromatography–Mass Spectrometry
GD	Glow Discharge
ICP	Inductively Coupled Plasma
LC–MS	Liquid Chromatography–Mass Spectrometry
LD	Laser Desorption
MALDI	Matrix-Assisted Laser Desorption Ionization
MS–MS	Mass Spectrometry–Mass Spectrometry
MW	Molecular Weight
NCI	Negative Ion Chemical Ionization
PA	Proton Affinity
PD	Plasma Desorption
rf	Radiofrequency
SI	Spray Ionization
SID	Surface-Induced Dissociation
SIM	Selected Ion Monitoring
SIMS	Secondary Ion Mass Spectrometry
TOF	Time-of-Flight
TS	Thermospray Ionization

part V

INTEGRATED PROBLEMS

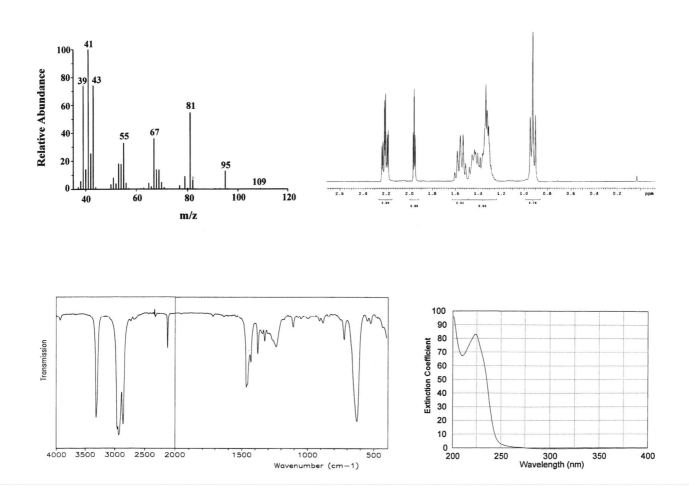

chapter 16

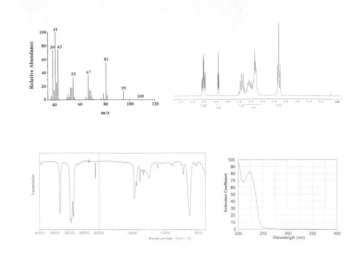

Integrated Problems

The preceding sections included spectral problems that focused on each separate area. In the following problems all four spectroscopic methods are to be used for the derivation of structures. The problems are arranged loosely in order of increasing difficulty. Elemental formulas are not provided. Both chemical ionization (CI with isobutane as the reagent gas) and electron impact (EI) mass spectra are given for nominal molecular weight. The EI spectra provide further information on fragmentation. Other types of ionization are utilized occasionally. The matrix-assisted laser desorption ionization (MALDI) technique was used for nonvolatile materials such as polymers, and the desorption ionization (DI) technique was used for salts, other nonvolatile materials, and compounds that are thermally unstable. Proton NMR spectra at 300 MHz are given for all exercises but one. Standard ^{13}C spectra at 75 MHz are given for all cases except when DEPT spectra are provided. The DEPT experiment includes a full ^{13}C spectrum at the bottom and a methine-only spectrum in the middle. In the top DEPT spectrum, methine and methyl carbons give positive peaks and methylene carbons give negative peaks. Infrared spectra of liquids were recorded from neat films between KBr windows, and spectra of solids were obtained from KBr discs. A Raman spectrum is given in one case. Ultraviolet–visible spectra are given for all cases, and circular dichroism spectra for those that are optically active.

The suggested procedure for approaching these problems is as follows. Obtain the nominal molecular weight from various mass spectra. Analyze fragmentation processes and isotopic clusters for additional structural information. Examine the ^{1}H and ^{13}C NMR spectra to determine the relative numbers of hydrogens and carbons and compare this result with that from MS. Suggest substructures from these data. When the DEPT spectra are given, sort out methyl, methylene, and methinyl groups. Consider the possibility of heteroatoms such as nitrogen, sulfur, or silicon by examination of the mass spectrum. Suggest a molecular formula. Use the infrared spectrum to determine what functional groups are present, and use the electronic spectrum to decide what chromophores may be present. Calculate the unsaturation number from the proposed formula and compare with the groupings identified thus far. Use the COSY experiment to identify connectivities between protons, and the HETCOR experiment to identify connectivities between protons and carbons. Circular dichroism spectra should be used to determine absolute configuration where appropriate.

Finally, bring the various proposed structural units together in possible structures and reexamine all the data to determine whether they are consistent with the structures. If you reach an impasse after due effort, we provide tables of molecular formulas and of functional groups as hints at the end of the problems. Refer to them only as a last resort. The full structures are given in the Solutions Manual. The solution to the first problem is presented here in detail.

Use the spectra as further exercises on the principles developed in the text. Try to identify every peak in the ^{1}H and ^{13}C NMR spectra, and justify chemical shifts and coupling constants. Use HETCOR to assign every hydrogen to an attached carbon, and identify all the COSY cross peaks. Find infrared absorptions for all the functional groups and hydrocarbon components in the molecules. Rationalize the location of the wavelength maxima and the intensities in the electronic spectra. Explain the differences among the various mass spectra (EI, CI, DI) and identify important fragmentation pathways.

PROBLEM 16-1

Mass spectrum (CI)

Mass spectrum (EI)

Proton NMR spectrum (CDCl$_3$)

DEPT spectra

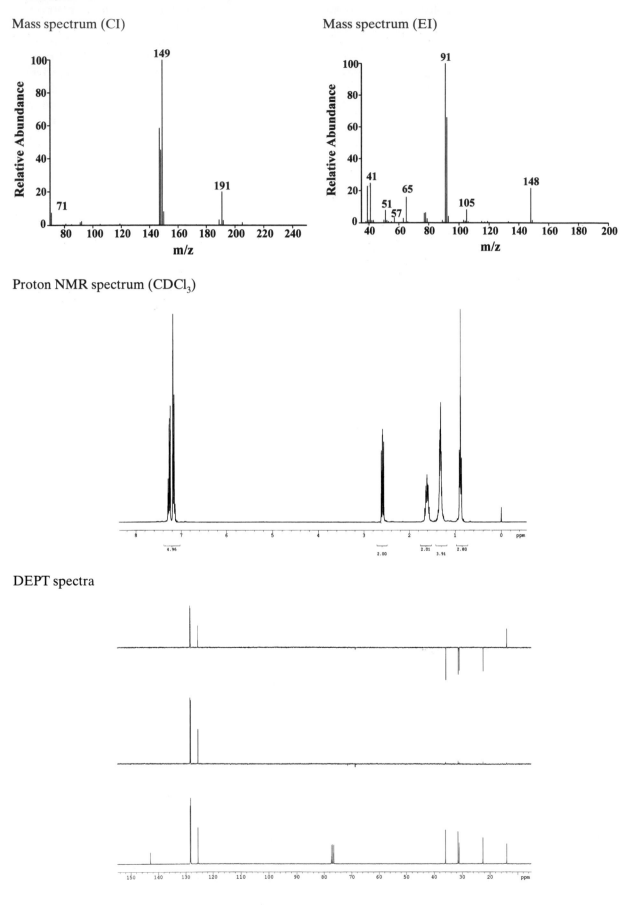

COSY spectrum

HETCOR spectrum

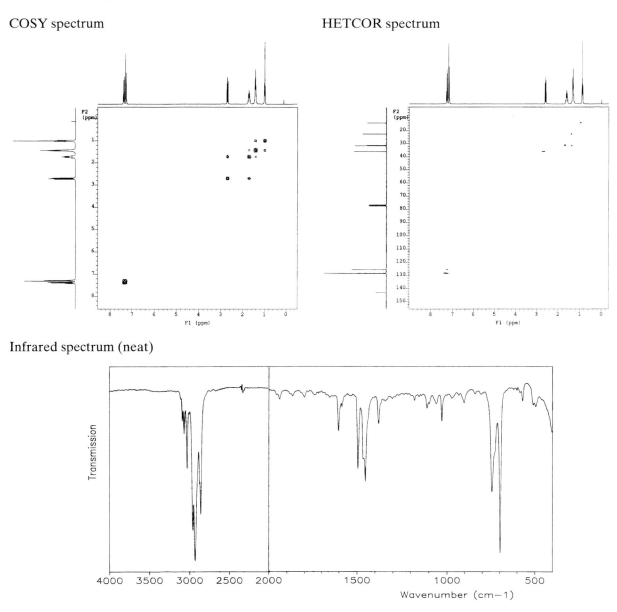

Infrared spectrum (neat)

Ultraviolet–visible spectrum (EtOH, ε[209] 8400, ε[243] 76, ε[248] 122, ε[253] 174, ε[262] 216, ε[268] 171)

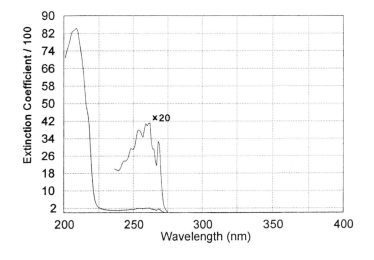

PROBLEM 16-2

Mass spectrum (CI)

Mass spectrum (EI)

Proton NMR spectrum (CDCl₃)

Carbon-13 NMR spectrum

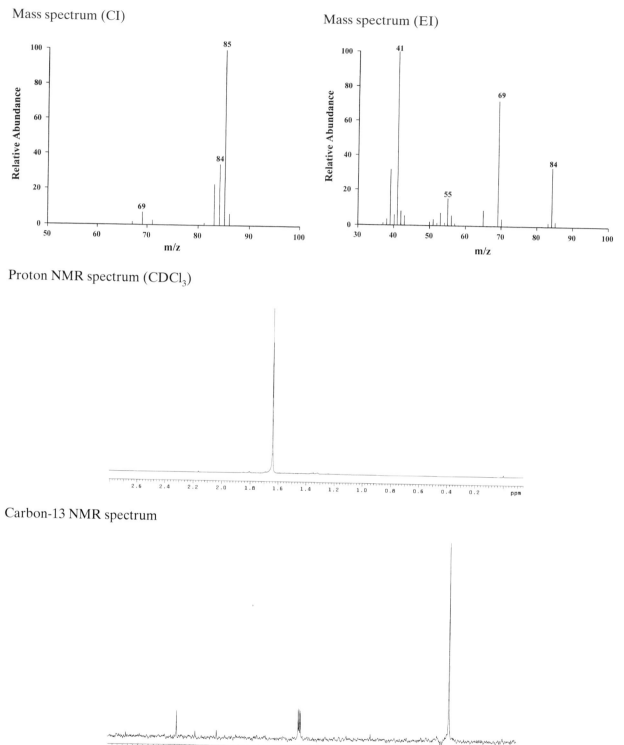

HETCOR spectrum

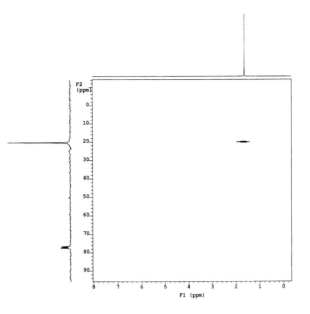

Infrared spectrum (neat)

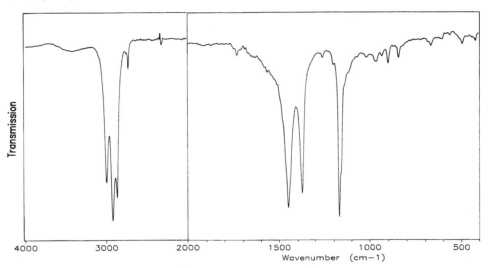

Raman spectrum (neat)

Ultraviolet–visible spectrum (EtOH)

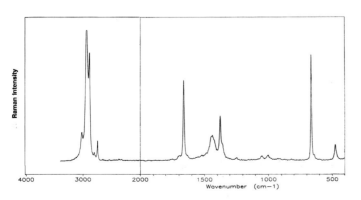

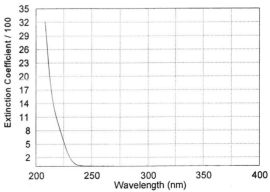

PROBLEM 16-3

Mass spectrum (CI)

Mass spectrum (EI)

Proton NMR spectrum (CDCl$_3$)

Carbon-13 NMR spectrum

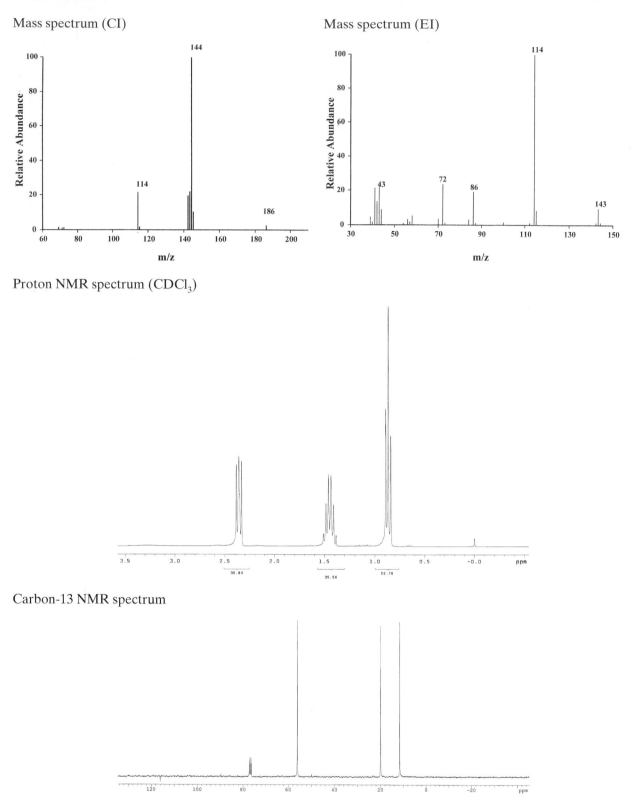

COSY spectrum   HETCOR spectrum

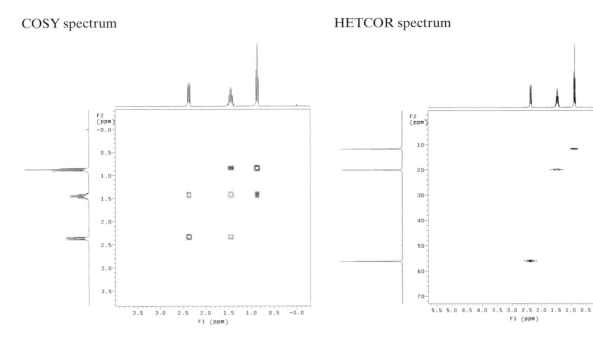

Infrared spectrum (neat)

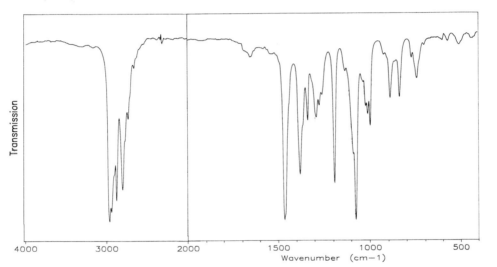

Ultraviolet–visible spectrum (EtOH)

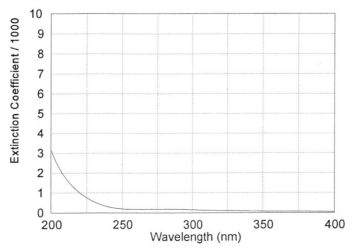

PROBLEM 16-4

Mass spectrum (CI)

Mass spectrum (EI) *[Note: The molecular ion is absent. Because EI is a hard method of ionization, loss of multiple hydrogen atoms sometimes occurs, that is, H loss followed by one or more H$_2$ losses.]*

Proton NMR spectrum (CDCl$_3$)

Carbon-13 NMR spectrum

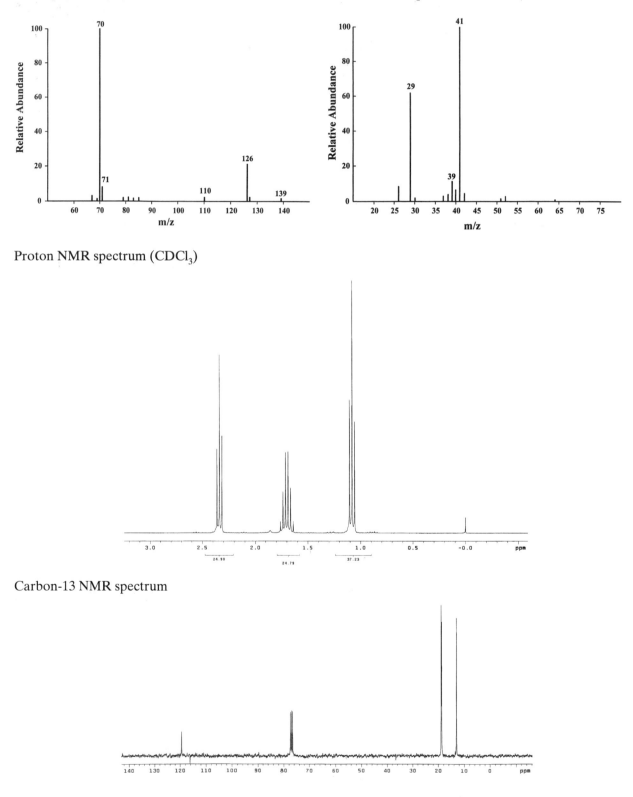

COSY spectrum   HETCOR spectrum

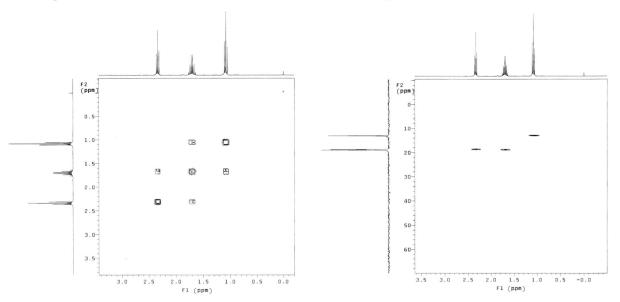

Infrared spectrum (neat)

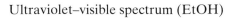

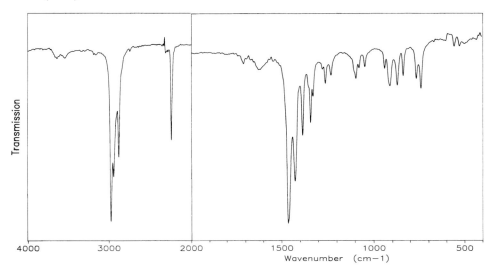

Ultraviolet–visible spectrum (EtOH)

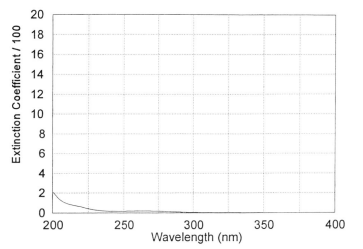

PROBLEM 16-5

Mass spectrum (CI)   Mass spectrum (EI)

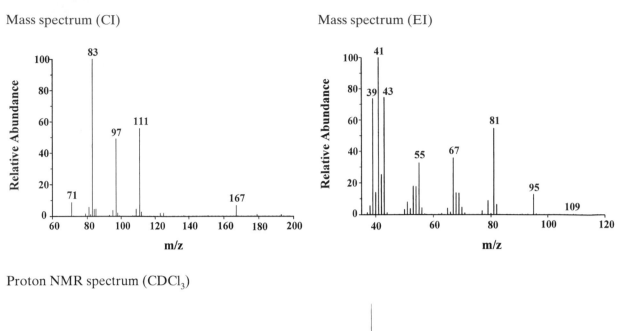

Proton NMR spectrum (CDCl₃)

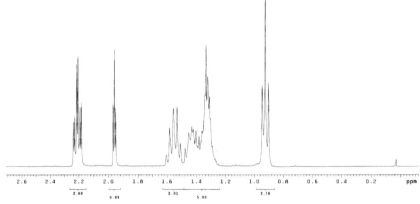

Carbon-13 NMR spectrum

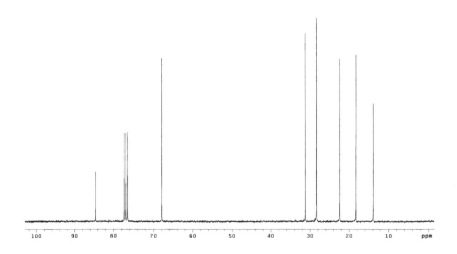

COSY spectrum   HETCOR spectrum

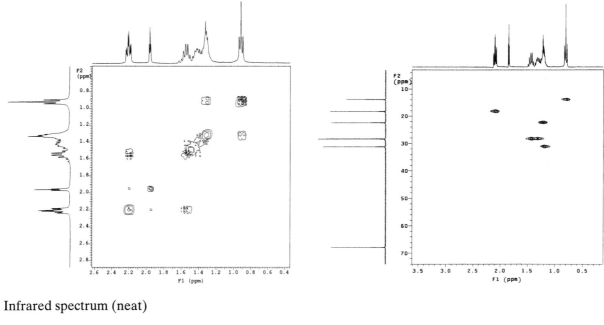

Infrared spectrum (neat)

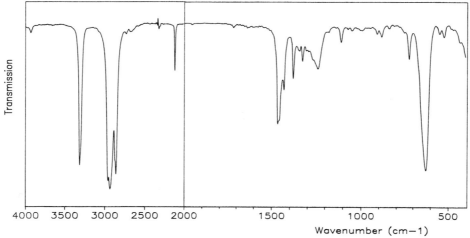

Ultraviolet–visible spectrum (EtOH, ε[224] 83)

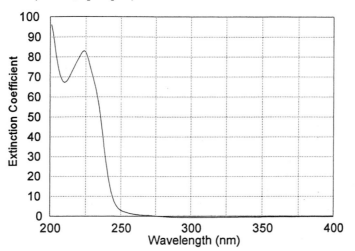

PROBLEM 16-6

Mass spectrum (negative ion CI) *[Note: This unknown contains halogen. Such compounds often lose halogen and give no molecular ion. Halogenated compounds are often best studied with negative rather than positive ionization.]*

Mass spectrum (EI)

Proton NMR spectrum (CDCl$_3$)

Carbon-13 NMR spectrum

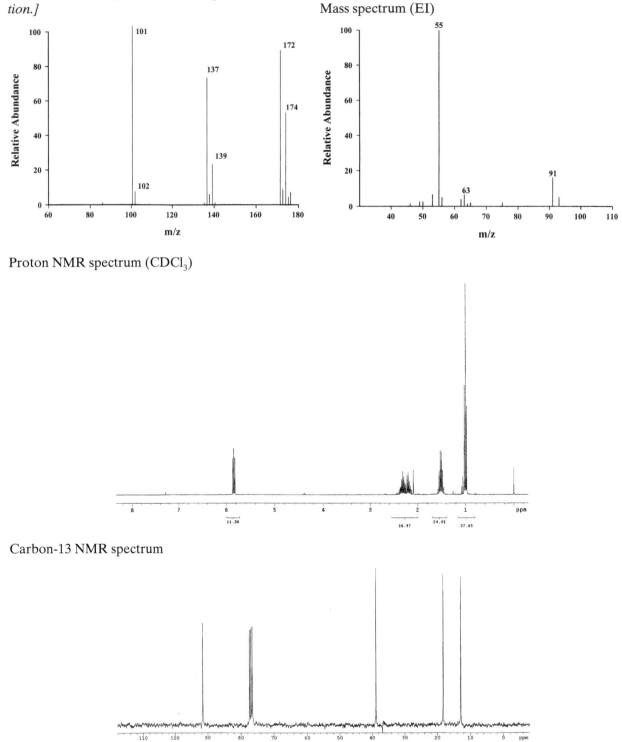

COSY spectrum   HETCOR spectrum

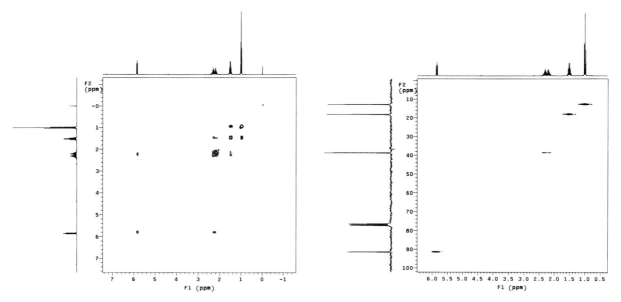

Infrared spectrum (neat)

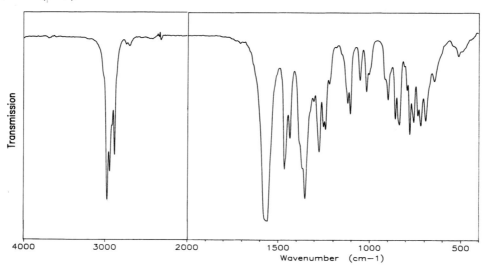

Ultraviolet–visible spectrum (EtOH, ignore the shoulder at 280 nm)

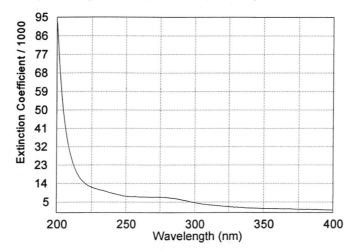

PROBLEM 16-7

Mass spectrum (EI)

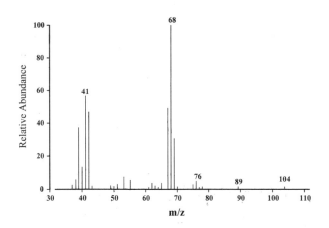

Proton NMR spectrum (CDCl₃)

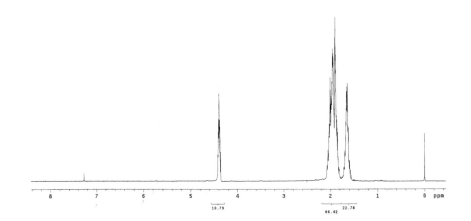

Carbon-13 spectrum

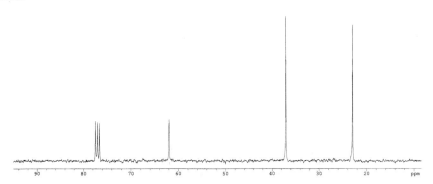

COSY spectrum

HETCOR spectrum

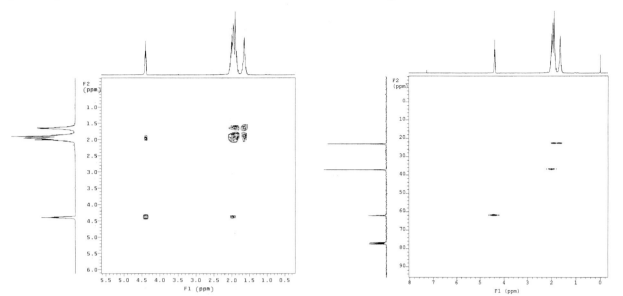

Infrared spectrum (neat)

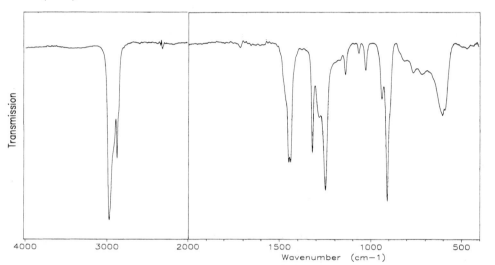

Ultraviolet–visible spectrum (EtOH)

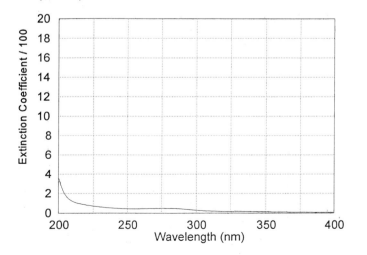

PROBLEM 16-8

Mass spectrum (CI)

Mass spectrum (EI)

Proton NMR spectrum with expansion (CDCl$_3$)

DEPT spectra

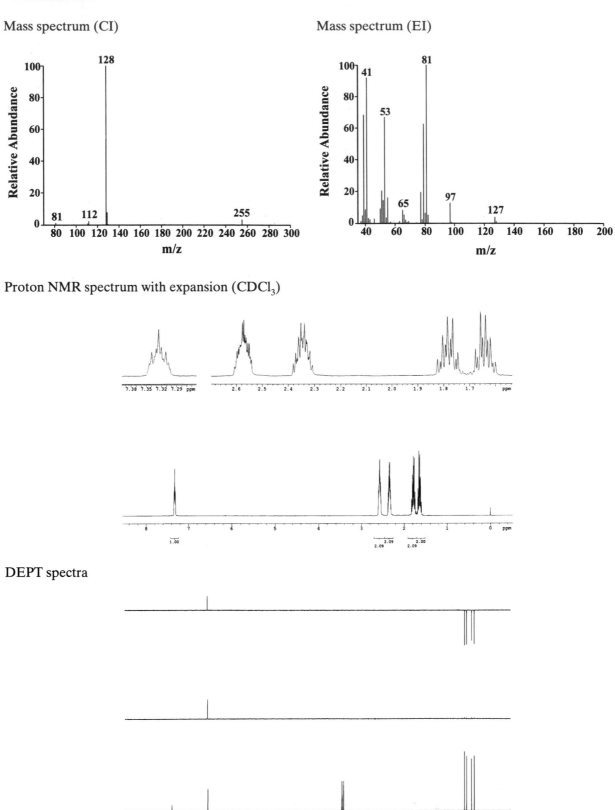

COSY spectrum   HETCOR spectrum

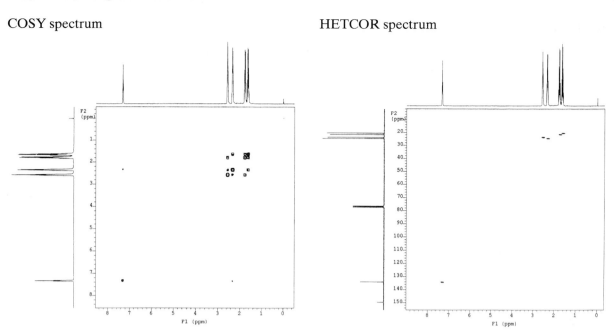

Infrared spectrum (neat)

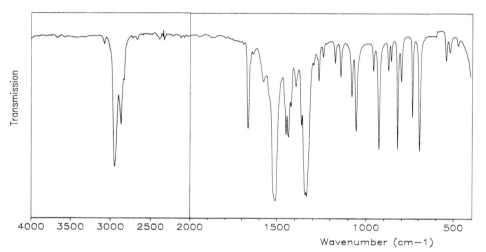

Ultraviolet–visible spectrum (EtOH, ε[225] 4300, ε[255] 5600, ε[322] 73)

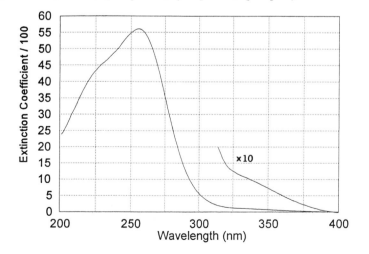

PROBLEM 16-9

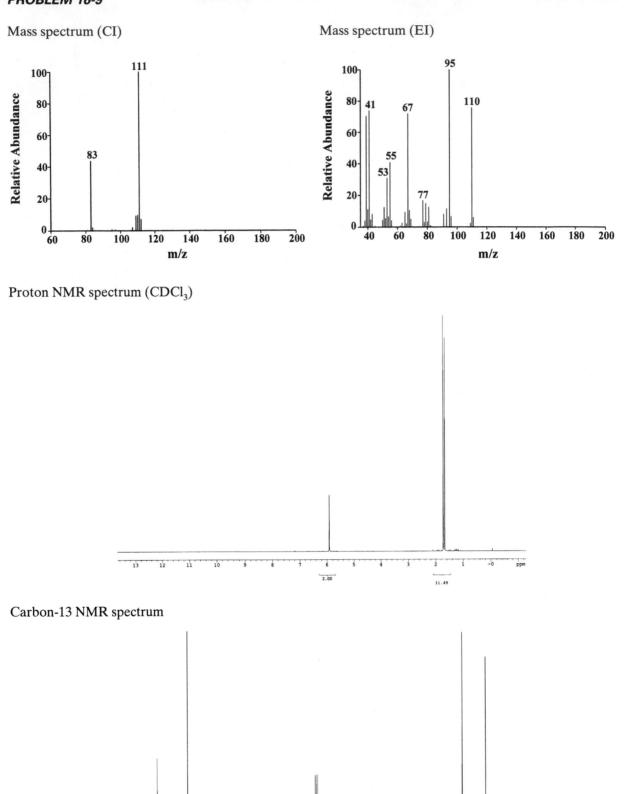

Mass spectrum (CI)

Mass spectrum (EI)

Proton NMR spectrum (CDCl₃)

Carbon-13 NMR spectrum

COSY spectrum   HETCOR spectrum

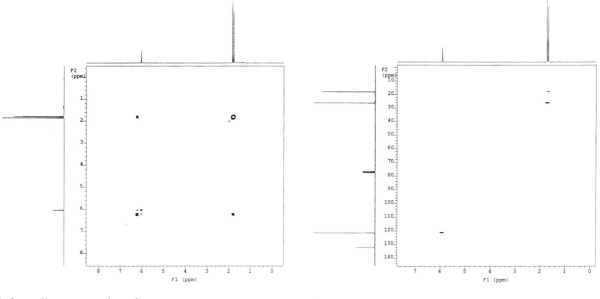

Infrared spectrum (neat)

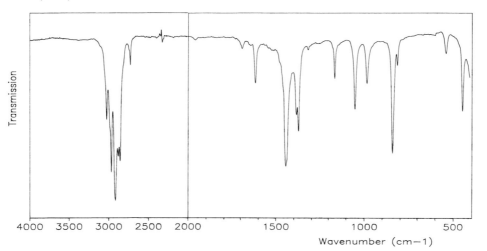

Ultraviolet–visible spectrum (EtOH, ε[243] 25,370)

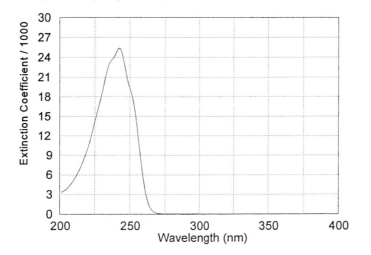

PROBLEM 16-10

Mass spectrum (CI)

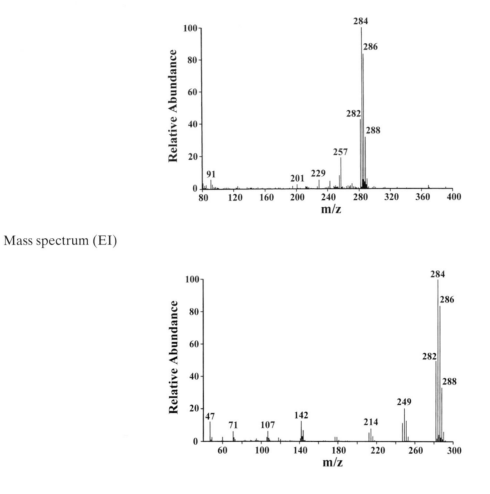

Mass spectrum (EI)

Carbon-13 NMR spectrum (tetrahydrofuran/acetone-d_6)

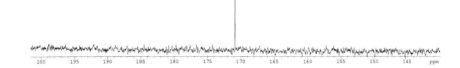

Infrared spectrum (KBr disc)

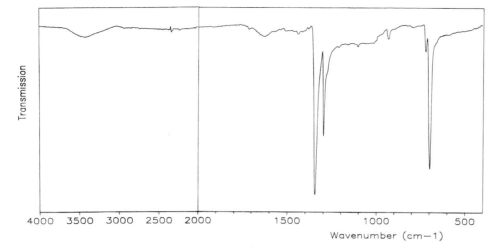

Ultraviolet–visible spectrum (EtOH ε[217] 75,000, ε[232] 15,000, ε[240] 7000, ε[251] 3200, ε[280] 530, ε[291] 261, ε[301] 211)

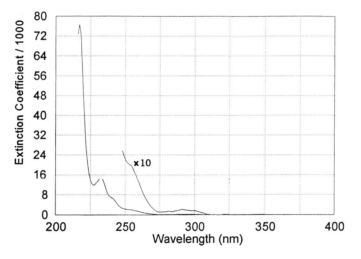

PROBLEM 16-12

Mass spectrum (CI)

Mass spectrum (EI)

Proton NMR spectrum (CDCl$_3$)

Carbon-13 NMR spectrum

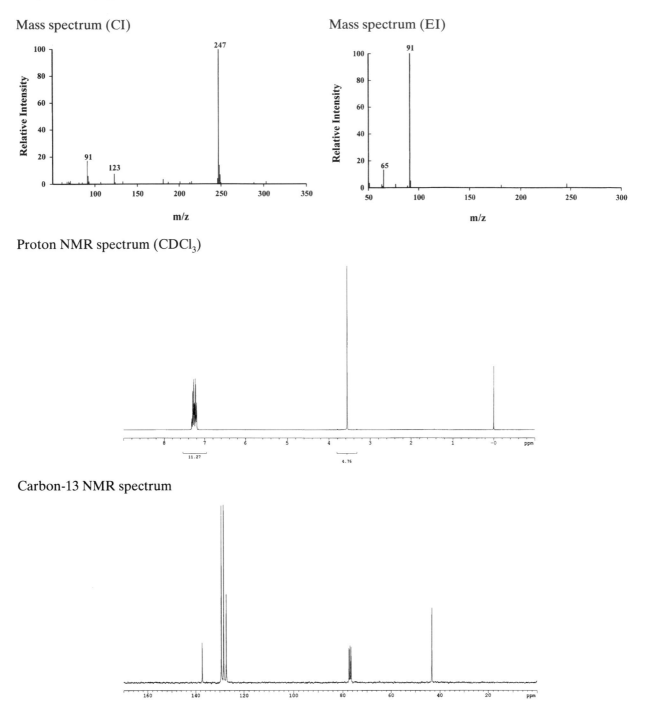

COSY spectrum HETCOR spectrum

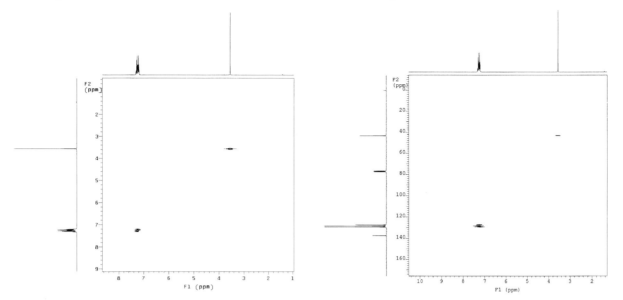

Infrared spectrum (KBr disc)

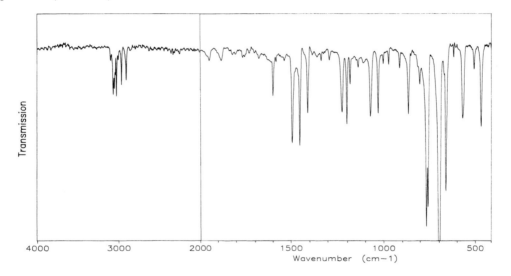

Ultraviolet–visible spectrum (EtOH, ε[220] 16,299, ε[261] 1652, ε[267] 1162)

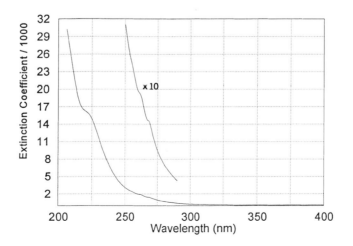

PROBLEM 16-13

Mass spectrum (desorption EI)

Mass spectrum (CI)

Mass spectrum (EI)

Proton NMR spectrum (D₂O)

Carbon-13 NMR spectrum

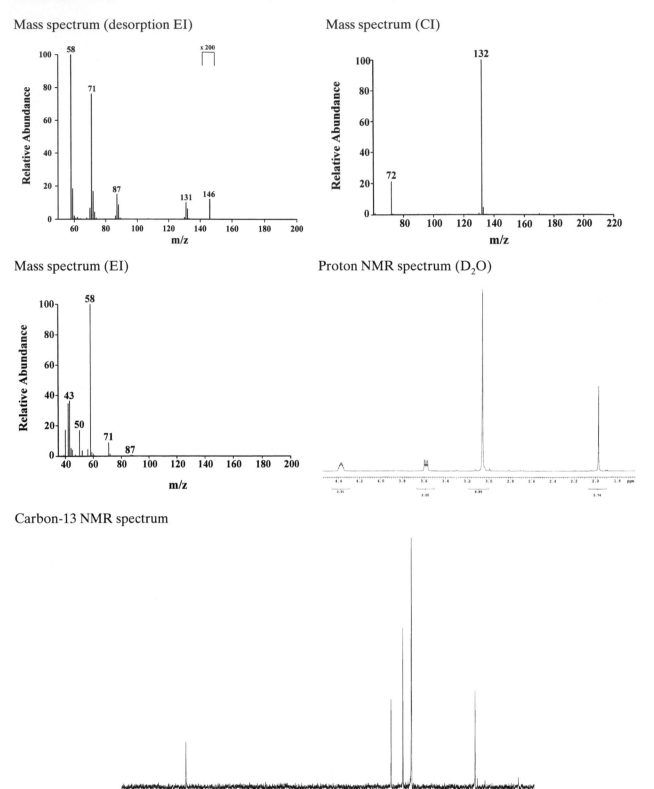

COSY spectrum

HETCOR spectrum

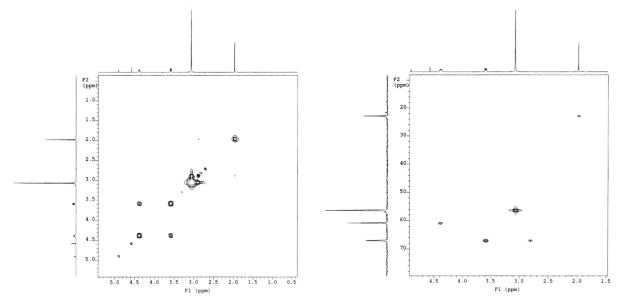

Infrared spectrum (KBr disc)

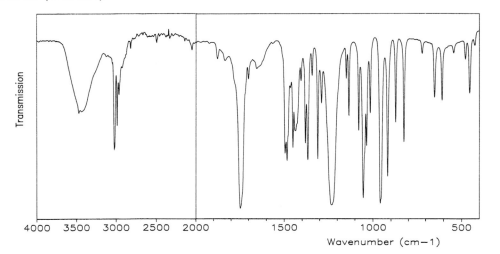

Wavenumber (cm−1)

Ultraviolet–visible spectrum (EtOH, ε[210] 82)

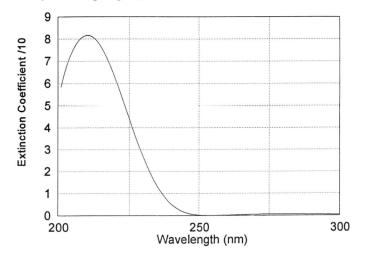

Wavelength (nm)

PROBLEM 16-14

Mass spectrum (CI)

Mass spectrum (EI)

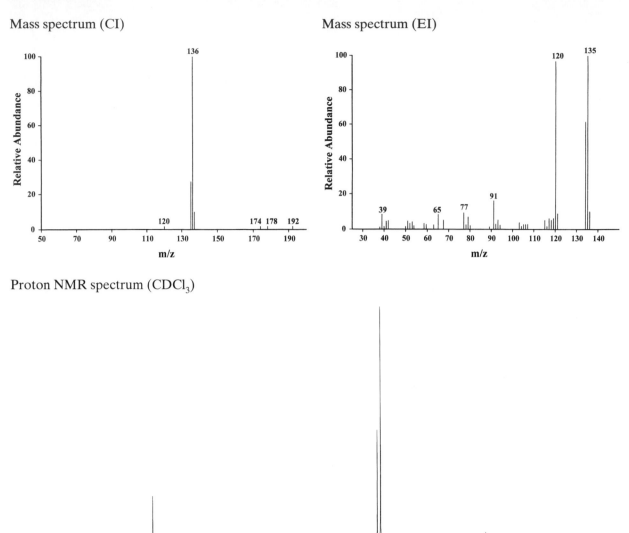

Proton NMR spectrum (CDCl₃)

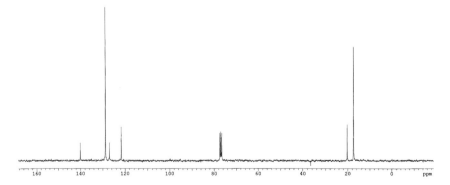

Carbon-13 NMR spectrum

COSY spectrum   HETCOR spectrum

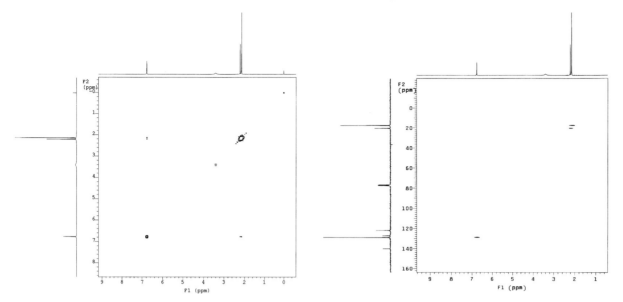

Infrared spectrum (neat)

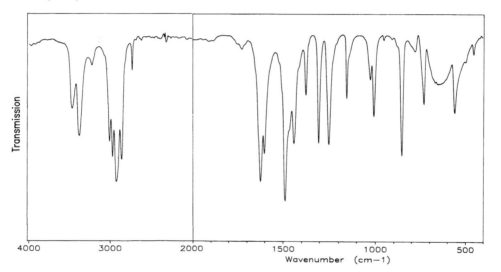

Ultraviolet–visible spectrum (EtOH, ε[236] 8123, ε[289] 2140)

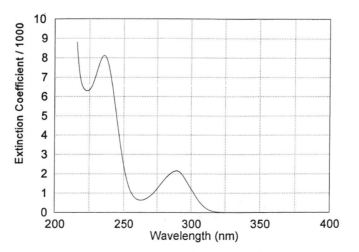

PROBLEM 16-15

Mass spectrum (CI)

Mass spectrum (EI)

Proton NMR spectrum (D_2O)

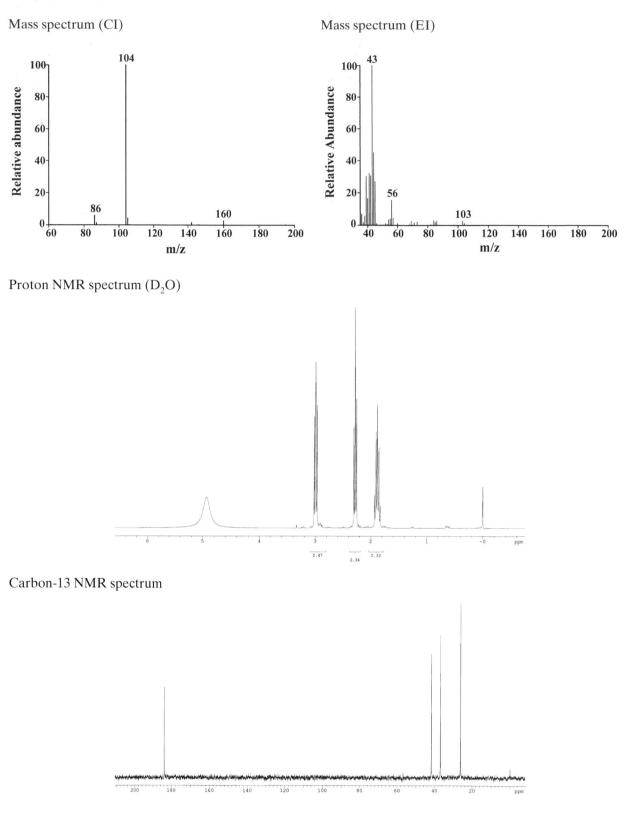

Carbon-13 NMR spectrum

COSY spectrum

HETCOR spectrum

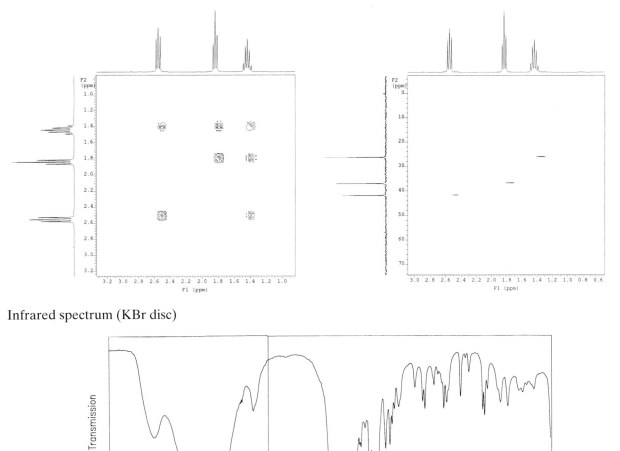

Infrared spectrum (KBr disc)

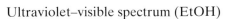

Ultraviolet–visible spectrum (EtOH)

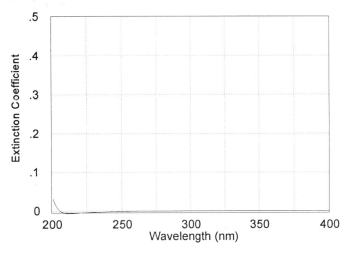

PROBLEM 16-16

Mass spectrum (CI)

Mass spectrum (EI)

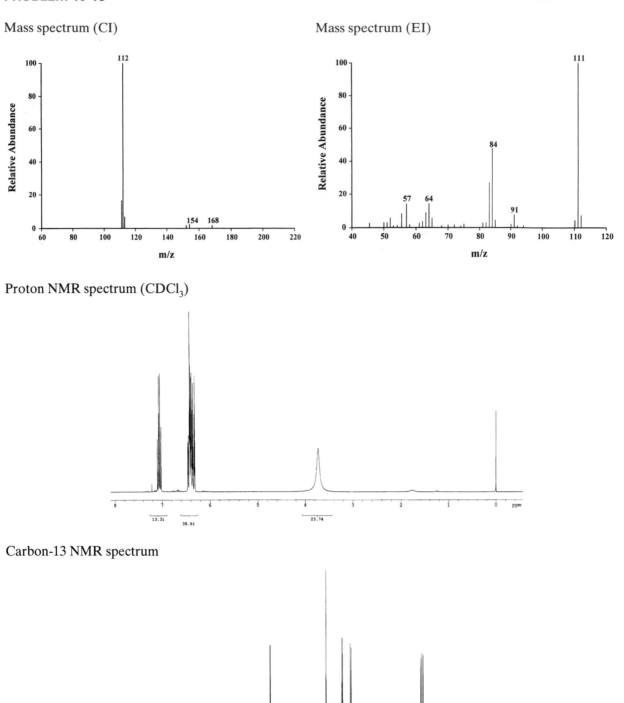

Proton NMR spectrum (CDCl₃)

Carbon-13 NMR spectrum

COSY spectrum

HETCOR spectrum

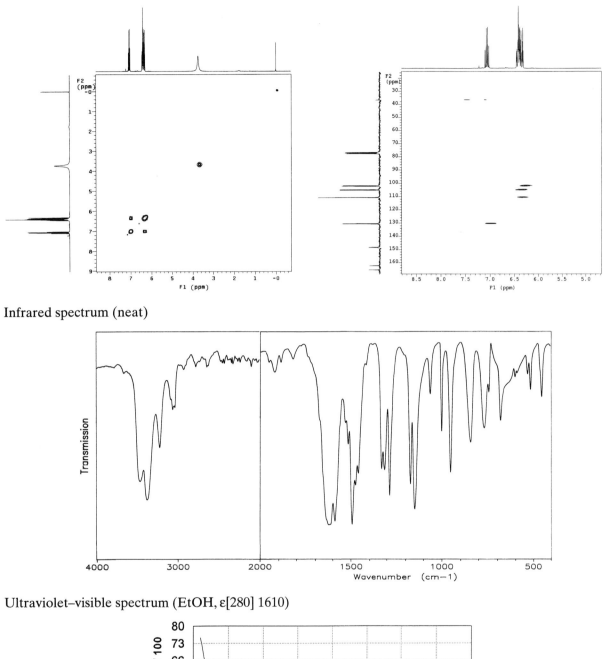

Infrared spectrum (neat)

Ultraviolet–visible spectrum (EtOH, ε[280] 1610)

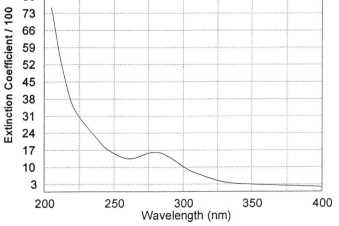

PROBLEM 16-17

Mass spectrum (CI)

Mass spectrum (EI)

Proton NMR spectrum (CDCl₃)

DEPT spectra

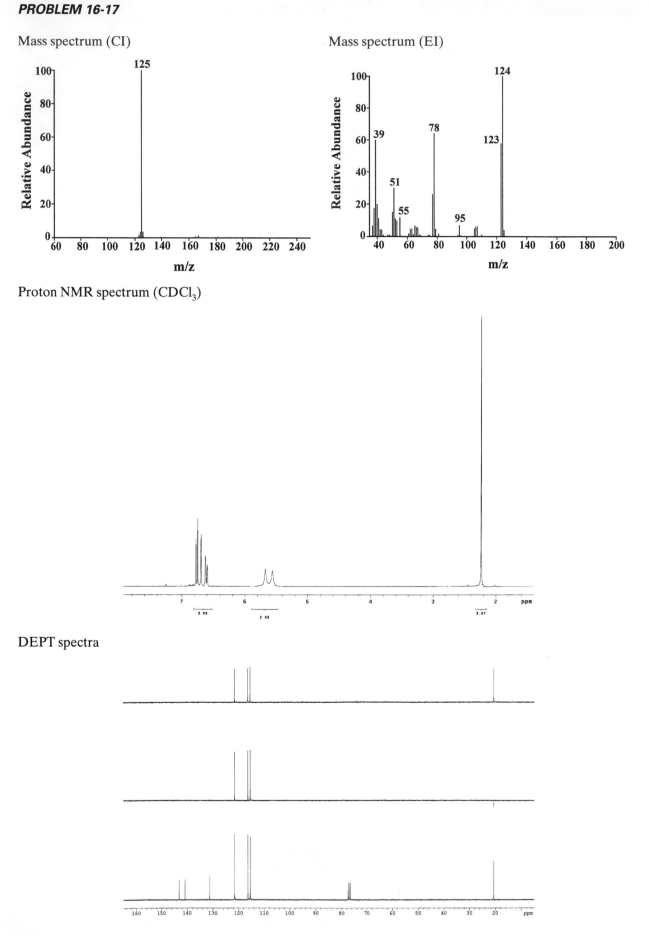

COSY spectrum   HETCOR spectrum

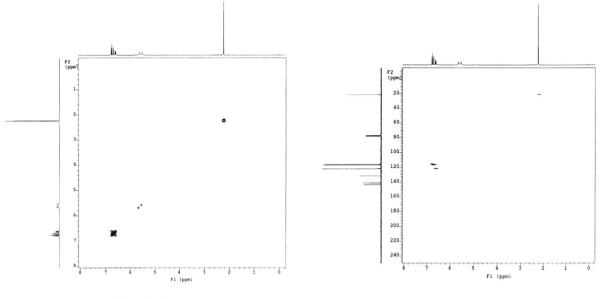

Infrared spectrum (KBr disc)

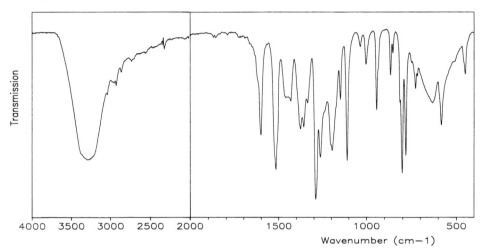

Ultraviolet–visible spectrum (EtOH, ε[203] 27,760, ε[218] 5336, ε[282] 2538)

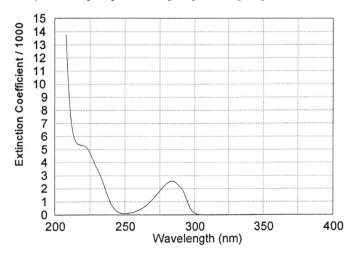

PROBLEM 16-18

Mass spectrum (CI)   Mass spectrum (EI)

Proton NMR spectrum (CDCl$_3$)

DEPT spectra

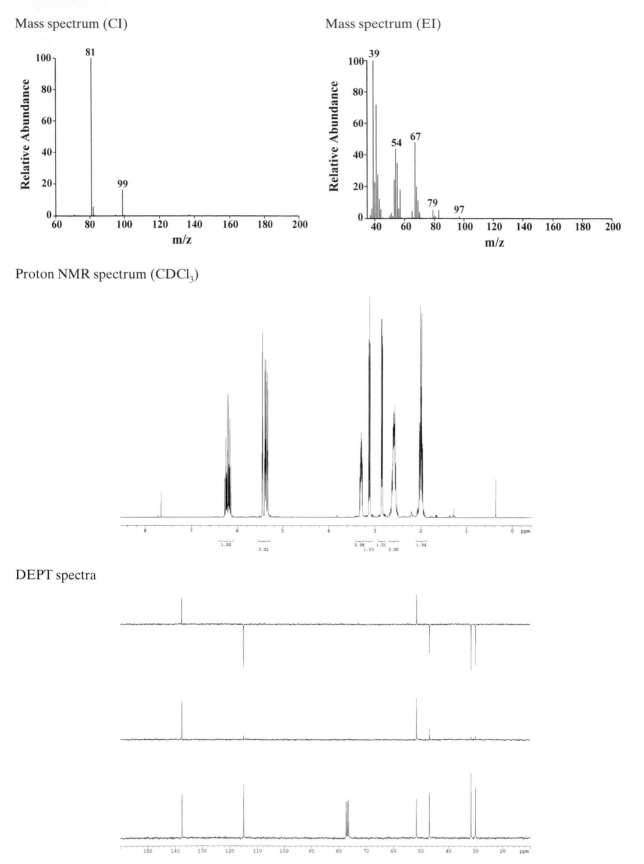

COSY spectrum

HETCOR spectrum

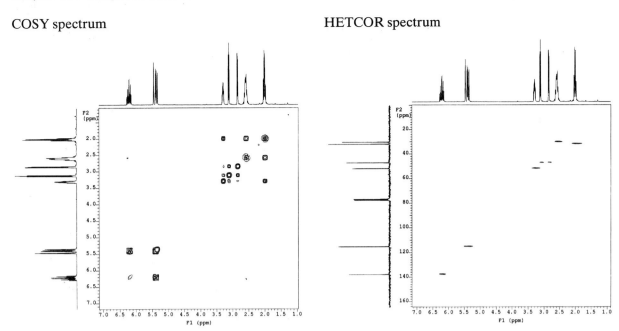

Infrared spectrum (neat)

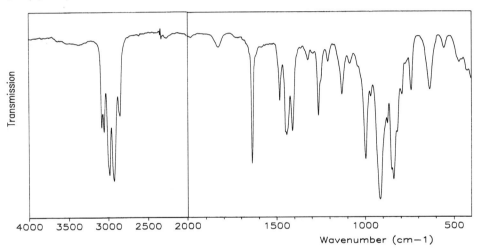

Ultraviolet–visible spectrum (EtOH, ε[202] 49, ε[211] 13, ε[271] 14)

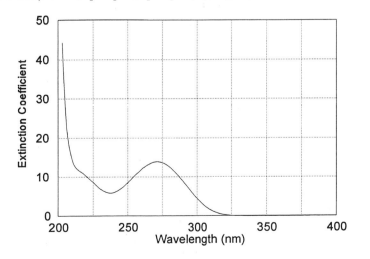

PROBLEM 16-19

Mass spectrum (CI)   Mass spectrum (EI)

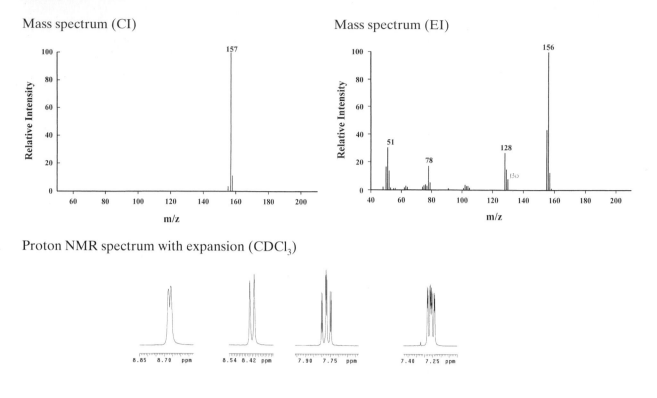

Proton NMR spectrum with expansion (CDCl$_3$)

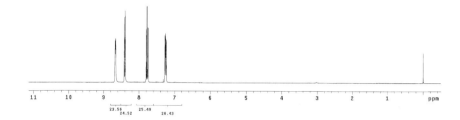

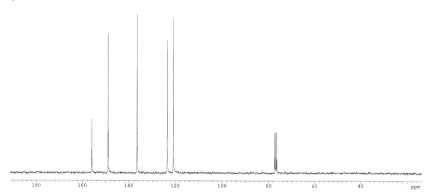

Carbon-13 NMR spectrum

COSY spectrum

HECTOR spectrum

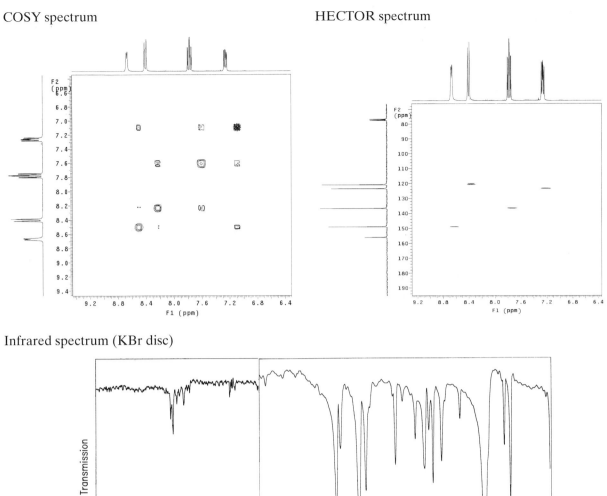

Infrared spectrum (KBr disc)

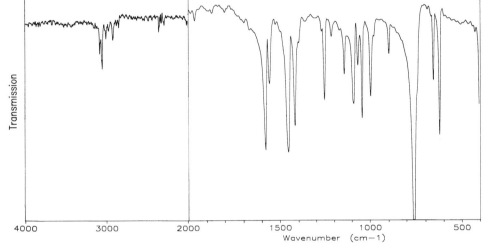

Ultraviolet–visible spectrum (EtOH, ε[236] 10,245, ε[242] 8595, ε[282] 13,419)

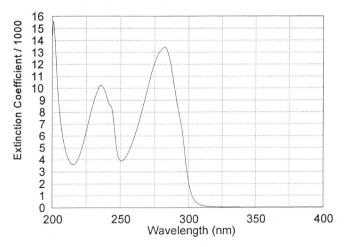

PROBLEM 16-20

Mass spectrum (CI)   Mass spectrum (EI)

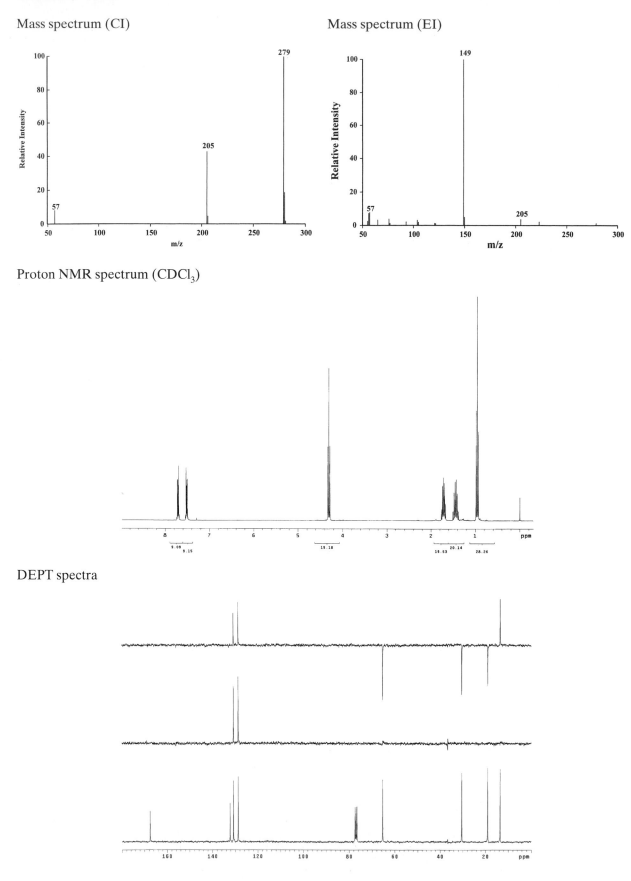

Proton NMR spectrum (CDCl$_3$)

DEPT spectra

COSY spectrum

HETCOR spectrum

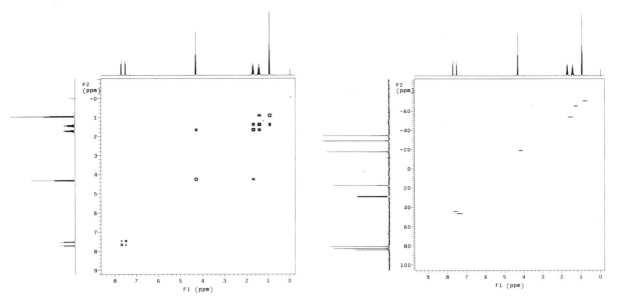

Infrared spectrum (neat)

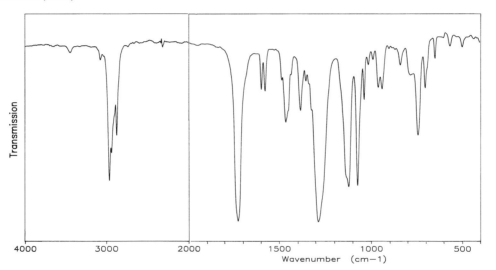

Ultraviolet–visible spectrum (EtOH, ε[225] 7313, ε[275] 1141, ε[280] 1037)

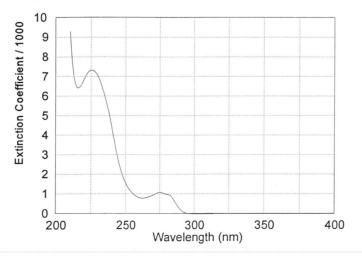

PROBLEM 16-21

Mass spectrum (CI)   Mass spectrum (EI)

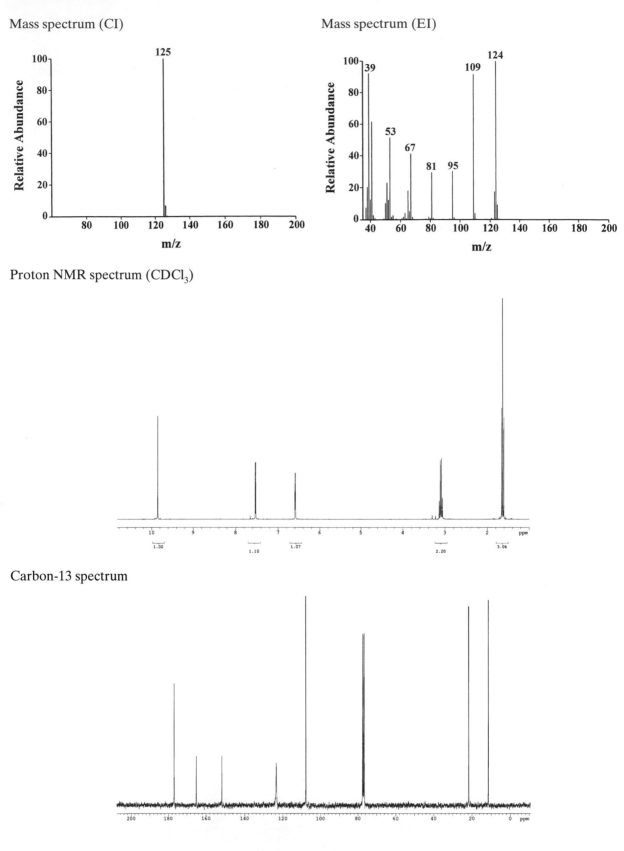

Proton NMR spectrum (CDCl₃)

Carbon-13 spectrum

COSY spectrum   HETCOR spectrum

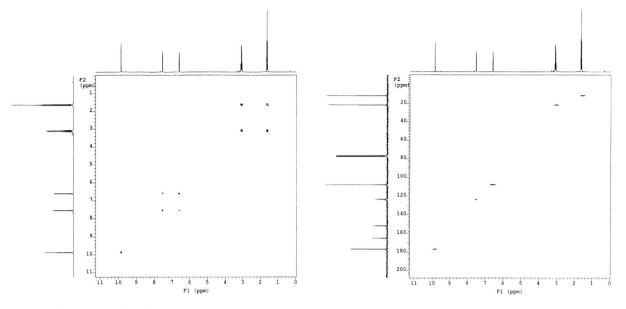

Infrared spectrum (neat)

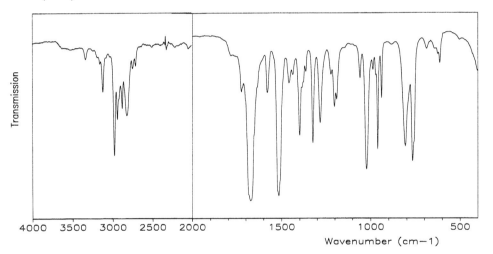

Ultraviolet–visible spectrum (EtOH, ε[225] 2525, ε[285] 16,421, ε[373] 39)

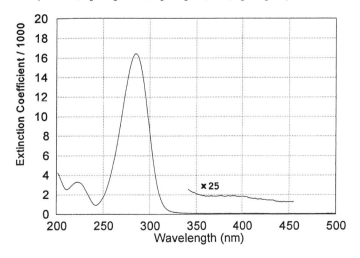

PROBLEM 16-24

Mass spectrum (MALDI in 2,5-dihydroxybenzoic acid)

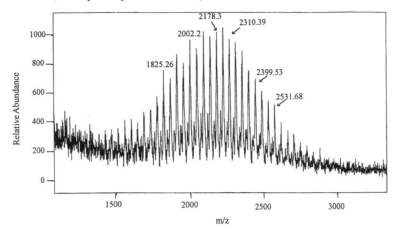

Proton NMR spectrum (CDCl$_3$)

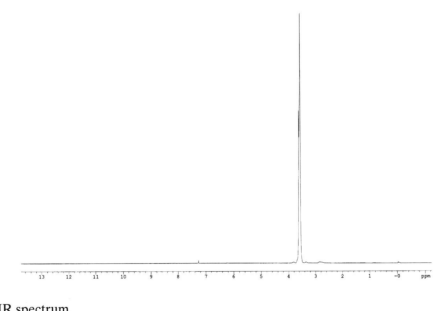

Carbon-13 NMR spectrum

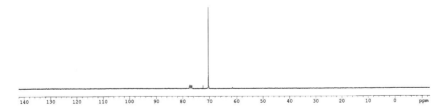

Infrared spectrum (KBr disc)

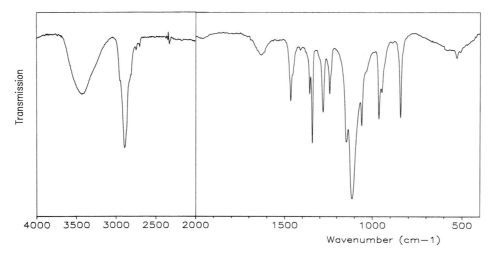

Ultraviolet–visible spectrum (EtOH)

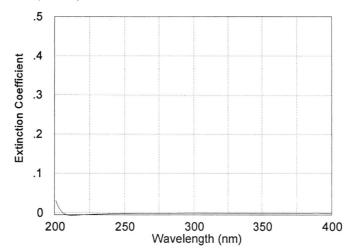

PROBLEM 16-25

Mass spectrum (CI)

Mass spectrum (EI)

Proton NMR spectrum with expansion (CDCl₃)

DEPT spectra

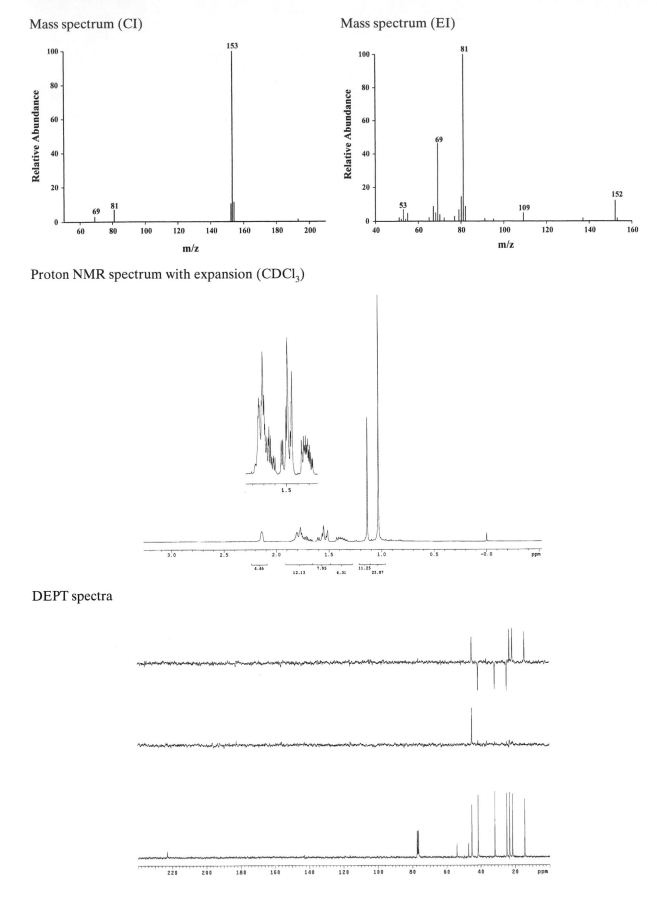

COSY spectrum

HETCOR spectrum

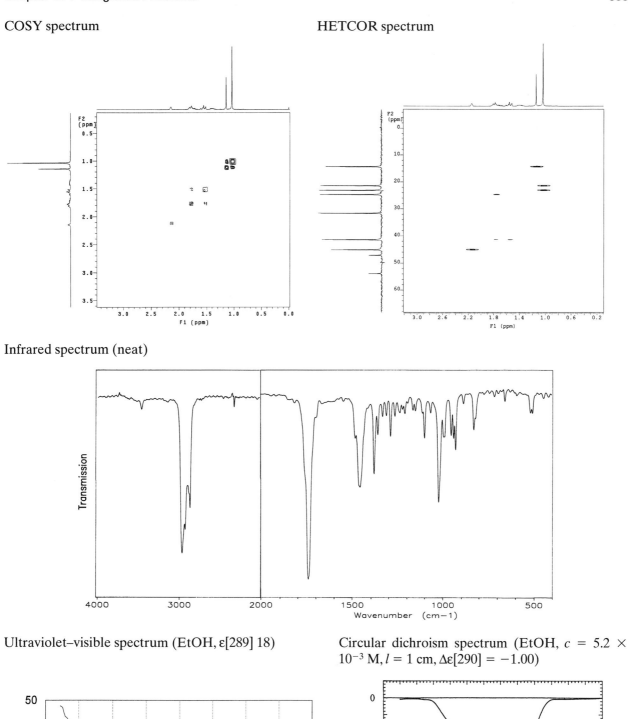

Infrared spectrum (neat)

Ultraviolet–visible spectrum (EtOH, ε[289] 18)

Circular dichroism spectrum (EtOH, $c = 5.2 \times 10^{-3}$ M, $l = 1$ cm, $\Delta\varepsilon[290] = -1.00$)

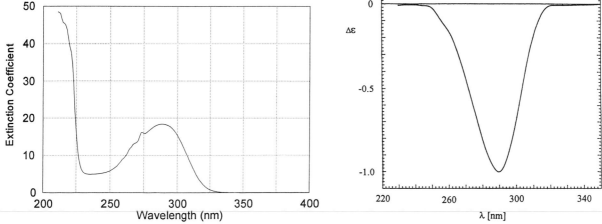

PROBLEM 16-26

Mass spectrum (CI)

Mass spectrum (EI)

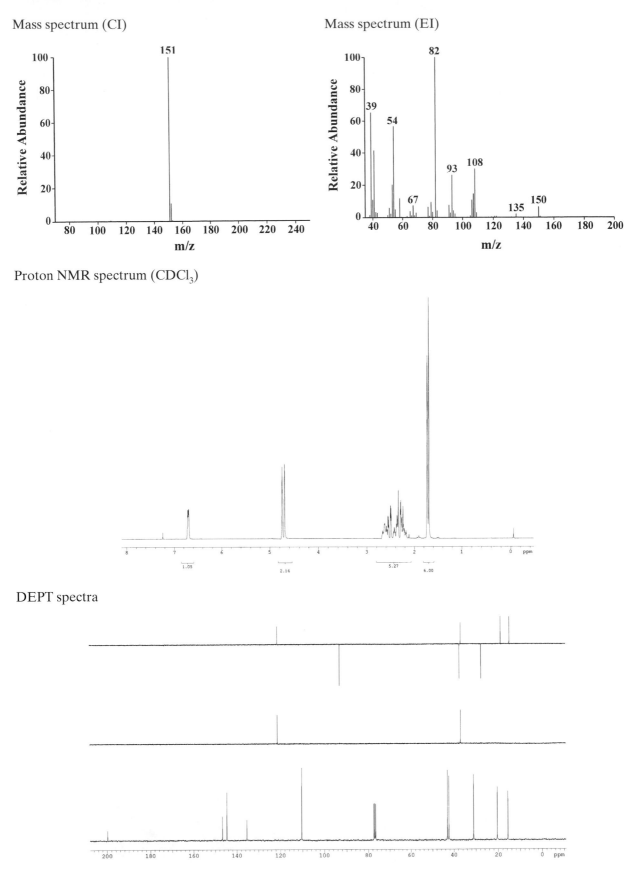

Proton NMR spectrum (CDCl₃)

DEPT spectra

COSY spectrum

HETCOR spectrum

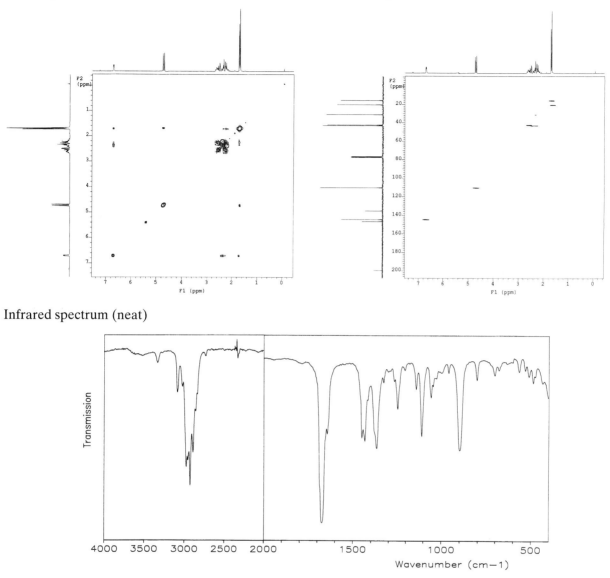

Infrared spectrum (neat)

Ultraviolet–visible spectrum (EtOH, ε[231] 43, ε[235] 9144)

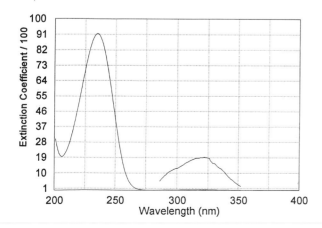

Circular dichroism spectrum (EtOH, $c = 1.277 \times 10^{-4}$ M, $l = 1$ cm, $\Delta\varepsilon[220] = -3.35$, $\Delta\varepsilon[247] = +1.96$, $c = 6.384 \times 10^{-3}$ M, $l = 1$ cm, $\Delta\varepsilon[308] = +0.15$, $\Delta\varepsilon[353] = -0.057$).

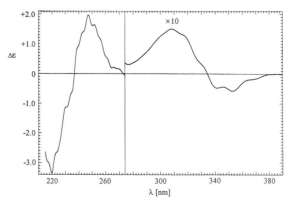

PROBLEM 16-28

Mass spectrum (CI)

Mass spectrum (EI)

Proton NMR spectrum (CDCl₃)

Carbon-13 NMR spectrum

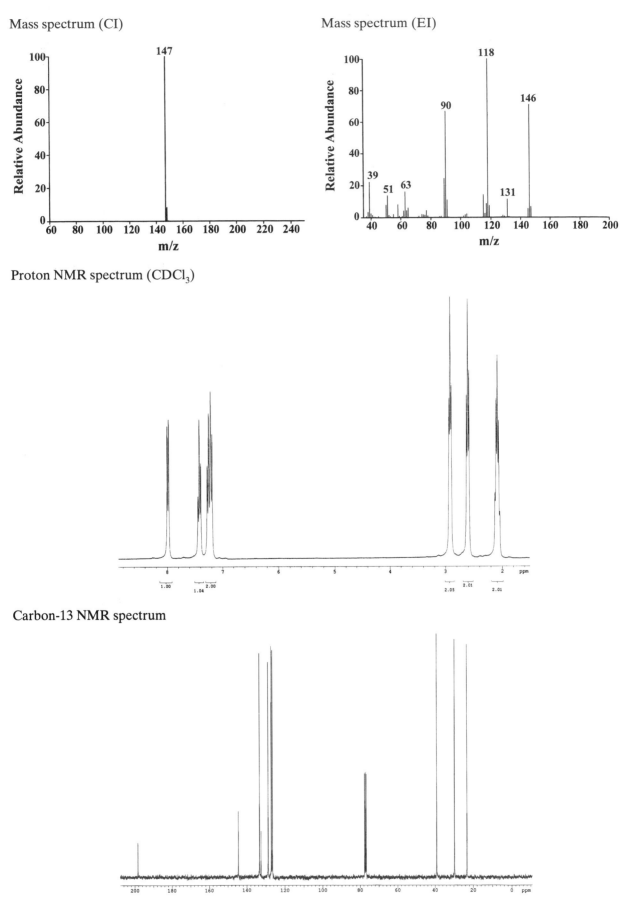

COSY spectrum   HETCOR spectrum

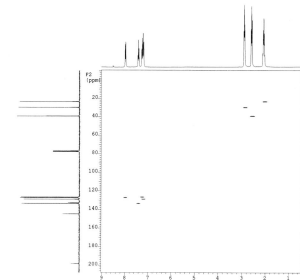

Infrared spectrum (neat)

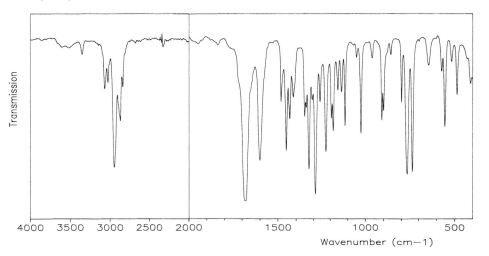

Ultraviolet–visible spectrum (EtOH, ε[206] 24,500, ε[248] 12,200, ε[291] 2200)

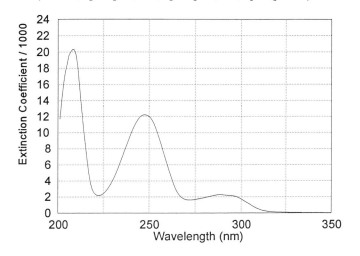

PROBLEM 16-29

Mass spectrum (CI)

Mass spectrum (EI)

Proton NMR spectrum (CDCl₃)

DEPT spectra

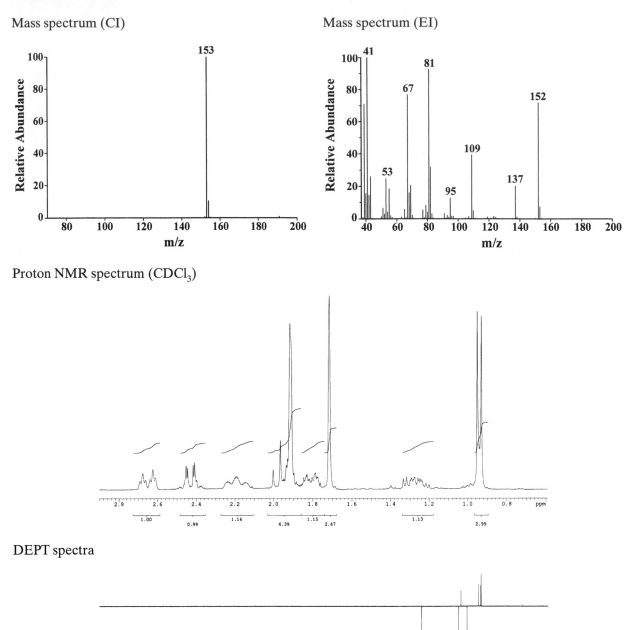

COSY spectrum

HETCOR spectrum

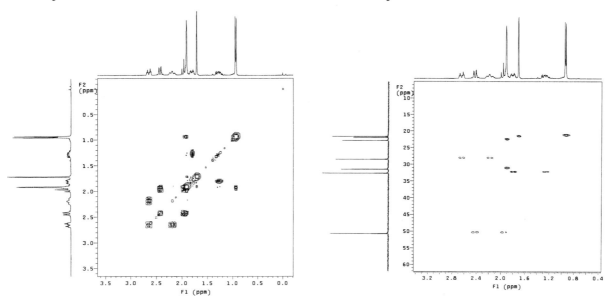

Infrared spectrum (neat)

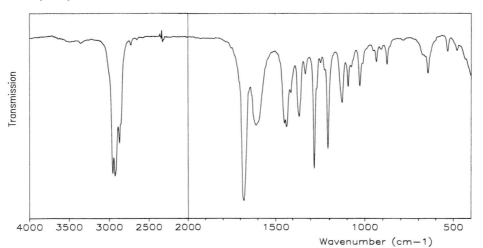

Ultraviolet–visible spectrum (EtOH, $\varepsilon[252]$ 7040, $\varepsilon[311]$ 72)

Circular dichroism spectrum (EtOH, $c = 6.149 \times 10^{-3}$ M, $l = 1$ cm, $\Delta\varepsilon[245] = -2.42$, $\Delta\varepsilon[319] = +0.66$)

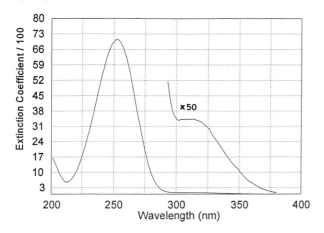

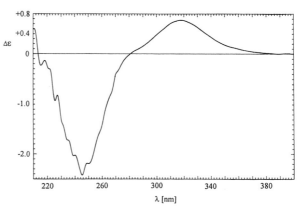

PROBLEM 16-30

Mass spectrum (CI)   Mass spectrum (EI)

Proton NMR spectrum with expansion (CD$_3$(SO)CD$_3$)

Carbon-13 NMR spectrum

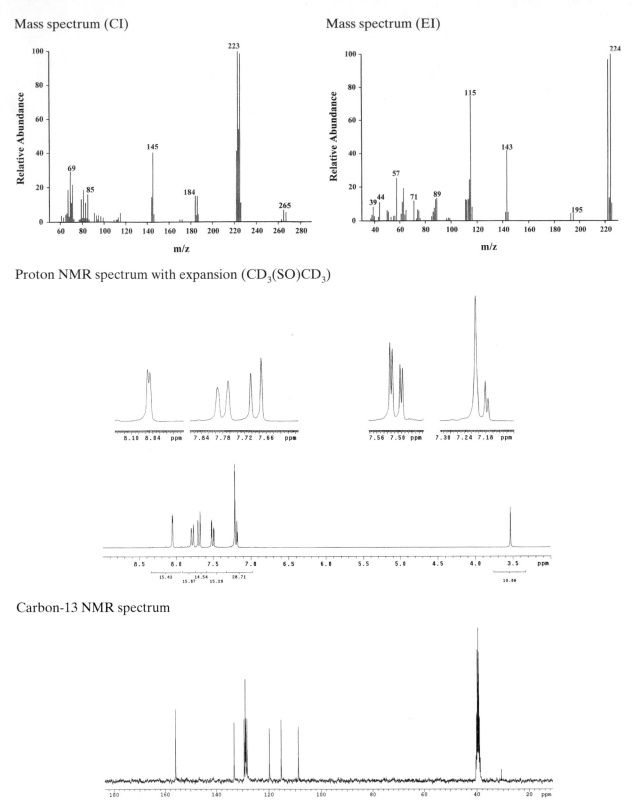

COSY spectrum

HETCOR spectrum

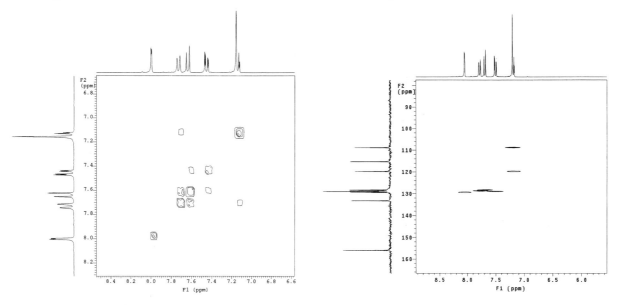

Infrared spectrum (KBr disc)

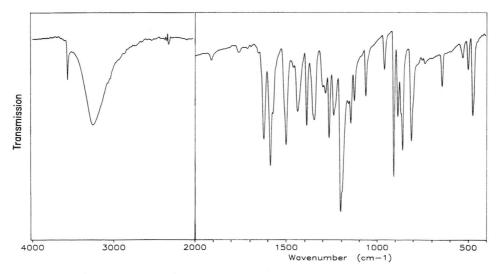

Ultraviolet–visible spectrum (EtOH, ε[232] 80,627, ε[276] 10,880, ε[286] 7669, ε[328] 1872, ε[340] 2056)

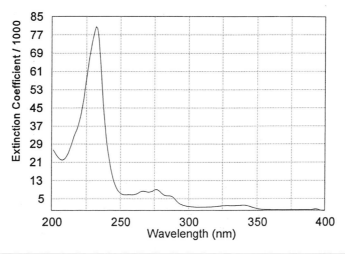

PROBLEM 16-31

Mass spectrum (CI)   Mass spectrum (EI)

Proton NMR spectrum with expansion (CDCl$_3$)

DEPT spectra

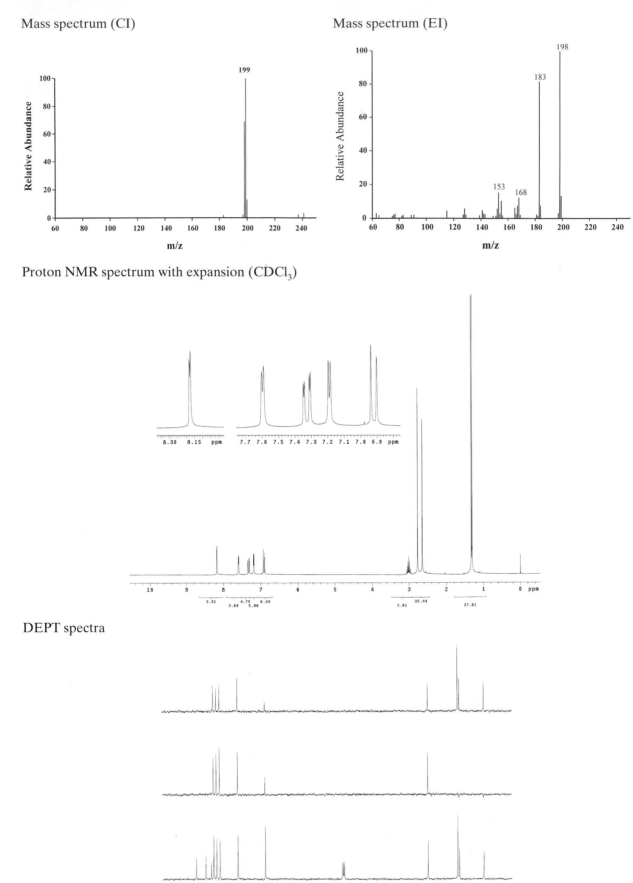

COSY spectrum

HETCOR spectrum

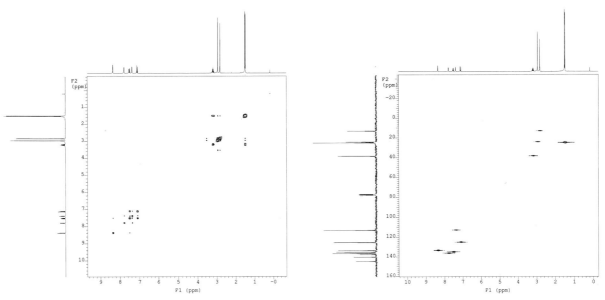

Infrared spectrum (KBr disc)

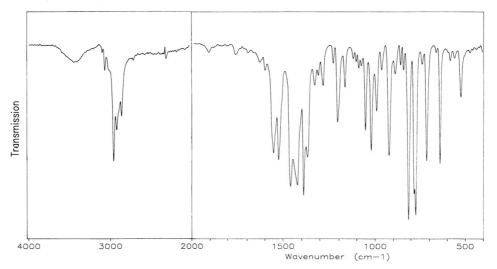

Ultraviolet–visible spectrum (EtOH, ε[214] 13,069, ε[244] 27,837, ε[284] 48,370, ε[289] 47,400, ε[304] 11,848, ε[349] 4664, ε[366] 3212, ε[602] 455, ε[649] 378, ε[720] 133)

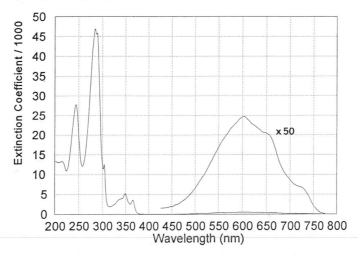

PROBLEM 16-32

Mass spectrum (desorption EI)

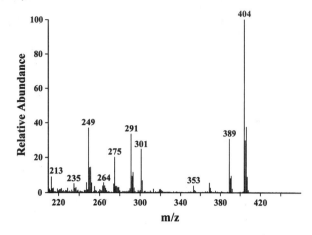

Proton NMR spectrum (CDCl$_3$)

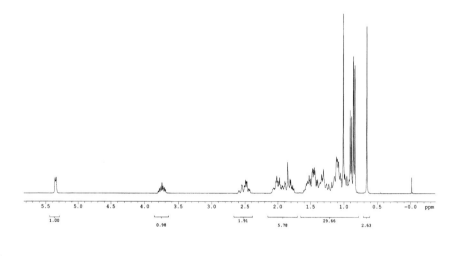

DEPT spectra

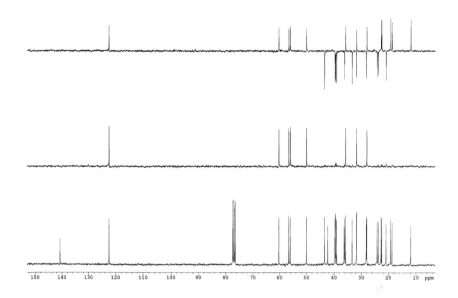

COSY spectrum

COSY spectrum with expansion

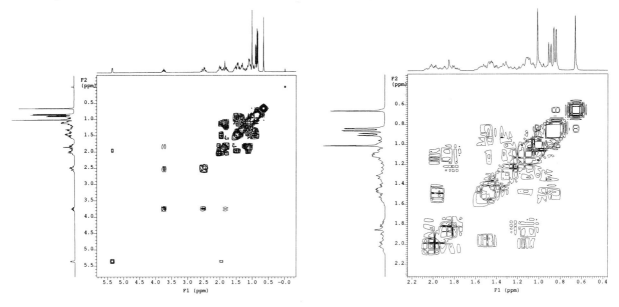

HETCOR spectrum

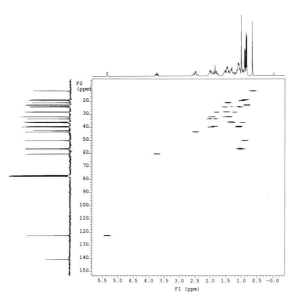

Infrared spectrum (KBr disc)

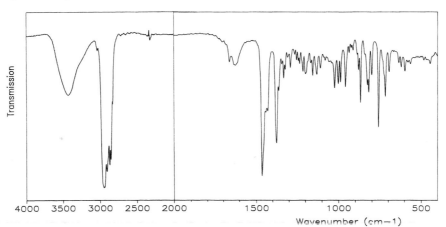

Continued

PROBLEM 16-32, cont'd

Ultraviolet–visible spectrum (EtOH, ε[203] 5500)

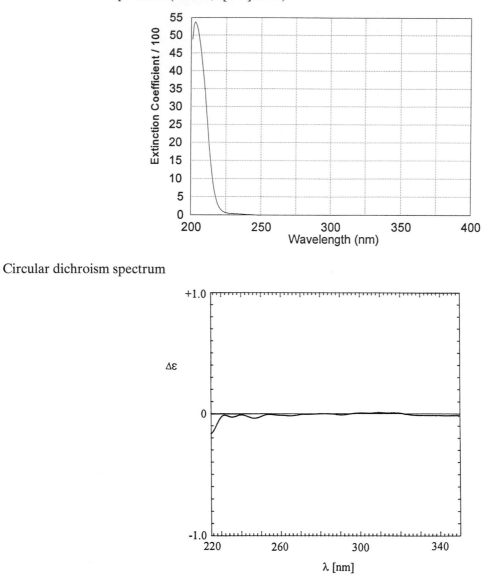

Circular dichroism spectrum

PROBLEM 16-33

Mass spectrum (CI)

Mass spectrum (EI)

Mass spectrum (desorption EI)

Proton NMR spectrum (CDCl₃)

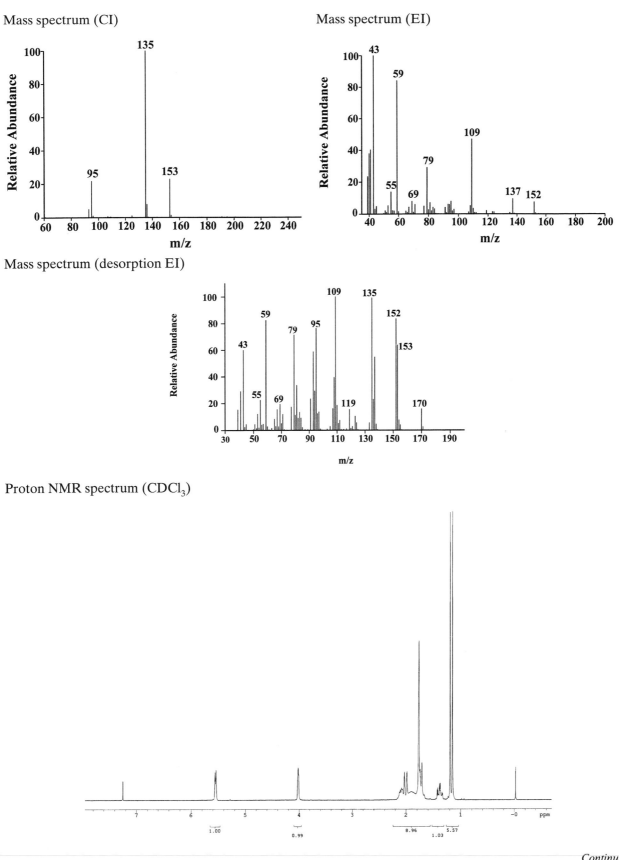

Continued

PROBLEM 16-33, cont'd

Carbon-13 NMR spectrum

COSY spectrum   HETCOR spectrum

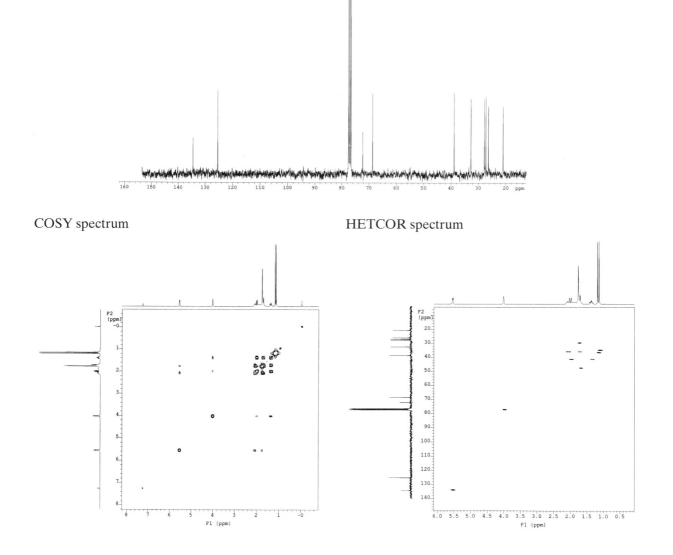

Infrared spectrum (KBr disc)

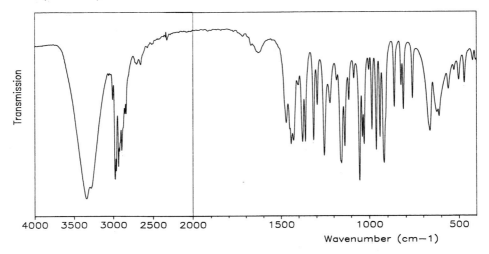

Ultraviolet–visible spectrum (EtOH)

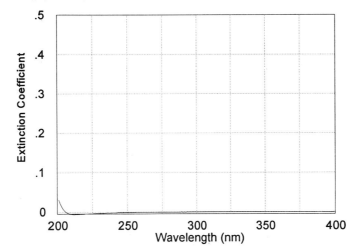

PROBLEM 16-34

Mass spectrum (MALDI in α-cyano-4-hydroxysinapinic acid)

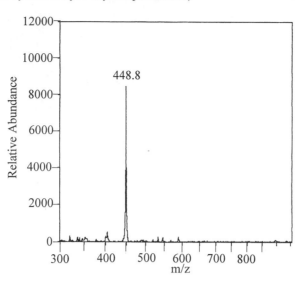

Proton NMR spectrum (CF$_3$CO$_2$D/CD$_3$(CO)CD$_3$)

DEPT spectra

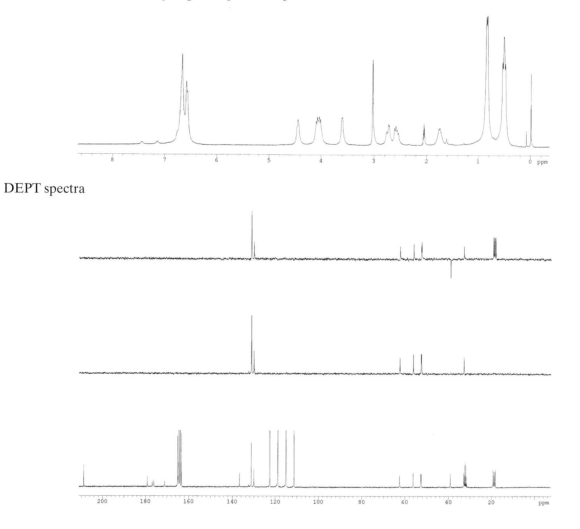

COSY spectrum

HETCOR spectrum

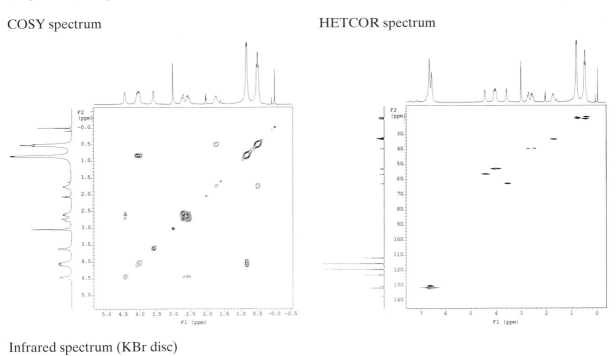

Infrared spectrum (KBr disc)

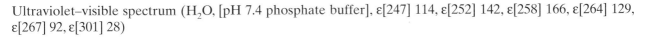

Ultraviolet–visible spectrum (H₂O, [pH 7.4 phosphate buffer], ε[247] 114, ε[252] 142, ε[258] 166, ε[264] 129, ε[267] 92, ε[301] 28)

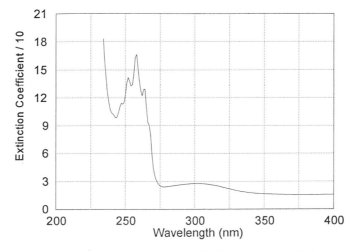

PROBLEM 16-35

Mass spectrum (MALDI in α-cyano-4-hydroxysinapinic acid)

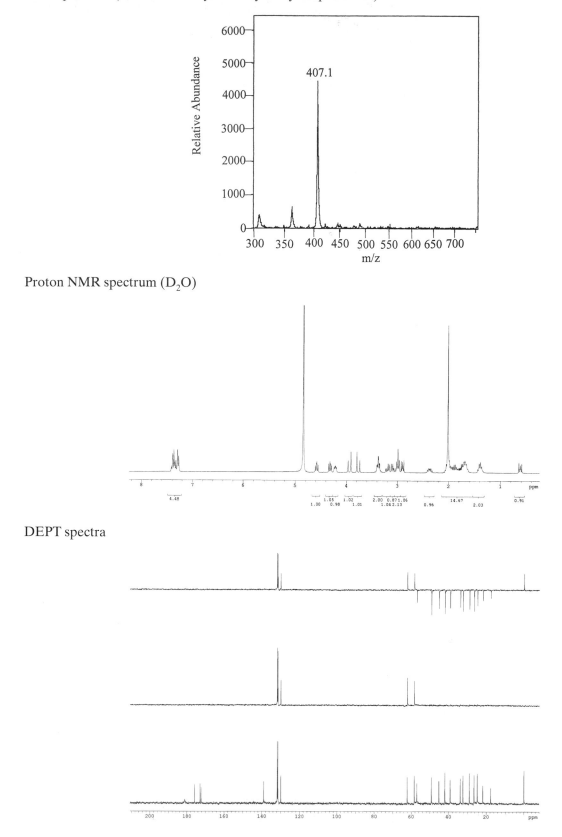

Proton NMR spectrum (D₂O)

DEPT spectra

COSY spectrum

HETCOR spectrum

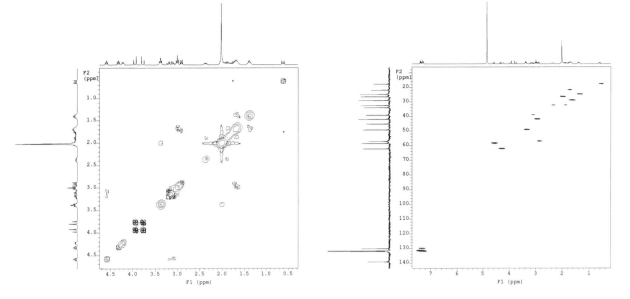

Infrared spectrum (KBr disc)

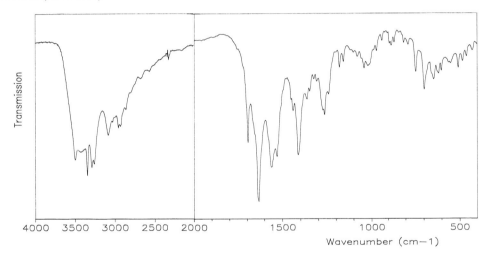

Ultraviolet–visible spectrum (H$_2$O, [pH 7.4 phosphate buffer], ε[247] 143, ε[252] 175, ε[258] 206, ε[264] 164, ε[267] 120, ε[312] 94)

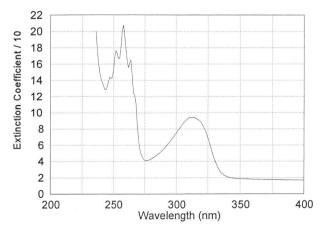

Elemental Formulas

1. $C_{11}H_{16}$
2. C_6H_{12}
3. $C_9H_{21}N$
4. C_4H_7N
5. C_8H_{14}
6. $C_4H_8NO_2Cl$
7. C_5H_9Cl
8. $C_6H_9NO_2$
9. C_8H_{14}
10. C_6Cl_6
11. $C_7H_{12}O_4$
12. $C_{14}H_{14}S_2$
13. $C_7H_{16}NO_2Cl$
14. $C_9H_{13}N$
15. $C_4H_9NO_2$
16. C_6H_6NF
17. $C_7H_8O_2$
18. $C_6H_{10}O$
19. $C_{10}H_8N_2$
20. $C_{16}H_{22}O_4$
21. $C_7H_8O_2$
22. C_6H_7NO
23. $C_6H_{10}O$
24. $(C_2H_4O_2)_n$
25. $C_{10}H_{16}O$
26. $C_{10}H_{14}O$
27. $C_{20}H_{14}O_2$
28. $C_{10}C_{10}O$
29. $C_{10}H_{16}O$
30. $C_{10}H_7OBr$
31. $C_{15}H_{18}$
32. $C_{27}H_{45}Cl$
33. $C_{10}H_{18}O_2$
34. $C_{20}H_{30}N_4O_5$
35. $C_{22}H_{33}N_5O_5 \cdot C_2H_4O_2$

Functional Groups

1. Monosubstituted benzene
2. Alkene
3. Amine
4. Nitrile
5. Alkyne
6. Chloronitroalkane
7. Chlorocycloalkane
8. Nitroalkene
9. Conjugated diene
10. Chlorinated aromatic
11. Carboxylic acid
12. Aromatic sulfide
13. Ester, quaternary ammonium chloride
14. Aromatic amine
15. Carboxylic acid, primary amine
16. Fluorinated aromatic amine
17. Substituted phenol
18. Alkene, epoxide
19. Monosubstituted pyridine
20. Aromatic ester
21. Furan, aldehyde
22. Methoxypyridine
23. Cyclic ketone
24. Polymeric ether
25. Bicyclic ketone
26. Unconjugated cyclic diene, α, β-unsaturated ketone
27. Hydroxynaphthalene
28. Fused aryl ketone
29. α, β-Unsaturated cyclic ketone
30. Bromohydroxynaphthalene
31. Nonalternate aromatic
32. Polycyclic chloroalkene
33. Cyclic alkene, dialcohol
34. Aromatic, peptide, amino acid
35. Aromatic, pyrrolidine, peptide, amino acid, acetate

Index

Proton Chemical Shift Ranges

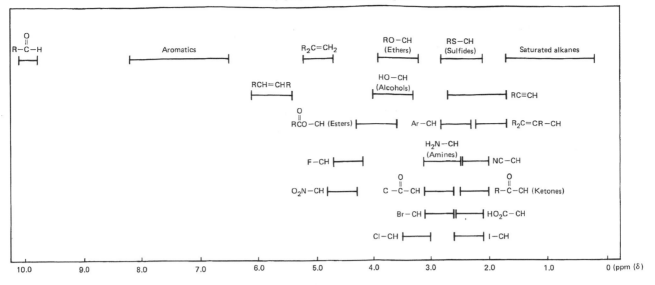

Carbon Chemical Shift Ranges

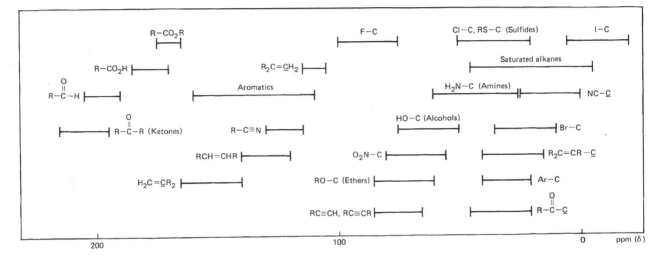

Ionization Methods

Method	Ionizing Agent	Sample	Molecular Ion
Electron ionization (EI)	Electrons	Vapor	$M^{+}\cdot$
Chemical ionization (CI)	Gaseous ions	Vapor	$(M+H)^{+}$, $(M-H)^{-}$, etc.
Desorption ionization (DI)	Photons, energetic particles	Solid	$(M+Na)^{+}$, $(M\ H)^{-}$, etc.
Spray ionization (SI)	Electric field, heat	Aqueous solution	$(M+H)^{+}$, $(M-nH)^{n-}$, $(M+nH)^{n+}$, etc.

Characteristic Fragment Ions as the Lowest Member of Homologous Series

Mass	Ion	Possible Functionality	Mass	Ion	Possible Functionality
15	CH_3^{+}	Methyl, alkane	50	$C_4H_2^{+}\cdot$	Aryl
29	$C_2H_5^{+}$, HCO^{+}	Alkane, aldehyde	51	$C_4H_3^{+}$	Aryl
30	$CH_2{=}NH_2^{+}$	Amine	77	$C_6H_5^{+}$	Phenyl
31	$CH_2{=}OH^{+}$	Ether or alcohol	83	$C_6H_{11}^{+}$	Cyclohexyl
39	$C_3H_3^{+}$	Aryl	91	$C_7H_7^{+}$	Benzyl
43	$C_3H_7^{+}$, CH_3CO^{+}	Alkane, ketone	105	$C_6H_5C_2H_4^{+}$	Substituted benzene
45	CO_2H^{+}, CHS^{+}	Carboxylic acid, thiophene		$CH_3C_6H_4CH_2^{+}$	Disubstituted benzene
				$C_6H_5CO^{+}$	Benzoyl
47	CH_3S^{+}	Thioether